D1083071

Stochastic Modelling and Applied Probability
formerly: Applications of Mathematics

(continued after index)

Søren Asmussen   Peter W. Glynn

Stochastic Simulation: Algorithms and Analysis

 Springer

Authors

Søren Asmussen
Department of Theoretical Statistics
Department of Mathematical Sciences
Aarhus University
Ny Munkegade
DK–8000 Aarhus C, Denmark
asmus@imf.au.dk

Peter W. Glynn
Department of Management Science
 and Engineering
Institute for Computational and
 Mathematical Engineering
Stanford University
Stanford, CA 94305–4026
glynn@stanford.edu

Managing Editors

B. Rozovskii
Division of Applied Mathematics
182 George St.
Providence, RI 02912
USA
rozovski@dam.brown.edu

G. Grimmett
Centre for Mathematical Sciences
Wilberforce Road, Cambridge CB3 0WB,
UK
G.R.Grimmett@statslab.cam.ac.uk

Mathematics Subject Classification (2000): 65C05, 60-08, 62-01, 68-01

Library of Congress Control Number: 2007926471

ISSN: 0172-4568
ISBN-13: 978-0-387-30679-7
e-ISBN-13: 978-0-387-69033-9

© 2007 Springer Science+Business Media, LLC
All rights reserved. This work may not be translated or copied in whole or in part without the written permission of the publisher (Springer Science+Business Media, LLC, 233 Spring Street, New York, NY 10013, USA), except for brief excerpts in connection with reviews or scholarly analysis. Use in connection with any form of information storage and retrieval, electronic adaptation, computer software, or by similar or dissimilar methodology now known or hereafter developed is forbidden.
The use in this publication of trade names, trademarks, service marks, and similar terms, even if they are not identified as such, is not to be taken as an expression of opinion as to whether or not they are subject to proprietary rights.

Printed on acid-free paper.

9 8 7 6 5 4 3 2 1

springer.com

Preface

Sampling-based computational methods have become a fundamental part of the numerical toolset of practitioners and researchers across an enormous number of different applied domains and academic disciplines. *This book is intended to provide a broad treatment of the basic ideas and algorithms associated with sampling-based methods, often also referred to as* Monte Carlo algorithms *or as* stochastic simulation. The reach of these ideas is illustrated here by discussing a wide range of different applications. Our goal is to provide coverage that reflects the richness of both the applications and the models that have found wide usage.

Of course, the models that are used differ widely from one discipline to another. Some methods apply across the entire simulation spectrum, whereas certain models raise particular computational challenges specific to those model formulations. As a consequence, the first part of the book focuses on general methods, whereas the second half discusses model-specific algorithms. The mathematical level is intended to accommodate the reader, so that for models for which even the model formulation demands some sophistication on the part of the reader (e.g., stochastic differential equations), the mathematical discussion will be at a different level from that presented elsewhere. While we deliver an honest discussion of the basic mathematical issues that arise in both describing and analyzing algorithms, we have chosen not to be too fussy with regard to providing precise conditions and assumptions guaranteeing validity of the stated results. For example, some theorem statements may omit conditions (such as moment hypotheses) that, while necessary mathematically, are not key to understanding the practical domain of applicability of the result. Likewise, in some arguments, we have provided an outline of the key mathematical steps necessary to understand (for example) a rate of convergence issue, without giving all the mathematical details that would serve to provide a complete and rigorous proof.

As a result, we believe that this book can be a useful simulation resource to readers with backgrounds ranging from an exposure to introductory probability to a much more advanced knowledge of the area. Given the wide range of examples and application areas addressed, our expectation is that students, practitioners, and researchers in statistics, probability, operations research, economics, finance, engineering, biology, chemistry, and physics will find the book to be of value. In addition to providing a development of the area pertinent to each reader's specific interests, our hope is that the book also serves to broaden our audience's view of both Monte Carlo and stochastic modeling, in general.

There exists an extensive number of texts on simulation and Monte Carlo methods. Classical general references in the areas covered by this book are (in chronological order) Hammersley & Handscombe [173], Rubinstein [313], Ripley [300], and Fishman [118]. A number of further ones can be found in the list of references; many of them contain much practically oriented discussion not at all covered by this book. There are further a number of books dealing with special subareas, for example Gilks et al. [129] on Markov chain Monte Carlo methods, Newman & Barkema [276] on applications to statistical physics, Glasserman [133] on applications to mathematical finance, and Rubinstein & Kroese [318] on the cross-entropy method.

In addition to standard journals in statistics and applied probability, the reader interested in pursuing the literature should be aware of journals like *ACM TOMACS* (*ACM Transactions of Modeling and Computer Simulation*), *Management Science*, and the *IEEE* journals. Of course, today systematic scans of journals are to a large extent replaced by searches on the web. At the end of the book after the References section, we give some selected web links, being fully aware that such a list is likely to be outdated soon. These links also point to some important recurrent conferences on simulation, see in particular $[w^3.14]$, $[w^3.16]$, $[w^3.17]$, $[w^3.20]$.

The book is designed as a potential teaching and learning vehicle for use in a wide variety of courses. Our expectation is that the appropriate selection of material will be highly discipline-dependent, typically covering a large portion of the material in Part A on general methods and using those special topics chapters in Part B that reflect the models most widely used within that discipline. In teaching this material, we view some assignment of computer exercises as being essential to gaining an understanding and intuition for the material. In teaching graduate students from this book, one of us (SA) assigns a computer lab of three hours per week to complement lectures of two hours per week. Exercises labeled (A) are designed for such a computer lab (although whether three hours is sufficient will depend on the students, and certainly some home preparation is needed). We have also deliberately chosen to not focus the book on a specific simulation language or software environment. Given the broad range of models covered, no single programming environment would provide a good universal fit. We prefer to let the user or teacher make the software choice herself. Finally, as a matter of teaching philosophy, we do not believe that programming should take a central role in a course taught from this book. Rather, the focus should be on understanding the intuition underlying the algorithms described here, as well as their strengths and weaknesses. In fact, to avoid a focus on the programming per se, we often hand out pieces of code for parts that are tedious to program but do not involve advanced ideas. Exercises marked (TP) are theoretical problems, highly varying in difficulty.

Since the first slow start of the writing of this book in 1999, we have received a large number of useful comments, suggestions, and corrections on earlier version of the manuscript. Thanks go first of all to the large number of students who have endured coping with these early versions. It would go too far to mention all the colleagues who have helped in one way or another. However, for a detailed reading of larger parts it is a pleasure to thank Hansjörg Albrecher, Morten Fenger-Grøn, Pierre L'Ecuyer, Thomas Mikosch, Leonardo Rojas-Nandayapa, and Jan Rosiński. At the technical level, Lars Madsen helped with many problems that were beyond our LaTeX ability.

A list of typos will be kept at [w^3.1], and we are greatful to be informed of misprints as well as of more serious mistakes and omissions.

Aarhus and Stanford   Søren Asmussen
February 2007   Peter W. Glynn

Contents

Notation

Internal Reference System

The chapter number is specified only if it is not the current one. As examples, Proposition 1.3, formula (5.7) or Section 5 of Chapter IV are referred to as IV.1.3, IV.(5.7) and IV.5, respectively, in all chapters other than IV where we write Proposition 1.3, formula (5.7) (or just (5.7)) and Section 5.

Special Typeface

d   differential like in dx, dt, $F(\mathrm{d}x)$; to be distinguished from a variable or constant d, a function $d(x)$ etc.

e   the base $2.71\ldots$ of the natural logarithm; to be distinguished from e which can be a variable or a different constant.

i   the imaginary unit $\sqrt{-1}$; to be distinguished from a variable i (typically an index).

$\mathbb{1}$ the indicator function, for example $\mathbb{1}_A$, $\mathbb{1}_{x \in A}$, $\mathbb{1}\{x \in A\}$, $\mathbb{1}\{X(t) > 0 \text{ for some } t \in [0,1]\}$.

O, o   the Landau symbols. That is, $f(x) . = O\big(g(x)\big)$ means that $f(x)/g(x)$ stays bounded in some limit, say $x \to \infty$ or $x \to 0$, whereas $f(x) = o\big(g(x)\big)$ means $f(x)/g(x) \to 0$.

π $3.1416\ldots$; to be distinguished from π which is often used for a stationary distribution or other.

$\mathcal{N}(\mu, \sigma^2)$ the normal distribution with mean μ and variance σ^2.

Probability, expectation, variance, covariance are denoted $\mathbb{P}$, $\mathbb{E}$, $\mathbb{V}ar$, $\mathbb{C}ov$. The standard sets are $\mathbb{R}$ (the real line $(-\infty, \infty)$), the complex numbers $\mathbb{C}$, the natural numbers $\mathbb{N} = \{0, 1, 2, \ldots\}$, the integers $\mathbb{Z} = \{0, \pm 1, \pm 2, \ldots\}$. Matrices and vectors are most often denoted by bold typeface, $\boldsymbol{C}$, $\boldsymbol{\Sigma}$, $\boldsymbol{x}$, $\boldsymbol{\alpha}$ etc., though exceptions occur. The transpose of $\boldsymbol{A}$ is denoted $\boldsymbol{A}^{\mathsf{T}}$.

Miscellaneous Mathematical Notation

$\stackrel{\text{def}}{=}$	a defining equality.
$\stackrel{\text{a.s.}}{\rightarrow}$	a.s. convergence
$\stackrel{\mathbb{P}}{\rightarrow}$	convergence in probability
$\stackrel{\mathscr{D}}{\rightarrow}$	convergence in distribution
$\stackrel{\mathscr{D}}{=}$	equality in distribution
$\longleftarrow$	an assignment in an algorithm (not used throughout)
$|\cdot|$	in addition to absolute value, also used for the number of elements (cardinality) $|S|$ of a set S, or its Lebesgue measure $|S|$.
$\mathbb{E}[X; A]$	$\mathbb{E}[X \mathbb{1}_A]$.
$\sim$	usually, $a(x) \sim b(x)$ means $a(x)/b(x) \to 1$ in some limit like $x \to 0$ or $x \to \infty$, but occasionally, other possibilities occur. E.g. $X \sim \mathcal{N}(\mu, \sigma^2)$ specifies X to have a $\mathcal{N}(\mu, \sigma^2)$ distribution.
$\approx$	a different type of asymptotics, often just at the heuristical level.
$\stackrel{\mathscr{D}}{\approx}$	approximate equality in distribution.
$\propto$	proportional to.
$\widehat{F}[\cdot]$	the m.g.f. of a distribution F. Thus $\widehat{F}[is]$ is the characteristic function at s. Sometimes $\widehat{F}[\cdot]$ is also used for the probability generating function of a discrete r.v.

The letter U is usually reserved for a uniform$(0, 1)$ r.v., and the letter z for a quantity to be estimated by simulation, Z for a r.v. with $\mathbb{E}Z = z$. As is standard, Φ is used for the c.d.f. of $\mathcal{N}(0, 1)$ and $\varphi(x) \stackrel{\text{def}}{=} e^{-x^2/2}/\sqrt{2\pi}$ for the density.. z_α often denotes the α-quantile of $\mathcal{N}(0, 1)$. A standard Brownian motion is denoted B and one with possibly drift $\mu \neq 0$ and/or variance σ^2 by W. Exceptions to all of this occur occasionally.

Conventions for a few selected standard distributions are given in A1.

Abbreviations

A-R	acceptance-rejection
BM	Brownian motion
c.g.f.	cumulant generating function (the log of the m.g.f.)
c.d.f.	cumulative distribution function, like $F(x) = \mathbb{P}(X \le x)$
CIR	Cox-Ingersoll-Ross
CLT	central limit theorem
CMC	crude Monte Carlo
ECM	exponential change of measure
fBM	fractional Brownian motion
FD	finite differences
FIFO	First-in-first-out
GBM	geometric Brownian motion
GSMP	generalized semi-Markov process
GW	Galton-Watson
i.i.d.	independent identically distributed
i.o.	infinitely often
IPA	infinitesimal perturbation analysis
l.h.s.	left hand side
LLN	law of large numbers
LR	likelihood ratio
MAP	Markov additive process
MCMC	Markov chain Monte Carlo
MH	Metropolis-Hastings
m.g.f.	moment generating function
MSE	mean square error
NIG	normal inverse Gaussian
ODE	ordinary differential equation
O-U	Ornstein-Uhlenbeck
PDE	partial differential equation
QMC	quasi Monte Carlo
RBM	reflected Brownian motion
r.h.s.	right hand side
r.v.	random variable
s.c.v.	squared coefficient of variation
SDE	stochastic differential equation
TAVC	time average variance constant
VaR	Value-at-Risk
w.l.o.g.	without loss of generality
w.p.	with probability

Chapter I
What This Book Is About

1 An Illustrative Example: The Single-Server Queue

We start by introducing one of the classical models of applied probability, namely the single-server queue. Queuing models are widely used across an enormous variety of application areas, and arise naturally when resource contention among multiple users creates congestion effects. We shall use the single-server queue as a vehicle for introducing some of the key issues that a simulator may need to confront when using simulation as a numerical tool; a parallel area illustrative for this purpose, option pricing, will be introduced in Section 3.

Consider a single-server queue possessing an infinite capacity waiting room and processing customers according to a "first-in–first out" (FIFO) queue discipline. Let A_n, D_n, and W_n be the arrival time, departure time, and waiting time (exclusive of service) for the nth customer to enter the queue. The FIFO discipline then clearly implies that

$$W_{n+1} = \left[D_n - A_{n+1} \right]^+,$$

where $[x]^+ \overset{\text{def}}{=} \max(0, x)$ ($\overset{\text{def}}{=}$ means a defining equality). Also, it is evident that $D_n = A_n + W_n + V_n$, where V_n is the service time of customer n, and hence

$$W_{n+1} = \left[W_n + V_n - T_n \right]^+ \tag{1.1}$$

(the *Lindley recursion*), where $T_n \overset{\text{def}}{=} A_{n+1} - A_n$ is the time between the arrivals of customers n and $n + 1$, $n = 0, 1, 2, \ldots$ Suppose that $\{V_n\}_{n \geq 0}$

and $\{T_n\}_{n \geq 0}$ are independent sequences of independent and identically distributed (i.i.d.) random variables (r.v.'s). Then the single-server queue model that we have described is known as the GI/G/1 queue.

Despite the simplicity of this model, it presents significant mathematical and computational challenges; in fact, thousands of papers have been devoted to the GI/G/1 queue and its applications. For example, consider computing the distribution of W_n. Even when $X_n \overset{\text{def}}{=} V_{n-1} - T_{n-1}$ has a distribution that can be computed explicitly (e.g., X_n is Gaussian or, more generally, infinitely divisible), the distribution of W_n can typically not be computed in closed form.

It follows from (1.5) below that

$$\mathbb{P}(W_n > x) = \int_{B_n(x)} \prod_{k=0}^{n-1} \mathbb{P}(V_k \in dv_k)\mathbb{P}(T_k \in dt_k), \qquad (1.2)$$

where

$$B_n(x) \overset{\text{def}}{=} \left\{(v_0, t_0), \ldots, (v_{n-1}, t_{n-1}) : \max_{k=0,\ldots,n-1} \sum_{j=k}^{n} (v_j - t_j) > x\right\},$$

so that $\mathbb{P}(W_n > x)$ can be computed as a $2n$-dimensional integral. Because of the high dimensionality, such a $2n$-dimensional numerical integration presents a significant challenge from a computational point of view. We shall return to this point in Chapter IX.

The distribution of W_n is an example of a *transient* characteristic, as opposed to *steady-state* or *stationary* characteristics, which are defined by taking the limit as $n \to \infty$. For example, an r.v. W_∞ having the limit distribution of W_n as $n \to \infty$ (provided such a limit exists as a probability measure on $\mathbb{R}$) is said to be the *steady-state waiting time*. Note that $\{W_n\}_{n \in \mathbb{N}}$ is a Markov chain with state space $[0, \infty)$. Therefore the theory of Markov chains with a discrete (i.e., finite or countable) state space suggests that under conditions corresponding to positive recurrence and aperiodicity,[1] W_∞ will exists, and that the Markov chain $\{W_n\}_{n \in \mathbb{N}}$ itself will obey the law of large numbers (LLN)

$$\frac{1}{N} \sum_{n=0}^{N-1} f(W_n) \overset{\text{a.s.}}{\to} \mathbb{E}f(W_\infty), \quad N \to \infty. \qquad (1.3)$$

This relation is one of the main reasons, if not the main one, for the interest in steady-state characteristics. Say we are interested in the average delay $N^{-1}\sum_0^{N-1} W_n$ of the first N customers. If N is large, (1.3) then asserts that $\mathbb{E}W_\infty$ should be a good approximation.

---

[1]The condition required for (1.3) and $W_\infty < \infty$ is that the load be strictly smaller than the offered service. This is expressed as $\rho < 1$, where $\rho = \mathbb{E}V/\mathbb{E}T$ is the *traffic intensity*; see [16].

Further transient characteristics of interest are first passage time quantities such as the time $\inf \{n : W_n > x\}$ until a customer experiences a long delay x, the number $\sigma \overset{\text{def}}{=} \inf \{n > 0 : W_n = 0\}$ of customers served in a busy period (recall $W_0 = 0$), and the total length $V_0 + \cdots + V_{\sigma-1}$ of the busy period. For both transient and steady-state characteristics, it is also of obvious interest to consider other stochastic processes associated with the system, such as the number $Q(t)$ of customers in system at time t (including the one being presently served), and the workload $V(t)$ (time to empty the system provided no new arrivals occur).

One of the key properties of the single-server queue is its close connection to random walk theory, which as a nice specific feature allows representations of steady-state distributions (equation (1.6) below) as well as transient characteristics (equation (1.5) below) in terms of an associated random walk. To make this connection precise, write as above $X_k = V_{k-1} - T_{k-1}$ and $S_n \overset{\text{def}}{=} X_1 + \cdots + X_n$ (with $S_0 = 0$), and note that if customer 0 enters an empty queue at time 0, then by (1.1),

$$
\begin{aligned}
W_1 &= \max(X_1, 0) = \max\big(S_1 - S_0, S_1 - S_1\big), \\
W_2 &= \max(W_1 + X_2, 0) \\
&= \max\big(\max(S_1 - S_0, S_1 - S_1) + S_2 - S_1, S_2 - S_2\big) \\
&= \max\big(S_2 - S_0, S_2 - S_1, S_2 - S_2\big) = S_2 - \min\big(S_0, S_1, S_2\big),
\end{aligned}
$$

and in general,

$$
W_n = S_n - \min_{k=0,\ldots,n} S_k = \max_{k=0,\ldots,n} [S_n - S_k]. \tag{1.4}
$$

Under our basic assumption that $\{V_n\}_{n\geq 0}$ and $\{T_n\}_{n\geq 1}$ are independent sequences of i.i.d. r.v.'s, $\{S_n\}_{n\geq 0}$ is a classical random walk, and (1.4) makes clear the connection between the GI/G/1 queue and the random walk.

Whereas (1.4) is a sample-path relation, a time-reversion argument translates (1.4) into a distributional relation of a simpler form. Indeed, using that

$$
\begin{aligned}
&\big(S_n - S_n, S_n - S_{n-1}, S_n - S_{n-2}, \ldots, S_n - S_1, S_n - S_0\big) \\
&= \big(0, X_n, X_n + X_{n-1}, \ldots, X_n + \cdots + X_2, X_n + \cdots + X_1\big) \\
&\overset{\mathscr{D}}{=} \big(0, X_1, X_1 + X_2, \ldots, X_1 + \cdots + X_{n-1}, X_1 + \cdots + X_n\big) \\
&= \big(S_0, S_1, S_2, \ldots, S_{n-1}, S_n\big),
\end{aligned}
$$

where $\overset{\mathscr{D}}{=}$ denotes equality in distribution, we get

$$
W_n = \max_{k=0,\ldots,n} [S_n - S_k] \overset{\mathscr{D}}{=} \max_{k=0,\ldots,n} S_k \overset{\text{def}}{=} M_n. \tag{1.5}
$$

As a consequence, $W_n \overset{\mathcal{D}}{\to} W_\infty$ as $n \to \infty$, where

$$W_\infty \overset{\mathcal{D}}{=} M \overset{\text{def}}{=} \max_{k \geq 0} S_k . \tag{1.6}$$

It follows that if $\rho = \mathbb{E}V_0/\mathbb{E}T_0 < 1$ (i.e., the mean arrival rate $1/\mathbb{E}T_0$ is smaller than the service rate $1/\mathbb{E}V_0$), then W_∞ is a proper r.v. (so that the steady state is well defined), and

$$\mathbb{P}(W_\infty > x) = \mathbb{P}(M > x) = \mathbb{P}(\tau(x) < \infty), \tag{1.7}$$

where $\tau(x) \overset{\text{def}}{=} \min\{n > 0 : S_n > x\}$.

It is easily seen that $\mathbb{P}(W_\infty > x)$ satisfies the integral equation

$$\mathbb{P}(W_\infty > x) = \int_0^\infty \mathbb{P}(W_\infty \in dy)\mathbb{P}(X_1 > x - y) \tag{1.8}$$

(the analogue of the stationarity equation for a Markov chain). One possible means of computing the distribution of W_∞ is therefore to numerically solve (1.8). However, rewriting (1.8) in the equivalent form

$$\mathbb{P}(W_\infty \leq x) = \int_0^\infty \mathbb{P}(W_\infty \leq x - y)\mathbb{P}(X_1 \in dy)$$

shows that (1.8) is of Wiener–Hopf type, and such equations are known to be numerically challenging.

The analytically most tractable special case of the GI/G/1 queue is the M/M/1 queue, where both the interarrival time and the service time distribution are exponential,[2] say with rates (inverse means) λ, μ. Then $\rho = \lambda/\mu$, and the distribution of W_∞ is explicitly available,

$$\mathbb{P}(W_\infty \leq x) = 1 - \rho + \rho(1 - e^{-\gamma x}),$$

where $\gamma \overset{\text{def}}{=} \mu - \lambda$; the probabilistic meaning of this formula is that the probability $\mathbb{P}(W_\infty = 0)$ that a customer gets served immediately equals $1 - \rho$, whereas a customer experiences delay w.p. ρ, and conditionally upon this the delay has an exponential distribution with rate parameter γ. Further, this is also the distribution of the steady-state workload $V(\infty)$, and the steady-state queue length $Q(\infty)$ is geometric with success parameter $1 - \rho$ (cf. A1), i.e., $\mathbb{P}(Q(\infty) = n) = (1 - \rho)\rho^n$, $n \in \mathbb{N}$. There are also explicit formulas for a number of transient characteristics such as the busy period density and the transition probabilities $p_{ij}^t = \mathbb{P}(Q(s + t) = j \mid Q(s) = i)$ of the Markov process $\{Q(t)\}_{t \geq 0}$, but the expressions are complicated and involve Bessel functions, even infinite sums of such, cf. [16, Section III.9].

Beyond the M/M/1 queue, the easiest special case of the GI/G/1 queue is GI/M/1 (exponential services), where steady-state quantitites have an

---

[2]M stands for Markovian or Memoryless. Note that the arrival process is just a Poisson process with rate λ.

explicit distribution given that one has solved the transcendental equation $\mathbb{E}e^{\gamma(V-T)} = 1$. Also, M/G/1 (Poisson arrivals) simplifies considerably, as discussed in Example 5.15 below.

2   The Monte Carlo Method

Given the stochastic origin of the integration problem (1.3), it is natural to consider computing $\mathbb{P}(W_n > x)$ by appealing to a sampling-based method. In particular, suppose that we could implement algorithms (on one's computer) capable of generating two independent sequences of i.i.d. r.v.'s $\{V_n\}_{n \geq 0}$ and $\{T_n\}_{n \geq 0}$ with the appropriate service-time and interarrival-time distributions. Then, by recursively computing the W_k according to the Lindley recursion (1.1), we would thereby obtain the r.v. W_n. By repeatedly drawing additional V_k and T_k, one could then obtain R i.i.d. copies $W_{1n}, \ldots, W_{Rn}$ of W_n. The probability $z \overset{\text{def}}{=} \mathbb{P}(W_n > x)$ could then be computed via the sample proportion of the W_{rn} that are greater than x, namely via the estimator

$$\widehat{z} \overset{\text{def}}{=} \widehat{z}_R \overset{\text{def}}{=} \frac{1}{R} \sum_{r=1}^{R} \mathbb{1}\{W_{rn} > x\}$$

($\mathbb{1}$ = indicator function). The LLN (1.3) guarantees, of course, that the algorithm converges to $z = \mathbb{P}(W_n > x)$ as the number R of independent replications tends to ∞.

This example exposes a basic idea in the area of stochastic simulation, namely to simulate independent realizations of the stochastic phenomenon under consideration and then to compute an estimate for the probability or expectation of interest via an appropriate estimator obtained from independent samples.

To be more precise, suppose that we want to compute $z = \mathbb{E}Z$. The idea is to develop an algorithm that will generate i.i.d. copies $Z_1, \ldots, Z_R$ of the r.v. Z and then to estimate z via the sample-mean estimator

$$\widehat{z} \overset{\text{def}}{=} \widehat{z}_R \overset{\text{def}}{=} \frac{1}{R} \sum_{r=1}^{R} Z_r \,. \qquad (2.1)$$

In other words, one runs R independent computer experiments replicating the r.v. Z, and then computes z from the sample. Use of random sampling or a method for computing a probability or expectation is often called the *Monte Carlo method*. When the estimator $\widehat{z}$ of $z = \mathbb{E}Z$ is an average of i.i.d. copies of Z as in (2.1), then we refer to $\widehat{z}$ as a *crude Monte Carlo* (CMC) estimator.

Note that an LLN also holds for many dependent and asymptotically stationary sequences, see, for example, the discussion surrounding (1.3) for

Markov chains and Chapter IV. As a consequence, one can compute characteristics like $\int f \mathrm{d}\pi$ of the limiting stationary distribution π by averaging over a long simulation-time horizon. For example, the integral equation (1.8) can be solved numerically in this way.

3   Second Example: Option Pricing

Financial mathematics has in recent years become one of the major application areas of stochastics. It draws on the one hand from a body of well-established theory, and on the other, it raises new problems and challenges, both from the theoretical and computational points of view.

The most classical problem in financial mathematics is option pricing, which we use here to parallel the single-server queue as a vehicle for illustrating some of the key issues arising in simulation. The prototype of an option is a European call option, which gives the holder the right (but not the obligation) to buy a certain amount of a given asset at the price K at time T (the *maturity time*). For example, the asset can be crude oil, and the option then works as an insurance against high oil prices. More precisely, if $S(T)$ is the market price[3] of the specified amount of oil at time T and $S(T) < K$, the holder will not exercise his right to buy at price K, but will buy at the market at price $S(T)$. Conversely, if $S(T) > K$, the holder will buy and thereby make a gain of $S(T) - K$ compared to buying at the market price. Thus, $\left[S(T) - K\right]^{+}$ is the value of exercising the option relative to buying at market price. What price Π should one pay for acquiring this option at time $t = 0$?

Let r be the continuously compounded interest rate (the short rate), so that the value at time 0 of receiving one monetary unit at time t is e^{-rt}. Traditional thinking would just lead to the price $\mathrm{e}^{-rT}\mathbb{E}\left[S(T) - K\right]^{+}$ of the option because the LLN predicts that this on average will favor neither the buyer nor the seller, assuming that both trade a large number of options over time. This is indeed how, for example, actuaries price their insurance contracts. Economists take a different starting point, the principle of *no arbitrage*, which states that the market will balance itself in such a way that there is no way of making money without risk (no *free lunches*). The "market" needs specification in each case, but is often a world with only two objects to invest in at time $0 \leq t < T$: the bank account (which may have a negative value) yielding a risk-free return at short rate r and the underlying asset priced at $S(t)$. Indeed, this leads to a different price of the option, as we will now demonstrate via an extremely simple example.

---

[3] Since $S(T)$ is not known at time $t = 0$, this is a random quantity, and we use $\mathbb{P}, \mathbb{E}$ to refer to this; for example, the expected gain of holding the option is $\mathrm{e}^{-rT}\mathbb{E}\left[S(T) - K\right]^{+}$.

Example 3.1 We will consider a binomial European call option with pay-off $\left(S(1) - K\right)^{+}$, where $S(1)$ is thought of as a stock price after one unit of time. We will assume that there are only two possible values for $S(1)$ and (w.lo.g.) that $S(0) = 1$. That is, $\mathbb{P}\left(S(1) = u\right) = p$, $\mathbb{P}\left(S(1) = d\right) = q \stackrel{\text{def}}{=} 1 - p$ (up and down), where $d < u$.

An investor with an initial wealth of w_0 is given only two possibilities, to invest at $t = 0$ in the asset (referred to as the *stock* in the following) or in a bank account at fixed interest rate r. If a_1 is the volume of stocks he invests in and a_2 the amount he puts in the bank, we thus have

$$w_0 = a_1 S(0) + a_2 = a_1 + a_2, \quad w_1 = a_1 S(1) + a_2 e^r,$$

where w_1 is the wealth at $T = 1$. We allow a_1, a_2 to be non-integer and to attain negative values, which corresponds to borrowing and short selling.[4] The pair a_1, a_2 is referred to as a *portfolio*.

A portfolio is said to *hedge* the option if its return w_1 is exactly the payout $\left(S(1) - K\right)^{+}$ of the option no matter how $S(1)$ comes out. An investor would therefore be indifferent whether he puts his money in the portfolio or in the option, so that we should have $\Pi = w_0$. Writing $V_u \stackrel{\text{def}}{=} (u - K)^{+}$ (the payoff if the stock goes up) and $V_d \stackrel{\text{def}}{=} (d - K)^{+}$, the requirement $w_1 = \left(S(1) - K\right)^{+}$ then means that $V_u = a_1 u + a_2 e^r$, $V_d = a_1 d + a_2 e^r$. This is two linear equations with two unknowns a_1, a_2, and the solution is $a_1 = (V_u - V_d)/(u - d)$, $a_2 = (uV_d - dV_u)/(u - d)e^r$. Thus

$$\Pi = w_0 = \frac{V_u - V_d}{u - d} + \frac{uV_d - dV_u}{(u - d)e^r}. \tag{3.1}$$

Probably the most surprising feature of this formula is that p does not enter. Intuitively, one feels that the option of buying a stock for a price of K at $T = 1$ is more attractive the larger p is. But this is not reflected in (3.1).

The market is said to *allow arbitrage* if it is possible to choose a_1, a_2 such that $\mathbb{P}(w_1 \geq w_0) = 1$ and $\mathbb{P}(w_1 > w_0) > 0$. It is easy to see that $d < e^r < u$ is a necessary condition for the market to be free of arbitrage.[5] Thus introducing $p^* \stackrel{\text{def}}{=} (e^r - d)/(u - d)$, $q^* \stackrel{\text{def}}{=} 1 - p^* = (u - e^r)/(u - d)$, we have $0 < p^* < 1$ in an arbitrage-free market, and it is easy to see that (3.1) can be rewritten

$$\Pi = w_0 = e^{-r}\left[p^* V_u + q^* V_d\right] = \mathbb{E}^*\left[e^{-r}(S(1) - K)^{+}\right], \tag{3.2}$$

---

[4] The meaning of $a_2 < 0$ is simply that $-a_2$ is borrowed from the bank, whereas $a_1 < 0$ means that the investor has taken the obligation to deliver a volume of $-a_1$ stocks at time $T = 1$ (short selling).

[5] For example, if $d > e^r$, an arbitrage opportunity is to borrow from the bank and use the money to buy the stock.

where $\mathbb{E}^*$ denotes expectation with p replaced by p^*. Thus, the price has the same form as the naive guess $\mathbb{E}[e^{-r}(S(1) - K)^+]$ above, but under a different probability specification. One can further check that $\mathbb{E}^* S(1) = S(0)e^r = e^r$. That is, in $\mathbb{E}^*$-expectation the stock behaves like the bank account. For this reason, p^* is referred to as the *risk-neutral probability*. Equivalently, $\{e^{-rt}S(t)\}_{t=0,1}$ is a martingale under the risk-neutral measure.   □

The same procedure applies to more than one period and to many other types of options and other models for $\{S(t)\}$. If $\Phi(S_0^T)$ is the payout at maturity where $S_0^T \overset{\text{def}}{=} \{S(t)\}_{0 \le t \le T}$, the principle is:

1. Compute risk-neutral parameters such that $\{e^{-rt}S(t)\}$ is a martingale under the changed distribution $\mathbb{P}^*$ and such that[6] $\mathbb{P}$ and $\mathbb{P}^*$ are equivalent measures on the space of paths for $\{S(t)\}$.[7]

2. Set the price as $\Pi = \mathbb{E}^*[e^{-rT}\Phi(S_0^T)]$.

In some cases, such a risk-neutral P^* exists and is unique; in other cases, there may be many or none, obviously creating some difficulties for this approach to option pricing.

Example 3.2 For continuous-time formulations, the plain vanilla model for the asset price process $\{S(t)\}_{0 \le t \le T}$ is geometric Brownian motion (GBM), i.e., $\{\log S(t)\}$ is Brownian motion (BM) with drift μ and variance constant σ^2, say; often this model is also called the *Black–Scholes model*. Here the risk-neutral measure is unique and is another GBM with changed drift $\mu^* \overset{\text{def}}{=} r - \sigma^2/2$ and the same variance constant σ^2. Equivalently, under $\mathbb{P}^*$, $\{S(t)\}$ is the solution of the stochastic differential equation (SDE)

$$\mathrm{d}S(t) = r\,\mathrm{d}t + \sigma\,\mathrm{d}B(t), \qquad (3.3)$$

where $\{B(t)\}$ is standard Brownian motion. The price therefore is just an integral in the normal distribution,

$$\begin{aligned}
\Pi &= e^{-rT}\int_{-\infty}^{\infty}\big[S(0)\exp\{\mu^*T + \sigma T^{1/2}x\} - K\big]^+\varphi(x)\,\mathrm{d}x \\
&= S(0)\Phi(d_1) - e^{-rT}K\Phi(d_2), \qquad (3.4)
\end{aligned}$$

where $\varphi(x) \overset{\text{def}}{=} e^{-x^2/2}/(2\pi)^{1/2}$ is the density of the standard normal distribution, Φ its c.d.f., and

$$d_1 \overset{\text{def}}{=} \frac{1}{\sigma T^{1/2}}\big[\log(S(0)/K) + (r + \sigma^2/2)T\big], \quad d_2 \overset{\text{def}}{=} d_1 - \sigma T^{1/2}.$$

Formula (3.4) is known as the *Black–Scholes formula*.   □

---

[6]We do not explain the reasons for this requirement here!

[7]Usually taken as the Skorokhod space $D[0,T]$ of real-valued functions on $[0,T]$, which are right-continuous and have left limits.

In view of the Black–Scholes formula, Monte Carlo simulation is not required to price a European call option in the GBM model. The need arises when one goes beyond this model, e.g., in the case of basket options (see below) or Lévy models, which have recently become popular, see Cont & Tankov [75] and Chapter XII.

Further examples of options include:

Asian (call) options, where the payoff $\Phi(S_0^T)$ is $(A - K)^+$, where A is an arithmetic average, either $T^{-1} \int_0^T S(t)\, dt$ (continuous sampling) or $N^{-1} \sum_1^N S(nT/N)$ (discrete sampling). In the example where the underlying asset is the oil price and the buyer an oil consumer, the Asian option serves to ensure a maximal spending of NK in $[0, T]$.

Put options, where the options gives the right to sell rather than to buy. For example, an European put option has payoff $\Phi(S_0^T) = \big[K - S(T) \big]^+$ and ensures a minimum price of K for the asset when sold at time T. Similar remarks apply to Asian put options with $\Phi(S_0^T) = (A - K)^+$.

Barrier options, where in addition to the strike price K one specifies a lower barrier $L < K$ and the payout depends on whether the lower barrier is crossed before T. There are now a variety of payoff functions, e.g., $\Phi(S_0^T) = \big(S(T) - K\big)^+ \mathbb{1}\{\tau > T\}$ ("down and out"), where $\tau \stackrel{\text{def}}{=} \inf\{t : S(t) < L\}$.

Basket options, where, as one example, $S(t)$ can be the weighted sum of several stock prices, that is, the value of a portfolio of stocks.

For the GBM Black–Scholes model, there are explicit formulas for the price of many types of options. Even so, Monte Carlo simulation is required to price some options in this model, e.g., a discretely sampled Asian option. For most other models, explicit price formulas are typically not available even for European options. In addition to alternative assumptions on the dynamics of $\{S(t)\}$, examples of model extensions beyond Black–Scholes include stochastic volatility and stochastic interest rate models, where the σ in (3.3) (the volatility) and/or the short rate r are replaced by stochastic processes $\{\sigma(t)\}$, $\{r(t)\}$.

4   Issues Arising in the Monte Carlo Context

We turn next to a discussion of some of the computational issues that arise in the Monte Carlo context.

Issue 1: *How do we generate the needed input random variables?*

Our above discussion for the GI/G/1 queue assumes the ability to generate independent sequences $\{V_n\}$ and $\{T_n\}$ of i.i.d. r.v.'s with appropriately de-fined distributions. Similarly, in the example of European options, we need to be able to generate $S(T)$ under the distributional assumption under con-sideration. The simplest examples are the M/M/1 queue, where the V_k, T_k are exponential, and the the Black-Scholes model where $S(T)$ is lognor-mal or, equivalently, $\log S(T)$ is normal, but more complicated models can occur, see, for example, Chapter XII.

As will be seen in Chapter II, all such algorithms work by transforming a sequence of uniform i.i.d. r.v.'s into the appropriate randomness needed for a given application. Thus, we shall need to discuss the generation of i.i.d. uniform r.v.'s (II.1 = Section 1 of Chapter II) and to show how such uniform randomness can be transformed into the required nonuniform randomness needed by the simulation experiment (II.2–II.6).

In the option-pricing example, we have the additional difficulty that we cannot simulate the entire path $\{S(t)\}_{0 \le t \le T}$. At best, we can simulate $\{S(t)\}_{0 \le t \le T}$ only at a finite number of discrete time points. Furthermore, because $\{S(t)\}$ typically evolves continuously through dynamics specified via an SDE, the same difficulties can arise here as in the solution of ordinary (deterministic) differential equations (ODEs). In particular, the forward simulation via the process with a finite-difference approximation induces a systematic bias into the simulated process. This is comparable to the error associated with finite-difference schemes for ODEs; see further Chapter X.

Issue 2: *How many computer experiments should we do?*

Given that we intend to use the Monte Carlo method as a computational tool, some means of assessing the accuracy of the estimator is needed. *Sim-ulation output analysis* comprises the body of methods intended to address this issue (Chapter III). One standard way of assessing estimator accuracy is to compute a (normal) confidence interval for the parameter of interest, so a great deal of our discussion relates to computing such confidence in-tervals. Special difficulties arise in:

(a) the computation of steady-state expectations, as occurs in analyzing the steady-state waiting time r.v. W_∞ of the GI/G/1 queue (the limit in distribution of W_n as $n \to \infty$), see Chapter IV;

(b) quantile estimation, as occurs in value-at-risk (VaR) calculations for portfolios; see Example 5.17 and III.4a.

A necessary condition for approaching item (a) at all is of course that we be able to deal with the following issue.

Issue 3: *How do we compute expectations associated with limiting station-ary distributions?*

For the GI/G/1 queue, the problem is of course that the relation $W_n \xrightarrow{\mathscr{D}} W_\infty$ and the possibility of generating an r.v. distributed as W_n does not allow us to generate an r.v. distributed as W_∞. In other words, we are facing an *infinite-horizon problem*, in which an ordinary Monte Carlo experiment cannot provide an answer in finite time. This is also clearly seen from the formula (1.7), stating that $\mathbb{P}(W_\infty > x) = \mathbb{P}(M > x) = \mathbb{P}(\tau(x) < \infty)$, where $\tau(x) = \min\{n > 0 : S_n > x\}$ and $M = \max_{n=0,1,2,\ldots} S_n$. Indeed, simulating a finite number n of steps of the random walk S_n cannot determine the value of M but only of M_n, and furthermore, one can get an answer only to whether the event $\{\tau(x) \leq n\}$ occurs but not whether $\{\tau(x) < \infty\}$ does.

Of course, one possible approach is to use a single long run of the W_n and appeal to the LLN; cf. the concluding remarks of Section 2. Here, the complication is that error assessment will be challenging because the observations are serially correlated.

Option-pricing examples typically involve only a finite horizon. However, a currently extremely active area of steady-state simulation with a rather different flavor from that of queuing models is *Markov chain Monte Carlo methods* (MCMC), which we treat separately in Chapter XIII; some key application areas are statistics, image analysis, and statistical physics.

Issue 4: *Can we exploit problem structure to speed up the computation?*

Unlike the sampling environment within which statisticians work, the model that is generating the sample is completely known to the simulator. This presents the simulator with an opportunity to improve the computational efficiency (i.e., the convergence speed as a function of the computer time expended) by exploiting the structure of the model.

For example, in a discretely sampled Asian option, the geometric average $A_{\text{geom}} \overset{\text{def}}{=} \left[S(T/N) \cdot S(2T/N) \cdots S(T)\right]^{1/N}$ has a lognormal distribution with easily computable parameters in the Black–Scholes model. Thus $\mathbb{E}A_{\text{geom}}$ is known, and for any choice of $\lambda \in \mathbb{R}$,

$$(A - K)^+ - \lambda(A_{\text{geom}} - \mathbb{E}A_{\text{geom}})$$

is again an estimator having mean $\mathbb{E}(A - K)^+$. Hence, in computing $\mathbb{E}(A - K)^+$, the simulator can choose λ to minimize the variance of the resulting estimator, thereby improving the efficiency of the naive estimator.

Similarly, in the context of the GI/G/1 queue, $\mathbb{E}X_1$ is known, so that for any choice of $\lambda \in \mathbb{R}$,

$$W_n - \lambda(S_n - n\mathbb{E}X_1)$$

is again an estimator having mean $\mathbb{E}W_n$.

The common idea of these two examples is just one example of *variance reduction* (and is the one specifically known as the method of *control variates*). Use of good variance-reduction methods can significantly enhance the efficiency of a simulation (and is discussed in detail in Chapter V).

Issue 5: *How does one efficiently compute probabilities of rare events?*

Suppose that we wish to compute the probability that a typical customer in steady state waits more than x prior to receiving service, i.e., $\mathbb{P}(W_\infty > x)$. If x is large, $z \stackrel{\text{def}}{=} \mathbb{P}(W_\infty > x)$ (or, equivalently, $z = \mathbb{P}(\tau(x) < \infty)$) is the probability of a *rare event*. For example, if $z = 10^{-6}$, the fraction of customers experiencing waiting times exceeding x is one in a million, so that the simulation of 10 million customers provides the simulator with (on average) only 10 samples of the rare event. This suggests that accurate computation of rare-event probabilities presents significant challenges to the simulator.

Chapter VI is devoted to a discussion of several different methods for increasing the frequency of rare events, and thereby improving the efficiency of rare-event simulation. The topic is probably more relevant for queuing theory and its communications applications than for option pricing. However, at least some option-pricing problems exhibit similar features, say a European call option that is out-of-the-money ($\mathbb{P}(S(T) > K)$ is small), and rare events also occur in VaR calculations, in which a not uncommon quantile to look for is the 99.97% quantile, cf. McNeil et al. [252].

Issue 6:   *How do we estimate the sensitivity of a stochastic model to changes in a parameter?*

The need to estimate such sensitivities arises in mathematical finance in hedging. More precisely, in a hedging portfolio formed by a bank account and the asset underlying the option one tries to hedge, it holds rather generally (Björk [45]) that the amount to invest in the asset at time t should equal what is called the the *delta*, the partial derivative of the expected payout at maturity w.r.t. the current asset price. For another example, assume that in the Black–Scholes model the volatility σ is not completely known but provided as an estimate $\widehat{\sigma}$. This introduces some uncertainty in the option price Π, for the quantification of which we need $\partial\Pi/\partial\sigma$, the *vega*.

In the GI/G/1 queue, suppose we are uncertain about the load the system will face. In such circumstances, it may be of interest to compute the sensitivity of system performance to changes in the arrival rate. More precisely, suppose that we consider a parameterized system in which the arrival epochs are given by the sequence $\{\lambda^{-1}A_n\}$, so that $\lambda/\mathbb{E}A_1$ is the arrival rate. The sensitivity of the system performance relating to long delays is then $(\mathrm{d}/\mathrm{d}\lambda)\mathbb{P}(W_\infty > x)$.

Chapter VII discusses the efficient computation of such derivatives (or, more generally, gradients).

Issue 7: *How do we use simulation to optimize our choice of decision parameters?*

Suppose that in the GI/G/1 queue with a high arrival rate (say exceeding λ_0), we intend to route a proportion p of the arriving customers to an alternative facility that has higher cost but is less congested, in order to ensure that customers are served promptly. There is a trade-off between increasing customer satisfaction (increasing p) and decreasing costs (decreasing p). The optimal trade-off can be determined as the solution to an optimization problem in which the objective function is computed via simulation.

Also, the pricing of some options involves an optimization problem. For example, an American option can be exercised at any stopping time $\tau \leq T$ and then pays $\left(S(\tau) - K\right)^+$. Optimizing τ for American options is a high-dimensional optimization problem, which due to its complexity and special nature is beyond the scope of this book (we refer to Glasserman [133, Chapter 8]), but is mentioned here for the sake of completeness.

A discussion of algorithms appropriate to simulation-based optimization is offered in Chapter VIII.

5   Further Examples

The GI/G/1 queue and option pricing are of course only two among many examples in which simulation is useful. Also, in some cases, other applications involve other specific problems than those discussed in Section 4, and the evaluation of an expected value is far from the only application of simulation. The following examples are intended to illustrate these points as well as to introduce some of the models and problems that serve as recurrent examples throughout the book.

Example 5.1 Let $f : (0,1)^d \to \mathbb{R}$ be a function defined on the d-dimensional hypercube and assume we want to compute

$$z \stackrel{\text{def}}{=} \int_{(0,1)^d} f\left(u_1,\ldots,u_d\right) du_1 \cdots du_d = \int_{(0,1)^d} f(\boldsymbol{u}) \, d\boldsymbol{u} \,,$$

where $\boldsymbol{u} \stackrel{\text{def}}{=} \left(u_1,\ldots,u_d\right)$. We can then write $z = \mathbb{E}f\left(U_1,\ldots,U_d\right) = \mathbb{E}f(\boldsymbol{U})$, where $U_1,\ldots,U_d$ are i.i.d. r.v.'s with a uniform$(0,1)$ distribution and $\boldsymbol{U} \stackrel{\text{def}}{=} \left(U_1,\ldots,U_d\right)$, and apply the Monte Carlo method as for the GI/G/1 queue to estimate z via

$$\widehat{z} \stackrel{\text{def}}{=} \frac{1}{R}\left(f(\boldsymbol{U}_1) + \cdots + f(\boldsymbol{U}_R)\right),$$

where $\boldsymbol{U}_1,\ldots,\boldsymbol{U}_R$ are i.i.d. replicates of $\boldsymbol{U}$; these can most often be easily generated by taking advantage of the fact that uniform$(0,1)$ variables are built in as standard in most numerical software packages (a totality of dR such variables is required).

Of course, the nonrandomized computation of the integral z has been the topic of extensive studies in numerical analysis, and for small d and smooth integrands f, standard quadrature rules perform excellently and are in general superior to Monte Carlo methods (see further Chapter IX).

More generally, let $f : \Omega \rightarrow \mathbb{R}$ be defined on a domain $\Omega \in \mathbb{R}^d$. We can then choose some reference density $g(\boldsymbol{x})$ on Ω and write

$$z = \int_\Omega f(\boldsymbol{x}) \, \mathrm{d}\boldsymbol{x} = \int_\Omega \frac{f(\boldsymbol{x})}{g(\boldsymbol{x})} g(\boldsymbol{x}) \, \mathrm{d}\boldsymbol{x} = \mathbb{E}\left[\frac{f(\boldsymbol{X})}{g(\boldsymbol{X})}\right],$$

where $\boldsymbol{X}$ is an r.v. with density $g(\boldsymbol{x})$. Thus replicating $\boldsymbol{X}$ provides a Monte Carlo estimate of z as the average of the $f(\boldsymbol{X}_r)/g(\boldsymbol{X}_r)$. For example, $g(\boldsymbol{x})$ could be the $\mathcal{N}(\boldsymbol{0}, \boldsymbol{I}_d)$ density or, if $\Omega \subseteq (0, \infty)^d$, the density $\mathrm{e}^{-x_1 - \cdots - x_d}$ of d i.i.d. standard exponentials. Another procedure in the case of a general Ω is to use a transformation $\varphi : \Omega \rightarrow (0, 1)^d$, where φ is 1-to-1 onto $\varphi(\Omega)$. Then

$$z = \int_\Omega f(\boldsymbol{x}) \, \mathrm{d}\boldsymbol{x} = \int_{\varphi(\Omega)} f(\psi(\boldsymbol{u})) J(\boldsymbol{u}) \, \mathrm{d}\boldsymbol{u} = \mathbb{E}\big[f(\psi(\boldsymbol{U})) J(\boldsymbol{U})\big],$$

where $\psi \stackrel{\mathrm{def}}{=} \varphi^{-1}$ and $J(\boldsymbol{u})$ is the Jacobian, that is, the absolute value of the determinant of the matrix

$$\left(\frac{\partial \boldsymbol{x}}{\partial \boldsymbol{u}}\right) = \begin{pmatrix} \frac{\partial \psi_1(\boldsymbol{u})}{\partial u_1} & \cdots & \frac{\partial \psi_d(\boldsymbol{u})}{\partial u_1} \\ \vdots & & \vdots \\ \frac{\partial \psi_1(\boldsymbol{u})}{\partial u_d} & \cdots & \frac{\partial \psi_d(\boldsymbol{u})}{\partial u_d} \end{pmatrix}.$$

We return to integration in Chapter IX. □

Example 5.2 An example of a similar flavor as the last part of Example 5.1 occurs in certain statistical situations in which the normalizing constant $z(\theta)$ determining a likelihood $f_\theta(x)/z(\theta)$ is not easily calculated. That is, we have an observation x that is the outcome of a r.v. X taking values in Ω (where Ω is some arbitrary space) with distribution F_θ governed by a parameter θ, such that the density w.r.t. some reference measure μ (independent of θ) on Ω is $f_\theta(y)/z(\theta)$, where thus $z(\theta) \stackrel{\mathrm{def}}{=} \int_\Omega f_\theta(y) \, \mu(\mathrm{d}y)$. However, there are situations in which $z(\theta)$ is not easily calculated, even if the form of $f_\theta(y)$ is simple.

If r.v.'s $X_1, \ldots, X_R$ with density $f_\theta(x)/z(\theta)$ are easily simulated, we may just for each θ calculate $z(\theta)$ as the average of the $Z \stackrel{\mathrm{def}}{=} f_\theta(X_r)$. However, often this is not the case, whereas generation from a reference distribution G with density $g(x)$ is easy. One may then write

$$z(\theta) = \int_\Omega \frac{f_\theta(y)}{g(y)} g(y) \, \mu(\mathrm{d}y) = \mathbb{E}_G Z(\theta),$$

where $Z(\theta) \stackrel{\mathrm{def}}{=} f_\theta(X)/g(X)$, and determine $z(\theta)$ by Monte Carlo simulation, where X is generated from G. A suitable choice of G could often be $G = F_{\theta_0}$ for some θ_0 such that F_{θ_0} has particularly simple properties.

The procedure is a first instance of *importance sampling*, to which we return in V.1.

Consider, for example, point processes in a bounded region Ω in the plane (or in $\mathbb{R}^d$), which are often specified in terms of an unnormalized density f_θ w.r.t. the standard Poisson process. One example is the *Strauss process*. A point x in the sample space can be identified with a finite (unordered) collection $a_1, \ldots, a_m$, where $m = m(x)$ is the number of points and a_i the position of the ith. Then $f_\theta(x) = \lambda^{m(x)}\eta^{t(x)}$ for the Strauss process, where

$$\theta = (\lambda, \eta) \in (0, \infty) \times [0, 1], \quad t(x) \overset{\text{def}}{=} \sum_{i \neq j} \mathbb{1}\{d(a_i, a_j) < r\}$$

(Euclidean distance) and $0 < r < \infty$ is given. For $\eta = 1$, this is just the Poisson(λ) process, whereas $\eta = 0$ corresponds to a so-called *hard-core model*, in this case a Poisson process conditioned to have no pair of points at distance less than r; the interpretation of the intermediate case $0 < r < 1$ is some repulsion against points being closer than r. For simulation of $z(\theta)$, one would obviously take $\theta_0 = (\lambda, 1)$ since a Poisson process is easy to simulate. See further Møller & Waagepetersen [265]. □

Example 5.3 A further related example is counting the number of elements $z \overset{\text{def}}{=} |S|$ of a finite but huge and not easily enumerated set S. Assume that $S \subseteq T$ with T finite and that some r.v. X on T is easily simulated, say the point probabilities are $p(x) \overset{\text{def}}{=} \mathbb{P}(X = x)$. We can then write

$$z = \sum_{x \in T} \mathbb{1}_{x \in S} = \sum_{x \in T} \frac{1}{p(x)} \mathbb{1}_{x \in S}\, p(x) = \mathbb{E}Z\,,$$

where $Z \overset{\text{def}}{=} \mathbb{1}_{X \in S}/p(X)$, and apply the Monte Carlo method.

For example, consider the graph-coloring problem in which S is the set of possible colorings of the edges $\mathscr{E}$ of a graph $G = (V, \mathscr{E})$ (V is the set of vertices) with colors $1, \ldots, q$ such that no two edges emanating from the same vertex have the same color. Say V is a set of countries on a map and two vertices are connected by an edge if the corresponding countries have a piece of common border. An obvious choice is then to take T as the set of all possible colorings (the cardinality is $n \overset{\text{def}}{=} |E|^q$) and X as uniform on T such that $p(x) = 1/n$.

This is an instance of a *randomized algorithm*, that is, an algorithm that involves coin-tossing (generated randomness) to solve a problem that is in principle deterministic. There are many examples of this, with Motwani & Raghavan [263] being a standard reference. We return to some in XIV.1. Other main ones like QUICKSORT, which sorts a list of n items, are not discussed because the involved simulation methodology is elementary.

Note that the suggested algorithm for the graph-coloring problem is somewhat deceiving because in practice $|T|$ is much bigger than $|S|$, making the algorithm inefficient (for efficiency, $p(\cdot)$ should be have substantial mass on S, cf. Exercise V.1.3). We return to alternative strategies in XIV.3. □

Example 5.4 An important area of application of randomized algorithms is optimization, that is, finding and locating the minimum of a deterministic function $H(x)$ on some feasible region S.

One approach involves the *Boltzmann distribution* with density $f(x) \overset{\text{def}}{=} e^{-H(x)/T}/C$ w.r.t. some measure $\mu(dx)$, where $C = C(T)$ is a normalizing constant ensuring $\int f \, d\mu = 1$. Here one refers to H as the *Hamiltonian* or *energy function*, and to T as the *temperature* (sometimes the definition also involves Planck's constant, which, however, conveniently can be absorbed into T). At high temperatures, the distribution is close to μ; at low, it concentrates its mass on x-values with a small $H(x)$. By choosing T small and generating an r.v. X according to f, the outcome of $H(X)$ will therefore with high probability be close to the minimum.

We return to this and further examples in XIV.1.   □

Example 5.5 *Filtering* or *smoothing* means giving information about an unobserved quantity X from an observation Y. Often, the joint density $f(x, y)$ of X, Y is readily available, but the desired conditional density

$$f(x|y) \;=\; \frac{f(x, y)}{\int f(x', y) \, dx'} \tag{5.1}$$

of X given Y involves a normalization constant obtained by integrating x' out in $f(x', y)$, a problem that is often far from trivial. Examples of filtering are reconstructing the current location of a target from noisy observations and reconstructing a signal or an image from blurred observations; see further XIII.2d and A5. Here often the observations arrive sequentially in time and X is a segment of a Markov chain with possibly nondiscrete state space. The problem is essentially the same as that of Bayesian statistics (see XIII.2a), where X plays the role of an unknown parameter (usually denoted by θ) with a prior $\pi^0(x)$, so that the joint density of (X, Y) is $f(x, y) = \pi^0(x)f(y \,|\, x)$, where $f(y \,|\, x)$ is the likelihood, and (5.1) is the posterior on x.

In the setting of filtering, Bayesian statistics, likelihood computations as in Example 5.2, the Bolztmann distribution, a uniform distribution on a finite set S with an unknown number S of elements (cf. Example 5.3), and many other situations with a density $f(x)$ only known up to a constant, most standard methods for generating r.v.'s from $f(x)$ fail. For example, how do we generate a realization of the Strauss process, or a point pattern that at least approximately has the same distributional properties?

One approach is provided by *Markov chain Monte Carlo methods*, a topic of intense current activity, as further discussed in Chapter XIII. Another is resampling methods and particle filtering, as surveyed briefly in XIV.2.   □

Example 5.6 A *PERT net* is a certain type of directed graph with one intial node ◁ and one terminal node ▷. It models a production process, whose stages are represented by the edges of the graph (PERT = Project

Evaluation and Review Technique). As an example, assume (following Winston [366] p. 410) that we want to produce an item assembled from two parts $1, 2$, and that the stages are A: train workers, B: purchase materials, C: produce 1, D: produce 2, E: test 2, F: assemble $1, 2$. Then C, D cannot be performed before A,B are finished, which is expressed by saying that A, B are predecessors of C, D. Similarly, C, E are predecessors of F and D is a predecessor of E. The traditional graph representation of this is given in Figure 5.1. The edge 0 is dummy (representing a task that is carried out instantaneously), and is added to conform with the convention that there is at most one edge connecting two nodes.

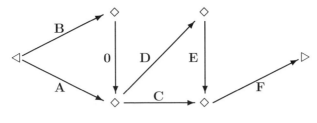

FIGURE 5.1

The set of all possible paths $i_1 i_2 \ldots i_N$ from $\triangleleft$ to $\triangleright$ is denoted by $\mathscr{P}$. Thus, in Figure 5.1 $\mathscr{P} = \{ACF, ADEF, B0CF, B0DEF\}$. Associated with edge i is a r.v. $Y_i \geq 0$, with the Y_i independent for different i but in general with different distributions, and a main quantity of interest is the length

$$L \stackrel{\text{def}}{=} \max\left(Y_{i_1} + Y_{i_2} + \cdots + Y_{i_N} : i_1 i_2 \ldots i_N \in \mathscr{P}\right)$$

of the maximal path; note that L is simply the time needed until the completion of the process. The Monte Carlo method may be used, for example, to evaluate $\mathbb{E}L$, $\mathbb{P}(L > \ell)$ or the probability that a given path such as ADEF is actually the maximal one. □

Example 5.7 *Buffon's needle experiment* (1777) for numerical evaluation of π is often mentioned as the first use of simulation. The basic observation is that if a needle of length $L < 1$ is thrown at random on a floor of parallel boards, each one unit wide, the needle will intersect one of the cracks with probability $p = 2L/\pi$. Thus if $\widehat{p}$ is the fraction of times an intersection is obtained in R replications, $\widehat{\pi} = 2L/\widehat{p}$ is a simulation-based estimator of π.

For more detail on the history, see Newman & Barkema [276, pp. 24–25]. Simulation aspects are discussed in more detail in Ripley [300].

For illustrative purposes, we will often consider an alternative estimator $Z \stackrel{\text{def}}{=} 4\mathbb{1}\{U_1^2 + U_2^2 \leq 1\}$ of π, where U_1, U_2 are independent and uniform$(0, 1)$; note that $\mathbb{P}(U_1^2 + U_2^2 \leq 1) = \pi/4$ is the probability that a uniform point in $(0, 1)^2$ falls in the quarter-disk

$$\{u_1 \geq 0,\, u_2 \geq 0,\, u_1^2 + u_2^2 \leq 1\}.$$

□

Example 5.8 Another historically important example is the determination of the critical mass K_c of the A-bomb. Assume that the bomb is a ball of radius K, that spontaneous emissions of neutrons occur according to a Poisson process in space and time, and that emitted neutrons may hit other nuclei and cause further emissions. Neutrons exiting the ball are lost. The total number of emitted neutrons in such a cascade will be infinite if $K = \infty$, in which case the bomb will explode, and finite in the other extreme $K = 0$. The problem for the simulator is to give an estimate of a number K_c such that a cascade is finite w.p. (with probability) 1 if $K < K_c$ and infinite with positive probability if $K > K_c$, which can be done by inspecting the sample path behavior of a number of simulated cascades.

Similar ideas have been used for dimensioning nuclear reactors. See further Harris [175], Hammersley & Handscombe [173], and XIV.10.   □

Example 5.9 The penetrability of a medium to say water is often modeled by physicists as a configuration of open links on an infinite lattice, say $\mathbb{Z}^2$. Two lattice points (x_1, x_2), (y_1, y_2) are *neighbors* if

$$|x_1 - y_1| + |x_2 - y_2| = 1,$$

and one defines a graph on $\mathbb{Z}^2$ by connecting neighbors independently with probability p for each pair. The model *percolates* if an infinite connected component exists. It was long conjectured that this occurs with a probability $z = z(p)$ which is 0 or 1, with 1 precisely when $p \geq p_c$ for a certain critical value p_c, and that more precisely $p_c = 1/2$. This was settled by Kesten around 1980 (see Grimmett [169]), but simulation played an important role in developing the conjecture and still does in higher dimensions and similar problems in statistical physics.

To obtain information on p_c, one possible approach is to simulate the percolation process on a finite lattice $\{-L, \ldots, L\}^2$ for different L-values, for each L let $z(L)$ denote the probability that a path from the left $\{-L\} \times \{-L, \ldots, L\}$ to the right $\{L\} \times \{-L, \ldots, L\}$ exists, and use what appears to be the limit z as L increases of the estimates $\widehat{z}(p) = \widehat{z}(p, L)$ of $z(p)$.   □

Example 5.10 It is worth pointing out that procedures such as in Example 5.9 do have their pitfalls. One example in which a misleading result occured is *bootstrap percolation* (a rather alogical terminology!), in which the sites of $\mathbb{Z}^2$ are initially independently marked w.p. p. In the first step, an unmarked site with two marked neighbors becomes marked, the procedure is repeated in the second step, and so on. Since the configuration of marked sites is increasing, a limit will exist, and the question is, what is the probability $z = z(p)$ that this limit is the whole of $\mathbb{Z}^2$?

Simulations by physicists long gave the impression that a critical value p_c exists and is strictly positive. However, later a reasonably simple proof was given that in fact, $p_c = 0$! The problem is roughly that certain configurations of marks may drift into regions very far away, so that simulations on a finite lattice will always be misleading.

At a philosophical level, one may of course argue that infinite lattices do not exist in the real world, and that a reliable simulation estimate of z (or the occupancy density of marks in the final configuration) for a realistically large lattice is better than a limit result as $L \to \infty$. However, "realistically large" may well be beyond the capacity of even today's computers!   □

Example 5.11 The *lilypond model* describes a pattern of disks in the plane. Initially, points (germs) are placed according to a Poisson process with unit intensity. Around each point, disks (grains) start to grow at a unit rate. When two disks meet, they both stop growing (one may have stopped already because of interaction with a third). Thus fixation occurs after a finite time in any finite region in the plane.

A quantity of interest is the *volume fraction*, that is, the fraction of the plane occupied by disks. Another is the expected area $\mathbb{E}Z$ of a typical connected component, defined as a maximal union of disks that meet another member of the union. None of these are explicitly available, so it is relevant to provide Monte Carlo estimates.

See further Exercise 6.6 and Daley et al. [79].   □

Example 5.12 Asymptotic methods are widely used in statistics. For example, when testing a simple hypothesis $\mathscr{H}$ using the test statistic $T \stackrel{\text{def}}{=} -2 \log Q$, where Q is the likelihood ratio,[8] such that $\mathscr{H}$ is rejected for large values of T (i.e. when $T > t_0$), one would ideally reject at the α level by choosing the critical value t_0 such that $\mathbb{P}_{\mathscr{H}}(T \le t_0) = 1 - \alpha$. However, the $\mathscr{H}$-distribution of T is often unknown, and one uses an approximation, typically that T has an approximate χ^2-distribution with an appropriate number f degrees of freedom as the number N of observations goes to ∞. Then t_0 is chosen approximately as the χ^2-quantile rather than as the $\mathscr{H}$-quantile, i.e., by $\mathbb{P}_{\chi^2}(T \le t_0) = 1 - \alpha$. The obvious question is how large the error of this procedure is for a given finite N, that is, how far $\mathbb{P}_{\mathscr{H},N}(T > t_0)$ is from $\mathbb{P}_{\chi^2}(Q > t_0) = \alpha$. This may be assessed by a Monte Carlo experiment with $Z = \mathbb{1}_{T>t_0}$.

For example, let X be binomial (N, θ). The traditional statistical test for $\theta = \theta_0$ rejects for large values of $T = -2 \log Q$, where

$$Q = \frac{\theta_0^X (1-\theta_0)^{N-X}}{\widehat{\theta}^X (1-\widehat{\theta})^{N-X}}.$$

and $\widehat{\theta} \stackrel{\text{def}}{=} X/N$ is the maximum likelihood estimator. The χ^2 approximation to T has $f = 1$, so that one rejects at the 95% level if $T \ge 3.84$, the 95%

---

[8]in the sense of statistics, not importance sampling as occuring otherwise widely throughout the book!

quantile in the χ^2 distribution with one degree of freedom; rejects at the 99% level if $T \geq 6.63$, the 99% quantile; etc. The traditional criterion for n being large enough is that both $N\theta_0 \geq 5$ and $N(1 - \theta_0) \geq 5$. Monte Carlo simulation may help in assessing whether this criterion is too crude or too conservative.

In addition to the level, a main quantity of interest in hypothesis testing is the *power*, which for a given alternative $\mathscr{H}^*$ is defined as $\mathbb{P}_{\mathscr{H}^*}(T > t_0)$ (the probability of rejecting a wrong hypothesis). Again, this may be evaluated by Monte Carlo simulation.   □

Example 5.13 The EM algorithm is one of the standard tools for maximum likelihood estimation of a parameter θ in the presence of incomplete observations. It is introduced in more detail in A4 (Section A4 of the Appendix), but at this stage it suffices to say that it involves the computation of certain conditional expectations that may not be readily available. For example, in Exercise VIII.5.1 one faces the evaluation of conditional expectations of the form

$$\mathbb{E}_\theta\left[X_1^2 \mid \lfloor X_1 \rfloor, \lfloor X_2 \rfloor\right] \quad \mathbb{E}_\theta\left[X_2^2 \mid \lfloor X_1 \rfloor, \lfloor X_2 \rfloor\right]$$

($\lfloor X \rfloor$ = integer part) for a bivariate $\mathscr{N}(\mathbf{0}, \boldsymbol{\Sigma})$ random vector (X_1, X_2), which is explicitly available only for special values of $\boldsymbol{\Sigma}$. The Monte Carlo method is an obvious tool, provided one can generate r.v.'s with the required conditional distributions, a problem to be discussed in II.2.4.
□

Example 5.14 The *Cramér–Lundberg model* is the most classical model for insurance risk, and assumes that claims arrive according to a Poisson(β) process $\{N(t)\}$, that the claim sizes $V_1, V_2, \ldots$ are i.i.d. and independent of $\{N(t)\}$, and that premiums come in at a continuous rate c. Thus, the aggregate claims in one unit of time (say a year) are distributed as

$$A \stackrel{\text{def}}{=} \sum_{i=1}^{N(1)} V_i \,,$$

and a classical object of study is the probability $z \stackrel{\text{def}}{=} \mathbb{P}(A > x)$ of a large loss (here x is some large number). The distribution of A is basically available only for exponential claims V_i, so simulation is one possible numerical method. Since x is typically large, we are again in a rare-event setting.   □

Example 5.15 Another basic quantity associated with the Cramér–Lundberg model is the infinite-horizon *ruin probability* $\psi(x)$ with initial reserve x. Let $S(t)$ denote the claim surplus at time t, that is, the amount by which the claims exceed the premiums. Then $S(t) = \sum_1^{N(t)} V_i - ct$, and ruin occurs when this quantity for the first time exceeds x. That is, $\psi(x) = \mathbb{P}(\tau(x) < \infty)$, where $\tau(x) \stackrel{\text{def}}{=} \inf\{t > 0 : S(t) > x\}$. One can prove (cf. IV.7) that if $c = 1$ (which w.l.o.g. may be achieved by a simple change of time), then

$$\psi(x) = \mathbb{P}(W_\infty > x) = \mathbb{P}\big(V(\infty) > x\big), \tag{5.2}$$

where $W_\infty, V(\infty)$ are the steady-state waiting time and workload in the M/G/1 queue with the same arrival rate and the service times distributed as the insurance claims. That such a connection exists is of course also strongly suggested by (1.7).

In both the insurance and M/G/1 models, one is faced with an infinite-horizon problem when evaluating (5.2) via simulation. The naive procedure is to choose some large T and use the approximate estimate $\mathbb{1}\{S(t) > x$ for some $t \leq T\}$. This is probably the best one can do in many similar situations, but in the Cramér–Lundberg and M/G/1 setting there is a more satisfying solution given by the *Pollaczeck–Khinchine formula*

$$\psi(x) = \mathbb{P}\big(Y_1 + \cdots + Y_N > x\big), \tag{5.3}$$

where the Y_k are i.i.d. r.v.'s with density $\mathbb{P}(V > y)/\mathbb{E}V$ (the *integrated tail distribution*) and N an independent geometric r.v. with $\mathbb{P}(N = n) = (1 - \rho)\rho^n$, $n \in \mathbb{N}$. Thus, one simply generates N and $Y_1, \ldots, Y_N$ to get the estimator $Z \overset{\text{def}}{=} \mathbb{1}\{Y_1 + \cdots + Y_N > x\}$.   □

Example 5.16 Example 5.14 and formula (5.3) both involve the distribution of a random sum. This comes up in many other examples. One is credit risk, where N is the number of those who have defaulted in a certain period and Y_n the loss on the nth default. An alternative representation is to let N be the total number of debtors and let $Y_n = 0$ if default does not occur. In both cases, $L = Y_1 + \cdots + Y_N$ represents the portfolio loss. The same interpretation can be given for other sums. For example, an investment company may be holding N assets (stocks, bonds, options of various types, etc.) and Y_n is the loss on the nth asset, that is, the (random) value at the end of the period under consideration minus the (known) value at the start.

A specific feature in the calculation of portfolio loss is that most often dependence between the Y_n cannot be neglected. Such dependence will typically arise from a random economic environment common to the individual risks. We touch upon ways to deal with rare-event calculations in a dependent setting in XIV.11.   □

Example 5.17 A variant of the problem of calculating $\mathbb{P}(L > x)$ in a portfolio setting is to asses the *value-at-risk* (VaR) x_α at a specific level α, defined as the α-quantile. That is, x_α satisfies $\mathbb{P}(L \leq x_\alpha) = \alpha$, with a typical value of α being 99%, 99.9% or 99.97%, see McNeil et al. [252].

VaR calculations are extensively performed by financial institutions and are in fact often required by legislators and regulators. A naive simulation scheme is to estimate $z(x) \overset{\text{def}}{=} \mathbb{P}(L > x)$ for selected values of x and estimate

x_α by the x for which the estimated $z(x)$ is closest to $1-\alpha$. A more satisfying procedure is to generate R replications $L_1, \ldots, L_R$ of L and estimate x_α by the empirical α-quantile, as discussed in III.4a. $\square$

Example 5.18 The *Galton–Watson process* $\{X_n\}_{n\in\mathbb{N}}$ is a model for a one-sex population. With $Z_{n,i}$ denoting the number of children of individual i in generation n, the $Z_{n,i}$ are assumed i.i.d. with mean $m \stackrel{\text{def}}{=} \mathbb{E}Z_{n,i}$ (the *offspring mean*). If $m < 1$, it is easy to see that the population eventually becomes extinct, i.e., $T \stackrel{\text{def}}{=} \inf\{n : X_n = 0\}$ is finite w.p. 1. It is known that the tail of the extinction time T decays geometrically at rate m under weak regularity conditions, i.e., $\mathbb{P}(T = n) \sim cm^n$ for some c, but c as well as further characteristics of T are not readily available. Thus, one may want to use simulation to provide Monte Carlo estimates of $c, \mathbb{E}T, \mathbb{V}ar\, T$, etc., or to provide histograms of simulated values to give a rough idea of the shape of the distribution of T.

We return to branching processes in XIV.10. $\square$

Example 5.19 The *Wright–Fisher model* (e.g., Ewens [115]) in population genetics is an example of a somewhat similar flavor as the Galton–Watson process. A diploid population is assumed to have a fixed size N and two alleles a, A of a gene are possible at a certain locus. The genes in generation $n + 1$ are assumed to be obtained from those in generation n by sampling with replacement. That is, if $X_n \in \{0, 1, \ldots, 2N\}$ is the total number of A genes in generation n, then X_{n+1} has a binomial $(2N, p_n)$ distribution given X_n, where $p_n = X_n/2N$.

The process $\{X_n\}$ is a Markov chain and will eventually end up in one of the two absorbing states $0, 2N$. Thus, with $T \stackrel{\text{def}}{=} \inf\{X_n = 0 \text{ or } 2N\}$ and $X_0 = x$, we may ask for the fixation probabilities $\mathbb{P}_x(X_T = 0)$, $\mathbb{P}_x(X_T = N)$ and for properties of the fixation time T such as $\mathbb{E}_x T, \mathbb{V}ar_x T$, or $\mathbb{E}_x[T; X_T = 0], \mathbb{E}_x[T; X_T = 2N]$. For the simple Wright-Fisher model, $\{X_n\}$ is a martingale, which easily yields

$$\mathbb{P}_x(X_T = 0) = 1 - x/2N, \quad \mathbb{P}_x(X_T = N) = x/2N.$$

However, adding features such as selection and mutation destroyes the martingale property, and simulation may be needed. Also, properties of T are not equally as easy as the fixation probabilities. For example, various approximations for $\mathbb{E}_x T$ are known when N is large, but one may ask for the accuracy of the approximations and for the behavior of T and the fixation probabilities in more complicated models.

A particular challenge for simulation the Wright–Fisher model is that N often is very large, say of order 10^6 to 10^9, so that we are dealing here with one more case in which speeding up the simulation is relevant. $\square$

Example 5.20 Consider two species such as lynxes and hares (or snakes and lizards, etc.), of which one is the predator and the other the prey. The growth of the lynx population is positively correlated to the size of the hare population, and the growth of the hare population is negatively correlated to the size of the lynx population. The classical deterministic model is the coupled ODEs

$$\dot{\ell}(t) = \ell(t)\big(a + bh(t)\big), \quad \dot{h}(t) = h(t)\big(c - d\ell(t)\big),$$

where typically $a < 0$ (the lynx population would decrease in the absence of hares) and $c > 0$ (the hare population would grow in the absence of lynxes). The solutions move around on closed curves in $(0, \infty)^2$, corresponding to phase-shifted periodic fluctuations of the two populations; cf. Figure 5.2.

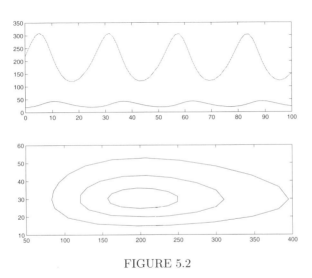

FIGURE 5.2

In Figure 5.2, the upper panel is a set of solutions (the upper curve represents the hares and the lower the lynxes) corresponding to $a = -0.2$, $b = 0.001$, $c = 0.3$, $d = 0.01$. The lower panel gives the level sets of the solutions, with the intermediate level set corresponding to the upper panel.

In fact, a classical set of data from Canada shows reasonable agreement with Figure 5.2, but of course also some stochastic deviations. In modeling the stochastic component, it is of course crucial that the proposed model have a behavior not too far from the data. This may be a difficult question to answer theoretically, but simulations may give a clue. Simulation may further help clarify questions such as whether the noisiness introduced in the model will eventually disturb the cyclical behavior, such that the process or one of its two species eventually becomes absorbed at 0, or that it drifts off to ∞. See further Exercise X.7.1.   □

Example 5.21 The *renewal equation* is the convolution equation $Z = z + F * Z$ on $[0, \infty)$, i.e.,

$$Z(x) = z(x) + \int_0^x Z(x - y) \, F(dy), \quad x \geq 0,$$

where Z is the unknown function, z a known function, and F a known positive measure (not necessarily a probability distribution). It is of fundamental importance in a variety of applied probability problems, cf. [16].

The solution is easily seen to be $Z = U * z$, where $U \overset{\text{def}}{=} \sum_0^\infty F^{*n}$ (sum of convolution powers) is the renewal measure. The difficulty in applying this formula is that U can easily be evaluated in explicit form in only a few special cases. But assuming the total mass c of F to be finite, $G \overset{\text{def}}{=} F/c$ is a probability distribution on $[0, \infty)$, and we can let $X_1, X_2, \ldots$ be i.i.d. with distribution G and $S_n \overset{\text{def}}{=} X_1 + \cdots + X_n$ (here $S_0 = 0$). Then $Z = U * z$ means

$$Z(x) = \mathbb{E} \sum_{n=0}^\infty c^n z(x - S_n)$$

(with the convention $z(x) = 0$, $x < 0$). So fixing $x_0 < \infty$ and letting $\tau \overset{\text{def}}{=} \inf \{n : S_n > x_0\}$, we can estimate $\big(Z(x)\big)_{x \leq x_0}$ by computing

$$\left(\sum_{n=0}^\tau c^n z(x - S_n) \right)_{0 \leq x \leq x_0}$$

over R replications and averaging. This provides an alternative to numerical methods for integral equations, which typically amount to discretization and recursive schemes.[9]   □

Example 5.22 Consider the partial differential equation (PDE)

$$0 = a f_{xx}(t, x) + \mu f_x(t, x) - f_t(t, x) - g(x) f(t, x), \tag{5.4}$$

where $f(t, x)$ is a function of two variables $x \in \mathbb{R}$, $t \geq 0$, and $f_t \overset{\text{def}}{=} \partial f / \partial t$, $f_x \overset{\text{def}}{=} \partial f / \partial x$, $f_{xx} \overset{\text{def}}{=} \partial^2 f / \partial x^2$. The primitive g is an arbitrary smooth function. Assume w.l.o.g. that $a > 0$. It is then straightforward to show (e.g., Steele [346]) that a solution is

$$f^*(x, t) \overset{\text{def}}{=} \mathbb{E}_x \exp \left\{ - \int_0^t g(X_s) \, ds \right\}, \tag{5.5}$$

where $\{X_s\}$ is a Brownian motion with drift μ and variance constant $\sigma^2 = 2a$, and $\mathbb{E}_x$ means $X_0 = x$. If we look for a solution satisfying

---

[9]The choice $G = F/c$ is made here mainly for expository purposes. Actually, an exponential change of measure is most often the more natural alternative.

the same initial condition $f(x, 0) = 1$ as f^*, we therefore obtain a numerical scheme for evaluating $f(t, x)$ using the Monte Carlo method with $Z(t) \overset{\text{def}}{=} \exp\left\{-\int_0^t g(X_s)\,\mathrm{d}s\right\}$ with independent Brownian paths $\{X_s\}_{0 \le s \le t}$ for the different replications. This provides an alternative to numerical methods for PDEs, which again typically amount to discretization and recursive schemes.

The connection between (5.4) and (5.5) is often referred to as a *Feynman–Kac formula*. In a suitable discrete time setting, this also incorporates the filtering problem discussed in Example 5.5. See further Del Moral [85] and XIV.2.   □

6   Introductory Exercises

We give here some exercises that may used at the start of a course to assist the reader in becoming acquainted with software and hardware and to get some first practical experience with simulation, but that do not require any of the theory to be developed later in the book (the software used should include, however, routines for generating r.v.'s from the standard distributions).

Exercises

6.1 (A) In a soccer tournament with 16 teams, the first 8 games are 1-2 (i.e., team 1 versus team 2), 3-4, ..., 15-16, the next four (the quarterfinals) the winner of 1-2 versus the winner of 3-4 and so on until the final. Ties are decided by sudden death (the winner is the first to score). Team i is characterized by three parameters λ_i, p_i, γ_i, such that normally the number of major possibilities for a goal it creates is Poisson(λ_i) and the probability of preventing the opponent from scoring on one of their major possibilities is p_i. However, a bad day may occur w.p. γ_i and then λ_i, p_i are reduced to 2/3 of their values. The parameters are as follows:

	1	2	3	4	5	6	7	8
λ	4.2	6.2	6.4	4.9	6.2	3.2	6.6	6.2
p	0.65	0.80	0.82	0.66	0.78	0.82	0.47	0.53
γ	0.36	0.23	0.23	0.32	0.42	0.19	0.37	0.41
	9	10	11	12	13	14	15	16
λ	4.2	4.1	8.7	3.3	6.8	0.7	4.1	4.9
p	0.65	0.60	0.88	0.55	0.72	0.50	0.74	0.69
γ	0.36	0.30	0.23	0.19	0.30	0.38	0.32	0.29

Explain that in a match between teams i and j, the score is N_{ij} goals to N_{ji}, where N_{ij} is Poisson(μ_{ij}) given two independent r.v.'s Y_i, Y_j that are 2/3 w.p. γ_i, γ_j and 1 otherwise, and $\mu_{ij} \overset{\text{def}}{=} \lambda_i Y_i (1 - p_j Y_j)$, and similarly for N_{ji}. Show also that the conditional probability that team i wins a possible draw is $\mu_{ij}/(\mu_{ij} + \mu_{ji})$. Give next a table over estimated values of the probabilities of the different teams to win the tournament.

For quantitative stochastic methods applied to sports, see further Albert et al.
[6].

6.2 (A) Consider an Asian call option with the underlying asset following the
GBM Black–Scholes model with yearly volatility 0.25, maturity $T = 3$ years, short
rate $r = 4\%$, and strike price $K = 100 = S(0)$, and sampling at N equidistant
time points.

Give a point estimate of the price for (a) half-yearly sampling $N = 6$, (b) monthly
sampling $N = 36$, (c) weekly sampling $N = 150$, and (d) daily sampling $N = 750$.
An obvious conjecture is that the price as a function of N approaches the price

$$e^{-rT}\mathbb{E}^*\left[\frac{1}{T}\int_0^T S(t)\,dt - K\right]^+$$

of the continuously sampled option at rate $N^{-\alpha}$ for some $\alpha > 0$. Check this
conjecture graphically, and give a rough estimate of α.

6.3 (A) A company holding a European call option with maturity $T = 1$ year
and strike price K delta hedges every week. Assuming the log-returns to follow
the GBM Black–Scholes model with (yearly) volatility $\sigma = 0.25$, $S(0) = K = 1$,
and $r = 4\%$ per year, investigate how good the delta hedge replicates the payout
$[S(T) - K]^+$, say by plotting 25 values of the pairs $([S(T) - K]^+, w(T))$, where
$w(t)$ is the value of the hedging portfolio at time t.

Explanation. Let $\Pi(x, t, T)$ be the price at time t of the option given $S(t) = x$
and $\Delta(t) = (\partial/\partial x)\Pi(x, t, T)\big|_{x=S(t)}$ (which can be computed by straightforward
differentiation of (3.4)). The portfolio that is delta hedging the option at times
$0 = t_0 < t_1 < \cdots < t_n < T$ invests in $a_1(t_i) = \Delta(t_i)$ units of the underlying asset,
whereas the amount $a_2(t_i)$ put in the bank is chosen so as to make the portfolio
self-financing, i.e., one should have

$$a_1(t_{i-1})S(t_i) + a_2(t_{i-1})e^{r(t_i - t_{i-1})} = a_1(t_i)S(t_i) + a_2(t_i).$$

The initial weights $a_1(0), a_2(0)$ are chosen such that $w(0-) = \Pi(S(0), 0, T)$, and
the terminal value of the hedge becomes $w(T) = a_1(t_n)S(T) + a_2(t_n)e^{T-t_n}$. See
further Björk [45].

6.4 (A) Consider the binomial testing problem in Example 5.12. The assignment
is to investigate via simulation whether the criterion that both $N\theta_0 \geq 5$ and
$N(1 - \theta_0) \geq 5$ be large is too crude or too conservative in the specific case
$\theta_0 = 0.1$ and $N = 10, 25, 50, 100$.

To do this for a given N, simulate R replications of X from $\mathbb{P}_{\theta_0}$, and use these
to estimate the actual level of the test and the 95% and 99% quantiles of $T = -2\log Q$.

6.5 (A) A gene in a population of size n occurs in two alleles A, a, such that the
number of individuals of genotypes AA, Aa, aa are N_{AA}, N_{Aa}, N_{aa}. The hypothesis $\mathscr{H}$ of *Hardy–Weinberg equilibrium* asserts that the corresponding frequencies
should be of the form $\theta^2, 2\theta(1 - \theta), (1 - \theta)^2$.

The maximum likelihood estimator of θ is $\widehat{\theta} = (2N_{AA} + N_{Aa})/n$ and the
likelihood ratio statistic for $\mathscr{H}$ is

$$Q = \frac{\widehat{\theta}^{2N_{AA}}\left[\widehat{\theta}(1-\widehat{\theta})\right]^{N_{Aa}}(1-\widehat{\theta})^{2N_{aa}}}{(N_{AA}/n)^{N_{AA}}(N_{Aa}/n)^{N_{Aa}}(N_{aa}/n)^{N_{aa}}}.$$

Investigate by simulation the validity of the usual χ^2 approximation with 1 degree of freedom for $-2\log Q$, taking $n = 5, 10, 20, 50$ and $\theta = 0.05, 0.1, 0.2, 0.5$.

6.6 (A) In the lilypond model, Example 5.11, give a Monte Carlo estimate of the expected area $z = \mathbb{E}Z$ of a typical connected component.

It is easy to see from properties of the Poisson process that Z can be simulated by placing an additional germ at the origin and measuring the area of the connected component containing it. In the practical implementation, restrict to the square $[-4, 4]^2$, which has proven large enough that boundary effects do not play a role.

6.7 (A) When applying for a secretarial position, n candidates line up in random order for interviews, where the manager after each interview has to decide to take or decline that candidate. His assessment is qualitative in the sense that at interview $k \geq 2$, he observes only $Y_{k,n}$, the indicator that candidate k is better than candidates $1, \ldots, k-1$, and his goal is to find a strategy whose probability of selecting the best candidate is close to the maximal value p_n^*.

It is easy to see that the optimal strategy is to select candidate

$$\tau_{k,n} \overset{\text{def}}{=} \inf\left\{\ell \geq k : Y_{\ell,n} = 1\right\} \wedge n$$

for some $k \in \{2, \ldots, n\}$. Let k_n^* denote the optimal such k. Plot simulation estimates of k_n^* and p_n^* for $n = 5, \ldots, 50$, and compare with the known asymptotics $k_n^* \sim n/e$, $p_n^* \sim 1/e$.

6.8 (TP) Which r.v.'s would you need to generate for Buffon's needle experiment?

6.9 (A) A small supermarket gets a batch of Q chickens each Monday and have to trash the unsold ones on Saturday. So far, the manager has taken $Q = 70$. Having a bachelor's degree in industrial engineering, she has collected some statistics and knows that the average number N asked for per week is 60, and now wants to use her background to make a rational choice of Q.

The weekly holding cost (electricity for cooling, interest rate on the cooling facility etc.) of ordering Q chickens is $a + bQ^{3/4}$ and the delivery costs $c + dQ^{1/2}$, where $a = 30, b = 5, c = 50, d = 3$ (the monetary unit is DKR; $1\,\text{USD} \approx 6\,\text{DKR}$). Each chicken costs $e = 18$ and is sold for $f = 25$.

The manager's first obvious idea is to assume that N is Poisson with $\lambda = 60$. What is then the optimal Q?

You will have to simulate for several values $q_1, \ldots, q_m$ of Q, and to do R replications (each corresponding to a week) for each. Thus each q_i requires R Poisson(60) variables. Try both to generate these as different for different i and to use the same for all i. Advantages and disadvantages?

After a while, it occurs to the manager that chickens are not always sold in units of 1 but some customers ask for 2 and a few for 3. She models this by $N = N_1 + 2N_2 + 3N_3$, where N_1 is Poisson(31), N_2 is Poisson(10), and N_3 is Poisson(3). Does this make a difference to the choice of Q and the expected profit?

For recent and more advanced treatments of the theory of inventory and storage, see Axsäter [33], Silver et al. [344], and Zipkin [368].

Part A:
General Methods and Algorithms

Chapter II
Generating Random Objects

1 Uniform Random Variables

The basic vehicle in the area of (stochastic) simulation is a stream $u_1, u_2, \ldots$ of numbers produced by a computer, which is treated as the outcome of a sequence $U_1, U_2, \ldots$ of i.i.d. r.v.'s with a common uniform distribution on $(0, 1)$. Many software libraries contain routines for generating such streams, such as a command of the type `u1:=random; u2:=random; ...` in the language C^{++} or `u=rand(1,n)` in Matlab (creating a row vector containing $u_1, \ldots, u_n$).

The algorithms in practical implementations are all deterministic (typically using recursion) and can therefore at best mimic properties of i.i.d. uniform r.v.'s;[1] for this reason, the sequence of outputs is called a sequence of *pseudorandom numbers*. Earlier generations of computers and software had quite a few examples of random number generators with unfortunate properties. However, it is our view that the situation is much improved now and that the development of algorithms for generation of pseudorandom numbers (as well as sequences of quasirandom numbers for similar uses, see IX.3) is now largely a specialist's topic: the typical reader of this book will do well with existing software (which is fast, certainly much faster than home-made high-level language routines) and will seldom be able to improve it substantially. We point out in this connection a common beginner's pitfall, to blame apparently erroneous simulation output

---

[1] For brevity reasons, called uniforms in the following.

on deficiencies in the generator. The (much) more common problem is an error of a different type, such as a programming error or a conceptual misunderstanding of the problem under study. The purpose of the discussion to follow is therefore largely an attempt to provide some common background and knowledge of random number generation.

A classical reference on uniform random number generators is Ripley's 1987 book [300], which covered many important aspects of the area at that time. More recent and up-to-date treatments are in Knuth [220], Gentle [127], and L'Ecuyer [232] where also an extensive bibliography can be found. Also the websites [w^3.7], [w^3.9] may be useful for getting an impression of the present state of the art.

1a   Physical Devices

In the early era of computer simulation. there were attempts to use physical devices for random number generation. A widely used idea is to exploit properties of radioactive decay: the times at which particles are emitted from a radioactive source is a classic example of a Poisson process, as has been verified empirically and which also has strong theoretical support. In particular, since individual atoms split in an unpredictable way, independent of one another, and the number of atoms is astronomical, the Poisson approximation to the binomial distribution is highly accurate. Thus, the interevent times can be viewed as i.i.d. exponentials. They can be transformed to other distributions such as the uniform.

Such physically based generators were slow, generating at best only hundreds of random numbers per second. The lack of replicability can also limit one's ability to apply certain variance-reduction techniques. Of course, the rapid decrease in the cost of high-speed computer memory offers the opportunity to prerecord huge quantities of such physically generated random numbers, thereby reducing the impact of the last two issues raised.

A more fundamental objection to the use of such physical devices to generate random numbers is that most of them produce numbers with a systematic (and difficult to quantify) bias. For example, while one could flip a coin to generate a sequence of i.i.d. Bernoulli(1/2) r.v.'s, it is likely that the coin exhibits a (small) systematic preference to either heads or tails. In the setting of radioactive counters, the device typically locks for a certain period of time subsequent to a particle registration, thereby eliminating the possibility of short interevent times. While deterministic mathematical algorithms also generally exhibit biases, such biases can often be studied mathematically, so that their possible effect is better understood.

There are situations in which physically generated "truly random" numbers are preferable (at a minimum as seeds) to pseudorandom numbers, in particular lotteries, gambling machines, and cryptography. A number of commercial software physical generators have recently come on the market, for example [w^3.10], [w^3.11]. Recent tests (Pierre L'Ecuyer, personal communication) indicate that these generators are reasonably reliable statistically. However, they remain slower and much more cumbersome than algorithmic random number generators.

1b Deterministic Recursive Algorithms

Instead of using physical devices, pseudorandom numbers $u_1, u_2, \ldots$ are typically produced by deterministic recursive algorithms. A general framework covering virtually all such generators occurring in practice and suggested by L'Ecuyer [229] is a quadruple (E, μ, f, g). Here E is the (finite) set of states, the state of the random number generator evolves according to the recursion $s_n = f(s_{n-1})$; and the random number stream produced is $u_n = g(s_n)$, where g maps S into $[0, 1]$. The initialization is determined by μ, which is a probability measure selecting x_1 (the *seed*).

Since E is finite, the range of g is not all of $[0, 1]$ but only the finite subset $g(E)$. In practice, the generator is modified so that the values 0 and 1 cannot occur (say 0 is replaced by ε, the smallest nonzero value of g or the smallest positive number representable on the computer). This is to avoid problems in using the sequence; say one needs division or to take logarithms. Finiteness also implies that there is a d such that $x_{\ell+d} = x_d$ for some ℓ; the *period* is the smallest d for which this happens. After ℓ, the algorithm will produce replicates of cycles of length d or smaller. One difficulty with generators having short periods is that the gaps in the sequence may not be evenly distributed.

To reduce the problems associated with the finiteness of $g(E)$ and the period, a necessary (but not sufficient!) condition is obviously that E be large.

Example 1.1 A class of simple algorithms of this type is that of the *linear congruential generators*, which were popular for many years but are now somewhat outdated. Such algorithms have the form

$$u_n = \frac{s_n}{M}, \quad \text{where} \quad s_{n+1} = (As_n + C) \bmod M. \tag{1.1}$$

The difficulty is in choosing a large M and associated A, C such that the period is large, preferably M when $C \neq 0$ or $M - 1$ when $C = 0$ (this property is called *full period*). Number-theoretic considerations provide verifiable conditions under which linear congruential generators are of full period. This has led to certain popular parameter choices for A, C, and M. A dominant one in earlier generations of computers and software was $M = 2^{31} - 1 = 2{,}147{,}483{,}647$, $A = 7^5 = 16{,}807$, $C = 0$. This choice has the nice property that its period is (very) close to the number of machine-representable integers in a 32-bit computer. Another example is $M = 2{,}147{,}483{,}563$, $A = 40{,}014$. □

Example 1.2 A generalization of linear congruential generators is obtained by replacing $x_{n+1} = (Ax_n + C) \bmod M$ in (1.1) by

$$x_n = (A_1 x_{n-1} + \cdots + A_k x_{n-k}) \bmod M. \qquad (1.2)$$

Thus, in the above formalism, $s_n = (x_n, x_{n-1}, \ldots, x_{n-k+1})$.

When M is a prime number, it is possible to achive the maximal period $M^k - 1$; see Knuth [220] for more detail. The typical choice of M is a large prime number close to the largest integer representable on the computer, e.g., $2^{31} - 1$ on a 32-bit computer. The implementation then requires some programming technicalities (long integers, etc.) because $A_1 x_{n-1} + \cdots + A_k x_{n-k}$ is not directly representable. $\qquad \square$

Example 1.3 Certain numerically demanding simulations require enormous numbers of random numbers. The congruential generators mentioned above have periods only on the order of billions. Much larger periods can be obtained by mixing generators. For example, many software packages (e.g., Arena, Kelton et al. [211] p. 501, or SAS) use the algorithm

1. $x_n \longleftarrow (A_1 x_{n-2} - A_2 x_{n-3}) \bmod M_1$,

2. $y_n \longleftarrow (B_1 y_{n-1} - B_2 y_{n-3}) \bmod M_2$,

3. $z_n \longleftarrow (x_n - y_n) \bmod M_1$,

4. If $z_n > 0$, return $u_n = z_n/(M_1 + 1)$; else return $u_n = M_1/(M_1 + 1)$,

where $M_1 = 4{,}294{,}967{,}087$, $M_2 = 4{,}294{,}944{,}443$, $A_1 = 1{,}403{,}580$, $A_2 = 810{,}728$, $B_1 = 527{,}612$, $B_2 = 1{,}370{,}589$ (the seed is the first three x's and the first three y's).

The algorithm originates with L'Ecuyer [230], where the above parameters were determined by a computer search for certain criteria determining the properties of the algorithm to be met. $\qquad \square$

Example 1.4 Instead of performing algebra modulo M for some huge integer, it is tempting to exploit the binary representation of the computer to look for algorithms using algebra modulo 2. A general scheme is

$$\boldsymbol{x}_n = \boldsymbol{A} \boldsymbol{x}_{n-1}, \quad \boldsymbol{y}_n = \boldsymbol{B} \boldsymbol{x}_n, \quad u_n = y_{n,1} 2^{-1} + \cdots + y_{n,\ell} 2^{-\ell}, \qquad (1.3)$$

where $\boldsymbol{x}_n \in \{0,1\}^k$, $\boldsymbol{y}_n \in \{0,1\}^\ell$, the matrices $\boldsymbol{A}, \boldsymbol{B}$ are $k \times k$ and $\ell \times k$, and all operations are performed modulo 2. A survey of this class of generators is given in L'Ecuyer & Panneton [234].

If the characteristic polynomial $\det(z\boldsymbol{I} - \boldsymbol{A})$ is written as $z^k - \alpha_1 z^{k-1} - \cdots - \alpha_{k-1} z - \alpha_k$, one can prove that the output sequence follows the recursion

$$u_n = \alpha_1 u_{n-1} + \cdots + \alpha_k u_{n-k} \bmod 2. \qquad (1.4)$$

Special cases include *Tausworthe generators*, which have the form

$$z_n = a_1 z_{n-1} + \cdots + a_k z_{n-k} \bmod 2, \quad u_n = z_{nt+1} 2^{-1} + \cdots + z_{nt+r} 2^{-k}.$$
(1.5)

A popular choice has been $z_n = z_{n-102} + z_{n-249}$, which has the obvious advantage in terms of speed that most of the a_i equal zero (the period is $2^{500} - 1$). Another example is the fairly recent *Mersenne twister* of Matsumoto & Nishimura [249], which has period $2^{19937} - 1$, excellent equidistributional properties and is still very fast (but see the caveat in Panneton et al. [283]).

L'Ecuyer [231] discusses how to combine such generators of mod 2 type to improve quality.   □

1c   Statistical Tests

The generators used in practice typically have the property that the marginal empirical distribution of the U_n is uniform (up to rounding errors) and that observations look independent within a narrow time range. These are also the properties that most statistical tests concentrate on confirming. However, it should be noted that (being deterministic) no stream of pseudorandom numbers is truly random. With knowledge of its basic mechanism, one can always design a statistical test that will reveal this.

Tests for deviations from a realization of a sequence of truly random numbers are basically just standard statistical goodness-of-fit tests. Considering first uniformity of the one-dimensional marginals, a useful graphical method is the Q-Q plot, which plots the empirical quantiles of the set $u_1, \ldots, u_n$ relative to the uniform quantiles $q_\alpha = \alpha$. Two standard significance tests are the χ^2-test and the Kolmogorov–Smirnov statistic. The χ^2-test splits $I = (0, 1)$ into subintervals $I_1, \ldots, I_K$, typically just $(0, 1/K), [1/K, 2/K), \ldots$ For subinterval k, O_k (O for observed) is defined as the number of $u_1, \ldots, u_n$ taking values in I_k and $E_k = n|I_k|$ (Lebesgue measure) as the expected number. The χ^2-statistic is $\sum_1^K (O_k - E_k)^2 / O_k$ and is approximately χ^2-distributed with $f = K - 1$ degrees of freedom provided the E_k are not too small. The Kolmogorov–Smirnov statistic is $\max_x |\widehat{F}_n(x) - F(x)|$, where $\widehat{F}_n$ is the empirical c.d.f. of $u_1, \ldots, u_n$ and F the theoretical c.d.f. (in the present case, $F(x) = x$); the asymptotic distribution of the statistic is complicated, but the quantiles have been tabulated and are available from statistical software packages.

Testing independence is a more complicated matter. One way is through a goodness-of-fit approach, testing that d-blocks $\left(u_{k+1}, \ldots, u_{k+d}\right)$ are uniformly distributed in $(0, 1)^d$. The χ^2 test (adapted in an obvious way) applies to this, but a difficulty is that the E_k quickly become too small for the asymptotic results as d grows, and one should note also that the independence assumptions are violated, since neighboring d-blocks are dependent (a way out is to base the test on blocks $k = 0, d, 2d, \ldots$

only). Another idea is to use correlations, say by a normal transformation $x_k = \Phi^{-1}(u_k)$ (see the discussion of inversion in Section 2), where the empirical correlations at lags $k = 1, 2, \ldots$ are possible test statistics. There are also quite a few ad hoc methods around: the gap test based on the fact that the spacings between the (random) indices k for which $u_k \in J \subset (0, 1)$ should behave like i.i.d. geometric r.v.'s with success parameter $|J|$; the coupon collector's test, where one records the successive lengths of minimal sequences containing all K values of $\lfloor K u_n \rfloor$ for a suitable integer K and compares to the theoretical distribution; and so on.

We point out that care must be taken w.r.t. the interpretation of multiple tests. For example, if 20 tests are each performed at a 5% rejection level in a traditional statistical setting, the expected number of rejections under the null hypothesis is one. It is therefore a delicate matter to assess when one should make an overall rejection in a multiple-test setting, and dependence further complicates the matter. The situation in testing random number generators is somewhat different, since one would usually not fix a significance level like 5% or 1%, but continue the test until the test comes out with a clear conclusion.

1d   Randomness Criteria

At the more philosophical level, one may ask, what are the criteria that $u_1, u_2, \ldots$ should satisfy in order to permit interpretation as a realization of an i.i.d. uniform sequence? Obviously, each u_k should be in some way "random". However, this concept is not at all well defined. For example, if a fair die is thrown 8 times, the sequences 33333333 and 14522634 have the same probability, so why should the second be declared more random than the first? See further Knuth [220].

A desirable property is obviously that the empirical distribution of d-blocks (the distribution on $(0, 1)^d$ giving mass $1/n$ to each of the n d-blocks in $u_1, \ldots, u_{n+d-1}$) should converge to the uniform distribution on $(0, 1)^d$ as $n \to \infty$, for any $d = 1, 2, \ldots$; this is sometimes referred to as the sequence being ∞-distributed (also the terminology "normal number" is used). It is also reasonable to require unpredictability. One formulation of this is to require all subsequences of $u_1, u_2, \ldots$ to be ∞-distributed, and one then talks of a *von Mises–Church collective*.

Another possible approach is via specific measures of deviations, in particular *discrepancies*; we discuss these in connection with quasirandom numbers in IX.3, but the relevance is not limited to this setting.

Exercises

1.1 (TP) Explain how Tausworthe generators fit into the framework (1.3).

2   Nonuniform Random Variables

Accepting the U_n as i.i.d. uniform, the next step is to use them to produce an r.v. X with a prescribed distribution F, say Poisson, exponential, normal, etc. As for uniform random numbers, the user will typically use standard routines for this purpose. However, occasionally one may come across a nonstandard distribution that is not avaliable in this way. We will meet at least two nonartificial examples in this book: conditional distributions occuring in Gibbs sampling (e.g., XIII.5.3); and generation of r.v.'s from truncated and normalized Lévy measures (see XII.3.4). In general, it seems reasonable (and, we hope, interesting) to have some insight into how nonuniform random numbers are produced. Comprehensive references in the area are Devroye [88], Ripley [300], Gentle [127], and Hörmann et al. [190].

Standard routines are usually designed with considerable attention to speed. In cases in which the user for one reason or another writes his/her own routine for generating random numbers, optimal speed is, however, rarely a concern, and the algorithms that we describe below are not necessarily the fastest ones; the average user may want to use a naive but easily programmed method rather than invoking more sophisticated methods or various library packages. A rule of thumb is that it is much faster to generate a bunch of uniform random numbers, say 5–10, than to evaluate even a single special function such as the logarithm, the exponential, or a trigonometric function.

In the following, $U, U_1, U_2, \ldots$ are used without further explanation for independent r.v.'s from the uniform$(0, 1)$ distribution.

Example 2.1 (SIMPLE DISCRETE R.V.'S)   A simple case is a Bernoulli r.v., $\mathbb{P}(X = 1) = 1 - \mathbb{P}(X = 0) = p$, where one can just let $X \stackrel{\text{def}}{=} \mathbb{1}_{U \leq p}$ (the indicator function, $= 1$ if $U \leq p$ and $= 0$ otherwise). This construction generalizes in a straightforward way to distributions with finite support. In particular, if we want to generate X as being uniform on $\{1, \ldots, N\}$, we may take $X = 1$ if $U \leq 1/N$, $X = 2$ if $1/N < U \leq 2/N$, and so on. This is often written as $X = \lceil U/N \rceil$, where $\lceil a \rceil$ is the smallest integer greater than or equal to a (many programming languages have $\lceil \cdot \rceil$ as a standard function).

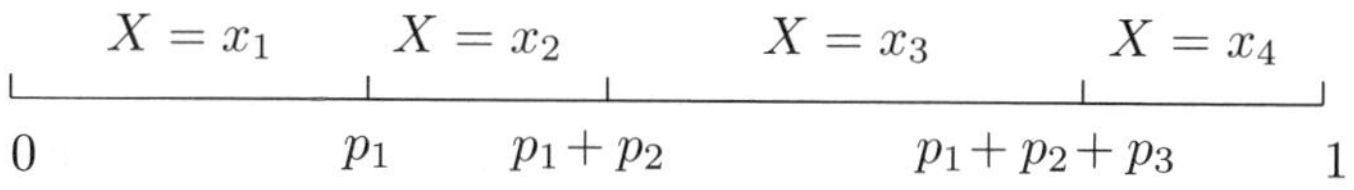

FIGURE 2.1

If instead the target distribution is given by $\mathbb{P}(X = x_i) = p_i$, where $p_1 + \cdots + p_N = 1$, $p_1, \ldots, p_N \geq 0$, we may let $X = x_1$ if $U \leq p_1$, $X = x_2$ if $p_1 < U \leq p_1 + p_2$ and so on; see Figure 2.1 for $N = 4$. This is a special case of the inversion algorithm described below. Note that the efficiency depends on how rapidly the search for the subinterval straddling U can be accomplished; for an efficient algorithm, see Example 2.15 below.

The search can be avoided when $\mathbb{P}(X = i) = k_i/n$ for integers $k_i \geq 0, n \geq 1$, at the cost of memory. One stores the integer 1 in the first k_1 memory locations, the integer 2 in the next k_2, and so on. Upon generating U, one then returns the integer in memory location $\lceil nU \rceil$.   □

For a continuous r.v. with a density $f(x)$, there are two general methods: inversion and acceptance–rejection (A–R), and a variety of more ad hoc methods that use special properties of the target distribution. We proceed next to a more detailed exposition of these ideas.

2a Inversion

For inversion, one defines $F^{\leftarrow}$ as the inverse of the c.d.f. F of X. The simplest case is that in which F is strictly increasing and continuous; then $x = F^{\leftarrow}(u) = F^{-1}(u)$ is the unique solution of $F(x) = u, 0 < u < 1$. For distributions with nonconnected support or jumps, more care is needed, and one then needs the more general definition of the inverse, where we here have chosen the left-continuous version

$$F^{\leftarrow}(u) \overset{\text{def}}{=} \inf\{x : F(x) \geq u\} = \min\{x : F(x) \geq u\}, \quad 0 < u < 1;$$

that the minimum is attained follows since $\{x : F(x) \geq u\}$ is an interval (infinite to the right) that must contain its left endpoint by right-continuity of F.

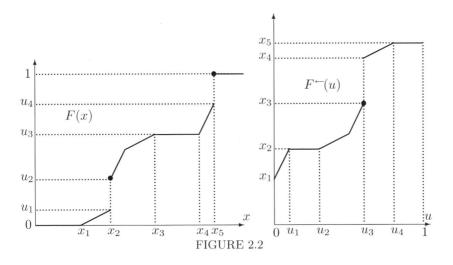

FIGURE 2.2

Proposition 2.2 (a) $u \leq F(x) \iff F^{\leftarrow}(u) \leq x$;
(b) *if U is uniform$(0,1)$, then $F^{\leftarrow}(U)$ has c.d.f. F;*
(c) *if F is continuous, then $F(X)$ is uniform$(0,1)$.*

Proof. Part (a) follows directly from the definition of $F^{\leftarrow}$ (it is clear that $F(x) \geq u$ for $x \geq F^{\leftarrow}(u)$ but not for any $x < F^{\leftarrow}(u)$), and (b) follows trivially from (a), which yields

$$\mathbb{P}\big(F^{\leftarrow}(U) \leq x\big) \;=\; \mathbb{P}\big(U \leq F(x)\big) \;=\; F(x).$$

Finally, for (c) we have to prove $\mathbb{P}\big(F(X) \geq u\big) = 1-u$, $0 < u < 1$. However, by (a) and continuity of F,

$$\mathbb{P}\big(F(X) \geq u\big) \;=\; \mathbb{P}\big(X \geq F^{\leftarrow}(u)\big) \;=\; \mathbb{P}\big(X > F^{\leftarrow}(u)\big) \;=\; 1 - F\big(F^{\leftarrow}(u)\big),$$

so we must prove $F\big(F^{\leftarrow}(u)\big) = u$. Here $F\big(F^{\leftarrow}(u)\big) \geq u$ is clear from (a) with $x = F^{\leftarrow}(u)$. Let $y_n < F^{\leftarrow}(u)$, $y_n \uparrow F^{\leftarrow}(u)$. Then $F(y_n) < u$ by (a), and so by continuity of F, $F\big(F^{\leftarrow}(u)\big) \leq u$. $\qquad\square$

The part that is mainly used in simulation is (b), which allows us to generate X as $F^{\leftarrow}(U)$ (the most common case is an F that is continuous and strictly increasing on an interval).

Example 2.3 (EXPONENTIAL R.V.'S) Let F be exponential with rate δ (the rate is the inverse of the mean, that is, the density is $f(x) = \delta e^{-\delta x}$). Then $F^{\leftarrow}(x) = -\log(1-x)/\delta$, so inversion means that $X = -\log(1-U)/\delta$ (in practice, one often uses $X = -\log U/\delta$ rather than $X = -\log(1-U)/\delta!$).

Standard packages usually use more elaborate but faster algorithms that avoid the time-consuming evaluation of the log.

From exponential r.v.'s, one can build Gamma(p, δ) r.v.'s with integer shape parameter p (the density is $\delta^{p-1}/(p-1)! \, e^{-\delta x}$, cf. A1) by simply adding p independent copies. A Poisson r.v. N with rate λ ($\mathbb{P}(X = n) = e^{-\lambda}\lambda^n/n!$) can be constructed using the relation between the exponential distribution and the Poisson process: generate $X_1, X_2, \ldots$ as i.i.d. exponentials with rate λ one at a time until the random time at which the sum exceeds 1, say $X_1 + \cdots + X_N < 1 < X_1 + \cdots + X_{N+1}$. Then N is the desired Poisson r.v. To reduce the number of evaluations of special functions, use

$$N \;=\; \max\Big\{n \geq 0 : \prod_{i=1}^{n} U_i > e^{-\lambda}\Big\}. \tag{2.1}$$

However, although such relations between standard distributions can be pushed even further, there are usually more efficient methods available. $\quad\square$

Inversion applies to many other examples, but (in addition to being slow) a main limitation is that quite often $F^{\leftarrow}$ is not available in explicit form, for example when $F = \Phi$, the standard normal c.d.f. Sometimes approximations are used. For example, the following rational polynomial approximation is standard, simple, and accurate for the normal distribution:

$$\Phi^{-1}(u) \approx y + \frac{p_0 + p_1 y + p_2 y^2 + p_3 y^3 + p^4 y^4}{q_0 + q_1 y + q_2 y^2 + q_3 y^3 + q^4 y^4}, \quad 0.5 < u < 1, \qquad (2.2)$$

where $y = \sqrt{-2\log(1-u)}$ and the p_k, q_k are given by the following table:

k	p_k	q_k
0	-0.322232431088	0.099348462606
1	-1	0.588581570495
2	-0.342242088547	0.531103462366
3	-0.0204231210245	0.10353775285
4	-0.0000453642210148	0.0038560700634

(the case $0 < u < 0.5$ is handled by symmetry).

Remark 2.4 A particularly convenient feature of inversion is that whenever $F^{\leftarrow}$ is available, one can simulate not only a r.v. distributed as X but also one from certain conditional distributions. In particular, if X has a density and $a < b$, then an r.v. distributed as X given $X \in (a, b)$ has c.d.f. $\big(F(x) - F(a)\big)/\big(F(b) - F(a)\big)$ and can therefore be generated as $F^{\leftarrow}\big(F(a)(1 - U) + F(b)U\big)$. Similarly, an r.v. distributed as the overshoot $X - a$ given $X > a$ can be generated as $F^{\leftarrow}\big(U\overline{F}(a)\big) - a$, etc.   □

2b   Simple Acceptance–Rejection

The idea is to start from an r.v. Y with a density $g(x)$ (the proposal), which is easily simulated and has the property $f(x) \le Cg(x)$, where $f(x)$ (the target) is the density of X and $C < \infty$ is a constant; see Figure 2.3.

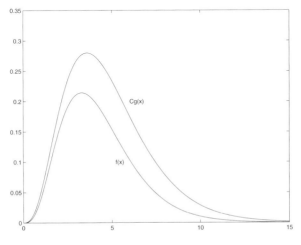

FIGURE 2.3

Given $Y = x$, one accepts Y and lets $X = Y$ w.p. $f(x)/Cg(x)$. Otherwise, a new Y is generated, and one continues until eventual acceptance. Algorithmically:

1. Generate Y from the density $g(x)$ and U as uniform(0,1).

2. If $U \le f(Y)/Cg(Y)$, let $X \longleftarrow Y$. Otherwise return to 1.

That this produces an r.v. with the desired density $f(x)$ follows from

$$\mathbb{P}(X \in dx) = \mathbb{P}(Y \in dx \,|\, A) = \frac{\mathbb{P}(Y \in dx;\, A)}{\mathbb{P}A}$$

$$= \frac{g(x)\,dx \cdot f(x)/Cg(x)}{\displaystyle\int_{-\infty}^{\infty} g(y) \cdot f(y)/Cg(y)\,dy} = \frac{f(x)\,dx}{\displaystyle\int_{-\infty}^{\infty} f(y)\,dy} = f(x),$$

where $A \overset{\text{def}}{=} \{U \le f(Y)/Cg(Y)\}$ is the event of acceptance.

Remark 2.5 The efficiency of A–R is determined by the speed of generation from $g(\cdot)$ and by the acceptance probability

$$\mathbb{P}\big(U \le f(Y)/Cg(Y)\big) = \mathbb{E}\big[\mathbb{P}\big(U \le f(Y)/Cg(Y) \,\big|\, Y\big)\big]$$

$$= \mathbb{E}\Big[\frac{f(Y)}{Cg(Y)}\Big] = \int \frac{f(y)}{Cg(y)} g(y)\,dy = \frac{1}{C}\int f(y)\,dy = \frac{1}{C}.$$

Obviously it is preferable to have C as close to 1 as possible, which in turn means that $g(\cdot)$ should look as much alike $f(\cdot)$ as possible.   □

Example 2.6 As a simple example, let $f(x)$ be a bounded density on $(0,1)$, say the Beta density $x^{\alpha-1}(1-x)^{\beta-1}/B(\alpha,\beta)$ with $\alpha,\beta > 1$. Then, we may take Y as uniform on $(0,1)$ and $C = \sup_{0<x<1} f(x)$.   □

Example 2.7 Let $f(x) = \sqrt{2/\pi}\,e^{-x^2/2}$ be the density of the absolute value $X = |Z|$ of a standard normal r.v. As trial density, we consider the standard exponential density $g(x) = e^{-x}$. Then $f(x)/g(x) = \sqrt{2/\pi}\,e^{x-x^2/2}$ is maximized for $x = 1$ and the maximum is $C = \sqrt{2e/\pi}$. The acceptance probability is $1/C \approx 0.76$.

The normal r.v. Z itself is easily simulated by assigning X a random sign.   □

It is rare that, as in this example, $f(x)/g(x)$ can easily be maximized analytically and thereby the optimal C determined. However, usually it is not worthwhile to look for the optimal C, and a rather crude bound determined numerically will do. This is illustrated in the following example:

Example 2.8 In Figure 2.3, the target density $f(\cdot)$ is the Gamma(α, λ) density with $\alpha = 4.3$, $\lambda = 1$.

In looking for a suitable proposal $g(x)$, it seems reasonable to exploit the fact that Gamma(p, λ) r.v.'s with integer p are easily simulated (as a sum

of p independent exponentials) and that the corresponding density $g(x)$ is
not too far from $f(x)$ if p is close to α and λ close to 1. The precise choice of
p, λ is restricted by boundary conditions: for the inequality $f(x) \leq Cg(x)$
to hold with a finite C for all $0 < x < \infty$, $g(x)$ must go more slowly
to 0 than $f(x)$ as x approaches either of the boundaries $0, \infty$. Now the
Gamma(α, λ) density is of order $x^{\alpha-1}$ as $x \to 0$ and of order $x^{\alpha-1}e^{-\lambda x}$
as $x \to \infty$. In chosing between $p = 4$ and $p = 5$, the $x \to 0$ requirement
therefore excludes $p = 5$. With $p = 4$ we must then take $\lambda < 1$ to meet the
$x \to \infty$ requirement since $x^{4.3-1}e^{-x}$ decays slower than $x^{4-1}e^{-\lambda x}$ when
$\lambda > 1$. The choice in Figure 2.3 is to take $\lambda = 1/1.2$. Thereafter one must
find C to satisfy $f(x) \leq Cg(x)$ for all x, which was done empirically from
the plot (check graphically that the inequality holds unless x is very close
to 0 or ∞, and rely on the boundary asymptotics in this range). It was
found that $C = 1.5$ works, and it is the value used in Figure 2.3. $\square$

Example 2.9 (LOG-CONCAVE DENSITIES) Let $f(x)$ be a log-concave decreasing density on $[0, \infty)$ with mode at $x = 0$; for r.v. generation, we
can w.l.o.g. assume $f(0) = 1$, since otherwise we can just proceed via
$\widetilde{X} \stackrel{\text{def}}{=} f(0)X$, which has a density $\widetilde{f}(x)$ satisfying the given assumptions.
Then

$$f(x) \leq \min(1, e^{1-x}). \qquad (2.3)$$

Indeed, assume $f(x_0) > e^{1-x_0}$ for some x_0 (necessarily > 1). By log-
concavity, $\log f(x)$ is above the line connecting $(0,0)$ and $(x_0, \log f(x_0))$
for $x \in (0, x_0)$, which yields

$$\int_0^\infty f(x)\,dx > \int_0^{x_0} e^{x(1-x_0)/x_0}\,dx = \frac{x_0}{x_0-1}(1 - e^{1-x_0}) > 1,$$

a contradiction. The r.h.s. of (2.3) can be interpreted as $2g(x)$, where $g(x)$
is the density of an r.v. Y with a distribution that is a mixture with equal
weights $1/2$ of a uniform$(0, 1)$ distribution and a standard exponential distribution shifted rightward by 1. It is straightforward to simulate such a Y
(and thereby X itself) by A–R, or as an r.v. from a mixture distribution.

The advantage of this algorithm is the wide occurrence of log-concave
densities, the simplicity, and the relatively high acceptance probability of
$1/2$. $\square$

Example 2.10 Assume still that f is log-concave, not necessarily concentrated on $(0, \infty)$, but now we want to generate an r.v. conditioned to (a, b)
where $-\infty < a < b < \infty$. Letting $\beta = f'(a)/f(a)$, the density of such an
r.v. is bounded by

$$\frac{f(a)}{F(b) - F(a)}e^{\beta(x-a)} = Ch(x),$$

where h is the density proportional to $e^{\beta x}$ in (a, b) and $C > 1$ an easily
computed constant. We can therefore just sample from h by inversion and

obtain the desired conditioned r.v. by rejection. The condition $b < \infty$ can be omitted if $\beta < 0$.   □

Log-concave densities are further discussed in Devroye [88]. A notable algorithm for r.v. generation within the class is an adaptive A–R algorithm, which uses a piecewise linear upper bound that is adaptively improved; see, e.g., Gentle [127, Section 4.13].

2c Ad Hoc Algorithms

A number of methods for generating nonuniform r.v.'s are based on relations between different distributions, and designing them often is quite tricky. We present here a few of the most standard ones; further examples are the algorithms for stable distributions in XII.2a and the inverse Gaussian distribution in XII.5.

Example 2.11 (THE BOX–MULLER METHOD FOR NORMAL R.V.'S) Scaling an exponential r.v. by 2, one obtains a χ^2 with 2 degrees of freedom, which is the distribution of the squared radial part $R^2 \overset{\text{def}}{=} Y_1^2 + Y_2^2$ of independent $\mathcal{N}(0,1)$ r.v.'s Y_1, Y_2. Since the conditional distribution of Y_1, Y_2 given $R = r$ is uniform on the circle $\{(v_1, v_2) : v_1^2 + v_2^2 = r\}$ with radius r, we obtain the following algorithm for generating (pairs of) normal r.v.'s from (pairs of) uniforms:

$$Y_1 \longleftarrow \sqrt{-2\log U_1} \sin 2\pi U_2, \quad Y_2 \longleftarrow \sqrt{-2\log U_1} \cos 2\pi U_2.$$

Due to the many special functions, the Box–Muller (not Müller!) algorithm is slow, and standard packages use quite complex methods to achieve fast generation of normal r.v.'s.   □

Example 2.12 (MARSAGLIA'S POLAR METHOD) This is a variant of the Box-Muller algorithm, replacing the evaluation of the trigonometric functions by an A–R step. The basic observation behind is that if $V_1 = Z \cos \Theta$, $V_2 = Z \sin \Theta$ is the polar representation of a random pair (V_1, V_2) that is uniform on the unit disk $\{(v_1, v_2) : v_1^2 + v_2^2 \le 1\}$, then $W = Z^2$ and Θ are independent and uniform on $(0,1)$, respectively $(0, 2\pi)$. Thus $\sqrt{-2\log Z} = \sqrt{-\log W}$ and $(\cos \Theta, \sin \Theta) = (V_1/\sqrt{W}, V_2/\sqrt{W})$ are independent with an exponential distribution, respectively an uniform distribution on the unit circle, so that by the same argument as for Box–Muller a valid algorithm producing a pair of independent standard normals is

$$Y_1 \longleftarrow \sqrt{-\log W/W}\, V_1, \quad Y_2 \longleftarrow \sqrt{-\log W/W}\, V_2. \qquad (2.4)$$

In practice, (V_1, V_2) are produced by A–R (generate (U_1, U_2) and repeat until $U_1^2 + U_2^2 \le 1$).   □

Example 2.13 The geometric distribution exists in two variants, one with support $\{0, 1, \ldots\}$ and point probabilities $p_n = (1 - \rho)\rho^n$, and the other with support $\{1, 2, \ldots\}$ and point probabilities $p'_n = (1-\rho)\rho^{n-1}$ (see A1). For the corresponding r.v.'s X, X' one has $X' \overset{\mathcal{D}}{=} X + 1$, so it suffices to discuss how to generate X.

The most straightforward procedure is to use the interpretation of failures before a success. That is, if a coin coming up heads w.p. $1 - \rho$ and tails w.p. ρ is flipped until a head comes up, the number of tails before then is distributed as X. Thus, one simply generates Bernoulli(p) r.v.'s with $p \overset{\text{def}}{=} 1 - \rho$ until the first 1, and let X be the number of 0's before then.

If ρ is close to one, this procedure is slow. An alternative is to use that a geometric r.v. X has the same distribution as the integer part of an exponential r.v. Y with mean $-1/\log\rho$. Thus, one may let $X \overset{\text{def}}{=} \lfloor \log U / \log \rho \rfloor$ with U uniform(0,1). □

Example 2.14 (NORMAL VARIANCE MIXTURES) Some r.v.'s with a standard distribution can be simulated as $X = SY$, where Y is standard normal and $S > 0$.

Examples occur in particular in stochastic volatility models in mathematical finance; see XII.5 for examples. □

Example 2.15 (THE ALIAS METHOD) Let F be an $(n+1)$-point distribution on $x_1, \ldots, x_{n+1}$ with weight p_k for x_k. Naive r.v. generation consists in letting $X = x_K$, where K satisfies $p_1 + \cdots + p_{K-1} < U < p_1 + \cdots + p_K$ with U uniform. To avoid the search for K, the *alias method* consists in representing F as a mixture of n distributions $G_1, \ldots, G_n$ with equal weights $1/n$, such that G_ℓ is a two-point distribution on $x_{1,\ell}, x_{2,\ell} \in \{x_1, \ldots, x_{n+1}\}$ with weights $p_{1,\ell}, p_{2,\ell}$ for each. Given such a representation, one can then generate two uniforms U_1, U_2, let $M = \lceil U_1/n \rceil$, and take $X = x_{1,M}$ if $U_2 \leq p_{1,M}$, $X = x_{2,M}$ otherwise.

To find $G_1, \ldots, G_n$, we start by choosing i such that $p_i < 1/n$; that such i exists follows since otherwise $\sum p_i > 1$. Next, we can choose j such that $p_i + p_j \geq 1/n$, since otherwise one would have $p_j < 1/n$ for $j \neq i$ and therefore $\sum p_i < 1/n + (n-1)/n \leq 1$. We then let

$$x_{1,1} \overset{\text{def}}{=} x_i, \quad x_{2,1} \overset{\text{def}}{=} x_j, \quad p_{1,1} \overset{\text{def}}{=} np_i, \quad p_{2,1} \overset{\text{def}}{=} 1 - np_i.$$

For $k \neq i, j$ we then have $p_k \leq 1 - p_i - p_j < (n-1)/n$. Therefore, if we define H_1 as the n-point measure on $\{x_1, \ldots, x_{n+1}\} \setminus \{x_i\}$ with point probability $n/(n-1) \cdot p_k$ for $x_k \neq x_i, x_j$ and $n/(n-1) \cdot (p_i + p_j - 1/n)$ for x_j, then H_1 is a probability distribution satisfying $F = G_1/n + (n-1)/n \, H_1$. Decomposing H_1 in a similar way as $1/(n-1) \cdot G_2 + (n-2)/(n-1) \cdot H_2$, we have $F = G_1/n + G_2/n + (n-2)/n \cdot H_2$, and repeating the procedure a further $n-2$ times yields the desired decomposition.

Whether the alias method is worthwhile is a trade-off between the setup time and the number of X_k that need to be generated. The method generalizes in a straightforward way to a mixture $F = p_1 F_1 + \cdots + p_n F_n$ (here

$0 < p_k < 1$, $p_1 + \cdots + p_n = 1$), where it is easy to simulate from each F_i.
□

We stress again that for the average user, optimal efficiency is not necessarily a major concern, and that one may want to use a naive but easily programmed method rather than invoking more sophisticated methods or various library packages.

2d   Further Uses of Acceptance–Rejection Ideas

Remark 2.16 The A–R method allows some extension beyond the case in which $f(x)$ is a normalized density. More precisely, one can sometimes deal with the situation in which the target distribution F has density $f(x)/D$ w.r.t. some reference measure μ but D is unknown. If for some easily simulated density $h(x)$, one can find an upper bound C on $C^* \overset{\text{def}}{=} \sup_x f(x)/h(x)$, one can then proceed eactly as for ordinary A–R, simulating r.v.'s $Y_1, Y_2, \ldots$ from $h(x)$ and accepting w.p. $f(x)/Ch(x)$.

The situations with a known $f(x)$ but unknown D are, however, typically complicated. In fact, they constitute the typical area of problems handled by Markov chain Monte Carlo (MCMC, see Chapter XIII), and the problems one encounters there when attempting to apply A–R rather than the traditional MCMC methods are first of all that it may be difficult to find a suitable h. Next, even if an $h(x)$ with $C^* < \infty$ and allowing for easy r.v. generation can be found, the bound C may be difficult to evaluate and will typically be far from C^*, making the rejection rate huge. Note that computing C^* itself is a problem of the same order of difficulty as maximum likelihood estimation!   □

Example 2.17 (RATIO OF UNIFORMS) As a motivation, note that by (2.4) the ratio V_1/V_2 of a random pair with a uniform distribution on the unit disk has the same distribution as that of the ratio of a pair of standard normals which (by Box–Muller) is in turn that of $\tan(2\pi U)$, that is, the Cauchy distribution. The ratio-of-uniforms method replaces the unit disc by a more general region

$$\Omega \overset{\text{def}}{=} \Omega(f) \overset{\text{def}}{=} \left\{ (v_1, v_2) : 0 < v_1 \le \sqrt{h(v_2/v_1)} \right\},$$

where $h \overset{\text{def}}{=} cf$ with f the target density, samples (V_1, V_2) from the uniform distribution on Ω, and returns $X = V_2/V_1$. To see that this gives X the desired density, let $|\Omega|$ denote the area of Ω. Using the change of variables $(v_1, v_2) \to (x = v_2/v_1, y = v_1)$, the Jacobian becomes $J = v_1^{-1} = y^{-1}$. Thus

$$|\Omega| = \iint_\Omega \mathrm{d}v_1 \mathrm{d}v_2 = \int \mathrm{d}x \int_0^{\sqrt{h(x)}} y \, \mathrm{d}y = \int h(x)/2 \, \mathrm{d}x = c/2 < \infty,$$

and the density of (X, Y) is the density $|\Omega|^{-1}$ of (V_1, V_2) divided by J, so that the marginal density of X is

$$|\Omega|^{-1} \int_0^{\sqrt{h(x)}} y \, dy = \frac{h(x)}{2|\Omega|} = f(x).$$

In practice, one usually implements the sampling of (V_1, V_2) by determining a rectangle R containing Ω, generating uniforms on R, and accepting when they fall in Ω. Representing R as $(0, a) \times (b_-, b_+)$, a possible choice is

$$a = \left[\sup \{h(x) : x \in \mathbb{R}\} \right]^{1/2}, \tag{2.5}$$

$$b_+ = \left[\sup \{x^2 h(x) : x \geq 0\} \right]^{1/2},$$
$$b_- = -\left[\sup \{x^2 h(x) : x \leq 0\} \right]^{1/2} \tag{2.6}$$

(provided the sup's are finite). Indeed, let $(v_1, v_2) \in \Omega$. Then $0 < v_1 \leq a$ is obvious. If $v_2 > 0$, then $v_1^2 \leq h(v_2/v_1)$ or equivalently $v_2^2 \leq x^2 h(x) \leq b_+$ where $x = v_2/v_1$. That $v_2 \geq b_-$ when $v_2 \leq 0$ follows similarly.

For a simple but relevant example, consider the standard normal distribution. We can take $h(x) = e^{-x^2/2}$ and get $a = 1$, $b_+ = b_- = (2/e)^{1/2}$, whereas the acceptance condition $V_1 \leq \sqrt{h(V_2/V_1)}$ reduces to $V_2^2 \leq -4V_1^2 \log V_1$, i.e. $X^2 \leq -4 \log V_1$. □

Example 2.18 (VON NEUMANN, EXPONENTIAL) Let X be standard exponential and Y exponential conditioned to $(0, 1)$, that is, with density $e^{-y}/(1 - e^{-1})$ for $0 < y < 1$. Then X can be generated as $M + Y$, where M is independent of Y and geometric with $\mathbb{P}(N = n) = (1 - e^{-1})e^{-n}$.

To generate Y, consider sequential sampling of i.i.d. uniforms until

$$N \overset{\text{def}}{=} \inf \{n > 1 : U_1 > U_2 > \cdots > U_{n-1} < U_n\}$$

(the time at which the descending order of the U_i is broken). Since $U_1 > U_2 > \cdots > U_{n-1} > U_n$ corresponds to one of $n!$ orderings, we have $\mathbb{P}(N > n) = 1/n!$. Similarly, the extra requirement $U_1 \leq y$ imposes the conditioning $U_k \leq y$ for all $k \leq n$, so that

$$\mathbb{P}(N > n, U_1 \leq y) = \frac{y^n}{n!}, \quad \mathbb{P}(N = n, U_1 \leq y) = \frac{y^{n-1}}{(n-1)!} - \frac{y^n}{n!}, \quad n = 2, 3, \ldots$$

Summing over $n = 2, 4, \ldots$ yields $\mathbb{P}(N \text{ is even}, U_1 \leq y) = 1 - e^{-y}$. Thus repeating the experiment until an even N comes out, we can take Y as the U_1 in the final (i.e., the first accepted) trial.

The algorithm can be written in a number of variants, e.g., Ripley [300] pp. 63–64. □

Example 2.19 (SQUEEZING) The target density $f(x)$ may sometimes be difficult to evaluate, say the expression for $f(x)$ involves nonstandard special functions. A–R implemented via the upper bound $f(x) \le Cg(x)$ is then potentially slow because of the repeated evaluations of $f(x)$. The idea of *squeezing* is to supplement with an easily evaluated lower bound $h(x) \le f(x)$, which makes it possible to reduce the number of evaluations of $f(x)$. Then one first tests $U \le h(Y)$ and then accepts, so that the test $U \le f(Y)/Cg(Y)$ needs to be performed only when $U > h(Y)$. See Figure 2.4 for an illustration.

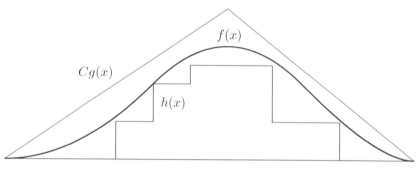

FIGURE 2.4

□

2e   Transform-Based Methods

In some cases, the density $f(x)$ or the c.d.f. $F(x)$ of a distribution $F(\mathrm{d}x)$ may not be available in explicit form, whereas the transform $\widehat{F}[s] = \int e^{sx} F(\mathrm{d}x)$ is. In others, $f(x)$ or $F(x)$ may be in principle computable, but is much more complicated than $\widehat{F}[s]$. One may therefore ask whether one can generate r.v.'s from F using $\widehat{F}[s]$ only. We just give a few basic remarks on this problem and refer to Devroye [88] for more detail.

For the following, note that $\psi(s) \overset{\text{def}}{=} \widehat{F}[is]$ viewed as function of a real variable s is the characteristic function of F.

Example 2.20 If $\psi(s)$ is Lebesgue integrable, the Fourier inversion formula states that

$$f(x) \; = \; \frac{1}{2\pi} \int_{-\infty}^{\infty} \psi(s) e^{-isx} \, \mathrm{d}s \,. \tag{2.7}$$

A simple approximate procedure is therefore the following:

1. evaluate $f(x_i), i = 1, \ldots, n$ from (2.7) for a finite set of selected points $x_1 < \cdots < x_n$;

2. as approximation to F, consider the discrete distribution $\widetilde{F}$ with point mass $p_i \overset{\text{def}}{=} f(x_i)/\big(f(x_1) + \cdots + f(x_n)\big)$ at x_i;

3. sample the desired r.v.'s from $\widetilde{F}$.   □

Example 2.21 (FFT) The method of Example 2.20 is particularly efficient when implemented via the FFT (fast Fourier transform; cf. A3) because of the speed of the FFT. For this, one needs to choose the support of $\widetilde{F}$ as

$$\{x_1, x_2 = x_1 + \varepsilon, x_3 = x_1 + 2\varepsilon, \ldots, x_n = x_1 + (n-1)\varepsilon\}.$$

The discrete Fourier transform of the corresponding (a priori unknown) approximate weights $p_1, p_2, \ldots, p_n$ is then $\widehat{\boldsymbol{p}} = \boldsymbol{F}\boldsymbol{p}/n$, where $\boldsymbol{F}$ is the finite Fourier matrix of order n and $\boldsymbol{p} \stackrel{\text{def}}{=} (p_1 \ldots p_n)^{\mathsf{T}}$. Element by element,

$$
\begin{aligned}
\widehat{p}_r &= \frac{1}{n} \sum_{s=1}^{n} p_s w^{rs} = \frac{1}{n} \sum_{s=1}^{n} p_s \exp\left\{ \frac{2\pi i r}{n} \left(1 + (x_s - x_1)/\varepsilon\right) \right\} \\
&= \frac{1}{n} \exp\left\{ \frac{2\pi i r}{n} (1 - x_1/\varepsilon) \right\} \sum_{s=1}^{n} p_s \exp\left\{ \frac{2\pi i r}{n\varepsilon} x_s \right\},
\end{aligned}
$$

where $w \stackrel{\text{def}}{=} e^{2\pi i/n}$ and we solved $x_s = x_1 + (s-1)\varepsilon$ for s in the second step. Here we can approximate the last expression by

$$\widehat{q}_r \stackrel{\text{def}}{=} \frac{1}{n} \exp\left\{ \frac{2\pi i r}{n} (1 - x_1/\varepsilon) \right\} \psi(2\pi r/n\varepsilon).$$

This suggests that a good choice of $\boldsymbol{p}$ is the inverse transform $\overline{\boldsymbol{F}}\widehat{\boldsymbol{q}}$, where $\widehat{\boldsymbol{q}} \stackrel{\text{def}}{=} (\widehat{q}_1 \ldots \widehat{q}_n)^{\mathsf{T}}$. However, this choice does not guarantee $\overline{\boldsymbol{F}}\widehat{\boldsymbol{q}}$ to be a probability distribution, so that one needs the modification

$$p_i^* \stackrel{\text{def}}{=} \left| (\overline{\boldsymbol{F}}\widehat{\boldsymbol{q}})_i \right|, \quad p_i \stackrel{\text{def}}{=} \frac{p_i^*}{p_1^* + \cdots + p_n^*}.$$

Sampling of $\widetilde{X}$ is then straightforward, since we are just dealing with a finite discrete distribution.

There are several variants of the algorithm, e.g., to return the continuous r.v. $\widetilde{X} + \varepsilon(U - 1/2)$ instead of $\widetilde{X}$.   □

Example 2.22 For a more sophisticated method, assume that F has a second moment, i.e., that ψ'' is integrable. Integration by parts and using (2.7) then yields

$$f(x) = \frac{1}{2\pi i x} \int_{-\infty}^{\infty} \psi'(s) e^{-isx} \, ds = -\frac{1}{2\pi i x^2} \int_{-\infty}^{\infty} \psi''(s) e^{-isx} \, ds.$$

Thus, in the tail $f(x)$ is of order at most $O(x^{-2})$, which suggests to use a proposal $g(x)$ proportional to $c \wedge kx^{-2}$ for suitable c, k.

As setup, the algorithm computes

$$c \stackrel{\text{def}}{=} \frac{1}{2\pi} \int_{-\infty}^{\infty} |\psi(s)| \, ds, \quad k \stackrel{\text{def}}{=} \frac{1}{2\pi} \int_{-\infty}^{\infty} |\psi''(s)| \, ds.$$

We then let $g(x) \stackrel{\text{def}}{=} (c \wedge kx^{-2})/A$, where $A \stackrel{\text{def}}{=} 4\sqrt{kc}$. Simulation of an r.v. Y from $g(\cdot)$ is easy, since $g(\cdot)$ is a mixture of a uniform distribution on $(-b,b)$ (where $b \stackrel{\text{def}}{=} \sqrt{k/c}$) and a shifted Pareto with $\alpha = 1$ equipped with a random sign, with weight $1/2$ for each. The algorithm then accepts Y w.p. $g(Y)/Af(Y)$, where $f(Y)$ has to be calculated by (2.7).   □

Exercises

2.1 (A$-$)[2] Write a routine for generation of a Weibull r.v. X with tail $\mathbb{P}(X > x) = e^{-x^\beta}$, $x > 0$, by inversion. Check the routine via a histogram of simulated values plotted against the theoretical density, say for $\beta = 1/2$.

2.2 (TP) In the uniform(0,1) distribution, derive the relevant formulas for generating $\max(U_1,\ldots,U_n)$ and $\min(U_1,\ldots,U_n)$ by inversion.

2.3 (TP) Show that $F^\leftarrow$ is left-continuous. Is the distinction between right or left continuity important for r.v. generation by inversion?

2.4 (TP) Let f be a density and $\mathscr{A}(f) = \{(v,x) : 0 \le v \le f(x)\}$. Show that if a bivariate r.v. (V,X) has a uniform distribution on $\mathscr{A}(f)$, then the marginal distribution of X has density f. More generally, show that the same is true if $\mathscr{A}(f,c) = \{(v,x) : 0 \le v \le cf(x)\}$ and (V,X) has a uniform distribution on $\mathscr{A}(f,c)$.

2.5 (TP) Verify the claims of Remark 2.4.

2.6 (TP) Verify the claim concerning the distribution of W in Example 2.12.

2.7 (TP) As a fast but approximate method for generating a r.v. X from $\mathscr{N}(0,1)$, it has been suggested to use a normalized sum of n uniforms, $X \stackrel{\text{def}}{=} (S_n - a_n)/b_n$, where $S_n \stackrel{\text{def}}{=} U_1 + \cdots + U_n$. Find a_n, b_n such that X has the correct two first moments. What are the particular features of the common choice $n = 12$?

2.8 (TP) Assume that the density of X can be written as $f(x) = cg(x)\overline{H}(x)$, where $g(x)$ is the density of an r.v. Y and $\overline{H}(x) = \mathbb{P}(Z > x)$ the tail of an r.v. Z. Show that X can be generated by sampling Y, Z independent and rejecting until $Y \le Z$.

2.9 (A) Produce 100,000 standard normal r.v.'s using each of the following methods: (a) inversion using the approximation (2.2), (b) Box–Muller, (c) Marsaglia polar, (d) A–R as in Example 2.7, (e) ratio-of-uniforms and (as comparison) (f) a routine of a standard package, say Matlab. The desired results are the CPU time needed for each of the methods.
Note that the results of course depend not only on hardware and software, but also on implementation issues such as the efficiency of the r.v. generation scheme for the exponential r.v.'s in (d).

2.10 (A) Write a routine for generation of an inverse Gaussian r.v. X with density (A1.2) by A–R when $\xi = c = 1$. Check the routine via confidence intervals for $\mathbb{E}X$, $\mathbb{E}X^2$, and (a little harder!) $\mathbb{V}ar\, X$, using the known formulas $\mathbb{E}X = c/\xi$, $\mathbb{V}ar\, X = c/\xi^3$.

---

[2]The $-$ indicates a somewhat lower degree of difficulty that the typical other assignments.

3   Multivariate Random Variables

3a   *Multivariate Normals*

A multivariate normal r.v. $\boldsymbol{X} = \boldsymbol{X}_{\boldsymbol{\mu},\boldsymbol{\Sigma}} \in \mathbb{R}^p$ is given by its mean vector $\boldsymbol{\mu}$ and its covariance matrix $\boldsymbol{\Sigma}$. Since $\boldsymbol{X}_{\boldsymbol{\mu},\boldsymbol{\Sigma}}$ can be generated just as $\boldsymbol{\mu} + \boldsymbol{X}_{\boldsymbol{0},\boldsymbol{\Sigma}}$, we assume $\boldsymbol{\mu} = \boldsymbol{0}$ in the following and write $\boldsymbol{X} \stackrel{\text{def}}{=} \boldsymbol{X}_{\boldsymbol{\Sigma}}$.

The case $p = 2$ is simple. We can write

$$\boldsymbol{\Sigma} = \begin{pmatrix} \sigma_1^2 & \rho\sigma_1\sigma_2 \\ \rho\sigma_1\sigma_2 & \sigma_2^2 \end{pmatrix},$$

where $\rho = \mathbb{C}orr(X_1, X_2)$, take Y_1, Y_2, Y_3 independent $\mathscr{N}(0,1)$, and let

$$X_1 = \sigma_1\big(\sqrt{1-|\rho|}\,Y_1 + \sqrt{|\rho|}\,Y_3\big), \quad X_2 = \sigma_2\big(\sqrt{1-|\rho|}\,Y_2 \pm \sqrt{|\rho|}\,Y_3\big),$$

where $+$ is for $\rho \geq 0$, $-$ for $\rho \leq 0$. The case $p > 2$ can in the same way be reduced to the i.i.d. case $\boldsymbol{\Sigma} = \boldsymbol{I}$ if we can find a square root of $\boldsymbol{\Sigma}$, that is, a matrix $\boldsymbol{C}$ satisfying $\boldsymbol{\Sigma} = \boldsymbol{CC}^{\mathsf{T}}$, since we can then generate $\boldsymbol{X}_{\boldsymbol{\Sigma}}$ as $\boldsymbol{CX}_{\boldsymbol{I}}$. Component by component,

$$X_i = \sum_{k=1}^{p} c_{ik}Y_k, \quad i = 1,\ldots,p, \tag{3.1}$$

where $Y_1,\ldots,Y_p$ are i.i.d. $\mathscr{N}(0,1)$ r.v.'s.

An obvious candidate for $\boldsymbol{C}$ is $\boldsymbol{\Sigma}^{1/2}$, the nonnegative definite (symmetric) square root of $\boldsymbol{\Sigma}$. This can be found from a diagonal form $\boldsymbol{\Sigma} = \boldsymbol{B}(\lambda_i)_{\text{diag}}\boldsymbol{B}^{\mathsf{T}}$ with $\boldsymbol{B}$ orthogonal by letting $\boldsymbol{\Sigma}^{1/2} \stackrel{\text{def}}{=} \boldsymbol{B}(\lambda_i^{1/2})_{\text{diag}}\boldsymbol{B}^{\mathsf{T}}$. Computationally, the following procedure is most often simpler:

Example 3.1 (CHOLESKY FACTORIZATION) This is an algorithm for writing a given symmetric $p \times p$ matrix $\boldsymbol{\Sigma} = (\Sigma_{ij})_{i,j=1,\ldots,p}$ as $\boldsymbol{\Sigma} = \boldsymbol{CC}^{\mathsf{T}}$, where $\boldsymbol{C} = (c_{ij})_{i,j=1,\ldots,p}$ is (square) lower triangular ($c_{ij} = 0$ for $j > i$). In the present simulation context, it works as follows. Suppose that we have already expressed the first $p-1$ components $X_1,\ldots,X_{p-1}$ of $\boldsymbol{X}$ as a lower triangular combination

$$X_1 = c_{11}Y_1, \quad X_2 = c_{21}Y_1 + c_{22}Y_2, \quad \ldots,$$
$$X_{p-1} = c_{(p-1)1}Y_1 + \cdots + c_{(p-1)(p-1)}Y_{p-1},$$

of i.i.d. $\mathscr{N}(0,1)$ r.v.'s. To generate the c_{pj} for which $X_p = \sum_1^p c_{pj}Y_j$, observe that for $1 \leq i \leq p$,

$$\Sigma_{pi} = \mathbb{C}ov(X_p, X_i) = \mathbb{C}ov\bigg(\sum_{j=1}^{p} c_{pj}Y_j, \sum_{k=1}^{i} c_{pk}Y_k\bigg) = \sum_{k=1}^{i} c_{pk}c_{ik},$$

leading to a system of p linear equations in $c_{p1}, \ldots, c_{pp}$. This has recursive solution

$$c_{pj} = \frac{1}{c_{jj}}\left(\Sigma_{pj} - \sum_{k=1}^{j-1} c_{pk}c_{jk}\right), \quad j < p, \quad c_{pp}^2 = \Sigma_{pp} - \sum_{k=1}^{p-1} c_{pk}^2. \quad (3.2)$$

We return to Cholesky factorization in connection with Gaussian processes in XI.2.   □

A number of special structures of particular interest allow for direct methods:

Example 3.2 (SYMMETRIC POSITIVE CORRELATIONS) This corresponds to $\Sigma_{ii} = \sigma^2$, $\Sigma_{ij} = \rho\sigma^2$ with $\rho \in (0, 1]$. We can take

$$X_k = \sigma\rho^{1/2}Z + \sigma(1-\rho)^{1/2}Y_k, \quad k = 1, \ldots, p,$$

where $Z, Y_1, \ldots, Y_p$ are i.i.d. standard normals.   □

Example 3.3 (SYMMETRIC NEGATIVE CORRELATIONS) This is like Example 3.2 except that now $\rho < 0$. Not all values $\rho < 0$ are, however, possible: for positive semidefiniteness, it is needed that $\rho \geq -1/(p-1)$. Assuming w.l.o.g. that $\sigma^2 = 1$, we can take

$$X_k = bY_k - a \sum_{\ell \in \{1,\ldots,p\}\setminus\{k\}} Y_\ell, \quad k = 1, \ldots, p,$$

where $Y_1, \ldots, Y_p$ are i.i.d. standard normals and

$$(p-1)a^2 + b^2 = 1, \quad (p-2)a^2 - 2ab = \rho.$$

In particular, the maximal attainable negative correlation is obtained by letting $X_k \stackrel{\text{def}}{=} (1 - 1/n)^{-1/2}\varepsilon_k$, where $\overline{Y} \stackrel{\text{def}}{=} (Y_1 + \cdots + Y_p)/p$ is the sample mean and $\varepsilon_k \stackrel{\text{def}}{=} Y_k - \overline{Y}$ the residual.   □

Example 3.4 Often, it may be reasonable to assume that $\mathbb{C}ov(X_k, X_\ell)$ depends only on $\ell - k$. That is,

$$\Sigma = \sigma^2 \begin{pmatrix} 1 & \rho_1 & \rho_2 & \cdots & \rho_{p-2} & \rho_{p-1} \\ & 1 & \rho_2 & \cdots & \rho_{p-3} & \rho_{p-2} \\ & & \ddots & & & \vdots \\ & & & 1 & \rho_1 \\ & & & & 1 \end{pmatrix},$$

where $\rho_\ell = \mathbb{C}orr(X_k, X_{k+\ell})$. A multivariate normal r.v. with such a covariance matrix is in particular obtained by sampling p consecutive values in a stationary Gaussian process, for which there are many specific candidates around, see Chapter XI.   □

Example 3.5 (MOVING AVERAGES) In Example 3.4, one will in many cases have a decreasing tendency of the ρ_ℓ. To obtain a particular desired pattern, one may for example consider an $MA(q)$ (moving average of order q) process where $\rho_\ell = 0$ for $\ell > q$, leading to

$$X_k = a_1 Y_k + a_2 Y_{k+1} + \cdots + a_q Y_{k+q-1}\,,$$

where $Y_1, Y_2, \ldots$ are i.i.d. standard normals, and play around with the equations

$$1 = a_1^2 + a_2^2 + \cdots + a_q^2,$$
$$\rho_1 = a_1 a_2 + a_2 a_3 + \cdots + a_{q-1} a_q,$$
$$\rho_2 = a_1 a_3 + a_2 a_4 + \cdots + a_{q-2} a_q,$$
$$\vdots$$
$$\rho_q = a_1 a_q.$$

$\square$

3b Other Parametric Multivariate Distributions

The area of multivariate distributions is extremely rich in examples; see Johnson & Kotz [202] and (for simulation aspects) also Johnson [201]. For example, there is an abundance of multivariate distributions with standard exponential marginals. We therefore constrain ourselves to a few basic examples. The first two deal with distributions on $\mathbb{R}^p$ with all marginal distributions equal to a given distribution F on the line and dependent components; the next three allow the marginals to vary in a parametric class.

Example 3.6 (MULTIVARIATE t) This is defined as the distribution of

$$\left(\frac{Y_1}{\sqrt{W}}, \ldots, \frac{Y_p}{\sqrt{W}} \right),$$

where $Y_1, Y_2, \ldots$ are i.i.d. standard normals, and W an independent χ_f^2/f r.v. (f = degrees of freedom). A generalization is to take $\boldsymbol{Y} \stackrel{\text{def}}{=} (Y_1 \ldots Y_p)^{\mathsf{T}}$ multivariate normal $(\boldsymbol{0}, \boldsymbol{\Sigma})$. $\square$

Example 3.7 (MULTIVARIATE LAPLACE) This is defined as the distribution of $(\sqrt{W} Y_1, \ldots, \sqrt{W} Y_p)$, where $Y_1, Y_2, \ldots$ are i.i.d. standard normals, and W an independent standard exponential r.v. For a financial application, see Huang & Shahabuddin [187]. $\square$

Example 3.8 (DIRICHLET DISTRIBUTION) This distribution has parameters $a_1, \ldots, a_p$ and is defined as the distribution of

$$\left(\frac{Y_1}{S}, \ldots, \frac{Y_p}{S} \right),$$

where $Y_1, \ldots, Y_p$ are independent and $\mathrm{Gamma}(a_1, 1), \ldots, \mathrm{Gamma}(a_p, 1)$ and $S \stackrel{\text{def}}{=} Y_1 + \cdots + Y_p$. The kth marginal is $\mathrm{Beta}(a_k, a)$, where $a \stackrel{\text{def}}{=} a_1 + \cdots + a_p$. □

Example 3.9 (MULTINOMIAL DISTRIBUTION) This is the distribution of the counts $(X_1, \ldots, X_k)$ of $N = X_1 + \cdots + X_k$ objects classified into k categories, with probability say p_j for the jth. A naive method is to go through the objects one by one and classify an object by sampling j w.p. p_j. A more efficient method (assuming that fast generation of binomial (M, p) variables is available also for large M) is to generate first X_1 as binomial(N, p_1), next X_2 as binomial$(N - X_1, p_2/(1 - p_1))$, and so on.

The algorithm uses sampling from the marginal distribution of X_1 and the conditional distributions of X_i given $X_1, \ldots, X_{i-1}$. This can in principle be used for any multivariate distribution, but the problem is that the conditional distributions seldom have an attractive form. □

Example 3.10 The *Marshall–Olkin bivariate exponential distribution* is defined as the distribution of (X_1, X_2), where $X_1 \stackrel{\text{def}}{=} T_1 \wedge T_{12}$, $X_2 \stackrel{\text{def}}{=} T_2 \wedge T_{12}$ with T_1, T_2, T_{12} independent exponentials with rates $\lambda_1, \lambda_2, \lambda_{12}$. We can think of X_i as the lifetime of an item which may potentially fail at a time T_i specific to the item and a time T_{12} common for both items, such that the actual failure occurs at the first of these potential failure times. The marginals are exponential with rates $\lambda_1 + \lambda_{12}$, respectively $\lambda_2 + \lambda_{12}$. □

Example 3.11 A completely different type of multivariate distributions is that of uniform distributions on a region Ω. The simplest algorithm is to find (if possible) a box $(a_1, b_1) \times \cdots \times (a_p, b_p)$ containing Ω, generate consecutive uniforms on the box in the obvious way from p uniform$(0, 1)$ r.v.'s, and accept the first in Ω.

A case not covered by this is $|\Omega| = 0$ (Lebesgue measure). The particular case in which Ω is the unit sphere $\{x : x_1 + \cdots + x_p^2 = 1\}$ in $\mathbb{R}^p$ is particularly important and easy: X can be generated as $(Y_1/\sqrt{R}, \ldots, Y_p/\sqrt{R})$, where $Y_1, \ldots, Y_p$ are i.i.d. standard normal and $R \stackrel{\text{def}}{=} Y_1^2 + \cdots + Y_p^2$. □

3c   Copulas

Whereas there is basically one and only one standard way to extend the normal distribution to multidimensions, that is not the case for other standard distributions. In a few cases (such as the exponential) there is a variety of suggestions around, but usually no clear picture of which one is the natural one.

However, in most cases it is hard to come up with even just one. Copulas then provide a possible approach, not least in examples in which one has a rather well-defined idea of the marginals but a rather vague one of the dependence structure.

A *copula* is defined as the distribution of a random vector $\boldsymbol{U} \stackrel{\text{def}}{=} (U_1 \ldots U_p)^{\mathsf{T}}$, where each U_i has a marginal distribution that is uniform$(0, 1)$ but the U_i may be dependent. For a general random vector $\boldsymbol{Y} \stackrel{\text{def}}{=} (Y_1 \ldots Y_p)^{\mathsf{T}}$ with a continuous marginal distribution, say F_i, of Y_i, the copula of $\boldsymbol{Y}$ is defined as the distribution of $\big(F_1(Y_1) \ldots F_p(Y_p)\big)^{\mathsf{T}}$. Some simple basic examples are *comonotonicity* (the completely positively dependent copula), where $U_1 = \cdots = U_p$; the completely negatively dependent copula for $p = 2$ where $U_1 = 1 - U_2$; Gaussian copulas corresponding to $\boldsymbol{Y}$ being multivariate normal, and multivariate t copulas.

In simulation, copulas are useful for generating multivariate dependent r.v.'s outside of standard situations such as the multivariate Gaussian. Say we want $X_1, \ldots, X_p$ to be dependent standard exponential. Then one can, for example, generate $X_1, \ldots, X_p$ according to a Gaussian copula, meaning that we take $\boldsymbol{Y} = (Y_1 \ldots Y_p)^{\mathsf{T}}$ to be $\mathcal{N}(\boldsymbol{0}, \boldsymbol{\Sigma})$, so that $U_i \stackrel{\text{def}}{=} \Phi\big(Y_i/(\mathbb{V}ar\, Y_i)^{1/2}\big)$ is uniform$(0, 1)$ and we can take $X_i = -\log(1 - U_i)$. Hence, the procedure is a way to export the Gaussian (or other) dependence structure to exponential marginals. In more general terms, one can describe copulas as a general tool to model dependence of whatever kind and to separate the dependence structure from the marginal distributions. Moreover, they help one to understand the deficiencies of quantifying dependence via correlation and suggest how to overcome the problem.

Copulas are nonstructural models in the sense that there is a lack of classes that on the one hand are characterized by a finite number of parameters so that statistical estimation is feasible, and on the other are flexible enough to describe a broad range of dependence structure. The way they are used is therefore most often as scenarios: for given marginals, experiment with different copulas and see how sensitive the quantity of interest is to the particular choice of copula. This approach is today widely used in areas such as multivariate loss distributions (loss from different lines of business), credit risk where losses from different debtors are highly dependent, market risk (multivariate asset returns), and risk aggregation where one combines market risk, credit risk, operational risk, etc. See in particular McNeil et al. [252]. Note, however, that copulas bring nothing more than a formalism for modeling dependence so that, as Mikosch [258] points out, the intense current interest in the concept may be somewhat overenthusiastic.

The example of exponential r.v.'s with a multivariate Gaussian copula explain the general procedure for simulation: first one generates $\boldsymbol{U} \in (0, 1)^p$ with the desired copula, and next one transforms the components of $\boldsymbol{U}$ with the respective inverse c.d.f.'s. Since inversion has been treated separately, we therefore concentrate on the first step, the generation of $\boldsymbol{U}$ outside settings such as the multivariate Gaussian or the multivariate t copula, where the problem is straightforward.

In the following, we take $p = 2$ for simplicity and write $C(u_1, u_2) \overset{\text{def}}{=} \mathbb{P}(U_1 \leq u_1, U_2 \leq u_2)$. If C is absolutely continuous on $(0,1)^2$, the corresponding density is

$$c(u_1, u_2) \overset{\text{def}}{=} \frac{\partial^2}{\partial u_1 \partial u_2} C(u_1, u_2).$$

A further relevant concept is the *tail copula*, which in terms of (U_1, U_2) is defined as the copula of $(1 - U_1, 1 - U_2)$.

In situations in which $c(u_1, u_2)$ is explicitly available in closed form, direct simulation may be feasible. Say that $c(u_1, u_2) \leq K$, where (necessarily) $K \geq 1$. Then we can simulate $\boldsymbol{Y}$ from the independence copula (that is, the components are i.i.d. uniforms) and use acceptance–rejection with acceptance probability $c(u_1, u_2)/K$. If $C(u_2 \mid u_1)$ is explicitly available, we may first simulate U_1 as uniform(0,1) and next U_2 as $C^{\leftarrow}(U_2 \mid u_1)$, where $u_1 = U_1$. In examples this procedure is, however, most often tedious, especially when $p > 2$.

In specific examples, different and more direct methods may be available. We next go through some of the more important copulas in the literature, from the point of view of both introducing these examples and to present simulation algorithms when available:

Example 3.12 (THE FARLIE–GUMBEL–MORGENSTERN COPULA) This corresponds to

$$C(u_1, u_2) = u_1 u_2 \big(1 + \varepsilon(1 - u_1)(1 - u_2)\big)$$

with $\varepsilon \in [-1, 1]$. The generation of an r.v. from this copula can be performed by drawing V_1, V_2 as uniform(0,1) and letting $U_1 \overset{\text{def}}{=} V_1$, $U_2 \overset{\text{def}}{=} 2V_2(a + b)$, where $a \overset{\text{def}}{=} 1 + \varepsilon(1 - 2V_1)$, $b \overset{\text{def}}{=} \big(a^2 - 4(a-1)V_2\big)^{1/2}$; see [201]. $\square$

Example 3.13 (ARCHIMEDEAN COPULAS) Such a copula is specified in terms of a *generator*, that is, a function $\phi : [0,1] \to [0, \infty]$ that is C^2 on $(0,1)$ with $\phi(1) = 0$ and $\phi'(u) < 0$, $\phi''(u) > 0$ for all $u \in (0,1)$. The bivariate c.d.f. of the copula is

$$\mathbb{P}(U_1 \leq u_1, U_2 \leq u_2) = \begin{cases} \phi^{-1}\big(\phi(u_1) + \phi(u_2)\big) & \text{if } \phi(u_1) + \phi(u_2) \leq \phi(0), \\ 0 & \text{otherwise.} \end{cases}$$

Some principal examples are the *Clayton copula*, where $\phi(u) = u^{-\alpha} - 1$ for some $\alpha > 0$; the *Frank copula*, where

$$\phi(u) = -\log \frac{e^{\alpha u} - 1}{e^\alpha - 1}$$

for some $\alpha \in \mathbb{R}$; and the *Gumbel–Hougaard copula*, where $\phi(u) = [-\log u]^\alpha$ for some $\alpha \geq 1$.

The generation of an r.v. from an Archimedean copula can be performed by drawing V_1, V_2 as uniform(0,1) and letting

$$U_1 \stackrel{\text{def}}{=} \phi^{-1}\big(U_1\phi(w)\big), \quad U_2 \stackrel{\text{def}}{=} \phi^{-1}\big((1-U_1)\phi(w)\big),$$

where $w \stackrel{\text{def}}{=} K^{-1}(U_2)$ with $K(t) \stackrel{\text{def}}{=} t - \phi(t)/\phi'(t)$; see [271] p. 134.   □

Example 3.14 (FRAILTY COPULAS) The idea is to create dependence via an unobservable r.v. Z. Taking U_1, U_2 to be conditionally independent given Z with

$$\mathbb{P}\big(U_1 \leq u_1 \,|\, Z = z\big) = u_1^z, \quad \mathbb{P}\big(U_2 \leq u_2 \,|\, Z = z\big) = u_2^z,$$

the bivariate c.d.f. of the copula becomes

$$\mathbb{P}(U_1 \leq u_1, U_2 \leq u_2) = \mathbb{E}\big[u_1 u_2\big]^Z = \widehat{Z}\big[\log u_1 + \log u_2\big],$$

where $\widehat{Z}[s] = \mathbb{E}e^{sZ}$ is the m.g.f. of Z. The Clayton copula is an example, see Exercise 3.4, and a frailty copula is Archimedean; see Exercise 3.6.   □

In some examples, one may even design one's own copula with a specific application in mind. Suppose, for example, that one is interested in bivariate tail dependence, that is, r.v.'s X_1, X_2 that are roughly independent at small values but strongly positively dependent at high. A relevant copula is then, for example, the one generated by an r.v. (X_1, X_2) that is uniform on the region in Figure 3.1.

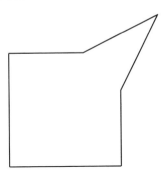

FIGURE 3.1

Example 3.15 A common way to illustrate the dependence structure of a bivariate copula is via a scatter-plot, that is, a plot of R simulated values (necessarily $(0,1)^2$-valued).

Six such scatter-plots with $R = 1{,}000$ are given in Figure 3.2. In lexicographic order, the first is from the independence copula (the uniform distribution on $(0,1)^2$), the second from a Gaussian copula with correlation 0.6, the third from a t copula with $f = 1$ and $\rho = 0$, the fourth from a t copula with $f = 1$ and $\rho = 0.6$, the fifth from the Marshall–Olkin exponential copula (cf. Example 3.10) with $\lambda_1 = 1$, $\lambda_2 = 3$, $\lambda_{12} = 5$, and the

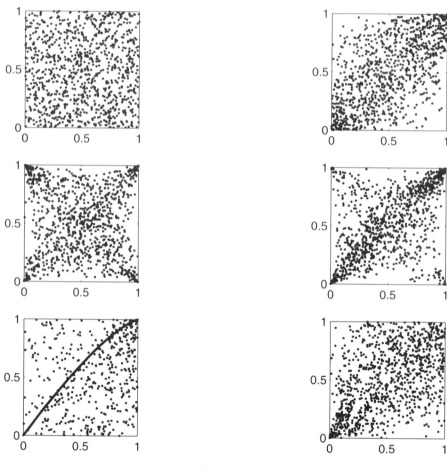

FIGURE 3.2

sixth from the Clayton copula with $\alpha = 1$. A huge number of scatter-plots for further copulas is given in Nelsen [271].

The singular component of the Marshall–Olkin copula of course corresponds to the case $X_1 = X_2$, which arises if $T_{12} < T_1 \wedge T_2$.
□

Nelsen [271] is a standard general text on copulas and also touches upon some simulation aspects. See also Joe [200], Johnson [201], and McNeil et al. [252].

3d   Tail Dependence

In the area of multivariate modeling, there is an abundance of concepts qualifying or quantifying dependence. We will not go into detail, since such discussion is a question of modeling rather than of simulation, but just give a few remarks on the bivariate case of a random vector $(X_1, X_2) \in \mathbb{R}^2$ with joint c.d.f. $F(x_1, x_2) \overset{\text{def}}{=} \mathbb{P}(X_1 \le x_1, X_2 \le x_2)$ with marginals $F_1(x_1) \overset{\text{def}}{=} F(x_1, \infty) = \mathbb{P}(X_1 \le x_1)$ and similarly for F_2.

Intuitively, one feels that given F_1, F_2, maximal positive dependence means that $F(x_1, x_2)$ is as large as possible for all x_1, x_2, and minimal positive dependence that $F(x_1, x_2)$ is as small as possible. A way of making this precise is to look for joint c.d.f.'s F such that the upper and lower *Fréchet–Hoeffding bounds*

$$\left[F_1(x_1) + F_2(x_2) - 1\right]^+ \;\le\; F(x_1, x_2) \;\le\; F_1(x_1) \wedge F_2(x_2) \qquad (3.3)$$

are attained. Define a set $S \subseteq \mathbb{R}^2$ to be *increasing* if $(x_1, x_2), (y_1, y_2) \in S$ and $x_1 < y_1$ implies $x_2 \le y_2$, and *decreasing* if $(x_1, x_2), (y_1, y_2) \in S$ and $x_1 < y_1$ implies $x_2 \ge y_2$.

Proposition 3.16 *A random vector $(X_1, X_2) \in \mathbb{R}^2$ attains the upper Fréchet–Hoeffding bound if and only if its support is an increasing set. Similarly, the lower Fréchet–Hoeffding bound is attained if and only if the support is a decreasing set.*

An obvious example of an increasing (decreasing) set is the graph of a nondecreasing (nonincreasing) function $f : \mathbb{R} \to \mathbb{R}$. In particular, taking $f(x) \overset{\text{def}}{=} \pm x$, we obtain the following corollary:

Corollary 3.17 *In the case of identical marginals, $F_1 = F_2$, the upper Fréchet–Hoeffding bound is attained for the comonotonic copula $X_1 = X_2$ a.s., and the lower Fréchet–Hoeffding bound is attained for the countermonotonic copula $X_1 = F^{\leftarrow}(U)$, $X_2 = F^{\leftarrow}(1 - U)$ with U uniform$(0, 1)$.*

A further relevant concept is *positive quadrant dependence*, meaning $F(x_1, x_2) \ge F_1(x_1)F_2(x_2)$ for all x_1, x_2, which is easily seen to be equivalent to $\mathbb{P}(X_1 > x_1, X_2 > x_2) \ge \mathbb{P}(X_1 > x_1)\mathbb{P}(X_2 > x_2)$. Somewhat related to this is the (upper) *tail dependence coefficient*

$$\lambda \overset{\text{def}}{=} \lim_{t \uparrow 1} \mathbb{P}\big(F_1(X_1) > t \,\big|\, F_2(X_2) > t\big) = 2 - \lim_{t \uparrow 1} \frac{1 - C(t, t)}{1 - t},$$

where C is the copula of (X_1, X_2). Note that the independence copula has $\lambda = 0$ and the comonotonic copula $\lambda = 1$. However, measuring tail dependence only via λ is somewhat rough. For example, if (X_1, X_2) are jointly Gaussian with correlation ρ and $\mathcal{N}(0, 1)$ marginals, then $\lambda = 0$ but $\mathbb{E}\big[X_1 \,\big|\, X_2 = x\big] = \rho x$ goes to ∞ as $x \to \infty$.

Exercises

3.1 (TP) Let $X_1, \ldots, X_p$ be continuous r.v.'s and $h_1, \ldots, h_p$ strictly increasing functions. Show that $X_1, \ldots, X_p$ and $h_1(X_1), \ldots, h_p(X_p)$ generate the same copula.

3.2 (TP) Show that the tail copula of the Marshall–Olkin bivariate exponential distribution is given by $C(u_1, u_2) = (u_1^{1-\alpha_1} u_2) \wedge (u_1 u_2^{1-\alpha_2})$, where $\alpha_i = \lambda_{12}/(\lambda_i + \lambda_{12})$.

3.3 (TP) Show that the Morgenstern copula depends only on ε but not F_1, F_2.

3.4 (TP) Show that the frailty copula with Z having the Gamma density $z^{1/\alpha - 1} e^{-z}/\Gamma(1/\alpha)$ reduces to the Clayton copula.

3.5 (TP) Show that the Clayton copula approaches comonotonicity as $\alpha \to \infty$ and independence as $\alpha \to 0$. Show more generally that the dependence is increasing in α in the sense that the c.d.f. is nondecreasing in α.

3.6 (TP) Show that a frailty copula is Archimedean with generator given by $\phi^{-1} = \widehat{Z}$.

3.7 (A) Your assignment is to use copulas to illustrate how robust a certain simple statistical test is to deviations from independence. Let $X_1, \ldots, X_n$ be i.i.d. Bernoulli(θ) r.v.'s and let $N \stackrel{\text{def}}{=} X_1 + \cdots + X_n$. If n is large, we can then test the hypothesis $\theta = 1/2$ via the normal approximation, so that the acceptance region at the 5% level is

$$\mathscr{A} \stackrel{\text{def}}{=} \left\{ \left| \frac{N - n/2}{\sqrt{n/4}} \right| \leq 1.96 \right\}.$$

Then $\mathbb{P}_{1/2}(\mathscr{A})$ is (approximately) 5% and the power of the test at $\theta \neq 1/2$ is $\mathbb{P}_\theta(\mathscr{A})$.

You can, for example, do as follows (there are many other ways!). Generate $X_1, \ldots, X_{100}$ with Bernoulli(θ) marginals, but having dependence corresponding to a Gaussian copula with weak dependence, say $(Y_1, \ldots, Y_{100})$ is a stationary autoregressive process $Y_{i+1} = \beta Y_i + \varepsilon_i$, where $\varepsilon_1, \ldots, \varepsilon_{99}$ are i.i.d. $\mathcal{N}(0, \sigma^2)$, $\sigma^2 \stackrel{\text{def}}{=} 1 - \beta^2$, and Y_1 is generated as $\mathcal{N}(0, 1)$. Then $\mathbb{V}ar\, Y_i = 1$ and the covariance function is $\rho_k = \mathbb{C}ov(Y_i, Y_{i+k}) = \beta^k$. Thus, taking β small and positive means weak positive dependence, and taking β small and negative means weak negative dependence. As measure of dependence, use $\tau^2 = \mathbb{V}ar\, N/25$ (then $\tau^2 = 1$ corresponds to independence). For a given value of β, simulate $X_1, \ldots, X_{100}$ with $\theta = 1/2$, observe whether the test accepts or rejects, and repeat 500 times to get an estimate of the level $\mathbb{P}_{1/2, \tau^2}(\mathscr{A})$. Repeat for different β to plot $\mathbb{P}_{1/2, \tau^2}(\mathscr{A})$ as function of τ^2 (evaluate τ^2 from the simulated values!). Similarly, for a given alternative like $\theta = 1/3$ plot the power $\mathbb{P}_{\theta, \tau^2}(\mathscr{A})$ as function of τ^2.

3.8 (A) A portfolio has 10 assets of current values

$$(x_1 \ x_2 \ \ldots \ x_{10}) = (5 \ 2 \ 10 \ 8 \ 8 \ 3 \ 15 \ 4 \ 1 \ 2).$$

The value of the holdings of asset i next week is $x_i e^{Y_i}$, where $Y_1, \ldots, Y_{10}$ are (possibly dependent) r.v.'s. Thus, the loss is

$$L \stackrel{\text{def}}{=} \sum_{i=1}^{10} x_i (1 - e^{Y_i}),$$

and the VaR (value at risk) is the 99% quantile in the distribution of L.

There is good statistical evidence that the marginal distributions of the Y_i are normal with mean zero (as is often the case for a short time horizon) and volatilities

$$4 \ 7 \ 1 \ 5 \ 3 \ 5 \ 2 \ 6 \ 4 \ 4$$

(in %), respectively (thus, for example, $Y_2 \sim \mathcal{N}(0,0049)$). However, the dependence structure is less well understood. Your assignment is to estimate the VaR for the given normal marginals and the following six dependence structures: the three $N_{10}(0,\boldsymbol{\Sigma})$ distributions where $\boldsymbol{\Sigma}$ corresponds to symmetric correlation 0, 0.3, 0.6, and the three Student $t_{10}(\boldsymbol{\Sigma},f)$ copulas with the same $\boldsymbol{\Sigma}$'s and $f=1$. Note that the c.d.f. of the t-distribution with $f=1$ is $1/2 + \arctan(x)/\pi$. The α-quantile of an r.v. L is estimated by simulating R replications $L_1,\ldots,L_R$, forming the order statistics $L_{(1)} < \cdots < L_{R)}$, and using the estimate $L_{(\alpha(R+1))}$ (thus, it is convenient to choose R such that $\alpha(R+1)$ is an integer); see further III.4a.

3.9 (TP) Show that the bivariate Gaussian copula is tail-independent ($\lambda=0$), but the bivariate t copula is not.

4   Simple Stochastic Processes

Example 4.1 (MARKOV CHAINS AND RECURSIONS)   Presumably the simplest process with dependence one could think of would be a (time-homogeneous) Markov chain $\{X_n\}_{n\in\mathbb{N}}$ with a finite or countable state space E, and presumably the first algorithm for generating a sample path by simulation one could think of would be based on the transition probabilities $p_{ij} = \mathbb{P}(X_{n+1} = j \mid X_n = i)$. That is, one would simply generate X_{n+1} from $X_n = i$ by generating an E-valued r.v. from the distribution $\boldsymbol{p}_{i\cdot} \overset{\text{def}}{=} (p_{ij})_{j\in E}$ with point probabilities p_{ij} by one of the parametric or nonparametric approaches surveyed in Section 2.

This is of course the path to follow if the primitives of the Markov chain $\{X_n\}$ are indeed the transition probabilities. However, most often the natural way to specify the time evolution of $\{X_n\}$ has a different form, as a recursion. This would then provide a natural simulation procedure that does not invoke the p_{ij}, and further, often would not be restricted to a discrete E. We have already seen one example of such a recursion, the Lindley recursion I.(1.1) for the waiting times of the GI/G/1 queue. Another example is an autoregressive process $X_{n+1} = aX_n + \varepsilon_n$ with the ε_n i.i.d., or ARMA processes as discussed briefly in XI.1b.[3] A generalization of autoregressive processes in a different direction is the GARCH model[4]

$$X_n = \sigma_n Z_n, \quad \sigma_n^2 = \alpha_0 + \alpha_1 X_{n-1}^2 + \beta\sigma_{n-1}^2 \tag{4.1}$$

with the Z_k i.i.d. The typical case in which the Z_k are standard normal has been designed to model mean-zero normals X_n with a variance σ_n^2 that fluctuates in a non-i.i.d. way, say stochastic volatility models in mathematical finance. The Markov chain in question is $\{\sigma_n^2, X_n\}$ with state space $(0,\infty) \times \mathbb{R}$.

---

[3] the ARMA(p,q) process is not itself Markov but can in a standard way be represented as a function of a higher-dimensional Markov chain.
[4] Generalized Autoregressive Conditionally Heteroscedatisc.

For autoregressive, ARMA, and GARCH processes, the expression for the transition kernel is so complicated compared to the recursion that simulation via the recursion is obviously the natural approach. The same remark applies to a huge number of other Markov chains and is in our view often overlooked. That is, we feel that the discussion in the literature on the simulation of Markov chains often is biased toward the transition probability approach.

At the formal level, any Markov chain satisfying weak regularity conditions can be represented via a recursion $X_{n+1} = \varphi(X_n, \varepsilon_n)$ for some deterministic function φ and some innovation sequence $\{\varepsilon_n\}$. Somewhat surprisingly, one can even take the ε_n as i.i.d. uniform$(0, 1)$. For example, if the state space is $\mathbb{R}$, one can define $F_x(y) \stackrel{\text{def}}{=} \mathbb{P}_x(X_1 \leq y)$ and $\varphi(x, u) \stackrel{\text{def}}{=} F_x^{\leftarrow}(u)$. But certainly this representation is rarely the most natural one!   □

Example 4.2 (INHOMOGENEOUS POISSON PROCESSES) The epochs σ_n of a standard Poisson process $\{N(t)\}$ with rate β are easily generated by noting that the interarrival times $T_n = \sigma_n - \sigma_{n-1}$ can be generated as i.i.d. exponential(β). However, in many situations it would be reasonable to assume that a given point process $\{N^*(t)\}$ is Poisson, but with a rate $\beta(t)$ depending on time. Assuming $\beta(t) \leq \beta$ for some constant $\beta < \infty$, $\{N^*(t)\}$ can then be constructed by thinning $\{N(t)\}$ with retention probability $\beta(t)/\beta$ at time t. That is, an epoch σ_n of $\{N(t)\}$ is accepted (retained) as an epoch σ_m^* of $\{N_t^*\}$ w.p. $\beta(\sigma_n)/\beta$. Algorithmically:

1. Let $n \longleftarrow 0$, $n^* \longleftarrow 0$, $\sigma \longleftarrow 0$, $\sigma^* \longleftarrow 0$.

2. Generate T as exponential with rate β;
   let $\sigma \longleftarrow \sigma + T$, $n \longleftarrow n + 1$.

3. Generate U as uniform$(0,1)$;
   if $U \leq \beta(\sigma)/\beta$, let $n^* \longleftarrow n^* + 1$, $\sigma^* \longleftarrow \sigma$.

4. Return to 2.

That this produces the correct intensity $\beta(t)$ for $\{N_t^*\}$ follows from

$$
\begin{aligned}
\beta^*(t)\,\mathrm{d}t \;\stackrel{\text{def}}{=}\; & \mathbb{P}\big(\sigma_m^* \in [t, t+\mathrm{d}t] \;\text{ for some } m = 0, 1, \ldots\big) \\
=\; & \mathbb{P}\big(\sigma_n \in [t, t+\mathrm{d}t] \;\text{ for some } n = 0, 1 \ldots\big) \cdot \frac{\beta(t)}{\beta} \\
=\; & \beta\,\mathrm{d}t \cdot \frac{\beta(t)}{\beta} = \beta(t)\,\mathrm{d}t.
\end{aligned}
$$

For nonbounded $\beta(\cdot)$, the procedure is easily modified by choosing different dominating β's on diferent intervals.

An alternative procedure, which may or may not be feasible depending on whether $B(t) \stackrel{\text{def}}{=} \int_0^t \beta(s)\,\mathrm{d}s$ and $B^{-1}(x)$ can be found in closed form, is to construct the first interarrival time T_1 as $B^{-1}(X_1)$, where X_1 is standard exponential; that this gives the correct distribution follows from

$$\mathbb{P}(T_1 > t) \;=\; \mathbb{P}\big(X_1 > B(t)\big) \;=\; \mathrm{e}^{-B(t)}.$$

The second interarrival time T_1 is then constructed from a second exponential r.v. as the solution of $B(T_2) - B(T_1) = X_2$, i.e., $T_2 = B^{-1}(X_1 + X_2)$, and so on.

The simulation of T_1 in an inhomogeneous Poisson process is also relevant in r.v. generation, where the interpretation of $\beta(t)$ is then the failure (hazard) rate at t. $\qquad\square$

Example 4.3 (MARKOV PROCESSES) Let $\{J(t)\}_{t\geq 0}$ be a Markov process with a finite state space E and intensity matrix $\mathbf{\Lambda} = (\lambda_{ij})_{i,j\in E}$. One can simulate $\{J(t)\}$ at transition epochs by noting that the holding time of state i is exponential with rate $\lambda_i \stackrel{\mathrm{def}}{=} -\lambda_{ii}$, and that the next state j is chosen w.p. λ_{ij}/λ_i:

1. Let $t \longleftarrow 0$, $J \longleftarrow i_0$.

2. Let $i \longleftarrow J$;
   generate T as exponential with rate λ_i and K with $\mathbb{P}(K = j) = \lambda_{ij}/\lambda_i$, $j \neq i$.

3. Let $t \longleftarrow t + T$, $J \leftarrow K$ and return to 2.

$\qquad\square$

Example 4.4 (UNIFORMIZATION OF MARKOV PROCESSES) In Example 4.3, the rate λ_i of an event being created depends on the current state $i = J(t)$. The uniformization algorithm creates instead events at a uniform rate η. A transition from i to $j \neq i$ then occurs w.p. λ_{ij}/η (thus, η should satisfy $\eta \geq \max_{i\in E} \lambda_i$) when the current state is $i = J(t)$; if $\eta > \lambda_i$, this leaves the possibility of a dummy transition $i \to i$ (t is rejected as transition epoch). Algorithmically:

1. Let $t \longleftarrow 0$, $J \longleftarrow i_0$.

2. Let $i \longleftarrow J$;
   generate T as exponential with rate η and K with $\mathbb{P}(K = j) = \lambda_{ij}/\eta$, $j \neq i$, and $\mathbb{P}(K = i) = \lambda_i/\eta$.

3. Let $t \longleftarrow t + T$, $J \leftarrow K$ and return to 2.

The algorithm makes the event times a homogeneous Poisson process with rate η, and the successive values of J a Markov chain with transition matrix $\mathbf{I} + \mathbf{\Lambda}/\eta$.

The method applies also to the countable case, provided $\sup_{i\in E} \lambda_i < \infty$. It is frequently promoted in the literature, but the efficiency gain compared to the more naive procedure in Example 4.3 is not clear to us. $\qquad\square$

Example 4.5 (MARKOV-MODULATED POISSON PROCESSES) Consider a Markov-modulated Poisson process with arrival rate β_i when $J(t) = i$ (here $\{J(t)\}$ is Markov with transition rates λ_{ij} as in Example 4.3). The intensity of an event (a transition $i \to j$ or a Poisson arrival) is $\lambda_i + \beta_i$ when $J(t) = i$. Thus, choosing $\eta \geq \max_{i \in E}(\lambda_i + \beta_i)$ and letting Δ be some point $\notin E$ (marking an arrival), we may generate the arrival epochs σ as follows:

1. Let $t \longleftarrow 0$, $J \longleftarrow i_0$, $\sigma \longleftarrow 0$.

2. Let $i \longleftarrow J$;
   generate T as exponential with rate η and K with

$$\mathbb{P}(K = j) = \begin{cases} \beta_i/\eta & j = \Delta, \\ \lambda_{ij}/\eta & j \in E, j \neq i, \\ (\eta - \lambda_i - \beta_i)/\eta & j = i. \end{cases}$$

   Let $t \longleftarrow t + T$.

3. If $K = \Delta$, let $\sigma \longleftarrow t$;
   otherwise, let $J \longleftarrow K$ and return to 2.

□

Exercises

4.1 (A) The *Gompertz–Makeham distribution* is given by its hazard rate $\beta(t) = a + be^{ct}$ and is widely used as lifetime distribution in life insurance. Write a routine generating an r.v. X having the distribution of the death time of a person of age 45, using the parameters $a = 5 \cdot 10^{-4}$, $b = 7.5858 \cdot 10^{-5}$, $c = \log(1.09144)$ from the so-called G82 base widely employed by Danish insurance companies (the time unit is years). Use the routine to compare the risks, as measured by the s.c.v. $\mathbb{E}Y^2/(\mathbb{E}Y)^2 - 1$, of the present values Y_1, Y_2 of the benefits to be paid by the company from (1) a life annuity paying a at a constant rate in $[65, X]$ and (2) a life insurance paying b at X, i.e.,

$$Y_1 = \int_{65}^{X} ae^{-rt}\, dt = \frac{a}{r}\left[e^{-65r} - e^{-rX}\right]\mathbb{1}_{X>65}, \quad Y_1 = be^{-rX},$$

where r is the interest rate taken to be $r = 5\%$.

5   Further Selected Random Objects

5a   Order Statistics

The aim is to generate the order statistics $X_{(1)} < \cdots < X_{(n)}$ or a subset, where $X_1, \ldots, X_n$ are i.i.d. with density f and c.d.f. F. The naive method (of complexity $\mathrm{O}(n \log n)$; cf. [293, Ch. 8]) is to generate $X_1, \ldots, X_n$ and sort them, but this method may be inefficient if just a few of the $X_{(k)}$ are needed or if the time needed for the sorting is much larger than that for the r.v. generation.

If $U_{(1)} < \cdots < U_{(n)}$ are the order statistics from the uniform$(0,1)$ distribution, $\big(F^{\leftarrow}(U_{(1)}), \ldots, F^{\leftarrow}(U_{(n)})\big)$ have the desired distribution. Thus, one can reduce to the uniform case *provided that inversion is feasible*. A simple algorithm, of complexity $O(n)$ exploits the connection to the Poisson process and standard properties of its relative spacings: we can just simulate $Z_1, \ldots, Z_{n+1}$ as standard exponentials and let $U_{(1)} \overset{\text{def}}{=} S_1/S_{n+1}, \ldots, U_{(n)} \overset{\text{def}}{=} S_n/S_{n+1}$, where $S_k \overset{\text{def}}{=} Z_1 + \cdots + Z_k$. Another algorithm uses $\mathbb{P}\big(U_{(k)} \le y \,\big|\, U_{(k+1)} = x\big) = y^k/x^k$, which implies that $U_{(1)} < \cdots < U_{(n)}$ can be generated by the downward recursion $U_{(k)} = U_k^{'1/k} U_{(k+1)}$ (with $U_1', U_2', \ldots$ i.i.d. uniforms) started from the fictitious value $U_{(n+1)} = 1$.

The uniform case can also be reduced to the exponential case by considering the tranformation $z \to u = 1 - \mathrm{e}^{-z}$, which may be useful, since the order statistics $Z_{(1)} < \cdots < Z_{(n)}$ from the standard exponential distribution are easily available by observing that the distribution of $Z_{(k+1)} - Z_{(k)}$ is that of the minimum of $n - k$ standard exponential r.v.'s, which in turn is exponential with $\lambda = n - k$. Thus, we can take $Z_{(k)} = \sum_1^k Z_i'/(n-i+1)$, where $Z_1', \ldots, Z_n'$ are i.i.d. standard exponentials.

In general, the most feasible alternative to inversion or sorting may be an A–R scheme, in which the trial density $g(y)$ is chosen such that its order statistics are easy to generate. One then chooses $m > n$ (for the optimal value, see Devroy [88, pp. 221–225], generates $Y_{(1)} < \cdots < Y_{(m)}$, accepts $Y_{(k)} = y$ w.p. $f(y)/Cg(y)$, and goes on choosing a new m-sample until the number N of accepted $Y_{(k)}$ is at least n. One then deletes $N - n$ random elements from this last sample.

The case of just the maximum $M_n \overset{\text{def}}{=} X_{(n)}$ (or the minimum) is of particular interest. If $F^{\leftarrow}$ is available, a variant of the inversion algorithm avoiding the generation of the complete sample is to generate M_n as $F^{\leftarrow}(U^{1/n})$ (this follows since M_n has c.d.f. $F^n(x)$). Other alternatives with complexity $O(\log n)$ in between the $O(1)$ for inversion and the $O(n \log n)$ for the naive method are the *quick elimination algorithm* of Devroye [88] and the record time approach described below. Quick elimination means that only a (typically small) number L of the X_i are actually simulated. Here L is generated as binomial(n,p), where $p \overset{\text{def}}{=} p_n \overset{\text{def}}{=} \overline{F}(x)$ for some suitably large $x \overset{\text{def}}{=} x_n$; one simulates $Y_1, \ldots, Y_L$ from F conditioned to (x, ∞) and lets $M_n \overset{\text{def}}{=} \max_{\ell \le L} Y_\ell$, with the modification that if $L = 0$, then M_n is taken as the maximum of n i.i.d. r.v.'s from F conditioned to $(-\infty, x]$. Basically, x_n should be chosen such that p_n is of order $\log n/n$.

A setting close to order statistics isthat of *records*, defined via the random times $1 = n_1 < \cdots < n_K$ at which $X_{n_k} > \max_{i < n_k} X_i$. The set

$$\big(n_1, X_{n_1}, \ldots, n_K, X_{n_K}\big)$$

can be generated by starting from $n_1 = 1$ and recursively generating $\big(n_{k+1} - n_k, X_{n_{k+1}}\big)$ as independent, such that $n_{k+1} - n_k$ follows the geometric distribution with success parameter $\overline{F}(X_{n_k})$, and $X_{n_{k+1}}$ has distribution F conditioned to (X_{n_k}, ∞). The procedure stops at the K with $n_K \leq n < n_{K+1}$. The computational effort is roughly the expected number $\sum_1^n i^{-1} \approx \log n$ of records in a sequence of length n (note that the probability of a record at i is i^{-1}).

5b   Random Permutations

The aim is for a given integer N, to generate a random permutation $(\sigma_1, \ldots, \sigma_N)$ of $(1, \ldots, N)$ such that all $N!$ permutations are equally likely.

One can represent a permutation by a set $(a_1, \ldots, a_N)$ of integers with $a_i \in \{1, \ldots, N-i+1\}$, such that σ_1 is element a_1 in the list $L_1 = (1, \ldots, N)$ and for $i > 1$, σ_i is element a_i in the list L_i obtained from L_{i-1} by deleting its element σ_{i-1} (note that L_N contains only one element, so that $a_N = 1$, as should be).

It is clear that in order to make every permutation equally likely, $a_1, \ldots, a_N$ should be chosen as the outcomes of independent r.v.'s $A_1, \ldots, A_N$ with A_i being uniform on $\{1, \ldots, N-i+1\}$. The A_i can be drawn one at a time, or starting from a uniform r.v. B_0 on $\{1, \ldots, n!\}$. Then

$$A_1 \stackrel{\text{def}}{=} \lceil B_0/(n-1)! \rceil, \quad B_1 \stackrel{\text{def}}{=} B_0 \bmod (n-1)!$$

are independent, with A_1 having the desired distribution and B_1 being uniform on $\{1, \ldots, (n-1)!\}$. Thus A_2 can be taken as $\lceil B_1/(n-1)! \rceil$, A_3 as $\lceil B_2/(n-1)! \rceil$, where $B_2 = B_0 \bmod (n-2)!$, and so on. It should be noted that the second algorithm quickly runs into overflow problems as N grows, because $N!$ (which is not needed for the first algorithm) explodes.

5c   Point Processes

Let $\Omega \subset \mathbb{R}^2$ be a region in the plane with finite area $|\Omega|$. A Poisson process on Ω with intensity λ can then be simulated by embedding Ω in a rectangle R, generating N as Poisson(λ/p), where $p \stackrel{\text{def}}{=} |\Omega|/|R|$), distributing N points uniformly within R, rejecting the ones not in Ω (the $*$'s in Figure 5.1) w.p. $1-p$, and retaining the rest (the $+$'s). The rectangle may be replaced by a disk when the shape of Ω makes this more natural; cf. Exercise 5.3.

Exercises

5.1 (TP) Consider sampling of n objects out of $N > n$ without replacement. That is, we want to sample a subset of size n uniformly among all such subsets. Give a rejection algorithm using uniform sampling from $\{1, \ldots, N\}$. Show next that an algorithm can also be obtained by examining $1, \ldots, N$ sequentially and incorporate k w.p. $(n-r)/(N-k)$ when r among $1, \ldots, k-1$ have already been chosen.

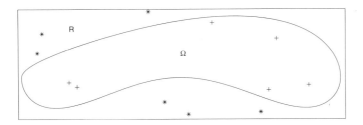

FIGURE 5.1

5.2 (A$-$) A line in the plane can be represented by its distance r from the origin and the (smallest) angle $\theta \in [0, \pi)$ it forms with the x-axis. A *Poisson line process* is a random collection of lines in the plane such that the (r, θ) form a Poisson process on $(0, \infty) \times [0, \pi)$ with intensity measure $\lambda \, dr \times \nu(d\theta)$ for some probability measure ν on $[0, \pi)$. Simulate the restriction of a Poisson line process to $[0, 1]^2$ for (a) ν uniform and (b) a ν giving preference to more vertical lines.

5.3 (TP) Let $0 < R_1 < R_2 < \cdots$ be the ordered radii of the points of a Poisson process on the disk $\Omega \stackrel{\text{def}}{=} \{(x_1, x_2); x_1^2 + x_2^2 < r\}$. Give algorithms for simulation of the homogeneous Poisson process on Ω by verifying and using that (a) the R_i are the points of an inhomogeneous Poisson process on $(0, r)$, (b) $R_1^2, R_2^2 - R_1^2, \ldots$ are i.i.d. exponentials.

6   Discrete-Event Systems and GSMPs

Discrete-event systems arise in many different areas of engineering, and a large number of special-purpose simulation packages have been commercially developed so as to serve this marketplace. In a nutshell, discrete-event dynamic systems describe the types of models that arise in queuing, production, communications, computer science, and inventory applications. These processes describe structures in which stochastic state transitions occur at a sequence of discrete transition epochs. Between such transitions, certain auxiliary state variables may change deterministically, but no stochastic behavior of any kind occurs. For example, in the single-server queue, one needs to "schedule" the next arrival (departure) at arrival (departure) epochs by drawing appropriate r.v.'s at those times from the interarrival (service) time distribution [the exception is the idle state, where an arrival requires scheduling both an arrival and a departure, and a departure requires no event scheduling whatsoever]. Between arrival and departure epochs, no random numbers are drawn.

Discrete-event systems are in close correspondence with the class of *generalized semi-Markov processes* (GSMPs; cf., e.g., Glynn [143]). As we shall see, a GSMP generalizes the concept of a discrete continuous-time Markov process as well as that of a semi-Markov process. To describe a GSMP, one starts by specifying the set S of (physical) states and the set E of possible events that can trigger state transitions; the events active in $s \in S$ are denoted by $E(s) \subseteq E$. For example, for the single-server queue, $S = \mathbb{N} = \{0, 1, 2, \ldots\}$ and $E = \{0, 1\}$, where 0 corresponds to an arrival event and 1 to a departure event. Here, $E(0) = \{0\}$ and $E(s) = E = \{0, 1\}$ for $s \geq 1$.

State transitions for a GSMP are detemined by "competing clocks". When an event $e \in E$ is scheduled, imagine that a corresponding clock is set. While the system is in physical state $s \in S$, the clock corresponding to e runs down at the deterministic rate $r_{s,e}$. When a clock reading hits zero, the event e^* corresponding to that clock occurs, and a state transition from the currently occupied state $s \in S$ to a new (randomly chosen) state s' occurs w.p. $p(s'; s, e^*)$. When the new state s' is entered, new events typically need to be rescheduled. If one denotes the set of such events to be rescheduled by $N(s'; s, e^*)$ (the new clocks), a clock reading for $e' \in N(s'; s, e^*)$ is then set independently according to the distribution $F(\cdot; s', e', s, e^*)$. On the other hand, the remaining clocks active in s', namely

$$O(s'; s, e^*) \stackrel{\text{def}}{=} E(s') \setminus N(s'; s, e^*)$$

(the old clocks), are permitted to continue running down in state s'.

For the GI/G/1 queue, $p(s + 1; s, 0) = 1$ and $p(s; s + 1, 1) = 1$ for $s \geq 0$; all other $p(s'; s, e)$ values are zero. The distributions $F(\cdot; s', e', s, e^*)$ depend only on e'; $F(\cdot; e')$ corresponds to the interarrival distribution if $e' = 0$ and the service time distribution if $e' = 1$. For this example, $r_{s,e} = 1$ for $s \in S$, $e \in E$.

If $S(t)$ is the state at time t, then $\{S(t)\}_{t \geq 0}$ is called a *generalized semi-Markov process*. Note that if all the distributions $F(\cdot; s', e', s, e^*)$ are exponential, then $\{S(t)\}$ is a continuous-time Markov process. If $E(s)$ is a singleton for each $s \in S$, then $\{S(t)\}$ is what is known as a semi-Markov process (cf. [16] VII.4).

Example 6.1 As a more complicated example, consider a processor-sharing queue in which all customers are served simultaneously at rate $1/s$ when $s \geq 1$ customers are present. We can again take $S = \mathbb{N}$ and use the convention that when $s > 1$, the customers are ordered in their order of arrival. We let $E = \mathbb{N}$, such that $e = 0$ is an arrival event and $e = i \geq 1$ the event that customer i departs. The clock speeds are $r_{s,0} = 1$ for all s and $r_{s,i} = 1/s$ for $s \geq 1$ and $i = 1, \ldots, s$, where $E(s) = \{0, 1, \ldots, s\}$. At an event of type $e^* = 0$, the state goes from s to $s + 1$, clock 0 is reset according to the interarrival time, clocks $1, \ldots, s$ remain the same, and clock $s + 1$ is initialized with a service time r.v. At an event of type $e^* = i \geq 1$, the state goes from s to $s - 1$, clocks $0, \ldots, i - 1$ remain the same, and clocks $i + 1, \ldots, s$ are shifted to positions $i, \ldots, s - 1$.   □

Because of the multiple clocks active in each state of a GSMP (and the memory the old clocks carry with them into the next state s' visited), GSMPs are typically analytically intractable. However, it should be clear that they can easily be simulated. In many practical situations, $E(s)$ can be large (in the dozens or hundreds). In such cases, efficient data structures must be used to represent the clock readings so as to make computation of the triggering event e^* as simple as possible. A common choice is to use an ordered linked list for this purpose, but more exotic data structures (such as trees) are also used.

Despite the fact that a GSMP is usually analytically intractable, it is important to recognize that it can be viewed as a Markov process. In particular, by appending the state vector $\boldsymbol{C}(t)$ of the clocks active in $S(t)$ to our state description, we arrive at a Markov process $\left\{\left(S(t), \boldsymbol{C}(t)\right)\right\}_{t \geq 0}$. Because the dynamics of this Markov process are deterministic between physical state transitions, one can alternatively use the embedded jump chain $\left\{\left(S(T_n), \boldsymbol{C}(T_n)\right)\right\}_{n \in \mathbb{N}}$ to study the GSMP (here T_n is the time of the nth event transition). The transition structures of $\left\{\left(S(t), \boldsymbol{C}(t)\right)\right\}$ and $\left\{\left(S(T_n), \boldsymbol{C}(T_n)\right)\right\}$ are messy to write down. Nevertheless, the very fact that GSMPs can be viewed as Markov processes is conceptually important from a simulation standpoint, since this means that any methodology developed for Markov processes on a general state space applies (at least in principle) to discrete-event simulations.

The concept of a discrete-event system and the formalism of GSMPs will not be used systematically in this book, but it is useful to keep in mind that a simulation program generating a dynamical system will typically have a structure that is implicitly that of a GSMP, even if this concept is not invoked explicitly; this intimate connection between program organization and GSMP representation is well illustrated by Example 6.1.

A closely related class of stochastic models that has essentially identical modeling power to that of GSMPs (in that there is essentially a 1-to-1 mapping between the two descriptions) is that of stochastic Petri nets; see, e.g., Haas [171]. Petri net representations of discrete-event systems have the advantage of offering graphical representations of the model that are often (much) more compact than comparable graph representations derived from the GSMP description, particularly when the model needs to resolve concurrency issues (as often occur in computer science applications).

Exercises

6.1 (TP) A system has N machines and $M < N$ repairmen. The operating time of a machine has distribution F and the repair time has distribution G. A failed machine is repaired immediately by a repairman if one is available, and otherwise it joins a queue of machines awaiting repair. Write up a GSMP representation of the system.

Chapter III
Output Analysis

1 Normal Confidence Intervals

As pointed out in Chapter I, substantial insight can often be obtained by taking advantage of visualization ideas (such as watching the sample path evolve). However, our emphasis in this book is on the use of simulation as a computational tool. In view of this, developing good methods for assessing an estimator's accuracy is of fundamental importance.

We start with the simplest possible situation, namely that of computing $z = \mathbb{E}Z$ in a context in which z is not available analytically but Z can be simulated. The Monte Carlo method, as introduced in I.2 in its basic version, then amounts to simulating R i.i.d. replicates $Z_1, \ldots, Z_R$ of Z and estimating z by the empirical mean $\widehat{z} \stackrel{\text{def}}{=} \widehat{z}_R \stackrel{\text{def}}{=} (Z_1 + \cdots + Z_R)/R$.

Clearly, assessment of the accuracy requires first studying the rate of convergence. Assuming $\sigma^2 \stackrel{\text{def}}{=} \mathbb{V}ar\, Z < \infty$, the central limit theorem (CLT) states that

$$\sqrt{R}\,(\widehat{z} - z) \stackrel{\mathscr{D}}{\to} \mathscr{N}(0, \sigma^2) \tag{1.1}$$

as $R \to \infty$. This can be interpreted as a statement that

$$\widehat{z} \stackrel{\mathscr{D}}{\approx} z + \frac{\sigma V}{\sqrt{R}}, \quad \text{where} \quad V \sim \mathscr{N}(0,1), \tag{1.2}$$

when R is large (where $\stackrel{\mathscr{D}}{\approx}$ can informally be interpreted as "has approximately the same distribution as"). This approximation has several implications:

1. The convergence rate is of order $R^{-1/2}$. This implies that if we want to add one significant figure of accuracy, we must increase R by a factor of 100. Relative to many other commonly used numerical methods, the rate $R^{-1/2}$ is slow (see, however, Chapter IX for additional explanation as to why simulation can be the method of choice for solving many problems). In view of this slow convergence rate, some type of accuracy assessment is critical if simulation is to be confidently used as a numerical method.

2. The error, for large R, is asymptotically normally distributed. This suggests the possibility of assessing accuracy via confidence intervals derived from the normal distribution.

3. The error, for large R, depends on the underlying problem's probability distribution through a single scalar measure, namely the standard deviation σ. In other words, a problem's difficulty is largely summarized by the magnitude of a single problem parameter, namely σ (or, equivalently, the variance σ^2).

Expanding on point 2 above, the implication of the CLT is that if z_α denotes the α-quantile of the normal distribution, i.e., $\Phi(z_\alpha) = \alpha$, then the asymptotic probability of the event

$$\left\{ z_{\alpha/2}\sigma/\sqrt{R} < \widehat{z} - z < z_{1-\alpha/2}\sigma/\sqrt{R}) \right\}$$

is $\Phi(z_{1-\alpha/2}) - \Phi(z_{\alpha/2}) = (1 - \alpha/2) - \alpha/2 = 1 - \alpha$. Equivalently, if I_α is the random interval[1]

$$\left(\widehat{z} - z_{1-\alpha/2}\sigma/\sqrt{R},\, \widehat{z} - z_{\alpha/2}\sigma/\sqrt{R} \right),$$

then $\mathbb{P}(z \in I_\alpha) \to 1 - \alpha$, which one expresses by saying that I_α is an *asymptotic $1 - \alpha$ confidence interval* for z. That is, I_α contains the true value w.p. $\sim 1 - \alpha$.

In practice, σ^2 is unknown and must be estimated. The traditonal estimator is the empirical (or sample) variance

$$s^2 \overset{\text{def}}{=} \frac{1}{R-1} \sum_{r=1}^{R} (Z_r - \widehat{z})^2 = \frac{1}{R-1} \left(\sum_{r=1}^{R} Z_r^2 - R\widehat{z}^2 \right) \tag{1.3}$$

[the occurrence of $R-1$ rather than R follows statistical tradition; in most Monte Carlo experiments, R is very large and the difference is minor]. Thus, replacing σ by s in I_α and noting that $s^2 \overset{\mathbb{P}}{\to} \sigma^2$, it follows that

$$\left(\widehat{z} - z_{1-\alpha/2}\, s/\sqrt{R},\, \widehat{z} - z_{\alpha/2}\, s/\sqrt{R} \right) \overset{\text{def}}{=} \widehat{z} \pm z_{1-\alpha/2}\, s/\sqrt{R} \tag{1.4}$$

---

[1]Whether one take such intervals open, as here, or closed is of course a matter of taste.

is an asymptotic $1 - \alpha$ confidence interval for z. This is the form in which the result of the simulation experiment usually not only is but should be reported, the philosophy being that the additional calculations needed for the whole confidence interval rather than just the point estimate $\hat{z}$ are minor compared to conducting the whole simulation experiment, and that an estimate with an error assessment is much more informative than one without.

Remark 1.1 The most standard choice is $1 - \alpha = 95\%$, corresponding to $z_{97.5\%} = -z_{2.5\%} = 1.96$ and the confidence interval $\hat{z} \pm 1.96 \, s/\sqrt{R}$. Other values are, of course, possible, say 99%, corresponding to $\hat{z} \pm 2.58 \, s/\sqrt{R}$, or 90%, corresponding to $\hat{z} \pm 1.64 \, s/\sqrt{R}$.

Also, one-sided confidence intervals may sometimes be relevant. Assume, for example, that Z is an indicator function telling whether a certain system failure occurs and z the corresponding failure probability. Then $\hat{z} + 1.64 \, s/\sqrt{R}$ is an upper 95% confidence limit for z.

For simplicity, we will often make statements such as, "the confidence interval based on the variance estimate s^2 is $\hat{z} \pm 1.96s/\sqrt{R}$", but of course such a statement is easily adapted to confidence levels other than 95% and equally to one-sided confidence bounds.   □

Example 1.2 As an illustration of the slow convergence rate of the Monte Carlo method, consider the estimation of $z = \pi = 3.1415\dots$ via the estimator $Z \stackrel{\text{def}}{=} 4\mathbb{1}\{U_1^2 + U_2^2 < 1\}$, where U_1, U_2 are independent uniform$(0, 1)$ r.v.'s. Here $\mathbb{1}\{U_1^2 + U_2^2 < 1\}$ is a Bernoulli r.v. with mean $\pi/4$, so the variance of Z is $4^2(\pi/4)(1 - \pi/4) \approx 2.70$. Thus, to get the leading 3 in π correct w.p. 95% we need $R \approx 1.96^2 2.70^2/0.5^2 \approx 112$, whereas for the next 1 we need $R \approx 11{,}200$, for the 4, $R \approx 1{,}120{,}000$, and so on.   □

Remark 1.3 (THE CANONICAL RATE) Being of the same dimension as the expectation z, the standard deviation σ is the natural measure of precision rather than the variance σ^2. The rate $R^{-1/2}$ is intrinsic for virtually any Monte Carlo experiment as the optimal rate which can be achieved. In fact, there are quite a few situations in which one gets slower rates such as $\mathrm{O}(R^{-(1/2-\varepsilon)})$ with $\varepsilon > 0$, e.g., in gradient estimation via finite differences (VII.1), or variance estimation via time series methods (IV.3). In view of these remarks, one often refers to $\mathrm{O}(R^{-1/2})$ as the *canonical Monte Carlo convergence rate*.

There are, however, a few isolated examples (but just a few!) in which $\mathrm{O}(R^{-1/2})$ can in fact be improved to supercanonical (faster) rates; see V.7.4 for examples.   □

Output analysis via normal confidence intervals can be and is in fact performed beyond the setting of i.i.d. replications. Let z be the unknown number to be estimated by simulation and consider a sampling-based scheme with sample size R (interpreted in an appropriate way depending on the context), in which $\widehat{z} \stackrel{\text{def}}{=} \widehat{z}_R$ is an estimator obeying a CLT of the form (1.1), i.e., $R^{1/2}(\widehat{z} - z) \stackrel{\mathscr{D}}{\to} \mathscr{N}(0, \sigma^2)$ for some σ^2. If an estimator $\widehat{\sigma}^2 \stackrel{\text{def}}{=} \widehat{\sigma}_R^2$ is avaliable such that $\widehat{\sigma}_R^2 \to \sigma^2$, an asymptotic $1 - \alpha$ confidence interval is then

$$\left(\widehat{z} - z_{1-\alpha/2}\widehat{\sigma}/\sqrt{R}, \, \widehat{z} + z_{\alpha/2}\widehat{\sigma}/\sqrt{R}\right) \stackrel{\text{def}}{=} \widehat{z} \pm z_{1-\alpha/2}\widehat{\sigma}/\sqrt{R}. \qquad (1.5)$$

We will meet many examples of this later in the chapter. Another example (see Chapters IV and XIII) is an ergodic Markov chain $\{X_n\}_{n\in\mathbb{N}}$. If f is a function on the state space, then under mild conditions $\widehat{z} \stackrel{\text{def}}{=} \left(f(X_0) + \cdots + f(X_{R-1})\right)/R$ has a limit, namely the expectation $z \stackrel{\text{def}}{=} \mathbb{E}_\pi f(X_0)$ w.r.t. the stationary distribution π. Further, $\widehat{z}$ obeys a CLT of the form (1.1). However, estimation of the variance σ^2 is more difficult than for the i.i.d. case and is discussed in Chapter IV. Note that in this example, one has both dependence and nonstationarity.

Exercises

1.1 (TP) One reason for $R - 1$ rather than R in (1.3) is that this choice makes s^2 unbiased. Verify this.

1.2 (TP) In Buffon's needle experiment, how should the length L of the needle be chosen for maximal efficiency (remember that we need $0 < L < 1$)? For this choice of L, perform a study of the run-length considerations parallel to Example 1.2.

1.3 (TP) A simulator wants to produce an unbiased estimate of $z \stackrel{\text{def}}{=} z'z''$, where $z' = \mathbb{E}Z'$, $z'' = \mathbb{E}Z''$ for r.v.'s Z', Z'' that can be generated by simulation. To this end, he simulates R replications $Z_1', \ldots, Z_R'$ of Z' and independently of this R replications $Z_1'', \ldots, Z_R''$ of Z''. He then has the following two estimators at his disposal:

$$\left(\frac{1}{R}\sum_{r=1}^{R} Z_r'\right)\left(\frac{1}{R}\sum_{r=1}^{R} Z_r''\right) \quad \text{and} \quad \frac{1}{R}\sum_{r=1}^{R} Z_r'Z_r''.$$

Show that the first one has the smaller variance.

2   Two-Stage and Sequential Procedures

In many computational settings, one wishes to calculate the solution to a desired level of accuracy. Two accuracy criteria are commonly used:

absolute accuracy: compute the quantity z to an accuracy ε;

relative accuracy: compute the quantity z to an accuracy $\varepsilon|z|$.

The confidence interval (1.4) suggests, in principle, what one should do to achieve such accuracies. In particular, for absolute precision ε, one should choose R such that the half-width $z_{1-\alpha/2}\sigma/\sqrt{R}$ is approximately equal to

ε, or, equivalently,

$$R \approx \frac{z_{1-\alpha/2}^2 \sigma^2}{\varepsilon^2}.$$

On the other hand, for relative precision ε, the appropriate number of simulations is $R \approx z_{1-\alpha/2}^2 \sigma^2 / \varepsilon^2 z^2$. As in the setting of confidence intervals, these run-length determination rules (i.e., rules for selecting R) cannot be implemented directly, since they rely on problem-dependent parameters such as σ^2 and z^2 that are unknown.

As a consequence, the standard means of implementing this idea involves first generating a small number of trial runs (say 50) to estimate σ^2 and z^2 by $\widehat{\sigma}_{\text{trial}}^2$ and $\widehat{z}_{\text{trial}}^2$. One then determines the number R of so-called *production runs* from either

$$R \approx \frac{z_{1-\alpha/2}^2 \widehat{\sigma}_{\text{trial}}^2}{\varepsilon^2} \quad \text{or} \quad R \approx \frac{z_{1-\alpha/2}^2 \widehat{\sigma}_{\text{trial}}^2}{\varepsilon^2 \widehat{z}_{\text{trial}}^2},$$

depending on whether absolute or relative precision is desired. The final confidence interval for z may be based on only the production runs or on both the production runs and the trial runs. If the latter method is used, one should ideally take into account the fact that the random nature of R and its dependence on the trial runs correlates all the observations in a nontrivial way. A significant literature has developed to address the construction of suitable confidence intervals for such two-stage procedures (trial runs, followed by production runs) that takes into account this dependency; see, for example, Nakayama [266].

When a specific level of accuracy is required, an alternative to using a two-stage procedure is to implement a sequential procedure. A sequential procedure essentially just monitors the confidence interval (1.4) until the precision criterion is met. For example, a sequential version of the absolute precision algorithm will end up by generating variates $Z_1, \ldots, Z_N$, where the (random) total sample size N is given by

$$N \stackrel{\text{def}}{=} \inf\{n > 1 : z_{1-\alpha/2} s_n \le \varepsilon\sqrt{n}\}. \tag{2.1}$$

In the above equation, we write s_n to make clear the fact that the sample standard deviation of the observations $Z_1, \ldots, Z_N$ must be recomputed after each additional sampling.

The main difficulty with a sequential procedure is that the standard deviation estimator s_n is highly unreliable at small values of n. If s_n is abnormally small, as might occur in estimating a probability $z = \mathbb{P}A$ via a sample proportion when no occurrences of A have yet been observed by time n (for then $s_n = 0$), this may cause the sequential procedure to shut off at too small a sample size, leading to a poor final estimator for z.

This problem of early termination is the most difficult challenge presented to users of sequential procedures. A second approach to addressing this issue is to modify (2.1) to

$$N' \stackrel{\text{def}}{=} \inf\{n \geq n_0 : z_{1-\alpha/2}s_n \leq \varepsilon\sqrt{n}\}, \tag{2.2}$$

so that at least n_0 ($n_0 = 50$, say) simulations are run before initialization of the sequential check of the precision criterion. In this way, one hopes that the standard deviation estimator has stabilized at something close to the correct value. The final $1 - \varepsilon$ confidence interval is then given by $\hat{z}_{N'} \pm \varepsilon$. Such confidence intervals can be shown to be asymptotically valid as $\varepsilon \downarrow 0$; see Chow & Robbins [71] and Glynn & Whitt [157].

Two implementation details are worth mentioning. The first is that computational effort must be expended on sequentially updating the estimators of σ and z; this can be mitigated by checking the precision criterion every $k > 1$ samples only (rather than after every sample is taken). Secondly, in sequentially updating s_n^2, one wishes to use an algorithm that is both numerically stable and can be updated in constant time (independent of n). Such an updating scheme can be found in Chan et al. [67].

Exercises

2.1 (A) Let $z = \mathbb{P}(U > u)$, where U is uniform$(0, 1)$ (i.e., $Z = \mathbb{1}_{U>u}$). Produce $M = 10{,}000$ 95% confidence intervals of the form (1.4) and compare the coverage (i.e., the frequency among the M confidence intervals in which $z \in \hat{z} \pm 1.96\,s/\sqrt{R}$) for $z = 0.5, 0.9, 0.95, 0.99$ and the following methods: (a) the two-stage method with 10, 100, and 1,000 trial runs, (b) the sequential method (2.2) with the same n_0 as in (a).

3   Computing Smooth Functions of Expectations

Although the problem of estimating $z = \mathbb{E}Z$ is a very general one, it does not subsume all computational problems of interest. In particular, suppose that $z = (z_1\, z_2\, \ldots\, z_d)^\mathsf{T}$ is a d-dimensional vector, in which each component z_i can be expressed as $\mathbb{E}Z(i)$ for some r.v. $Z(i)$. Some applications demand that one compute $f(z)$, where $f : \mathbb{R}^d \to \mathbb{R}$ is an explicitly known and smooth function. This is known as the "smooth function estimation problem" and occurs in traditional statistics as well as in the analysis of simulation output.

Example 3.1 The function $f(z) = 1/z$ occurs in Buffon's needle experiment I.5.7. For another example with this f, let $Z_1, Z_2, \ldots$ be i.i.d. failure times of some system. Then $f(z) = 1/z = 1/\mathbb{E}Z$ is the long-run failure intensity. □

Example 3.2 Suppose that we want to compute the standard deviation σ of an r.v. W. Then $\sigma = f(z)$, where $f(z) = (z_1 - z_2^2)^{1/2}$ and $Z(1) = W^2$, $Z(2) = W$. □

Example 3.3 Suppose that we want to compute the correlation ρ between two r.v.'s X, Y. Then $\rho = f(\mathbf{z})$, where

$$\mathbf{Z} = (X, Y, X^2, Y^2, XY), \quad f(\mathbf{z}) = \frac{z_5 - z_1 z_2}{(z_3 - z_1^2)^{1/2}(z_4 - z_2^2)^{1/2}}. \qquad \square$$

Example 3.4 Let z be the price of a European call option under a non-Black–Scholes model (say with stochastic volatility, stochastic interest rate, non-Gaussian log returns, etc.) and Z an unbiased simulation estimator. The *implied volatility* σ_{impl} is the σ giving the same price z in the Black–Scholes model. That is, $z = BS(r, K, T, \sigma_{\mathrm{impl}})$ (with BS being given in I.(3.4)), so we need $f(z)$, where $f = BS^{-1}(r, K, T, \cdot)$. $\qquad \square$

Example 3.5 A *perpetuity* is an r.v. of the form

$$Y = B_0 + A_1 B_1 + A_1 A_2 B_2 + A_1 A_2 A_3 B_3 + \cdots,$$

where all r.v.'s are independent, the B_n are i.i.d. with finite mean and the A_n are i.i.d. and positive with $\mathbb{E}\log A_n < 0$ [this ensures that Y is well-defined]. Then $\mathbb{E}Y$ may be computed by observing that $Y^* \overset{\text{def}}{=} B_1 + A_2 B_2 + A_2 A_3 B_3 + \cdots$ has the same distribution as Y, so that

$$\mathbb{E}Y = \mathbb{E}\big[B_0 + A_1 Y^*\big] = \mathbb{E}B + \mathbb{E}A \cdot \mathbb{E}Y = \frac{\mathbb{E}B}{1 - \mathbb{E}A},$$

which has the form $f(z_1, z_2)$.

Of course, in this simple setting $\mathbb{E}A$ and $\mathbb{E}B$ wil typically be known, so that simulation is not needed. An extension typically requiring simulation is that of discounted rewards of a finite Markov chain $\{X_n\}$, where A_n, B_n have distribution F_i, G_i when $X_n = i$ and one wants $\mathbb{E}_x Y$ for some x. One can then let $\tau \overset{\text{def}}{=} \inf\{n > 0 : X_n = x\}$,

$$Z(1) \overset{\text{def}}{=} \sum_{n=0}^{\tau-1} A_1 \ldots A_n B_n, \quad Z(2) \overset{\text{def}}{=} A_1 \ldots A_\tau$$

to conclude easily that

$$\mathbb{E}_x Y = \mathbb{E}_x\left[Z(1) + Z(2)\sum_{n=\tau}^{\infty} A_{\tau+1} \ldots A_n B_n\right]$$

$$= \mathbb{E}_x Z(1) + \mathbb{E}_x Z(2)\mathbb{E}_x Y = \frac{\mathbb{E}_x Z(1)}{1 - \mathbb{E}_x Z(2)}.$$

See Fox & Glynn [122] for further discussion of such discounted rewards. $\square$

We return to the general discussion of the smooth-function estimation problem. The obvious estimator of $f(\mathbf{z})$ is, of course, $f(\widehat{\mathbf{z}})$, where $\widehat{\mathbf{z}} \overset{\text{def}}{=} (\mathbf{Z}_1 + \cdots + \mathbf{Z}_R)/R$ and the $\mathbf{Z}_r$ are i.i.d. replicates of the random vector

$$\mathbf{Z} \overset{\text{def}}{=} (Z(1) \ldots Z(d))^{\mathsf{T}}.$$

If f is continuous at z, consistency (i.e., $f(\widehat{z}) \overset{\text{a.s.}}{\to} f(z)$) is guaranteed due to the consistency of $\widehat{z}$.

The more subtle issue is the rate of convergence of $f(\widehat{z})$ and the construction of associated confidence intervals for $f(z)$. The key is a development of a CLT for $f(\widehat{z})$. This CLT can easily be derived via an appropriate Taylor expansion. This approach for deriving CLTs is known in the statistical literature as the *delta method*; a short introduction can be found in van der Vart [356, Chapter 3].

In this context, the delta method just requires observing that if f is smooth enough (in the context, differentiability at z is enough), then

$$f(\widehat{z}) - f(z) = \nabla f(z)(\widehat{z} - z) + \mathrm{o}\big(\|\widehat{z} - z\|\big)$$

$$= \frac{1}{R}\big(V_1 + \cdots + V_R\big) + \mathrm{o}\big(\|\widehat{z} - z\|\big), \qquad (3.1)$$

where ∇f denotes the gradient (written as row vector) and $V_r \overset{\text{def}}{=} \nabla f(z)(Z_r - z)$. Note that $\mathbb{E}V_i = 0$. Furthermore, if the matrix $\Sigma \overset{\text{def}}{=} \mathbb{V}ar\, Z_1$ with ijth element $\mathbb{C}ov\big(Z(i), Z(j)\big)$ is well defined and finite, then $\|\widehat{z} - z\| = \mathrm{O}(R^{-1/2})$ by the CLT and

$$\sigma^2 \overset{\text{def}}{=} \mathbb{V}ar\, V_1 = \nabla f \Sigma (\nabla f)^{\mathsf{T}} = \sum_{i,j=1}^{d} f_{z_i}(z) f_{z_j}(z) \mathbb{C}ov\big(Z(i), Z(j)\big) < \infty.$$

Thus multiplying (3.1) by $R^{1/2}$, we obtain

$$R^{1/2}\big(f(\widehat{z}) - f(z)\big) \overset{\mathscr{D}}{\to} \mathscr{N}\big(0, \sigma^2\big). \qquad (3.2)$$

Note that if $d = 1$, we have just $\sigma^2 = f'(z)^2 \mathbb{V}ar\, Z$.

Given the CLT (3.2), much of our discussion of Section 1 continues to follow. However, there are two key differences. Firstly, the estimator $f(\widehat{z})$ is generally biased as an estimator of $f(z)$, in the sense that $\mathbb{E}f(\widehat{z}) \neq f(z)$. More precisely, note that if f is twice differentiable at z, then

$$f(\widehat{z}) - f(z) = \nabla f(z)(\widehat{z} - z) + \frac{1}{2}(\widehat{z} - z)^{\mathsf{T}} H(z)(\widehat{z} - z) + \mathrm{o}\big(\|\widehat{z} - z\|^2\big),$$

where $H(z) = \big(H_{ij}(z)\big)_{1 \le i,j \le d}$ is the Hessian (the $d \times d$ matrix of second derivatives) at z. Taking expectations, we formally arrive at

$$\mathbb{E}f(\widehat{z}) - f(z) = \frac{1}{2R} \sum_{i,j=1}^{d} \mathbb{C}ov\big(Z(i), Z(j)\big) H_{ij}(z) + \mathrm{o}(1/R) \qquad (3.3)$$

as $R \to \infty$. Under additional regularity conditions, a further Taylor expansion shows that the $\mathrm{o}(1/R)$ term in (3.3) is actually of the form $\beta/R^2 + \mathrm{o}(1/R^2)$ for some constant β.

Remark 3.6 The arguments leading to (3.3) must be viewed as being heuristic. A rigorous version of (3.3) requires conditions ensuring uniform integrability. $\square$

In view of (3.3), a possible means of reducing bias would be to use the modified estimator

$$f(\widehat{\boldsymbol{z}}) \; - \; \frac{1}{2R} \sum_{i,j=1}^{d} \widehat{\text{Cov}}(Z(i), Z(j)) H_{ij}(\widehat{\boldsymbol{z}}), \tag{3.4}$$

where $\widehat{\text{Cov}}(Z(i), Z(j))$ is the sample covariance between $Z(i)$ and $Z(j)$ based on $\boldsymbol{Z}_1, \ldots, \boldsymbol{Z}_R$. Because the difference between (3.4) and $f(\widehat{\boldsymbol{z}})$ is of order $1/R$, it follows that (3.4) satisfies precisely the same CLT as does $f(\widehat{\boldsymbol{z}})$. Consequently, the lower bias of (3.4) does not improve the asymptotic rate of convergence. Nevertheless, bias corrections are often implemented due to a widely held belief that such estimators exhibit better small-sample properties. To the extent that this belief is valid, it should be seen as a "rule of thumb", and it does not hold universally that such estimators behave better for small R.

Remark 3.7 To compute (3.4) requires calculating $(d + 1)d/2$ different Hessian entries, as well as estimating $(d+1)d/2$ different sample covariances. We will see in Section 5 how one can construct a bias-corrected estimator that avoids these steps. □

The second key difference in the CLT (3.2) relative to (1.1) is that estimating the variance σ^2 is harder. Note that we cannot observe the V_r, since the definition involves the unknown $\boldsymbol{z}$. Of course, if R is large, $\nabla f(\widehat{\boldsymbol{z}})(\boldsymbol{Z}_r - \widehat{\boldsymbol{z}})$ should be close to $\nabla f(\boldsymbol{z})(\boldsymbol{Z}_r - \boldsymbol{z})$, suggesting that we use the variance estimator

$$\widehat{\sigma}^2 \; \overset{\text{def}}{=} \; \frac{1}{R-1} \sum_{r=1}^{R} \left(\nabla f(\widehat{\boldsymbol{z}})(\boldsymbol{Z}_r - \widehat{\boldsymbol{z}}) \right)^2. \tag{3.5}$$

The challenge in computing $\widehat{\sigma}^2$, practically speaking, is the need to explicitly compute the gradient of f. For complex functions or high d, one may wish to avoid this calculation. Section 5 describes two different methods capable of producing suitable estimators that have this desirable characteristic. In any case, given a suitable estimator $\widehat{\sigma}^2$ with $\widehat{\sigma}^2 \to \sigma^2$ as $R \to \infty$, the generation of an asymptotic $1-\alpha$ confidence interval for $f(\boldsymbol{z})$ is straightforward (see (1.5) with $\widehat{\boldsymbol{z}}$ replaced by $f(\widehat{\boldsymbol{z}})$). Centering the interval on (3.4) instead of $f(\widehat{\boldsymbol{z}})$ would also produce an asymptotically valid procedure.

Exercises

3.1 (TP) Give a bivariate CLT for $\widehat{\boldsymbol{z}}$ and s^2 in the i.i.d. setting.

3.2 (TP) The skewness of an r.v. Z is defined as $\lambda \overset{\text{def}}{=} \mathbb{E}(Z - \mathbb{E}Z)^3 / (\mathbb{V}\text{ar}\, Z)^{3/2}$. Give a CLT for its empirical counterpart $\sum_1^R (Z_t - \widehat{z})^3 / Rs^{3/2}$.

3.3 (TP)   A system develops in i.i.d. cycles. According to whether a certain catastrophic event occurs or not within a cycle, the cycle is classified as failed or nonfailed. Denote by p the probability that a cycle is failed, by ℓ_1 the expected length of a cycle given that it does not fail, and by ℓ_2 the expected time until failure in a cycle given that it is failed. We are interested in ℓ, the expected time until a failure occurs.

1. Express ℓ in terms of p, ℓ_1, ℓ_2.
2. You are presented with statistics of 1,000 cycles, of which 87 failed. The nonfailed cycles had an empirical mean of 20.2 and an empirical variance of 18.6, and the average time until failure in the failed cycles was 5.4 with an empirical variance of 3.1. Give a confidence interval for ℓ.

3.4 (A) A man of age 45 signs a life annuity contract, according to which he has to pay premium at (continuous) rate π until age 67 or the time T of death, and if still alive at age 67, he will receive a payment at rate 1 until death. Assuming a constant interest rate r, the equation determining π is therefore

$$\mathbb{E} \int_{55}^{67 \wedge T} \pi e^{-rt} \, dt = \mathbb{E} \int_{T \vee 67}^{T} e^{-rt} \, dt$$

(the two sides are the present values of contributions, respectively benefits). Your assignment is to give a simulation estimate of π and an associated confidence interval, assuming $r = 4\%$ and that T follows the *Gompertz–Makeham distribution* (conditioned on $T > 55$), which is widely used as lifetime distribution in life insurance. As parameters, use the G82 base in Exercise II.4.1. The r.v. T may be generated by acceptance–rejection with a uniform$(55, 110)$ proposal (neglecting the possibility $T > 110$), using the fact that $f(t) \leq 0.035$ for all t.

4   Computing Roots of Equations Defined by Expectations

The simplest instance of the problem considered in this section is to find a root θ^* to an equation of the form

$$f(z, \theta) = 0, \tag{4.1}$$

where f is explicitly known and $z \overset{\text{def}}{=} \mathbb{E} \mathbf{Z} \in \mathbb{R}^d$, $\theta \in \mathbb{R}$. For certain functions f, the root θ^* to (4.1) can be written as $\theta^* = \zeta(z)$, where ζ is defined via the relation $f(z, \zeta(z)) = 0$. If ζ can be computed explicitly and is smooth, this reduces the root-finding problem to one of the type considered in the previous section.

When ζ cannot be computed explicitly, iterative algorithms can be used to find the empirical root $\widehat{\theta}$ to the equation

$$f(\widehat{z}, \theta) = 0, \tag{4.2}$$

where $\widehat{z}$ is the average of R replicates $\mathbf{Z}_1, \ldots, \mathbf{Z}_R$ of $\mathbf{Z}$.

Given that $\widehat{z} \overset{\text{a.s.}}{\to} z$ as $R \to \infty$, it is straightforward to argue that $\widehat{\theta} \overset{\text{a.s.}}{\to} \theta^*$ under mild additional regularity conditions on f. As in the preceding section, the key to deriving asymptotically valid confidence intervals is developing a CLT for $\widehat{\theta}$. Because $\widehat{\theta}$ and θ^* are respectively roots of (4.2) and (4.1), it is evident that

$$0 \;=\; f(\widehat{z}, \widehat{\theta}) - f(z, \theta^*) \;=\; f(\widehat{z}, \widehat{\theta}) - f(z, \widehat{\theta}) + f(z, \widehat{\theta}) - f(z, \theta^*).$$

The delta method establishes that if f is sufficiently smooth,

$$\nabla_z(z, \theta^*)(\widehat{z} - z) + f_\theta(z, \theta^*)(\widehat{\theta} - \theta^*) + o(\|\widehat{z} - z\|) \;=\; 0,$$

where f_θ is the partial derivative w.r.t. θ. Arguments similar to those of Section 3 yield the CLT

$$R^{1/2}(\widehat{\theta} - \theta^*) \;\overset{\mathscr{D}}{\to}\; \mathscr{N}(0, \sigma^2) \tag{4.3}$$

as $R \to \infty$ (provided $f_\theta(z, \theta^*) \neq 0$), where

$$\sigma^2 \;=\; \frac{\mathbb{V}ar(\nabla_z f(z, \theta^*)Z)}{f_\theta(z, \theta^*)^2}.$$

A sample estimate of σ^2 is therefore

$$\widehat{\sigma}^2 \;=\; \frac{\dfrac{1}{R-1}\displaystyle\sum_{r=1}^{R}[\nabla_z f(\widehat{z}, \widehat{\theta})(Z_r - z)]^2}{f_\theta(\widehat{z}, \widehat{\theta})^2},$$

from which the asymptotic confidence interval is computed in the obvious way, cf. (1.5) with $\widehat{z}$ replaced by $\widehat{\theta}$.

A still more demanding root-finding problem arises when the objective is to find the root θ^* of an equation of the form $z(\theta) = 0$, where $z(\theta) = \mathbb{E}Z(\theta)$ for each θ.

Example 4.1 (STOCHASTIC COUNTERPART METHOD) Suppose that we wish to numerically minimize a convex function $w(\theta)$ over $\theta \in \mathbb{R}$, where $w(\theta) = \mathbb{E}W(\theta)$. This requires computing the root θ^* to $z(\theta) = 0$, where $z(\theta) = \mathbb{E}W'(\theta)$. Here, we have assumed that $W(\theta)$ is smooth in θ and that the derivative interchange is valid (see also VII.2). □

The natural estimator θ^* is of course the root to $\widehat{z}(\widehat{\theta})$, where $\widehat{z}(\theta) \overset{\text{def}}{=} (Z_1(\theta) + \cdots + Z_R(\theta))/R$. Obviously, this assumes that an efficient numerical procedure for computing the root of the random function $\widehat{z}(\cdot)$ is used. Under suitable regularity conditions, $\widehat{\theta} \overset{\text{a.s.}}{\to} \theta^*$ as $R \to \infty$.

To derive a CLT for $\widehat{\theta}$, we suppose here that $Z(\theta)$ is smooth in θ and that the derivative interchange with the expectation is valid. Because θ^* and $\widehat{\theta}$ are the roots of their respective equations,

$$\widehat{z}(\widehat{\theta}) - \widehat{z}(\theta^*) \;=\; z(\theta^*) - \widehat{z}(\theta^*). \tag{4.4}$$

But $\widehat{z}(\widehat{\theta}) - \widehat{z}(\theta^*) = \widehat{z}'(\xi)(\widehat{\theta} - \theta^*)$, where ξ lies between $\widehat{\theta}$ and θ^*. If $z'(\theta^*) \neq 0$, then applying the CLT to the r.h.s. of (4.4) shows that

$$R^{1/2}(\widehat{\theta} - \theta^*) \;\overset{\mathscr{D}}{\to}\; \mathscr{N}(0, \sigma^2), \quad \text{where } \sigma^2 = \frac{\mathbb{V}ar\, Z(\theta^*)}{z'(\theta^*)^2}. \tag{4.5}$$

Here the natural estimator for σ^2 is

$$\widehat{\sigma}^2 = \frac{1}{\widehat{z}'(\widehat{\theta})^2(R-1)} \sum_{r=1}^{R} Z_r(\widehat{\theta})^2, \qquad (4.6)$$

from which the asymptotic confidence interval is computed in the obvious way, cf. (1.5) with $\widehat{z}$ replaced by $\widehat{\theta}$.

Remark 4.2 Assume in Example 4.1 that $W(\cdot)$ is twice continuously differentiable with the interchange valid. Then $\widehat{\sigma}^2$ takes the form

$$\widehat{\sigma}^2 = \frac{1}{R-1} \sum_{r=1}^{R} W_r'(\widehat{\theta})^2 \bigg/ \left(\frac{1}{R} \sum_{r=1}^{R} W_r''(\widehat{\theta})\right)^2. \qquad \square$$

The problem of computing a root to $z(\theta) = 0$ becomes more challenging when $Z(\theta)$ is not smooth in θ. We next illustrate this in a setting that is of particular application relevance (cf., for instance, VaR calculations in mathematical finance).

4a   Quantile Estimation

Suppose that we want to compute the α-quantile, denoted by $q = q_\alpha$, of a continuous r.v. Z with positive and continuous density. To this end, note that q is the root of $z(\theta) = 0$, where $z(\theta) = \mathbb{E}Z(\theta)$ and $Z(\theta) \overset{\text{def}}{=} \mathbb{1}_{Z \leq \theta} - \alpha$.

Starting from (4.4), the main difficulty is dealing with its l.h.s. If R is large, we expect that $\widehat{z}(\theta)$ will be uniformly close to $z(\theta)$, in which case we might hope that

$$\widehat{z}(\widehat{\theta}) - \widehat{z}(\theta^*) \approx z'(\theta^*)(\widehat{\theta} - \theta^*)$$

if z is smooth [this smoothness of $z(\cdot) = \mathbb{E}Z(\cdot)$ often follows from the smoothing of the $\mathbb{E}$ operator, even if $Z(\cdot)$ is not smooth]. When this approach is valid, (4.5) continues to hold.

There are two difficulties here, one theoretical and one practical.

The main practical issue is that, in contrast to the root-finding context explained above, producing a confidence interval for θ^* is no longer straightforward. The obstacle is that the $\widehat{\sigma}^2$ formula of (4.6) no longer applies, since $\widehat{z}'(\widehat{\theta})$ cannot be used to estimate $z'(\theta^*)$.

The above analysis suggests that $\widehat{q} = \widehat{\theta}$ satisfies the CLT

$$R^{1/2}(\widehat{q} - q) \overset{\mathscr{D}}{\to} \mathscr{N}(0, \sigma^2), \quad \text{where } \sigma^2 = \frac{\alpha(1-\alpha)}{f(q)^2}, \qquad (4.7)$$

and $f(\cdot)$ is the density of Z. An estimator of σ^2 must therefore address the issue of estimating $f(\cdot)$. This is a nontrivial issue that will be discussed further in Section 6.

The theoretical issue is the need to justify the above Taylor expansion when $z(\cdot)$ is not smooth. One means of addressing this is to apply ideas from the theory of Gaussian processes. Let z be a continuous real function defined on an interval I and having a unique zero θ^*, where $z'(\theta^*) > 0$. Assume that a simulation with run length R generates a process $\{Z_R(t)\}$, such that

$$\left\{ \sqrt{R}\left(Z_R(t) - z(\theta)\right) \right\}_{\theta \in I} \xrightarrow{\mathscr{D}} \{G(\theta)\}_{\theta \in I} ,$$

where $\{G(\theta)\}$ is a Gaussian process satisfying suitable continuity requirements, and define

$$\widehat{\theta}_R \overset{\text{def}}{=} \inf \{\theta \in I : Z_R(\theta) \geq 0\} .$$

Then $\widehat{\theta}_R \xrightarrow{\mathbb{P}} \theta^*$, and by similar calculations as those for (4.7),

$$\sqrt{R}\left(\widehat{\theta}_R - \theta^*\right) \xrightarrow{\mathscr{D}} \mathscr{N}\left(0, \tfrac{\mathbb{V}\text{ar}\, G(\theta^*)}{z'(\theta^*)^2}\right). \tag{4.8}$$

In the context of quantile estimation, we take as above $z(\theta) \overset{\text{def}}{=} F(\theta) - \alpha$, so that $z(\theta^*) = 0$, where $\theta^* \overset{\text{def}}{=} q$, and as $Z_R(t)$, we take the empirical c.d.f. centered around α,

$$Z_R(t) \overset{\text{def}}{=} \frac{1}{R} \sum_{r=1}^{R} \mathbb{1}\{Z_r \leq t\} - \alpha.$$

Since the sum is binomial $(R, F(\theta))$, the assumptions above hold with $\mathbb{V}\text{ar}\left(G(\theta)\right) = F(\theta)\left(1 - F(\theta)\right)$, $\theta^* = q$. The empirical α-quantile $Z_{(\alpha(R+1))}$ (the $\alpha(R+1)$th order statistic) is close to $\widehat{\theta}_R$ above, and so as for (4.7) we are led to the same conclusion that $\sqrt{R}\left(Z_{(\alpha(R+1))} - q\right)$ is asymptotically normal with variance $\alpha(1 - \alpha)/f(q_\alpha)^2$, where $f = F'$ is the density. This leads to the 95% confidence interval

$$Z_{(\alpha(R+1))} \pm \frac{1.96\sqrt{\alpha(1 - \alpha)}}{\widehat{f}(q)\sqrt{R}}$$

for $q = q_\alpha$, where $\widehat{f}(q)$ is an estimate of $f(q)$.

For more detailed and rigorous discussions of the material of the present section, see further van der Vart [356, Chapter 21], Serfling [333], or Borovkov [52].

Exercises

4.1 (TP) Find the bias of the quantile estimator $Z_{(\alpha(R+1))}$. [Hint: Consider first the case in which the Z_r are uniform.]

5   Sectioning, Jackknifing, and Bootstrapping

The discussions of Sections 3 and 4 illustrate the point that the construction of suitable confidence intervals (and bias-corrected estimators) can become nontrivial in some settings, not least because of a potential need to compute large numbers of partial derivatives in some problems. This section is concerned with providing alternative methods that seek to avoid some of these difficulties.

5a Sectioning. The Delta Method in Function Space

Starting at an abstract level, we note that all the problems described thus far in this chapter are special cases of a general problem, namely the computation of a quantity $\psi(F)$, where F is the distribution of some underlying random element Z and ψ is a real-valued functional on the space of probability distributions. For example, the problem of estimating $z = \mathbb{E}Z$ is the special case in which $\psi(F) = \int x \, F(\mathrm{d}x)$. Assuming that one can draw i.i.d. replicates $Z_1, \ldots, Z_R$ from F, the natural estimator for $\psi(F)$ is $\psi(\widehat{F}_R)$, where $\widehat{F}_R$ is the sample (or empirical) distribution defined by

$$\widehat{F}_R(\mathrm{d}x) \stackrel{\mathrm{def}}{=} \frac{1}{R} \sum_{r=1}^{R} \delta_{Z_r}(\mathrm{d}x),$$

where $\delta_{Z_r}(\cdot)$ is a unit point mass distribution at Z_r. If $Z \in \mathbb{R}$, the corresponding c.d.f. is

$$\widehat{F}_R(x) \stackrel{\mathrm{def}}{=} \frac{1}{R} \sum_{r=1}^{R} \mathbb{1}\{Z_r \leq x\},$$

but more general sets of values of Z are needed for many problems. This estimator is, of course, the one suggested in the examples of Sections 2 and 3. Under minor topological conditions, $\psi(\widehat{F}_R) \stackrel{\mathrm{a.s.}}{\to} \psi(F)$ as $R \to \infty$.

To develop confidence intervals for $\psi(F)$, one needs to apply a general version of the delta method to the functional ψ. More precisely, the program for rigorously verifying the requisite smoothness is to find a Banach space $(V, \|\cdot\|)$ of (signed) measures such that one is willing to assume that both F and $\widehat{F}_R$ are in V, and that ψ is defined on all of V and Fréchet or Hadamard differentiable at F. See, for example, Gill [130] or van der Vart [356, Chapter 20] for surveys.

When this program works, one is led to the CLT

$$R^{1/2}\big(\psi(\widehat{F}_R) - \psi(F)\big) \stackrel{\mathscr{D}}{\to} \mathscr{N}(0, \sigma^2)$$

as $R \to \infty$, where σ^2 depends on the (Fréchet or Hadamard directional) derivative $(\mathscr{D}\psi)(F)$ of ψ at F. In other words,

$$\psi(\widehat{F}_R) \stackrel{\mathscr{D}}{\approx} \psi(F) + Y, \quad \text{where } Y \sim \mathscr{N}\big(0, \sigma/\sqrt{R}\big) \tag{5.1}$$

for R large, but where σ^2 may be difficult to estimate directly via an estimator $\widehat{\sigma}^2$.

Suppose now that $R = NK$ and split the sample into N "sections", each of length K. Set

$$\widehat{F}_{n,K}(\mathrm{d}x) \overset{\text{def}}{=} \frac{1}{K} \sum_{r=(n-1)K+1}^{nK} \delta_{Z_r}(\mathrm{d}x),$$

and let $\psi(\widehat{F}_{n,K})$ be the estimator of $\psi(F)$ constructed from the nth section of the sample. According to (5.1), when K is large,

$$\psi(\widehat{F}_{n,K}) \overset{\mathscr{D}}{\approx} \psi(F) + \frac{\sigma}{\sqrt{K}} Y_n, \tag{5.2}$$

where the Y_i are i.i.d. $\mathscr{N}(0,1)$ r.v.'s. Furthermore, because the above Banach space machinery leads to the Taylor-type expansion

$$\psi(\widehat{F}_R) \approx \psi(F) + (\mathscr{D}\psi)(F)(\widehat{F}_R - F),$$

it follows that

$$R^{1/2}\left(\psi(\widehat{F}_R) - \frac{1}{N}\sum_{n=1}^{N}\psi(\widehat{F}_{n,K})\right) \overset{\mathbb{P}}{\to} 0 \tag{5.3}$$

as $R \to \infty$ with N fixed. It is evident from (5.2) and (5.3) that

$$N^{1/2}\left(\psi(\widehat{F}_R) - \psi(F)\right) \Big/ \sqrt{\frac{1}{N-1}\sum_{n=1}^{N}\left(\psi(\widehat{F}_{n,K}) - \psi(\widehat{F}_R)\right)^2}$$

$$\overset{\mathscr{D}}{\to} \sum_{n=1}^{N} Y_n \Big/ \sqrt{\frac{1}{N-1}\sum_{n=1}^{N}\left(Y_n - N^{-1}\sum_{m=1}^{N} Y_m\right)^2} \tag{5.4}$$

as $R \to \infty$ with N fixed. The limiting r.v. in (5.4) is both free of the difficult-to-estimate σ^2 and also has a standard distribution, namely that of a Student t with $f \overset{\text{def}}{=} N - 1$ degrees of freedom. Letting t_α be the α-quantile of this distribution and noting that the t distribution is symmetric, this sectioning idea therefore leads to the asymptotic confidence interval

$$\left(\psi(\widehat{F}_R) - t_{1-\alpha/2}\widehat{\sigma}/\sqrt{N},\ \psi(\widehat{F}_R) - t_{\alpha/2}\widehat{\sigma}/\sqrt{N}\right) \overset{\text{def}}{=} \psi(\widehat{F}) \pm t_{1-\alpha/2}\widehat{\sigma}/\sqrt{N}, \tag{5.5}$$

where

$$\widehat{\sigma}^2 \overset{\text{def}}{=} \frac{1}{N-1}\sum_{n=1}^{N}\left(\psi(\widehat{F}_{n,K}) - \psi(\widehat{F}_R)\right)^2$$

is the corresponding estimator for the variance.

Note that this method relies on K being large enough that the approximation (5.2) is reasonable. In view of the fact that a t distribution with $f \geq 30$ is very close to $\mathscr{N}(0,1)$, it is therefore recommended that N be chosen relatively small, and certainly not larger than 30.

5b   The Jackknife

We next turn to the jackknife method, and focus our attention on the smooth functions of expectations problem discussed in Section 3. One difficulty there was the significant potential effort in developing a bias-corrected estimator for $f(z)$ based on second-order Taylor expansions. We further argued that

$$\mathbb{E}f(\widehat{z}_R) \; = \; f(z) + \frac{c}{R} + \frac{\beta}{R^2} + o(1/R^2),$$

where we now choose to write $\widehat{z} = \widehat{z}_R$ to indicate the dependence on R, the number of replications. Because c is now troublesome to estimate, we follow the same idea as that used for sectioning. In particular, we seek a method that "cancels out" the constant c, eliminating our need to estimate it.

Note that

$$R\,\mathbb{E}f(\widehat{z}_R) \; = \; Rf(z) + c + \frac{\beta}{R} + o(1/R),$$

so that

$$\mathbb{E}\big[R\,f(\widehat{z}_R) - (R-1)\,f(\widehat{z}_{R-1})\big] \; = \; f(z) + \frac{\beta}{R^2} + o(1/R^2). \tag{5.6}$$

Of course, we can view $\widehat{z}_{R-1}$ as the estimator obtained by leaving out Z_R and averaging over the remaining $R-1$ replications. Since $Z_1, \ldots, Z_R$ are i.i.d., each of the quantities

$$\widehat{J}_{(r)} \; \overset{\text{def}}{=} \; R\,f(\widehat{z}_R) - (R-1)f\Big(\frac{1}{R-1}\sum_{r' \neq r} Z_{r'}\Big)$$

has expectation equal to the r.h.s. of (5.6). This leads to the jackknife estimator

$$\frac{1}{R}\sum_{r=1}^{R} \widehat{J}_{(r)}.$$

By virtue of (5.6), the estimator is bias-corrected, in the sense that its bias (of order R^{-2}) is of smaller order than the R^{-1} of $f(z)$. Of course, it comes at a significant additional cost (because computing the $\widehat{J}_{(r)}$ may be costly when R is large). If this is a concern, one way to mitigate this cost is to apply the jackknife idea to the sections described earlier, by averaging over each of the N values

$$N\,f(\widehat{z}_R) \; - \; (N-1)f\Big(\frac{1}{(N-1)K}\sum_{r' \notin \{nK+1,\ldots,(n+1)K\}} Z_{r'}\Big),$$

which again reduces the bias to order R^{-2} but with a substantially lower cost.

However, the jackknife estimator comes with other advantages. If one treats the $\widehat{J}_{(r)}$ as i.i.d. r.v.'s (they are, in fact, highly dependent!) and computes their sample variance

$$\frac{1}{R-1}\sum_{r=1}^{R}\left(\widehat{J}_{(r)} - \frac{1}{R}\sum_{r'=1}^{R}\widehat{J}_{(r')}\right)^2, \tag{5.7}$$

this sample variance estimator converges (surprisingly!) to the σ^2 of (3.2) as $R \to \infty$. Note that (5.7) is an estimator of σ^2 that avoids the need to compute ∇f and can be used in place of the $\widehat{\sigma}^2$ of (3.5) in the confidence interval.

5c   The Bootstrap

Bootstrapping (e.g., Davison & Hinkley [82]) is a general term for a set of methods for performing statistical inference by resampling from the empirical distribution $\widehat{F}_R$. We focus here on the implementation for producing confidence intervals for $\psi(F)$. As pointed out in Section 5a, the natural estimator for $\psi(F)$ is $\psi(\widehat{F}_R)$, where $\widehat{F}_R$ is the empirical distribution of $Z_1, \ldots, Z_R$. Suppose, for the moment, that the distribution of the r.v. $\psi(\widehat{F}_R) - \psi(F)$ is known, continuous, and strictly increasing. We can then compute the (unique) $\alpha/2$ and $1 - \alpha/2$ quantiles z_1, z_2 satisfying

$$\mathbb{P}\big(\psi(\widehat{F}_R) - \psi(F) < z_1\big) = \mathbb{P}\big(\psi(\widehat{F}_R) - \psi(F) > z_2\big) = \alpha/2,$$

so that

$$\mathbb{P}\Big(\psi(F) \in \big(\psi(\widehat{F}_R) - z_2, \psi(\widehat{F}_R) - z_1\big)\Big) = 1 - \alpha.$$

In other words,

$$\big(\psi(\widehat{F}_R) - z_2, \psi(\widehat{F}_R) - z_1\big)$$

is a $100(1-\alpha)\%$ confidence interval.[2] We conclude that the construction of an exact confidence interval is straightforward, given that quantiles of $\psi(\widehat{F}_R) - \psi(F)$ can be calculated.

To make clear the dependence of the distribution of $\psi(\widehat{F}_R) - \psi(F)$ on the underlying distribution F, we write $\mathbb{P}_F\big(\psi(\widehat{F}_R) - \psi(F) \in \cdot\big)$ for $\mathbb{P}\big(\psi(\widehat{F}_R) - \psi(F) \in \cdot\big)$. The bootstrap principle asserts that since $\widehat{F}_R \approx F$ when R is large, the sampling distribution of the estimator under $\widehat{F}_R$ should be close to that under F. Observe that the quantity $\psi(\widehat{F}_R)$ is the value of $\psi(\cdot)$ when the underlying distribution is $\widehat{F}_R$ rather than F. This suggests that

---

[2]Note that (when $\alpha < 1/2$) the upper quantile z_2 determines the left endpoint of the confidence interval and the lower z_1 the upper; this is disguised in the most traditional setting of normal confidence interval because the normal distribution is symmetric.

good approximations z_1^*, z_2^* to z_1, z_2 can be computed as approximative[3] quantiles. That is, we look for z_1^*, z_2^* satisfying

$$\mathbb{P}_{\widehat{F}_R}\left(\psi(\widehat{F}_R) - \psi(F) < z_1^*\right) \approx \mathbb{P}_{\widehat{F}_R}\left(\psi(\widehat{F}_R) - \psi(F) > z_2^*\right) \approx \alpha/2 \,.$$

It remains to describe how z_1^*, z_2^* can be efficiently computed. Sampling represents one obvious alternative. To be precise, generate $b = 1, \dots, B$ so-called *bootstrap samples* $Z_{1b}^*, \dots, Z_{Rb}^*$ by sampling the Z_{rb}^* with replacement from the set $Z_1, \dots, Z_R$. That is, $\mathbb{P}(Z_{rb}^* = Z_{r'}) = 1/R$ and the Z_{rb}, $b = 1, \dots, B$, $r = 1, \dots, R$ are i.i.d. Let

$$\widehat{F}_{Rb}^*(\cdot) \overset{\text{def}}{=} \frac{1}{R} \sum_{r=1}^{R} \mathbb{P}(Z_{rb} \in \cdot)$$

be the empirical distribution of the bth bootstrap sample, and compute z_1^*, z_2^* as the empirical quantiles of the B i.i.d. r.v.'s

$$\psi(\widehat{F}_{R1}^*) - \psi(\widehat{F}_R), \ \dots, \psi(\widehat{F}_{RB}^*) - \psi(\widehat{F}_R) \,. \tag{5.8}$$

The standard convention is to take the empirical β quantile of a set of B i.i.d. r.v.'s as the $\lfloor \beta(B+1) \rfloor$th order statistic, so one conveniently chooses B such that $\alpha(B+1)/2$ and $(1-\alpha/2)(B+1)$ are both integers (say $B = 999$ or $B = 9{,}999$ when $\alpha = 5\%$).

Remark 5.1 Note that the bootstrap confidence interval requires drawing RB i.i.d. r.v.'s from the empirical distribution $\widehat{F}_R$, and computing $\psi(\cdot)$ B times (for each of the B independent bootstrap samples of size R). Given that B must be large in order that the quantile estimators z_1^*, z_2^* be accurate (certainly, B of the order of hundreds is necessary), the construction of a bootstrap confidence interval for $\psi(F)$ can be expensive, rising to the point where the computer time expended in computing the confidence interval potentially dominates the computer time for generating the original R r.v.'s $Z_1, \dots, Z_R$ and $\psi(\widehat{F}_R)$. □

Remark 5.2 One way to minimize the problem raised in Remark 5.1 is by combining the bootstrap with sectioning. Suppose that $R = NK$, so that the sample $Z_1, \dots, Z_R$ may be thought of as N i.i.d. sections, each section n with its own empirical distribution $\widehat{F}_n$. We now draw a bootstrap sample of size R by sampling each of its N sections at random from the N sections of $Z_1, \dots, Z_R$. This reduces the amount of bootstrap sampling to BN. However, in contrast to Section 5a, one needs now N to be reasonably large in order that $(\widehat{F}_{1,K}^* + \cdots + \widehat{F}_{N,K}^*)/N$ look like a sample from $\widehat{F}_R$. □

Example 5.3 For quantile estimation, one has (in the setting of Section 1) that $\psi(F) =$ the α-quantile of F. One first draws i.i.d. replications $Z_1, \dots, Z_R$ from F and next $b = 1, \dots, B$ bootstrap samples, each of

---

[3]Since $\widehat{F}_R$ is a discrete distribution, exact quantiles typically are not unique.

which consists of R observations $Z^*_{1b}, \ldots, Z^*_{Rb}$ from the empirical distribution of $Z_1, \ldots, Z_R$. For each bootstrap sample, $\psi(\widehat{F}^*_{Rb})$ is computed as the empirical α-quantile $T^*_b \overset{\text{def}}{=} T(Z^*_{1b}, \ldots, Z^*_{Rb}) \overset{\text{def}}{=} Z^*_{(\alpha(R+1)),b}$, and one lets $\Delta^*_b \overset{\text{def}}{=} T^*_b - t$, where $t \overset{\text{def}}{=} Z_{(\alpha(R+1))}$ is the value computed from $Z_1, \ldots, Z_R$. The order statistics $\Delta^*_{(1)}, \ldots, \Delta^*_{(B)}$ are then computed and the desired confidence interval (say an equitailed 95% one, i.e., $\beta = 0.05$) is finally obtained as

$$\left(Z_{(\alpha(R+1))} - \Delta^*_{(0.975(N+1))}, Z_{(\alpha(R+1))} + \Delta^*_{(0.025(N+1))} \right) . \qquad \square$$

Exercises

5.1 (A) Complement your solution of Exercise II.3.8 by giving an upper 95% bootstrap confidence limit for the VaR. For simplicity, you may consider the multivariate Gaussian copula only.

5.2 (A) The *Kolmogorov–Smirnov test* for testing whether n observations $X_1, \ldots, X_n$ are i.i.d. with a given common c.d.f. F rejects for large values of $\|\widehat{F}_n - F\|$, where $\|\cdot\|$ is the supremum norm and $\widehat{F}_n$ is the empirical c.d.f. of n i.i.d. observations. More precisely, the rejection region at the 95% level is approximately $\{\|\widehat{F}_n - F\| > \sqrt{n}q\}$, where q is the 95% quantile of the maximum $M \overset{\text{def}}{=} \sup_{0 \le t \le 1} |B(t) - tB(1)|$ of the absolute value of the Brownian bridge (B is standard Brownian motion). Give a simulation estimate of q and an associated 95% bootstrap confidence interval.[4]

6   Variance/Bias Trade-Off Issues

We have seen earlier in this chapter that estimating the variance parameter σ^2 that arises in quantile estimation (see Section 4a) involves computing the density of the r.v. Z. Of course, density estimation is also of interest in its own right.

The density-estimation problem is representative of an important class of computational problems, in which the object to be calculated cannot be expressed directly in terms of the expectation of a simulatable r.v. Rather, in the density-estimation setting, the density is defined as a limit of simulatable r.v.'s, specifically

$$f(q) = \lim_{h \downarrow 0} \frac{\mathbb{E}\mathbb{1}\{q < Z \le q + h\}}{h} \overset{\text{def}}{=} \lim_{h \downarrow 0} \mathbb{E}D(h) . \tag{6.1}$$

More generally, there are many problems in which the quantity of interest can be expressed only as a limit (in expectation) of simulatable r.v.'s.

If W is an estimator of w, the quantity $\mathbb{E}W - w$ is called the *bias* of W. The estimator is said to be *unbiased* for w if the bias is zero; otherwise,

---

[4]The distribution of M is known, but the form is complicated, see [98] p. 335.

the estimator is *biased*. The bias measures the systematic error in W that cannot be eliminated through sampling i.i.d. copies of W. In the setting of (6.1), it is clear that $D(h)$ is biased for each $h > 0$. However, the bias disappears as $h \downarrow 0$, creating a preference for estimating $f(q)$ via repeated sampling of the r.v. $D(h)$ with h small.

On the other hand,

$$\mathbb{V}ar\, D(h) \;=\; \frac{1}{h^2}\big(F(q+h) - F(q)\big)\big(1 - F(q+h) + F(q)\big),$$

which is asymptotic to $f(q)/h$ as $h \downarrow 0$. Thus, the variance blows up as $h \downarrow 0$. This suggests that the right choice for h must involve some kind of "trade-off" between variance and bias. It is this variance/bias trade-off that is the key to the design of a good simulation scheme when one is presented with computing quantities that are expressible only as limits of "simulatable expectations".

Remark 6.1 Note that if the bias is analytically computable, one would subtract an estimator of the bias off the original estimator. On the other hand, if the bias cannot be estimated and the bias is of order $R^{-1/2}$ or more, then the estimator's standard deviation $\sigma/\sqrt{R}$ is dominated by the bias, so the only reasonable candidate $\widehat{z}_R \pm 1.96\widehat{\sigma}_R/\sqrt{R}$ for the confidence interval is centered at the wrong point (due to the bias), and the resulting coverage may be disastrous. Thus, one has the following general principle: *the standard deviation should dominate the bias* in order that confidence intervals be meaningful.   ☐

To make concrete the typical theoretical approach used to study the variance/bias trade-off, consider the density estimator $\sum_1^R D_r(h)/R$, where the $D_r(h)$ are i.i.d. replications of $D(h)$. For a given simulation budget (corresponding to the number of simulation runs one is willing to do), the goal is to select h optimally.

Of course, the notion of optimality depends on the objective that is chosen. The most analytically tractable measure of estimator quality is that of mean square error (MSE). A key to the tractability of MSE is that

$$\mathrm{MSE}(W) \;=\; \mathbb{V}ar\, W + \big(\mathrm{bias}(W)\big)^2.$$

In our density-estimation example,

$$\mathrm{MSE}\Big(R^{-1}\sum_1^R D_r(h)\Big) \;=\; \frac{1}{R}\mathbb{V}ar\, D(h) + \big[\mathrm{bias}\big(D(h)\big)\big]^2.$$

Note that the second term is unaffected by sending $R \to \infty$, thereby offering further support to our prior assertion that the bias measures the systematic error associated with the estimator. To proceed further, we need to study the bias of $D(h)$. Note that

$$\mathrm{bias}\big(D(h)\big) \;=\; \frac{F(q+h) - F(q) - hF'(q)}{h} \;\sim\; \frac{h}{2}F''(q) = \frac{h}{2}f'(q).$$

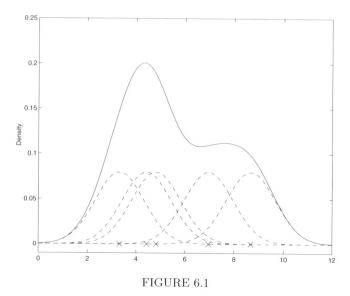

FIGURE 6.1

Given our analysis of $\text{Var}\, D(h)$, this suggests that

$$\text{MSE}\!\left(R^{-1}\sum_{1}^{R} D_r(h)\right) \;\approx\; \frac{f(q)}{hR} \;+\; \frac{h^2}{4}f'(q)^2\,. \tag{6.2}$$

For a given R, we should select h so that it roughly balances the terms on the r.h.s. of (6.2). More precisely, the function $a/h + bh^2$ of h is minimized for $h = (a/2b)^{1/3}$, so letting $a = f(q)/R$, $b = f'(q)^2/4$, it follows that (6.2) is minimized by taking h of order $R^{-1/3}$, in which case the MSE goes to 0 at rate $R^{-2/3}$. For further calculations of this type, see VII.1.

A frequently used alternative to the density estimator $D(h)$ above is a *kernel estimator*, defined as

$$\widehat{f_k}(x) \;=\; \frac{1}{hR}\sum_{r=1}^{R} k\big((Z_r - x)/h\big)\,,$$

where the kernel k is a probability density (often just the standard normal). An example of a kernel estimate for a density is given in Figure 6.1 for $R = 5$ observations and $h = 0.5$. Note that in practice, the choice of h is crucial for the quality of the density estimate; too large an h will yield an oversmoothed estimate, too large an h a too jaggy one.

7   Multivariate Output Analysis

Our discussion thus far has focused on a generic situation in which our goal is to compute some scalar quantity (e.g., $z = \mathbb{E}Z$). We briefly describe here the generalization to the setting in which we wish to compute a vector. For concreteness, suppose our goal is to compute the multivariate quantity $z = \mathbb{E}Z$, where $Z \in \mathbb{R}^d$.

From an estimation viewpoint, the problem is straightforward: we clearly should estimate the ith component z_i of z via $\widehat{z}_{iR}$, where $\widehat{z}_{iR} \stackrel{\text{def}}{=} R^{-1}(Z_{i1} + \cdots + Z_{iR})$ and $Z_r = (Z_{1r}, \dots, Z_{dr})$ is the rth copy of the r.v. Z generated. However, the construction of an asymptotic confidence region for z involves new ideas.

Let

$$\widehat{\Sigma} \stackrel{\text{def}}{=} \frac{1}{R-1} \sum_{r=1}^{R} (Z_r - \widehat{z})^{\mathsf{T}} (Z_r - \widehat{z})$$

be the sample covariance matrix associated with the R multivariate outputs $Z_1, \dots, Z_R$. Then $\widehat{\Sigma} \stackrel{\text{a.s.}}{\to} \Sigma$ as $R \to \infty$, where Σ is the $d \times d$ covariance matrix $\mathbb{E}[(Z - z)^{\mathsf{T}}(Z - z)]$. It follows that

$$(\widehat{z}_R - z)^{\mathsf{T}} \widehat{\Sigma}^{-1} (\widehat{z}_R - z) \stackrel{\mathscr{D}}{\to} \chi_d^2$$

as $R \to \infty$, so that if x is the $1 - \alpha$ quantile in the χ_d^2 distribution and E the ellipsoid

$$E \stackrel{\text{def}}{=} \{w : (\widehat{z}_R - w)^{\mathsf{T}} \widehat{\Sigma}^{-1} (\widehat{z}_R - w) \leq x\},$$

then

$$\mathbb{P}(z \in E) \to 1 - \alpha. \tag{7.1}$$

In other words, the ellipsoid E contains z with probability converging to $1 - \alpha$ as $R \to \infty$. This is the desired confidence region in the above multivariate context.

Remark 7.1 When sectioning is applied to the d-dimensional context, $\psi(F)$ is then d-dimensional, as is $\psi(\widehat{F}_R)$ and the $\psi(\widehat{F}_{n,K})$, $n = 1, \dots, N$. The analogue of (7.1) is then

$$\mathbb{P}(\psi(F) \in E_R) \to 1 - \alpha,$$

where E_R is the d-dimensional ellipsoid

$$E_R \stackrel{\text{def}}{=} \left\{ w : N(\psi(\widehat{F}_R) - w)^{\mathsf{T}} \widehat{S}_N^{-1} (\psi(\widehat{F}_R) - w) \leq d \frac{N-1}{N-d} f \right\},$$

f is the $1 - \alpha$ quantile of an F distribution with $(d, N - d)$ degrees of freedom, and

$$\widehat{S}_N \stackrel{\text{def}}{=} \frac{1}{N-1} \sum_{n=1}^{N} (\psi(\widehat{F}_{n,K}) - \psi(\widehat{F}_R))^{\mathsf{T}} (\psi(\widehat{F}_{n,K}) - \psi(\widehat{F}_R)). \qquad \square$$

8   Small-Sample Theory

The confidence intervals described in the preceding sections of this chapter are asymptotic in nature, and depend on the use of limit theorems. In particular, all the asymptotic confidence intervals of this chapter depend, in a fundamental way, on the central limit theorem.

Note that the normal approximation associated with the CLT is exact if the underlying distribution of the Z_r is normal. Thus, the error in the normal approximation to the distribution of $\widehat{z}$ can be attributed to nonnormality in the underlying distribution of the Z_r. As one might expect, the more nonnormal this distribution, the slower the convergence to the normal distribution in the CLT. This is quantified through the Edgeworth expansion ([116, p. 538])

$$\mathbb{P}\big(R^{1/2}(\widehat{z}_R - z) \le \sigma x\big)$$
$$= \Phi(x) + \frac{\mathbb{E}(Z - z)^3}{6\sigma^3\sqrt{R}}(1 - x^2)\varphi(x) + o(1/\sqrt{R}). \qquad (8.1)$$

Hence, the larger the skewness $\mathbb{E}(Z - z)^3/\sigma^3$, in absolute terms, the slower the rate of convergence. Such Edgeworth-type expansions hold for many complex estimators, including those that arise in computing a smooth function of expectations; see, for example, Battacharya & Ghosh [43]. Similar confidence-interval concerns arise there.

Some Monte Carlo computations give rise to large skewness coefficients, thereby leading to slow rate of convergence in the CLT. These slow rates manifest themselves, in the confidence interval context, through poor coverage probabilities. For example, it may be that a sample size R on the order of thousands is required in order that a 95% asymptotic confidence interval cover z more than 90% of the time (although, thankfully, usually smaller sample sizes suffice). An example is sampling from the Bernoulli(p) distribution with p small, for example of order 10^{-3} or less, as occurs in rare event simulation implemented via the crude Monte Carlo method; see Chapter VI.

Given the fact that the convergence rate in the CLT (and the associated coverage of the confidence intervals) is problem dependent, examining the degree of nonnormality in the underlying sampling population provides one means of assessing the magnitude of the problem. An alternative is to build a confidence-interval methodology with guaranteed performance (i.e., a 95% confidence interval is guaranteed to cover at least 95% of the time). Such intervals can be constructed when Z is a bounded r.v. (with known bounds) but typically leads to intervals that are significantly "fatter" than those based on asymptotic theory; see, for example, Fishman [118] and Exercise 8.1.

The bootstrap presents one more means of addressing this problem. While the bootstrap procedure described in Section 5c does not improve upon coverage error, a modified bootstrap procedure does so (in some settings), and often significantly improves upon coverage at a practical level as well. The idea is not to bootstrap the r.v. $\psi(\widehat{F}_R) - \psi(F)$ but instead to bootstrap the standardized (Studentized) r.v. $\left(\psi(\widehat{F}_R) - \psi(F)\right)/\widehat{\sigma}_R$. Thus, from a bootstrap sample $Z_{1b}^*, \ldots, Z_{Rb}^*$ one generates $\left(\psi(\widehat{F}_{Rb}^*) - \psi(\widehat{F}_R)\right)/\widehat{\sigma}_{Rb}^*$, where F_{Rb}^* is the empirical distribution of the bootstrap sample and $\widehat{\sigma}_{Rb}^*$ an estimate of the standard deviation for the problem based on $Z_{1b}^*, \ldots, Z_{Rb}^*$. When $\psi(F)$ is the mean of F (so that we are computing a confidence interval for $\mathbb{E}Z$), this procedure can be shown to improve asymptotic coverage relative to the standard normal confidence interval (1.4).

One also sometimes encounters the suggestion of using the t distribution as a basis for confidence intervals, leading to $\widehat{z} \pm s\,t_{1-\alpha/2}/\sqrt{R}$, where t_α is the α-quantile of the t distribution with $f = R - 1$ degrees of freedom. The motivation behind this is that the normalized average $(\widehat{z}_R - z)/s\sqrt{R}$ of normal r.v.'s is t_f distributed. However, since the sampling distribution of the Z_r is typically highly nonnormal, our view is that t confidence intervals very rarely improve upon normal ones. We will, however, encounter one very convincing example in connection with batch means; see IV.5.

Exercises

8.1 (TP) Explain how Chebyshev's inequality $\mathbb{P}\left(|\widehat{z} - z| > \varepsilon\right) \leq \mathbb{V}ar(Z)/R\varepsilon^2$ leads to the asymptotic 95% confidence $\widehat{z} \pm 4.47s/\sqrt{R}$.

9   Simulations Driven by Empirical Distributions

Most often, the r.v. Z in the CMC method is produced from other r.v.'s belonging to parametric distributions. Say that in a queuing simulation one uses a Gamma service time distribution. In this example, the exact form is, however, seldom known but some statistical observations are available, and it is then appealing not to fit a parametric distribution but to simulate directly from the empirical distribution of the observed service times. A terminology that is often used in such a situation is *trace-driven simulation* or *resampling*. In mathematical finance, one often talks about *historical simulation*; see, for example, McNeil et al. [252, Section 2.3.2].

To illustrate how to perform output analysis in such situations, we consider a simple case in which $x_1, \ldots, x_m$ are i.i.d. observations from an unknown distribution F. The aim is to estimate $\psi(F) = \mathbb{E}_F g(X_1, X_2, \ldots)$, where $X_1, X_2, \ldots$ are i.i.d. with distribution F, by simulation based on drawings from the empirical distribution $\widehat{F}_m$ of $x_1, \ldots, x_m$ ($\widehat{F}_m$ is the distribution putting mass $1/m$ at each x_k). Say g is the length of the busy period in a D/G/1 queue with service-time distribution F and constant interarrival times, or $\max_{1 \leq n \leq N} W_n$, where R is fixed and W_n the waiting time of customer n.

The naive procedure is to use Monte Carlo with

$$Z_r \overset{\text{def}}{=} g\big(x_{K(r,1)}, x_{K(r,2)}, \dots \big), \quad r = 1, \dots, R,$$

where the $K(r,j)$ are i.i.d. and uniform on $\{1, \dots, m\}$. The problem is that thereby one estimates $z \overset{\text{def}}{=} \mathbb{E}Z = \psi(\widehat{F}_m)$, so that the confidence interval produced in this way is a confidence interval for $\psi(\widehat{F}_m)$, not $\psi(F)$ as desired: the stochastic variation in $\psi(\widehat{F}_m)$ is ignored.

To resolve this problem, note that typically $\psi(\widehat{F}_m)$ has an asymptotic normal distribution with mean $\psi(F)$ and variance of the form ω^2/m for some ω^2 as $m \to \infty$. More precisely, as in Section 5a and references there, the program for rigorously verifying this is to find a Banach space $(V, \|\cdot\|)$ of (signed) measures such that one is willing to assume that $F \in V$, that $\widehat{F}_m \in V$, and that ψ is defined on all of V and Fréchet or Hadamard differentiable at F.

It follows that in the Monte Carlo setting, we can write

$$\widehat{z} = \psi(\widehat{F}_m) + \frac{\sigma_{\widehat{F}_m}}{\sqrt{R}} V_1 = \psi(F) + \frac{\sigma_{\widehat{F}_m}}{\sqrt{R}} V_1 + \frac{\omega}{\sqrt{m}} V_2,$$

where V_1, V_2 are independent and asymptotically standard normal when both R and m are large. To produce a confidence interval, we need therefore an additional estimate of ω^2. In some cases, ω^2 has been evaluated analytically, but typically, an estimate needs to be produced by simulation and to this end a variant of the CMC method is required.

The idea is similar to sectioning. We divide the m observations into groups, say k groups of size ℓ (assuming for convenience that m can be written as $m = k\ell$) and perform p (say) simulations within each group, using the empirical distribution $\widehat{F}_{\ell,i}$ in group i. For each group, we thus in a straightforward way obtain an estimate $\widehat{\psi}(\widehat{F}_{\ell,i})$ of $\psi(\widehat{F}_{\ell,i})$ and an associated estimate $\widehat{\sigma}^2_{\widehat{F}_{\ell,i}}$ of $\sigma^2_{\widehat{F}_{\ell,i}}$. The estimator of $\psi(F)$ is

$$\widehat{\psi} = \frac{1}{k}\left(\widehat{\psi}(\widehat{F}_{\ell,1}) + \cdots + \widehat{\psi}(\widehat{F}_{\ell,k}) \right).$$

We can write

$$\widehat{\psi}(\widehat{F}_{\ell,i}) = \psi(\widehat{F}_{\ell,i}) + \frac{\sigma_{\widehat{F}_{\ell,i}}}{\sqrt{p}} V_{1,i} = \psi(F) + \frac{\sigma_{\widehat{F}_{\ell,i}}}{\sqrt{p}} V_{1,i} + \frac{\omega}{\sqrt{\ell}} V_{2,i},$$

where the X_{ij} are independent and asymptotically standard normal when both p and ℓ are large. When ℓ is large, we can replace $\sigma^2_{\widehat{F}_{\ell,i}}$ by $\sigma^2 \overset{\text{def}}{=} \sigma^2_F$, and so the asymptotic variance of $\widehat{\psi}$ becomes

$$\frac{\sigma^2}{kp} + \frac{\omega^2}{k\ell}. \tag{9.1}$$

The natural estimates of σ^2, ω^2 are

$$\widehat{\sigma}^2 = \frac{1}{k} \left(\widehat{\sigma}^2_{F_{\ell,1}} + \cdots + \widehat{\sigma}^2_{F_{\ell,k}} \right), \quad \widehat{\omega}^2 = \frac{\ell}{k-1} \sum_{i=1}^{k} \left(\widehat{\psi}(F_{\ell,i}) - \widehat{\psi} \right)^2,$$

and so the confidence interval is

$$\widehat{\psi} \pm 1.96 \sqrt{\frac{\widehat{\sigma}^2}{kp} + \frac{\widehat{\omega}^2}{k\ell}}. \tag{9.2}$$

An obvious question is how k and p should be chosen based on a budget of c drawings from an empirical distribution. Clearly, $c = kp$, so since both c and $m = k\ell$ are fixed, (9.1) shows that in terms of minimizing the variance, the answer is that the choice is unimportant. However, consider next the variance of the variance estimator. Since $\mathbb{V}ar(\widehat{\sigma}^2) \sim c_1/kp = c_1/c$, $\mathbb{V}ar(\widehat{\omega}^2) \sim c_2\ell/k$,

$$\mathbb{V}ar\left(\frac{\widehat{\sigma}^2}{kp} + \frac{\widehat{\omega}^2}{k\ell} \right) \sim \frac{c_1}{c^3} + \frac{c_2\ell^2}{m^3}.$$

This indicates that choosing ℓ small or, equivalently, the number of groups k large is preferable. But note that the largest possible choice $k = m$ is not feasible because the asymptotics used in the arguments require that also $\ell = m/k$ be sufficiently large for the CLT for $\psi(F_\ell)$ to be in force.

Exercises

9.1 (A) The Pollaczeck–Khinchine formula for the expected waiting time w (say) in the M/G/1 queue states that

$$w = w(F) = \frac{\rho \mathbb{E}V^2}{2(1 - \rho)\mathbb{E}V},$$

where ρ is the traffic intensity and V an r.v. having the service-time distribution F.
For $\rho = 0.8$ and service-time density xe^{-x}, simulate m service times $v_{r,1}, \ldots, v_{r,m}$ for $r = 1, \ldots, R$, compute (without simulation!) the w_r corresponding to the empirical distributions $\widehat{F}_{r,m}$, and plot a histogram of the w_r. Experiment with different values of m to see how big m needs to be for the distribution of $w(\widehat{F}_m)$ to look reasonably normal.

10   The Simulation Budget

At first sight, it may appear that the variance is the universal measure of efficiency of an MC estimator. In particular, given the choice between two Monte Carlo schemes based on r.v.'s Z, Z' with variances $\sigma_Z^2, \sigma_{Z'}^2$, one should choose the one with smallest variance. However, this argument cheats because it does not take into account that the expected CPU times

T, T' required to generate one replicate may be very different. Instead, one can formulate the problem in terms of a constraint on the *simulation budget*: given that we are prepared to spend c units of CPU time for the simulation, will Z or Z' give the lower variance? The answer is that by renewal theory, the number of replications obtained within time c will be $R \sim c/T$, respectively $R' \sim c/T'$, and so the variances on the estimates are $\sigma_Z^2 T/c$, respectively $\sigma_{Z'}^2 T'/c$ (that the appropriate CLT holds also for such random sample sizes follows from Anscombe's theorem; cf. A2). Thus, we should prefer Z if $\sigma_Z^2 T < \sigma_{Z'}^2 T'$ and Z' otherwise.

The interpretation of $\sigma_Z^2 T$ is *variance per unit CPU time*, and the inverse $1/\sigma_Z^2 T$ may be taken as a measure of efficiency of the estimator Z. To quote Hammersley & Handscombe [173], "The efficiency of a Monte Carlo process may be taken as inversely proportional to the product of the sampling variance and the amount of labour expended in obtaining this estimate."

It is of course rarely the case that the simulator will specify c in advance. Nevertheless, considerations like the above are appealing from a conceptual point of view. A general decision-theoretic formulation was given by Glynn & Whitt [156] and involves the following ingredients in the estimation of a number z by simulation:

(i) A *loss function* $L(y, z)$, specifying the loss in estimating z by y;

(ii) the *experiment*, a stochastic process $\{(Y(t), C(t))\}_{t \geq 0}$ with t representing simulated time (*not* CPU time), $Y(t)$ the corresponding estimator of z, and $C(t)$ the cost of obtaining the estimator $Y(t)$;

(iii) a *budget constraint* c;

(iv) the *realized length* $T(c) \stackrel{\text{def}}{=} \sup\{t : C(t) \leq c\}$;

(v) the *budget-constrained estimator* $Y(T(c))$;

(vi) the *risk function* $R(c) \stackrel{\text{def}}{=} \mathbb{E} L(Y(T(c)), z)$;

(vii) the *efficiency* $e(c) \stackrel{\text{def}}{=} 1/R(c)$.

The idea is to take properties of $R(c)$ and $e(c)$ for large c as measures of performance of a given simulation scheme.

Example 10.1 The loss function is often the mean squared error $\mathbb{E}(Y(t) - z)^2$, which reduces to the variance in the unbiased case $\mathbb{E} Y(t) = z$, and the cost $C(t)$ is often the CPU time. In the i.i.d. Monte Carlo setting, $t = R$ is the number of replications, $C(t)$ is a random walk, $T(c)$ is roughly proportional to c, $T(c) \propto c$ by renewal theory, and hence $R(c) \propto 1/c$, $e(c) \propto c$. Different cases occur, for example, in Gaussian processes in which the cost to generate a sample path of length t is $O(t^3)$ for some algorithms and $O(t \log t)$ for others; see Chapter XI. Then $T(c) \propto c^{1/3}$, respectively $T(c) \propto t/\log t$ (which is not necessarily an indication that the $O(t \log t)$

algorithms are preferable: in fact, some will typically have a larger bias and hence a larger loss; see again Chapter XI). □

The large-c asymptotics will often be of the form

$$c^r R(c) \;\to\; 1/v \tag{10.1}$$

as $c \to \infty$, where $r, v > 0$ are constants. When comparing two estimators with different characteristics (r_1, v_1), (r_2, v_2), one therefore prefers the one with the larger r and, if $r_1 = r_2$, the one with the larger v. A statement of type (10.1) will typically be based on a functional CLT of the form

$$\left\{\varepsilon^{-\gamma}\big(Y(t/\varepsilon) - z\big)\right\}_{t \geq 0} \;\xrightarrow{\mathscr{D}}\; \{W(t)\}_{t \geq 0} \quad \text{in } D[0, \infty) \tag{10.2}$$

as $\varepsilon \downarrow 0$, where W is often Brownian motion but can also be another Gaussian process, a stable process, fractional Brownian motion, or another process. Similarly, it is often reasonable to assume that

$$t^{-\beta} C(t) \;\xrightarrow{\mathbb{P}}\; 1/\lambda \tag{10.3}$$

as $t \to \infty$ for some constant $0 < \lambda < \infty$. We then get

$$c \;\approx\; C\big(T(c)\big) \;\approx\; T(c)^\beta/\lambda, \quad T(c) \;\approx\; (\lambda c)^{1/\beta},$$

and, taking $\varepsilon = 1/T(c)$ in (10.2),

$$c^{\gamma/\beta}\big[Y\big(T(c)\big) - z\big] \;\xrightarrow{\mathscr{D}}\; \lambda^{-\gamma/\beta} W(1); \tag{10.4}$$

the rigorous proof requires the additional assumption that W is continuous at $t = 1$. Taylor expanding L around $y = z$ and making the (reasonable) assumptions that $L(z, z) = 0$, $L_y(z, z) = 0$ (partial derivative) then yields (10.1) with

$$r = 2\beta/\gamma, \quad v = \frac{2\lambda^{2\beta/\gamma}}{L_{yy}(z)\mathbb{E}W(1)^2}. \tag{10.5}$$

Exercises

10.1 (TP) What is your guess of the analogue of (10.5) for the $C(t) = O(t \log t)$ situation in Gaussian processes?

10.2 (TP) What is your guess of the analogue of (10.5) for $L(y, z) = |y - z|^p$?

Chapter IV
Steady-State Simulation

1 Introduction

Let $Y = \{Y(t)\}_{t \geq 0}$ be a stochastic process in continuous time.[1] Suppose that the time-average limit for Y exists, in the sense that there exists a (deterministic) constant z such that

$$\frac{1}{t} \int_0^t Y(s) \, ds \xrightarrow{\mathbb{P}} z \qquad (1.1)$$

as $t \to \infty$. The steady-state simulation problem is concerned with computing z.

Given the limit (1.1), the most natural estimator for z is to simulate Y up to time t, and to use the time-average

$$\overline{Y}(t) \stackrel{\text{def}}{=} \frac{1}{t} \int_0^t Y(s) \, ds$$

as the estimator for z. There are two principal difficulties inherent in this approach:

   (i) The estimator $\overline{Y}(t)$ is generally biased as a consequence of an initial transient that is induced by the fact that the process is usually initialized from a nonequilibrium distribution.

---

[1] The small differences that arise in discrete time are discussed in Section 6a.

(ii) Because $\overline{Y}(t)$ is based on a single realization of the process, computing variance estimates is problematic, thereby creating challenges in the construction of confidence intervals.

In many applications, $Y(t)$ can be represented as $Y(t) = f\big(X(t)\big)$, where f is a real-valued function defined on the state space E of a Markov process X. This is the setting on which we focus in this section and Sections 2, 4, and which is the one arising in most applications; we comment on the generality in Section 6. When X is suitably positive recurrent, (1.1) then typically follows as a consequence of a stronger ergodic theorem, asserting that there exists a (deterministic and independent of the initial distribution) probability distribution $\pi(\cdot)$ for which

$$\frac{1}{t}\int_0^t g\big(X(s)\big)\,\mathrm{d}s \;\xrightarrow{\;\mathbb{P}\;}\; \int_E g(x)\,\pi(\mathrm{d}x) \tag{1.2}$$

as $t \to \infty$, for all bounded and/or nonnegative functions g. Furthermore, the distribution π generally satisfies the integral equation

$$\pi(\mathrm{d}y) \;=\; \int_E \mathbb{P}\big(X(t) \in \mathrm{d}y \,\big|\, X(0) = x\big)\,\pi(\mathrm{d}x) \tag{1.3}$$

for $y \in E$ and $t \geq 0$. Because of (1.3), the distribution π is variously called the *equilibrium distribution*, the *invariant distribution*, and the *stationary distribution*. In addition, π is often called the *steady-state distribution*, in part because of its interpretations as a limit in (1.2).

In view of (1.2), the quantity z can be expressed as

$$z \;=\; \int_E f(x)\,\pi(\mathrm{d}x)\,.$$

It follows that if one can generate variates $X(\infty)$ from the distribution π, then $z = \mathbb{E}Z$, where $Z \overset{\text{def}}{=} f\big(X(\infty)\big)$, reducing the computation of z to the problem studied in Chapter III. If E is low-dimensional and π is known explicitly, the methods of Chapter II may be applicable to efficient random variate generation for $X(\infty)$. On the other hand, if π is known and E is high-dimensional, it may be that computing π through the time average (1.2) is more efficient than available algorithms based on direct variate generation from π itself.

More commonly, π is not known explicitly. Perhaps surprisingly, there exist algorithms (for certain classes of Markov chains) that offer the ability to generate variates from the distribution π based on one's ability to dynamically simulate paths of the process X. In Sections 8 and XIV.7b, we study the extent to which this is possible for Markov chains with a discrete state space. The answer is that algorithms exist for E finite ($|E| < \infty$) but not in general for E countably infinite.

The case $|E| < \infty$ is, however, special, and even there the algorithms are often prohibitively inefficient in terms of computer time. The most commonly used algorithm for computing such steady-state quantities z are therefore based on the LLN (1.1).

Remark 1.1 An alternative view of the of steady-state simulation is that the LLN (1.1) provides a simulation-based method for solving the integral equation (1.1), in particular for computing an "inner product" $\int_E f(x)\,\pi(\mathrm{d}x)$. □

In great generality, the LLN (1.1) is accompanied by a central limit theorem (CLT) taking the form

$$t^{1/2}\big(\overline{Y}(t) - z\big) \xrightarrow{\mathscr{D}} \mathscr{N}(0,\sigma^2), \tag{1.4}$$

where σ^2 is a (deterministic) constant, called the TAVC (time average variance constant) in the following. So,

$$\overline{Y}(t) \stackrel{\mathscr{D}}{\approx} z + \frac{\sigma}{\sqrt{t}} V,$$

where $V \sim \mathscr{N}(0,1)$, when t is large. Assuming appropriate uniform integrability,

$$\mathbb{V}ar\,\overline{Y}(t) \sim \sigma^2/t \quad \text{and} \quad \big(\text{bias}(\overline{Y}(t))\big)^2 = \mathrm{o}(1/t)$$

as $t \to \infty$. It follows that the bias of $\overline{Y}(t)$ is small relative to the standard deviation when t is large.

Remark 1.2 To get a better sense of the relative magnitude of the bias and the standard deviation, suppose that the underlying Markov process X is suitably aperiodic, so that

$$\mathbb{P}\big(X(t) \in \cdot\big) \xrightarrow{\mathscr{D}} \pi(\cdot)$$

as $t \to \infty$. Again, assuming suitable uniform integrability,

$$\mathbb{E}f\big(X(t)\big) \to z$$

as $t \to \infty$. Suppose that the convergence is rapid enough so that the integral

$$b \stackrel{\text{def}}{=} \int_0^\infty \big[\mathbb{E}f\big(X(s)\big) - z\big]\,\mathrm{d}s \tag{1.5}$$

converges absolutely. Then

$$\mathbb{E}\overline{Y}(t) - z = \frac{b}{t} - \frac{1}{t}\int_t^\infty \big[\mathbb{E}f\big(X(s)\big) - z\big]\,\mathrm{d}s = \frac{b}{t} + \mathrm{o}(1/t), \tag{1.6}$$

so that the bias is an order of magnitude smaller than the estimator's $\mathrm{O}(t^{-1/2})$ sampling variability.

Returning to the absolute convergence of the integral (1.5), note that the rate at which $\mathbb{E}\overline{Y}(t) - z$ converges to zero is related to the mixing rate of X (mixing rates are measures of the rate at which the autocorrelations decay). In fact, for Markov processes that mix exponentially rapidly,

$$\left| \mathbb{E}_x f\big(X(t)\big) - \mathbb{E} f\big(X(\infty)\big) \right| \; \leq \; c(x) \mathrm{e}^{-\lambda t} \tag{1.7}$$

for some function $c(\cdot)$ and positive λ, where $\mathbb{E}_x(\cdot) \stackrel{\mathrm{def}}{=} \mathbb{E}\big[\cdot \mid X(0) = x\big]$ (for precise conditions guaranteeing (1.7), see Meyn & Tweedie [255]). Hence,

$$\left| \mathbb{E}\overline{Y}(t) - z \right| \; \leq \; \frac{\mathrm{e}^{-\lambda t}}{\lambda t} \mathbb{E} c\big(X(0)\big), \tag{1.8}$$

so that

$$\mathrm{bias}\big(\overline{Y}(t)\big) \; = \; \frac{b}{t} + \mathrm{O}\big(\mathrm{e}^{-\lambda t}/t\big). \tag{1.9}$$

Thus, in contrast to the smooth estimation problem of III.3, the bias of $\overline{Y}(t)$, when written as a power series in $1/t$, has vanishing coefficients except for the first term. □

In the following, we write

$$f_c(x) \; \stackrel{\mathrm{def}}{=} \; f(x) - z \; = \; f(x) - \pi(f), \tag{1.10}$$

$$c(s) \; \stackrel{\mathrm{def}}{=} \; \mathbb{C}ov_\pi\big(Y(0), Y(s)\big) \; = \; \mathbb{E}_\pi\big[f_c\big(X(0)\big) f_c\big(X(s)\big)\big], \tag{1.11}$$

$$\rho(s) \; \stackrel{\mathrm{def}}{=} \; \mathbb{C}orr_\pi\big(Y(0), Y(s)\big) \; = \; \frac{\mathbb{E}_\pi\big[f_c\big(X(0)\big) f_c\big(X(s)\big)\big]}{\sqrt{\mathbb{E}_\pi f_c^2\big(X(0)\big) \mathbb{E}_\pi f_c^2\big(X(s)\big)}}$$

$$= \; \frac{\mathbb{E}_\pi\big[f_c\big(X(0)\big) f_c\big(X(s)\big)\big]}{\mathbb{E}_\pi f_c^2\big(X(0)\big)} \; = \; \frac{c(s)}{c(0)}, \tag{1.12}$$

$$\sigma_0^2 \; \stackrel{\mathrm{def}}{=} \; \mathbb{V}ar\big[f\big(X(\infty)\big)\big] \; = \; \mathbb{V}ar_\pi Y(0). \tag{1.13}$$

That is, $c(s)$ is the steady-state covariance at lag s and $\rho(s)$ the correlation. Returning again to the CLT (1.4), it seems reasonable to expect (in view of the fact that the initial transient affects only the initial segment of the process) that the same CLT should hold under any initial distribution for X, specifically its stationary distribution π. This leads to the following:

Proposition 1.3 *Assume* $\int_0^\infty |\rho(s)|\,\mathrm{d}s \; < \; \infty$ *and* $\sigma_0^2 < \infty$. *Then the TAVC* σ^2 *in* (1.4) *is given by*

$$\sigma^2 \; = \; \lim_{t \to \infty} t\, \mathbb{V}ar\,\overline{Y}(t) \; = \; 2\sigma_0^2 \int_0^\infty \rho(s)\,\mathrm{d}s.$$

Proof. Write $c(s, v) \stackrel{\text{def}}{=} \mathbb{C}ov_\pi\big(Y(s), Y(v)\big)$ and note that $c(s, v) = c(|s - v|)$. Hence

$$
\begin{aligned}
t \, \mathbb{V}ar_\pi \, \overline{Y}(t) &= \frac{1}{t} \, \mathbb{E}_\pi \left(\int_0^t f_c\big(X(s)\big) \, ds \right)^2 \\
&= \frac{1}{t} \, \mathbb{E}_\pi \int_0^t \int_0^t f_c\big(X(s)\big) f_c\big(X(v)\big) \, ds \, dv \\
&= \frac{1}{t} \int_0^t \int_0^t c(s, v) \, ds \, dv = \frac{2}{t} \int_0^t ds \int_0^s c(s, v) \, dv \\
&= \frac{2}{t} \int_0^t ds \int_0^s c(s - v) \, dv = 2 \int_0^t (1 - s/t) c(s) \, ds \\
&\to 2 \int_0^\infty c(s) \, ds .
\end{aligned}
$$

Since $c(s) = \sigma_0^2 \rho(s)$, the result follows. $\square$

Note that if we simulate X for $\sigma^2 t$ time units, then (1.4) establishes that $\mathbb{V}ar\big(\overline{Y}(\sigma^2 t)\big) \sim 1/t$ as $t \to \infty$. Similarly, if we had the ability to simulate $\lfloor \sigma_0^2 t \rfloor$ i.i.d. replicates of $f\big(X(\infty)\big)$, the resulting sample mean would have variance $1/t$. So, the factor

$$
\frac{\sigma^2}{\sigma_0^2} = 2 \int_0^\infty \rho(s) \, ds \tag{1.14}
$$

has an interpretation as the amount of time one must simulate X in order that the time average over $[0, \sigma^2/\sigma_0^2]$ contain as much information as a single sample of $f\big(X(\infty)\big)$. Therefore, the factor σ^2/σ_0^2 given by (1.14) is of great interest in the steady-state simulation context.

In addition, it is evident that the steady-state variance σ_0^2 can be easily estimated via the sample quantity

$$
\frac{1}{t} \int_0^t f\big(X(s)\big)^2 \, ds - \overline{Y}(t)^2 .
$$

Therefore, the entire difficulty in estimating the TAVC σ^2 (and hence in producing a confidence interval for z based on (1.4)) is the need to estimate the factor (1.14).

1a   The Relaxation Time

To get a handle on the steady-state autocorrelation function $\rho(\cdot)$, we take advantage of the fact that for Markov chains exhibiting exponentially rapid mixing, one can typically refine (1.7) to

$$
\mathbb{E}_x f\big(X(t)\big) - z \sim \widetilde{c}(x) t^{k-1} e^{-\gamma t} \tag{1.15}
$$

as $t \to \infty$, for some suitable function $\widetilde{c}(\cdot)$, some positive constant γ (that clearly must be larger than the λ of (1.7)), and some k. For example, when X is an irreducible finite-state continuous-time Markov process, the constant γ is typically the eigenvalue of the rate matrix $\boldsymbol{A}$ having the second-largest real part (the principal eigenvalue is 0) and k its multiplicity. In queuing, more precisely the GI/G/1 queue, one has

$$\gamma \;=\; \min_{s>0} \mathbb{E}e^{s(V-T)}, \quad k = -1/2\,,$$

where V, T are independent such that V follows the service-time distribution and T the interarrival-time distribution; see [16, Section XIII.2].

It follows that

$$
\begin{aligned}
\rho(t) \;&=\; \frac{\mathbb{E}_\pi\big[f_c\big(X(0)\big)f_c\big(X(t)\big)\big]}{\mathbb{E}_\pi f_c^2\big(X(0)\big)} \\[2mm]
&=\; \frac{\mathbb{E}_\pi\big[f_c\big(X(0)\big)\mathbb{E}\big[f\big(X(t)\big) - z \,\big|\, X(0)\big]\big]}{\mathbb{E}_\pi f_c^2\big(X(0)\big)} \\[2mm]
&\sim\; \frac{\mathbb{E}_\pi\big[f_c\big(X(0)\big)\widetilde{c}\big(X(0)\big)\big]}{\mathbb{E}_\pi f_c^2\big(X(0)\big)}\, t^k e^{-\gamma t}\,.
\end{aligned}
\tag{1.16}
$$

Neglecting the presence of the constant multiplier in (1.16) and t^k, we conclude that the factor (1.14) is roughly of order $1/\gamma$. In other words, $1/\gamma$ is a rough measure of the amount of time X must be simulated in order to provide roughly the same amount of statistical information as that contained in a single sample from the system in equilibrium.

Of course, the asymptotic (1.15) suggests that the amount of time $t = t(\varepsilon)$ required in order that the bias satisfy

$$\big|\mathbb{E}_x f\big(X(t)\big) - z\big| \;\leq\; \varepsilon$$

obeys the asymptotic $t(\varepsilon) \sim \gamma^{-1} \log(1/\varepsilon)$ as $\varepsilon \downarrow 0$. Hence, the quantity $1/\gamma$ not only is critical to determining the TAVC σ^2 but also governs the rate at which the initial transient dissipates. In view of these considerations, the quantity $1/\gamma$ is of great interest in the context of steady-state simulation, since this constant is intimately connected to both of the principal difficulties described at the outset of this chapter.

The quantity $1/\gamma$ is often called the *relaxation time* of the system, since it represents a rough measure of how long it takes to relax to the equilibrium state (in other words, to *decorrelate*). As indicated above, in the finite case γ is typically characterized as the second-largest eigenvalue of the rate matrix $\boldsymbol{A}$ for the process. Consequently, bounds on this second eigenvalue are of great interest for purposes of planning steady-state simulations.

In general, such bounds can be very difficult to compute. Eigenvalue estimates for a matrix generally take advantage of the Rayleigh–Ritz characterization of its eigenvalues and eigenvectors (see Parlett [285]). This tool is most straightforward to apply in the context of symmetric matrices. Given that reversibility of a Markov chain is closely related to self-adjointness of $\boldsymbol{A}$, it is not surprising that the great majority of analytic bounds on the second eigenvalue assume reversibility. Reversibility is natural in some application areas such as mathematical physics, but it rarely arises in others such as performance engineering. On the other hand, in the Markov chain Monte Carlo context, sampling algorithms are typically designed so as to generate reversibility, see XIII.3.2 for details. In this latter context, analytic bounds on the second eigenvalue are sometimes explicitly computable and can be quite useful in planning steady-state simulations. Such bounds are discussed, for example, in Brémaud [57], and we also later outline a case study in which they are essential, see XIV.3.

Remark 1.4 A pragmatic approach that is often used to mitigate the initial transient problem is to use a *burn-in* or *warm-up* period of lenght v (say), hoping that the bias have been substantially reduced by time v. Instead of $\overline{Y}(t)$, the estimator of z based on simulation in $[0, t]$ is then

$$\frac{1}{t-v}\int_v^t Y(s)\,\mathrm{d}s\,.$$

Knowledge of γ may of course help to provide guidelines for how large v needs to be. In absence of such knowledge often more ad hoc rules are used, for example setting $v = s/10$. $\qquad\square$

2   Formulas for the Bias and Variance

When $X = \{X(t)\}_{t\geq 0}$ is a finite Markov process, explicit formulas for the bias and TAVC are available. Let $\boldsymbol{A} = \big(A(x, y)\big)_{x,y\in E}$ be the rate matrix and $\boldsymbol{P}(t) \stackrel{\mathrm{def}}{=} \big(P(t, x, y)\big)_{x,y\in E}$ the lag-t transition matrix of X, i.e., $P(t, x, y) = \mathbb{P}_x\big(X(t) = y\big)$. Assuming that $\boldsymbol{A}$ is irreducible, a unique stationary distribution π exists. Representing π as a row vector $\boldsymbol{\pi} \stackrel{\mathrm{def}}{=} \big(\pi(x)\big)_{x\in E}$, π is uniquely characterized by the equation $\boldsymbol{\pi A} = \boldsymbol{0}$ together with the normalization $\boldsymbol{\pi 1}$, where $\boldsymbol{1}$ is the column vector of ones. Further, if $\boldsymbol{\Pi} \stackrel{\mathrm{def}}{=} \boldsymbol{1\pi}$ is the matrix in which each row is identical to $\boldsymbol{\pi}$, then $\boldsymbol{P} \to \boldsymbol{\Pi}$ (exponentially rapidly) as $t \to \infty$ and

$$\boldsymbol{\Pi P}(t) \;=\; \boldsymbol{P}(t)\boldsymbol{\Pi} \;=\; \boldsymbol{\Pi}^2 \;=\; \boldsymbol{\Pi}\,.$$

For these and other relevant facts on finite Markov processes, see [16, Chapter II].

The initial bias of $\overline{Y}(t)$ is given by the following result, where $(\boldsymbol{\Pi} - \boldsymbol{A})^{-1}$ is known as the *fundamental matrix* (the existence of the inverse follows from the proof below):

Proposition 2.1 *Let $\boldsymbol{\mu}$ be the row-vector representation of the initial distribution. Then the constant b in (1.6), (1.9) is given by $b = \boldsymbol{\mu}(\boldsymbol{\Pi} - \boldsymbol{A})^{-1}\boldsymbol{f}_c$, where $\boldsymbol{f}_c$ is the column-vector representation of $f_c(\cdot) = f(\cdot) - z$.*

Proof. It is well known that the $\boldsymbol{P}(t)$ satisfy the backward equation $\boldsymbol{P}'(t) = \boldsymbol{A}\boldsymbol{P}(t)$ subject to $\boldsymbol{P}(0) = \boldsymbol{0}$, so that

$$\boldsymbol{P}(t) - \boldsymbol{I} \;=\; \int_0^t \boldsymbol{A}\boldsymbol{P}(s)\,\mathrm{d}s\,.$$

Now, $t\big[\mathbb{E}_x\overline{Y}(t) - z\big]$ is the xth entry of the column vector $\int_0^t \boldsymbol{P}(s)\boldsymbol{f}_c\,\mathrm{d}s$. Observe that

$$(\boldsymbol{\Pi} - \boldsymbol{A})\Big(\int_0^t \big(\boldsymbol{P}(s) - \boldsymbol{\Pi}\big)\,\mathrm{d}s \;+\; \boldsymbol{\Pi}\Big) \;=\; \boldsymbol{\Pi} - \boldsymbol{P}(t) + \boldsymbol{I}$$

because $\boldsymbol{\Pi}\boldsymbol{A} = \boldsymbol{A}\boldsymbol{\Pi} = \boldsymbol{0}$. Since $\boldsymbol{P}(t) \to \boldsymbol{\Pi}$ exponentially rapid, this implies

$$(\boldsymbol{\Pi} - \boldsymbol{A})\Big(\int_0^\infty \big(\boldsymbol{P}(s) - \boldsymbol{\Pi}\big)\,\mathrm{d}s \;+\; \boldsymbol{\Pi}\Big) \;=\; \boldsymbol{I}\,,$$

so that $(\boldsymbol{\Pi} - \boldsymbol{A})^{-1}$ exists. So,

$$\int_0^t \boldsymbol{P}(s)\boldsymbol{f}_c\,\mathrm{d}s \;=\; (\boldsymbol{\Pi} - \boldsymbol{A})^{-1}(\boldsymbol{\Pi} - \boldsymbol{P}(t) + \boldsymbol{I})\boldsymbol{f}_c \;-\; \boldsymbol{\Pi}\boldsymbol{f}_c\,,$$

from which the result follows since $\boldsymbol{\Pi}\boldsymbol{f}_c = \boldsymbol{0}$.   □

Also, the TAVC can be easily computed from knowledge of the fundamental matrix (or, more precisely, from knowledge of the solution $\boldsymbol{g} = (\boldsymbol{\Pi} - \boldsymbol{A})^{-1}\boldsymbol{f}_c$ to Poisson's equation $\boldsymbol{A}\boldsymbol{g} = -\boldsymbol{f}_c$; see [16, Section II.4d]):

Proposition 2.2 *The TAVC is given by*

$$\sigma^2 \;=\; 2\sum_{x\in E} \pi(x)f_c(x)\big((\boldsymbol{\Pi} - \boldsymbol{A})^{-1}\boldsymbol{f}_c\big)(x)\,.$$

Proof. We have

$$\begin{aligned}
\sigma^2 &\;=\; 2\int_0^\infty \mathbb{E}_\pi\big[f_c\big(X(0)\big)f_c\big(X(t)\big)\big]\,\mathrm{d}t \\
&\;=\; 2\mathbb{E}_\pi\Big[f_c\big(X(0)\big)\int_0^\infty \big(\boldsymbol{P}(t)\boldsymbol{f}_c\big)\big(X(0)\big)\,\mathrm{d}t\Big] \\
&\;=\; 2\mathbb{E}_\pi\big[f_c\big(X(0)\big)\big((\boldsymbol{\Pi} - \boldsymbol{A})^{-1}\boldsymbol{f}_c\big)\big(X(0)\big)\,,
\end{aligned}$$

which is the same as the asserted expression.   □

When one can explicitly perform such calculations,[2] one can invariably also analytically (or numerically) compute z. Hence, the discussion of this section is not generally intended as a guide to computing either b or σ^2 for practical simulations. Rather, it is intended to provide the reader with a taste of the analytical theory that is bound up with computing such constants in general. Such formulas also turn out to be valuable in validating

---

[2]For analogues of Propositions 2.1, 2.2 for finite discrete-time Markov chains, see [16, Section I.7].

various steady-state simulation output analysis algorithms empirically for simple models in which the constants can be explicitly calculated.

Remark 2.3 There is, however, one useful application of such formulas. Suppose that one can explicitly compute the necessary constants for a model that is believed to be a good approximation to the system to be simulated. This opens the possibility of using the constant for the approximating model when planning steady-state simulations for the model under consideration. This approach can be nicely illustrated by considering a GI/G/1 queue in *heavy traffic*, i.e., a situation in which the traffic intensity ρ is close to 1. It can then be shown ([16, Section X.7], [361]) that such a queue satisfies a functional limit theorem as $\rho \uparrow 1$, in which the approximating limit process is a one-dimensional reflecting Brownian motion (RBM). In view of the abundance of explicit formulas and computational methods for Brownian motion, the quantitites γ, b, and σ^2 are explicitly available for RBM. This permits us to easily compute an approximation to the relaxation time of the queue, and to determine an approximation to the amount of simulated time t_0 required to reduce the half-width of a $100(1 - \delta)\%$ confidence interval to ε ($t_0 \approx \sigma^2 z^2/\varepsilon^2$, with $z \stackrel{\text{def}}{=} z_{1-\delta/2}$ the $1 - \delta/2$ quantile of the standard normal distribution; cf. (1.4)). See further Whitt [360] and Asmussen [12].   □

3   Variance Estimation for Stationary Processes

As discussed earlier, the rate $t^{-1/2}$ at which the sampling variability goes to zero is much slower than the rate t^{-1} at which the systematic error associated with the bias decays to zero. Furthermore, one has only a single realization of the process from which to attempt to estimate the effect of the initial transient, so that statistical approaches to initial transient determination are likely to be problematic. As a consequence, both theoretical and pragmatic considerations suggest that the dominant problem in steady-state simulation is the estimation of the TAVC σ^2 (and the associated construction of confidence intervals based on (1.4)).

Given that

$$\sigma^2 = 2 \int_0^\infty c(s) \, ds, \quad \text{where} \quad c(s) \stackrel{\text{def}}{=} \mathbb{E}_\pi \big[f_c\big(X(0)\big) f_c\big(X(s)\big) \big], \quad (3.1)$$

note that $c(s)$ can be estimated via

$$\widehat{c}(s) \stackrel{\text{def}}{=} \frac{1}{t-s} \int_0^{t-s} Y(u) Y(s+u) \, du \; - \; \overline{Y}(t)^2 .$$

The ergodic theorem guarantees that when Y is a square integrable process, $\widehat{c}(s) \to c(s)$ as $t \to \infty$. Of course, when the lag s corresponding to the covariance $c(s)$ is large, $\widehat{c}(s)$ involves an average over a relatively short time interval and hence $\widehat{c}(s)$ tends to be a noisy estimator of $c(s)$ when s is large. On the other hand, it is necessary that the lag s covariance with s large be included in the integral (3.1) in order that the bias for an estimator of σ^2 be made small. Therefore, the development of a good estimator for σ^2 depends on trading off variance against bias (see III.6). In particular, consider an estimator of the form

$$\widehat{\sigma}^2 = 2 \int_0^t w_t(u)\widehat{c}_t(u)\,\mathrm{d}u\,,$$

where $w_t(u)$ is a weighting function chosen to minimize mean-square error and having the properties that $w_t(u) \to 1$ as $t \to \infty$ and $w_t(u) \to 0$ as $u \uparrow t$. Such estimators arise as special cases of spectral density estimators (because σ^2 is a constant multiple of the spectral density of a stationary version of the process Y at frequency 0). With an appropriate choice of $w_t(\cdot)$, one obtains a root-mean-square convergence rate of $t^{-1/3}$; see Grenander & Rosenblatt [167].

An alternative variance estimation method is to fit a finite-order autoregressive process to Y, and to use the fact that the TAVC for such a process can be easily computed in closed form. One then uses the TAVC for the fitted process as an estimator for the simulated process. The theoretical justification for such a method rests on the Wold representation theorem (which asserts that a large class of stationary sequences can be represented as infinite-order autoregressions). A key complication in this approach is the need to determine the appropriate model order for autoregressive processes, which is a nontrivial problem. However, the Akaike information criterion (AIC) is one common tool for determining model order.

Exercises

3.1 (TP) Assume that we simulate kR values $X_1, \ldots, X_{kR}$ from a stationary process and in addition to the sample mean $\widehat{z}_{kR}$ also consider the alternative estimator $\widehat{z}_{kR}^{(k)} = (X_k + X_{2k} + \cdots + X_{kR})/R$, the idea being that the spacings typically will reduce dependence and thereby possibly the overall variance. Show that this is not the case, i.e., that $\mathbb{V}ar(\widehat{z}_{kR}) \leq \mathbb{V}ar(\widehat{z}_{kR}^{(k)})$.

4   The Regenerative Method

Suppose that $Y(t) = f(X(t))$, where X is positive recurrent and suitably irreducible. Let g be a solution to Poisson's equation $Ag = -f_c$, where A is the generator of the Markov process X. In great generality (see, e.g., [112]),

$$M(t) \stackrel{\mathrm{def}}{=} g(X(t)) + \int_0^t f_c(X(s))\,\mathrm{d}s$$

is then a martingale (adapted to the natural filtration of X). It follows that if the optional sampling theorem $\mathbb{E}M(T) = \mathbb{E}M(0)$ (see again [112]) can be applied at the stopping time T, then

$$\mathbb{E}g(X(T)) + \mathbb{E}\int_0^T (f(X(s)) - z)\,ds = \mathbb{E}g(X(0)).$$

Hence, the steady-state mean satisfies

$$z = \frac{1}{\mathbb{E}T}\left[\mathbb{E}g(X(T)) - \mathbb{E}g(X(0)) + \mathbb{E}\int_0^T f(X(s))\,ds\right].$$

Assume that we can find a time T at which X *regenerates* in the sense that $X(T)$ has the same distribution as $X(0)$.[3] In this case,

$$z = \frac{1}{\mathbb{E}T}\mathbb{E}\int_0^T f(X(s))\,ds. \tag{4.1}$$

Formula (4.1) is known as the *regenerative ratio formula* for z.

For discrete-state-space irreducible positive-recurrent Markov processes, determination of such regeneration times is straightforward. In particular, choose any state $x^* \in E$ as "return state." If we initialize X in x^* and choose $T \stackrel{\text{def}}{=} \inf\{t > 0 : X(t) = x, X(t-) \neq x^*\}$ to be the first return time to x^*, $X(T)$ clearly has the same distribution as $X(0)$ and (4.1) holds. We adhere to this framework for the rest of this section and discuss extensions in Section 6.

The regenerative ratio formula (4.1) shows that the quantity z, which in general depends on the entire infinite history of X, in the regenerative setting reduces to a ratio of two finite-horizon expectations. This suggests that z and an associated confidence interval can be computed by the smooth function estimation ideas of III.3.

Specifically, let

$$\tau \stackrel{\text{def}}{=} T, \quad \widetilde{Y} \stackrel{\text{def}}{=} \int_0^T f(X(s))\,ds,$$

and suppose that we simulate R i.i.d. copies $(\widetilde{Y}_1, \tau_1), \ldots, (\widetilde{Y}_R, \tau_R)$ of the random vector $(\widetilde{Y}, \tau)$. Thus, $z = \mathbb{E}\widetilde{Y}_1/\mathbb{E}\tau_1$ can be estimated via

$$\widehat{z}_R \stackrel{\text{def}}{=} \frac{R^{-1}(\widetilde{Y}_1 + \cdots + \widetilde{Y}_R)}{R^{-1}(\tau_1 + \cdots + \tau_R)} = \frac{\widetilde{Y}_1 + \cdots + \widetilde{Y}_R}{\tau_1 + \cdots + \tau_R}. \tag{4.2}$$

By recurrence, the epochs $T(1) \stackrel{\text{def}}{=} T < T(2) < \cdots$ of return to x^* form an infinite sequence of regeneration points, and the r.h.s. of (4.2) has the same distribution as the r.v.

$$\frac{1}{T(R)}\int_0^{T(R)} f(X(s))\,ds, \tag{4.3}$$

---

[3]See Section 6 for terminology and further discussion.

that is, a time average over the first R "regenerative cycles." The simulation can therefore be performed simply by simulating the process up to $T(R)$, i.e., by taking

$$\widetilde{Y}_r \overset{\text{def}}{=} T(r) - T(r-1), \quad \tau_r \overset{\text{def}}{=} \int_{T(r-1)}^{T(r)} f(X(s)) \, ds.$$

Moreover, on the scale of regenerative cycles, the bias in this ratio estimator manifests itself as the bias associated with estimating a nonlinear function of a vector-valued expectation. Here, $z = k(\mathbb{E}\widetilde{Y}, \mathbb{E}\tau)$ where $k(x_1, x_2) = x_1/x_2$. Because the bias on the regenerative cycles scale arises as nonlinear function estimation bias, the methods of Chapter III (e.g., III.(3.4) using Taylor expansions or jackknifing as in III.5b) apply directly here as a means of reducing bias. This serves to address one of the two principal difficulties associated with steady-state simulation and described at the outset of this chapter.

As for the other principal difficulty, note that the i.i.d. cycle structure permits one to obtain confidence intervals in a completely straightforward manner:

Proposition 4.1 *The estimator $\widehat{z}_R$ in (4.2) satisfies the CLT*

$$R^{1/2}(\widehat{z}_R - z) \overset{\mathcal{D}}{\to} \mathcal{N}(0, \eta^2)$$

as $R \to \infty$, where

$$\eta^2 \overset{\text{def}}{=} \frac{\mathbb{E}Z_1^2}{(\mathbb{E}\tau_1)^2}, \quad Z_1 \overset{\text{def}}{=} \widetilde{Y}_1 - z\tau_1.$$

Hence, an asymptotic $100(1-\delta)\%$ confidence interval is given by

$$\widehat{z}_R \pm z_{1-\delta/2}\widehat{\eta}_R/\sqrt{R}, \tag{4.4}$$

where $z_{1-\delta/2}$ is the $1-\delta/2$ quantile of $\mathcal{N}(0,1)$ and

$$\widehat{\eta}_R^2 \overset{\text{def}}{=} \frac{1}{R-1} \sum_{r=1}^{R} (\widetilde{Y}_r - \widehat{z}_R\tau_r)^2 \Big/ \Big(\frac{1}{R}\sum_{r=1}^{R}\tau_r\Big)^2.$$

The interval (4.4) is the basic confidence interval for z (in the time scale of regenerative cycles).

Proof. Define as above $k(x_1, x_2) \overset{\text{def}}{=} x_1/x_2$, $z_1 \overset{\text{def}}{=} \mathbb{E}\widetilde{Y}_1$, $z_2 \overset{\text{def}}{=} \mathbb{E}\tau_1$. Then $z = k(z_1, z_2)$, and the delta method, when applied to the ratio function $k(x_1, x_2) = x_1/x_2$, yields the asserted CLT with η^2 as the variance of the r.v.

$$(\widetilde{Y}_1 - z_1)k_{x_1}(z_1, z_2) + (\tau_1 - z_2)k_{x_2}(z_1, z_2)$$
$$= (\widetilde{Y}_1 - z_1)/z_2 - (\tau_1 - z_2)z_1/z_2^2 = \frac{1}{z_2}(\widetilde{Y}_1 - z_1 - z\tau_1 + z_1) = \frac{Z_1}{\mathbb{E}\tau_1}.$$

The remaining assertions are then obvious. $\qquad\square$

A closely related cousin to the estimator $\widehat{z}_R$ is the time average $\overline{Y}(t) = t^{-1} \int_0^t Y(s)\, ds$, in which the time horizon is specified in terms of the time scale of "simulated time." We have already established that the bias of $\overline{Y}(t)$ takes the form $b_1/t + b_2/t^2 + \cdots$, where $b_k = 0$ for $k \geq 2$ often holds. Of course, for producing confidence intervals, a CLT is needed.

Proposition 4.2 *The TAVC σ^2 is given by $\sigma^2 = \mathbb{E}Z_1^2/\mathbb{E}\tau_1$.*

Proof. We give a heuristic derivation and refer to [16, Section VI.3] for a rigorous proof. Observe that if

$$N(t) \stackrel{\text{def}}{=} \max\{n \geq 0 : T(n) \leq t\}$$

is the number of regenerative cycles completed by time t, then N is a renewal counting process. The LLN for such processes ([16, p. 140]) asserts that $N(t)/t \stackrel{\text{a.s.}}{\to} \lambda \stackrel{\text{def}}{=} 1/\mathbb{E}\tau_1$ as $t \to \infty$. Hence $t \approx \sum_1^{N(t)} \tau_j$, $\int_0^t Y(s)\, ds \approx \sum_1^{N(t)} \widetilde{Y}_j$, and so

$$t^{1/2}(\overline{Y}(t) - z) \approx t^{1/2}\left(\sum_{j=1}^{N(t)} (\widetilde{Y}_j - z\tau_j)\right) \Big/ \sum_{j=1}^{N(t)} \tau_j$$

$$\approx t^{-1/2} \sum_{j=1}^{\lfloor \lambda t \rfloor} Z_j \stackrel{\mathscr{D}}{\to} \mathscr{N}\big(0, \mathbb{E}Z_1^2/\mathbb{E}\tau_1\big).$$

$\square$

Remark 4.3 Because we expect the effect of the initialization to wear off after a period of time roughly proportional to the relaxation time, the asymptotic distribution of the time average should be independent of the initialization. It follows that z and σ^2 are quantities that do not depend on the state x^* chosen as a means of defining the cycles that play a role in the theoretical analysis associated with the CLT for the time average. Hence, the choice of the return state x^* has no effect on the asymptotic efficiency of the time-average estimator $\overline{Y}_t$ for the steady-state mean z. $\square$

Remark 4.4 In contrast, the choice of x^* has an effect on the asymptotic efficiency of the estimator for the TAVC σ^2. Specifically, let

$$\widehat{\sigma}^2(t) \stackrel{\text{def}}{=} \frac{1}{t} \sum_{j=1}^{N(t)} \big(\widetilde{Y}_j - \overline{Y}(t)\tau_j\big)^2 ; \tag{4.5}$$

then $\widehat{\sigma}^2(t) \to \sigma^2$ as $t \to \infty$. The definition of the TAVC estimator $\widehat{\sigma}^2(t)$ depends on the choice of x^* in a critical way, so (not surprisingly) its statistical efficiency as an estimator of σ^2 does so too. More precisely, one can use delta method ideas to establish a CLT of the form

$$t^{1/2}\big(\widehat{\sigma}(t) - \sigma\big) \stackrel{\mathscr{D}}{\to} \mathscr{N}\big(0, \nu^2\big),$$

see Glynn & Iglehart [151] for details. The variance constant ν^2 depends on the return state x^*. Examples in [151] show that ν^2 is not minimized by choosing the x^* with the smallest expected cycle duration $\mathbb{E}\tau_1$. Nevertheless, choosing the state x^* that maximizes frequency of regeneration is a rule of thumb that often works quite well empirically. In discrete time, this rule of thumb is equivalent to choosing x^* with maximal stationary probability; cf. the discussion around (6.3) below.   □

In any case, the regenerative method for producing an asymptotic $100(1 - \delta)\%$ steady-state confidence intervals (on the time scale of simulated time rather than cycles) takes the form

$$\overline{Y}(t) \pm z_{1-\delta/2}\widehat{\sigma}(t)/\sqrt{t},$$

where $\widehat{\sigma}^2(t)$ is defined as in (4.5).

For exercises in regenerative simulation, see the end of Section 6.

5   The Method of Batch Means

Perhaps the most straightforward and generally applicable approach to constructing confidence intervals is the method of *batch means*. This method requires only that the process Y satisfy a functional version of the CLT, namely

$$\varepsilon^{-1/2}t\{\overline{Y}(t/\varepsilon) - z\}_{t\geq 0} \xrightarrow{\mathscr{D}} \{\sigma B(t)\}_{t\geq 0} \tag{5.1}$$

as $\varepsilon \downarrow 0$ in the sense of weak convergence in $D[0, \infty)$, where B is standard Brownian motion. Such a functional CLT requires no Markovian assumption, and from a practical standpoint is typically applicable when the one-dimensional CLT (1.4) holds. The key idea is that under the assumption (5.1), the so-called *batch means*

$$\overline{Y}_k(t) \overset{\text{def}}{=} \frac{1}{t/m} \int_{(k-1)t/m}^{kt/m} Y(s)\,\mathrm{d}s,$$

$k = 1, \ldots, m$, are asymptotically i.i.d. with $\mathscr{N}(z, \sigma^2 m/t)$ marginals. It follows that the sectioning approach of III.5a is applicable in this context. In particular, since $\overline{Y}(t) = (\overline{Y}_1(t) + \cdots + \overline{Y}_m(t))/m$,

$$m^{1/2}\frac{\overline{Y}(t) - z}{s_m(t)} \xrightarrow{\mathscr{D}} T_{m-1} \tag{5.2}$$

as $t \to \infty$, where T_{m-1} is a Student t r.v. with $m - 1$ degress of freedom, and

$$s_m^2 \overset{\text{def}}{=} \frac{1}{m-1} \sum_{k=1}^{m} \left(\overline{Y}_k(t) - \overline{Y}(t)\right)^2.$$

Hence, if $t_{1-\delta/2}$ is the $1 - \delta/2$ quantile of T_{m-1}, an asymptotic $100(1-\delta)\%$ confidence interval for z is given by

$$\overline{Y}(t) \pm t_{1-\delta/2} \frac{s_m(t)}{\sqrt{m}}.$$

The quality of this confidence interval (as determined by how close the coverage probability is to $1 - \delta$) depends on the degree to which the batch means are i.i.d. normal r.v.'s. Each of these three characteristics (independence, identical marginals, Gaussianity) are improved by taking large batches. This suggests that one should choose the number m of batches to be reasonably small, say 5 to 30. Choosing m to be in the upper end of this range leads to confidence intervals with potentially poorer coverage but with half-width characteristics closer to those obtained when one attempts to consistently estimate σ (as in the methods of Section 3).

The key idea underlying the method of batch means is to cancel the unknown parameter σ appearing as a common scale factor in the asymptotic distribution of both the numerator and denominator in (5.2), thereby removing the need to estimate σ explicitly. The method of batch means is a special case of "cancellation methods" known as *standardized time series* methods; see Schruben [332] and Glynn & Iglehart [154] for details.

6   Further Refinements

6a   *Discrete Time Modifications*

A discrete-time sequence $\{Y_n\}_{n\in\mathbb{N}}$ can formally be embedded in continuous time by setting $Y(t) \stackrel{\text{def}}{=} Y_{\lfloor t \rfloor}$ for $t \geq 0$. However, the more natural approach is most often just to use a discrete-time interpretation of formulas for the continuous-time case, for example to interpret integrals as sums with appropriate conventions for when endpoints of the range of integration should be included or not.

One important such modification is that the TAVC σ^2 in Proposition 1.3 becomes

$$\sigma^2 = \operatorname{Var}_\pi Y_0 + 2 \sum_{n=1}^{\infty} \operatorname{Cov}_\pi(Y_0, Y_n) \qquad (6.1)$$

(this holds also beyond the Markov case). Another is that in the regenerative setting $X_0 = x^*$, $T \stackrel{\text{def}}{=} \{n > 0 : X_n = x^*\}$, one has to take $\widetilde{Y}_1 \stackrel{\text{def}}{=} \sum_0^{T-1} f(X_n)$ rather than $\sum_0^T$. Thus, for discrete Markov chains, formula (4.1) becomes

$$\pi(f) = \mathbb{E}_\pi f(X_0) = \frac{1}{\mathbb{E}_{x^*} T} \mathbb{E}_{x^*} \sum_{n=0}^{T-1} f(X_n) = \frac{1}{\mathbb{E}_{x^*} T} \mathbb{E}_{x^*} \sum_{n=1}^{T} f(X_n). \quad (6.2)$$

In particular, taking $f(y) \stackrel{\text{def}}{=} \mathbb{1}\{y = x\}$, this yields the standard regenerative representation of the stationary distribution,

$$\pi_x \; = \; \frac{1}{\mathbb{E}_{x^*} T} \mathbb{E}_{x^*} \sum_{n=0}^{T-1} \mathbb{1}\{X_n = x\} \; = \; \frac{1}{\mathbb{E}_{x^*} T} \mathbb{E}_{x^*} \sum_{n=1}^{T} \mathbb{1}\{X_n = x\}, \qquad (6.3)$$

which as a special case contains the formula $\pi_{x^*} = 1/\mathbb{E}_{x^*} T$.

6b   Regenerative Processes

The definition of regeneration that we have used so far, that $Y(0) \stackrel{\mathscr{D}}{=} Y(T)$ for a suitable stopping time T such as the first return to the initial state x^*, is more general than the traditional one, in which a stochastic process $Y = \{Y(t)\}_{t \geq 0}$ is called *regenerative* if there exists a sequence $0 = T(0) < T(1) < T(2) < \cdots < \infty$ of random times (the *regeneration points*) such that the *cycles*

$$\{Y(T(k) + s)\}_{0 \leq s < T(k+1) - T(k)} \qquad (6.4)$$

are independent for different k and have the same distribution for $k \geq 1$. The interpretation of the different distribution for $k = 0$ is as a *delay*, for example when Y is a function of a Markov chain, $T(1), T(2), \ldots$ are the returns of X to x^*, but $\mathbb{P}(X_0 = x^*) < 1$, say $X_0 = y \neq x^*$ or X_0 is chosen with the stationary distribution π. For expositions of the theory of regenerative processes, see [16, Chapters VI, VII] and [354].

In contrast, focusing on returns to a fixed state as regeneration points, as we have done for expository purposes, is of course more restrictive than the above definition.

A zero-delayed ($T(1) = 0$) regenerative process Y can be viewed as a function of a Markov process X: just define $A(t) \stackrel{\text{def}}{=} t - T(k)$ as the *age* of the cycle when $T(k) \leq t < T(k+1)$, and let $X(t) \stackrel{\text{def}}{=} (A(t), Y(t))$. In the delayed case the same is possible under certain conditions on cycle 0.

The existence of limiting time averages of functions f of a regenerative process holds without further conditions, provided only that the mean cycle length $\mathbb{E}(T(2) - T(1))$ is finite. The existence of a limiting distribution, in the sense of weak convergence or total variation convergence, is a more technical topic and involves conditions on the distribution of the cycle length $T(2) - T(1)$; see, e.g., [16]. However, for the purpose of steady-state simulation it is the existence of time averages that matters, not of a limit distribution.

Apart from the wider definition of regeneration, the methodology of Section 4 for simulation and variance estimation applies unchanged. However, beyond discrete-state Markov processes and chains, the identification of regeneration points is sometimes straightforward and sometimes requires more sophistication, as illustrated by the following examples. Further examples are in the exercises; see also Section 6d on Harris chains below.

Example 6.1 For the GI/G/1 queue in continuous time and $Y(t)$ the workload at time t or the queue length, one can let $T(0), T(1), \ldots$ be the epochs with a customer entering an empty system. Thus, cycles $1, 2, \ldots$ are simply the conventional busy cycles (a busy period followed by an idle period).

For the M/G/1 queue, one can alternatively let $T(0), T(1), \ldots$ be the epochs with a customer departing to leave the server idle. Thus, a cycle becomes an idle period followed by a busy period. This definition does not work for the GI/G/1 queue, because the evolution of the process after the start of an idle period depends on the residual arrival time at that instant (for M/G/1, one can just appeal to the memoryless property of the exponential distribution).   □

Example 6.2 Reflected Brownian $\{Y(t)\}$ can be viewed as a continuous-state queuing or storage model, so it is tempting to copy the busy period construction from Example 6.1 by (in the zero-delayed case $Y(0) = 0$) letting $T(0) = 0$, $T(1) \stackrel{\text{def}}{=} \inf\{t > 0 : Y(t) = 0\}$, $T(2) \stackrel{\text{def}}{=} \inf\{t > T(1) : Y(t) = 0\}$, and so on. However, sample path properties of Brownian motion imply that this definition yields the unusable $0 = T(0) = T(1) = T(2) = \cdots$.

Two among many alternatives are

$$T(k+1) \stackrel{\text{def}}{=} \inf\{t > T(k) + 1 : Y(t) = 0\},$$

$$T(k+1) \stackrel{\text{def}}{=} \inf\left\{t > T(k) : Y(t) = 0, \sup_{T(k) \leq s \leq t} Y(s) \geq 1\right\}$$

("up to 1 and back to 0").   □

6c   On the Necessity of Regenerative Structure

Our regenerative discussion in Section 4 has focused on discrete state space Markov processes, in which the regenerative structure is particularly transparent (as defined through the sequence of times at which the process enters a fixed state x^*). However, it turns out that regenerative structure must necessarily exist whenever the steady-state simulation problem is well posed in the following sense:

Definition 6.3 Let $X = \{X(t)\}_{t \geq 0}$ be a Markov process taking values in a separable metric space E. We say that the steady-state simulation problem for X is well posed if for each bounded measurable function $f : E \to \mathbb{R}$ there exists a number $z(f)$ such that for $y \in E$,

$$\mathbb{E}_y \frac{1}{t} \int_0^t f(X(s)) \, ds \rightarrow z(f), \quad t \to \infty.$$

The key point of this definition is that the time-average expectation converges to a quantity independent of the initial state y.

Theorem 6.4 *If the steady-state simulation problem is well posed, there exist randomized stopping times (relative to the filtration induced by X) $T(0) < T(1) < \cdots < \infty$ such that:*

(i) *The processes $\theta_{T(i)}X \stackrel{\text{def}}{=} \{X(T(i)+s)\}_{s\geq 0}$ have the same distribution for all i;*

(ii) *The cycles $\{X(T(i)+s)\}_{0\leq s<T(i+1)-T(i)}$ form a sequence of one-dependent random elements, i.e., the ith cycle is independent of the cycles j with $|j - i| \geq 2$.*

Proof. See Glynn [145].   □

Theorem 6.4 guarantees that $X(T(0))$ has the same distribution as $X(T(1))$. In view of the argument leading to (4.1), it follows that the steady-state mean z can be expressed in terms of the ratio estimation problem

$$z = \mathbb{E}\widetilde{Y}_1/\tau_1, \quad \text{where} \quad \tau_j \stackrel{\text{def}}{=} T(j) - T(j-1), \quad \widetilde{Y}_j \stackrel{\text{def}}{=} \int_{T(j-1)}^{T(j)} f(X(s))\,\mathrm{d}s.$$

As in the discrete-state-space setting, one can now compute z by generating i.i.d. copies of the cycle

$$\{X(T(0)+s)\}_{0\leq s<\tau_1}. \tag{6.5}$$

Remark 6.5 It is no longer the case, however, that generating R such i.i.d. copies is probabilistically equivalent to simulating X up to time $T(R)$, because the cycle structure is generally one-dependent rather than i.i.d. Therefore, an alternative implied estimation algorithm for computing z is obtained by simulating X to time $T(n)$ and using the time average over that time interval to estimate z. In view of the one-dependence, the CLT for the time average $\overline{Y}(t)$ takes the form

$$t^{1/2}(\overline{Y}(t) - z) \stackrel{\mathscr{D}}{\to} \mathscr{N}(0,\sigma^2), \quad t \to \infty,$$

where now

$$\sigma^2 \stackrel{\text{def}}{=} \frac{\mathbb{E}(Z_1^2 + 2Z_1Z_2)^2}{\mathbb{E}\widetilde{Y}_1}, \quad Z_j \stackrel{\text{def}}{=} \widetilde{Y}_j - z\tau_j, \ j \geq 1.$$

Given that we expect the cycle quantities Z_1 and Z_2 to be generally positively correlated, this analysis suggests that a variance reduction in computing z can be obtained by generating i.i.d. copies of the cycle (6.5) rather than by generating a single path of X. The construction of confidence intervals then follows the approach described earlier in this section for regenerative processes with i.i.d. cycle structure.   □

Of course, the algorithms just described rely heavily on the ability to computationally identify the cycle boundaries $T(0), T(1), \ldots$. While computing the regeneration times $T(0), T(1), \ldots$ is a triviality in discrete state space, we note that Theorem 6.4 does not provide a constructive definition of these random times in general. We therefore next discuss this issue in greater generality.

6d Regeneration in Harris Chains

We focus on the case of most practical interest, in which $X = \{X_n\}_{n \in \mathbb{N}}$ is a discrete time Markov chain (taking values in a general state space E). In this setting, Theorem 6.4 is established by showing that X is a positive recurrent Harris chain. Such chains are guaranteed to possess a set A for which one can find an integer $m \geq 1$, a $\lambda > 0$, and a probability $\varphi(\cdot)$ for which

$$\mathbb{P}_x(X_n \in A \text{ i.o.}) = 1, \qquad x \in E, \qquad (6.6)$$

$$\mathbb{P}_x(X_m \in \cdot) \geq \lambda \varphi(\cdot), \qquad x \in A. \qquad (6.7)$$

Note that condition (6.7) permits one to write the m-step transition probability for X on the set A in the form

$$\mathbb{P}_x(X_m \in \cdot) = \lambda \varphi(\cdot) + (1 - \lambda)Q(x, \cdot),$$

where

$$Q(x, \mathrm{d}y) \stackrel{\text{def}}{=} \frac{\mathbb{P}_x(X_m \in \mathrm{d}y) - \lambda \varphi(\mathrm{d}y)}{1 - \lambda}$$

is a Markov transition kernel. Hence, if $X_\ell \in A$, we may generate $X_{\ell+m}$ as a mixture of the distributions φ and $Q(x, \cdot)$. Conceptually, we may view the corresponding variate generation in terms of a coin flip, in which we distribute $X_{\ell+m}$ according to φ if the coin comes up heads (w.p. λ) and according to $Q(X_\ell, \cdot)$ otherwise. So, every time we visit A and observe a head on our coin toss, we regenerate m time units later (because the distribution at that time is φ, regardless of our position on A). [Condition (6.6) is a technical condition guaranteeing that infinitely many coin tosses will occur and therefore infinitely many successful ones, thereby producing an infinite sequence of regeneration times.]

Remark 6.6 Assume that $\mathbb{P}_x(X_m \in \cdot)$ has a transition density $p(x, \cdot)$ relative to a reference measure ϕ, i.e.,

$$\mathbb{P}_x(X_m \in \mathrm{d}y) = p_m(x, y)\phi(\mathrm{d}y).$$

In this case, natural candidates for φ and λ are

$$\varphi(\mathrm{d}y) \stackrel{\text{def}}{=} \frac{1}{\lambda} \inf_{x \in A} p_m(x, y)\, \phi(\mathrm{d}y), \quad \lambda \stackrel{\text{def}}{=} \int_E \inf_{x \in A} p_m(x, y)\, \phi(\mathrm{d}y). \qquad (6.8)$$

In most applications, the set A will be chosen as some compact subset of the state space. In the presence of modest continuity restrictions on p_m, λ will be positive and φ will indeed be a probability, so that the candidates in (6.8) do indeed satisfy (6.5). $\qquad\square$

There are two difficulties with direct implementation of the above idea computationally. The first problem is that we have said nothing about how to generate the intermediate path values $X_{\ell+1}, \ldots, X_{\ell+m-1}$ when $m > 1$. The most obvious mechanism for accomplishing this is to generate the set $(X_{\ell+1}, \ldots, X_{\ell+m-1})$ directly from its conditional distribution given X_ℓ and $X_{\ell+m}$. This variate generation approach is problematic, because generating high-dimensional random vectors (as would occur when m is large) is generally both theoretically and practically challenging. Further, the form of the conditional distribution given X_ℓ and $X_{\ell+m}$ may not be easily available. An alternative idea is to generate the path $(X_{\ell+1}, \ldots, X_{\ell+m})$ as usual, that is, using the one-step transition kernel or the one-step stochastic recursion underlying X. Having now generated $X_{\ell+m}$, we need to determine whether a regeneration has occurred there. We can do this with an acceptance–rejection step. In particular, let

$$ \omega(x,y) \;\overset{\text{def}}{=}\; \frac{\lambda\varphi(\mathrm{d}y)}{\mathbb{P}_x(X_m \in \mathrm{d}y)}. $$

Generate a uniform r.v. U. If $U \le \omega(X_\ell, X_{\ell+m})$, we may accept $X_{\ell+m}$ as having been distributed according to φ, so that a regeneration has occurred at the time $\ell+m$. If $U > \omega(X_\ell, X_{\ell+m})$, a rejection occurs and no regeneration is present at time $\ell+m$. This A-R implementation is clearly preferable to the direct coin toss idea described above.

The second implementation difficulty is more problematic from a practical standpoint. The formulas (6.8) for λ and φ rely on knowledge of p_m. While computing p_m is often straightforward when $m = 1$, it can be very challenging for $m > 1$. Unfortunately, there are many models for which it is known theoretically that $m^* > 1$, where m^* is the smallest m^* for which conditions (6.6), (6.7) hold. For example, in applying these ideas to the discrete-event simulations that can be characterized as GSMPs, m^* corresponds to the minimal number of events that are simultaneously scheduled at any given time t. As a result, the single-server queue, when viewed as a GSMP, has $m^* = 1$ (since only one arrival event is scheduled at times at which the server is idle), while multistation networks give rise to $m^* > 1$. The implementation of regenerative simulation methodology in problems in which $m > 1$ in (6.6), (6.7) is an area of ongoing research; see, for example, Henderson & Glynn [182].

6e   Stationary Processes. Palm Structure

A stochastic process $Y = \{Y(s)\}_{s \geq 0}$ is called *strictly stationary* or just *stationary* if the shift $\theta_t Y \overset{\text{def}}{=} \{Y(t+s)\}_{s \geq 0}$ has the same distribution as $Y = \theta_0 Y$ for all $t \geq 0$. A main example is Gaussian processes and derived processes. For example, let $W = \{W(t)\}$ be a stationary Gaussian process and assume that the goal of the simulation is to produce an estimate of

$$z \overset{\text{def}}{=} \mathbb{P}\left(\sup_{0 \leq t \leq 1} W(t) > 1 \right).$$

This could be performed by simulating W with the methods of Chapter XI and letting

$$Y(t) \overset{\text{def}}{=} \mathbb{1}\left\{ \sup_{t \leq s \leq t+1} W(s) > 1 \right\}, \quad \widehat{z} \overset{\text{def}}{=} \overline{Y}(t) = \frac{1}{t} \int_0^t Y(s) \, \mathrm{d}s.$$

Consistency, i.e. $\widehat{z} \to z$ as $t \to \infty$, in this and other stationary settings is usually established via the ergodic theorem, for which one in turn needs to verify that the shift-invariant σ-field of W is trivial. For the purpose of variance estimation, the representation of the TAVC σ^2 in Proposition 1.3 continues to hold for a stationary process, but in the Gaussian max example and many others, no regenerative structure is available, and so the TAVC has to be estimated via the methods of Sections 3 or 5.

Remark 6.7 Of course, the simulation of the steady-state mean $\mathbb{E}f(W(0))$ of a function of the Gaussian process W itself is uninteresting: if we are in position to generate W, we also know the mean and the variance of $W(0)$. The relevant examples are functionals of the entire path.

The setup of a general stationary process Y is not more general than a function of a time-homogeneous Markov process X: just define $X(t) \overset{\text{def}}{=} \{Y(s)\}_{-\infty < s \leq t}$, with the state space defined as an appropriate function space. This remark is formal rather than practically useful!   □

Another type of path functional deals with cycle properties. In the stationary process, one then assumes that X can be divided into cycles, as defined by (6.4), separated by $0 = T(0) < T(1) < T(2) < \cdots$. Cycle stationarity means that the cycles form a stationary sequence of random elements. Thus, we are dealing with a setup with a generality one level higher than the one-dependence in Harris chains, going under the name of *Palm theory* ([34], [341], [16, Section VII.6], or [354]), an area connecting time and cycle stationarity.

Example 6.8 A cycle-stationary version Y^* expresses properties of Y as viewed from a "typical" start of a cycle. Consider for example a stationary Gaussian process with differentiable sample paths, and let the $T(k)$ be the times of upcrossings of level 0, i.e., $Y(T(k)) = 0$, $Y'(T(k)) > 0$. Then the cycles of Y^* can be viewed as "typical" cycles of Y separated by the upcrossings, whereas the cycles of Y itself do not even form a stationary sequence. For a relevant example of the cycle-stationary point of view, see Aberg et al. [3] and Exercise XI.4.1 □

Given a cycle-stationary process Y^*, one can define a time-stationary version Y by the *Palm inversion formula*

$$\mathbb{P}(X \in F) = \frac{1}{\mathbb{E}T(1)} \mathbb{E} \int_0^{T(1)} \mathbb{1}\{\theta_t X^* \in F\} \, dt. \qquad (6.9)$$

Conversely, a cycle-stationary version can be constructed from a time-stationary one by

$$\mathbb{P}(X^* \in F) = \frac{1}{\lambda h} \mathbb{E} \sum_{k:\, 0 \le T(k) \le h} \mathbb{1}\{\theta_{T(k)} X \in F\}, \qquad (6.10)$$

where $\lambda = \mathbb{E}\#\{k : 0 \le T(k) \le 1\}$ is the rate of cycle starts (here $h = 1$ is a common choice, but any positive value is in principle possible).

In simulation, the ability to simulate a cycle-stationary version permits one to estimate $z = \mathbb{E}f(X(0))$ because

$$z = \frac{1}{\mathbb{E}T(1)} \mathbb{E} \int_0^{T(1)} f(X(t)) \, dt$$

by (6.9). One would thereby ideally generate i.i.d. cycles; cf. the discussion in Remark 6.5. However, we know of few implementations of this beyond one-dependence such as in Harris chains, and rather it is formula (6.10) that is the more useful one. For example, in the Gaussian process Example 6.8, the time-stationary version is readily simulated, but the cycle-stationary version is not, and one could use (6.10) to estimate the probability of exceedance of level x in a cycle by

$$\mathbb{P}\left(\max_{0 \le t < T(1)} Y^*(t) > x \right) = \frac{1}{\lambda h} \mathbb{E} \sum_{k:\, 0 \le T(k) \le h} \mathbb{1}\left\{ \max_{T(k) \le t < T(k+1)} Y(t) > x \right\}.$$

This means simulating Y in $[0, h]$, completing the cycle straddling h, if any, counting the number of cycles with exceedance of x, replicating the experiment R i.i.d. times, and averaging over the R copies. A competing estimator is the time-average, that is, the frequence of cycles exceeding x among all cycles initiated in $[0, Rh]$. Our i.i.d. estimator based on (6.10) enjoys two advantages. The first is that the i.i.d. structure permits confidence intervals to be easily computed. The second is that in the presence of positive correlation in the cycle exceedance sequence, this estimator can be expected to be more efficient than the one based on a single run of duration Rh.

Exercises

6.1 (TP) Consider a stochastic sequence $\{X_n\}_{n\in\mathbb{N}}$ given by the recursion

$$X_{n+1} = \sqrt{\beta X_n + Z_{n+1}},$$

where the Z_n are i.i.d. exponential r.v.'s with rate parameter λ. Use the memoryless property of the exponential distribution to identify regeneration points for $\{X_n\}$.

6.2 (A) A supermarket has on a given day n a demand for a Poisson(20) number Y_n of packages of raw oysters, and a contract with a supplier to deliver 20 packages daily. A package can be sold only on the day of supply or the following day, and the supermarket has a clever way of arranging the packages that ensures that one-day-old packages will be sold before the freshly delivered ones. Find a recursion for X_n, the number of one day old packages for sale, and use regenerative simulation to find the expected steady-state number of packages that have to be trashed per day.

6.3 (A) Consider an (s, S) inventory system in which the number of items stored at time t is $V(t)$. Demands arrive (one at a time) according to a Poisson process with intensity λ. When $V(t-) = s+1$, $V(t) = s$, an order of size $S-s$ is placed and arrives after a random time Z (the lead time). Demands arriving while $V(t) = 0$ are lost. It is assumed that $S - s > s$.

Write a program for regenerative simulation of p, the long-run probability that a demand is lost. Use whatever values of s, S, λ you like and whatever distribution of Z you prefer. Save your program for Exercise 6.4.

6.4 (A) Consider again the model of Exercise 6.3, but assume that λ is small compared to Z in the sense that the event that $V(t) = 0$ in a cycle is rare.

Improve your program for Exercise 6.3 by combining it with a variance reduction technique.

You can, for example, use a change of λ depending on the value of Z. Be also clever and use the fact that the expected number of demands lost in a cycle depends only on the residual lead time at the time at which $V(t-) = 1$, $V(t) = 0$.

See also Exercise VI.2.6.

7   Duality Representations

Our starting point is the following example :

Example 7.1 In the GI/G/1 queue, we have the representation $W \overset{\mathscr{D}}{=} M$, where W is the steady-state waiting time, $M = \sup_{n=0,1,\dots} S_n$, and $\{S_n\}$ is a random walk with increments X_k distributed as the difference between a service time and an independent interarrival time, cf. I.1. Assume that we want to estimate $z = \mathbb{P}(W \geq x)$. We can then write $z = \mathbb{P}(M \geq x) = \mathbb{P}(\tau(x) < \infty)$, where $\tau(x) \overset{\text{def}}{=} \inf\{n > 0 : S_n \geq x\}$. In this way the problem of simulating a stationary characteristic is converted to the problem of simulating a first-passage probability.

The difficulty is of course that CMC is not feasible, since $\mathbb{P}(\tau(x) < \infty) <$ 1 under the stability condition $\rho < 1$ (equivalent to $\mathbb{E}X < 0$) and so $\mathbb{1}\{\tau(x) < \infty\}$ cannot be generated by simulating $\{S_n\}$ up to a stopping time. However, in this example a classical importance sampling technique based on exponential change of measure exists and not only does resolve the infinite horizon problem but does in fact give extremely accurate estimates also for very large x. We return to this in VI.2a.   □

Here are two further less-standard examples:

Example 7.2 The $GI/G/1$ queue waiting-time process can be viewed as a random walk reflected at zero in view of the two equivalent representations $W_{n+1} = (W_n + X_n)^+$ (the Lindley recursion) and $W_n = S_n - \inf_{0 \le k \le n} S_k$. In many problems involving a finite buffer $K < \infty$, one has instead a random walk reflected both at 0 and K,

$$V_{n+1} = \min\big(K, \max(0, V_n + X_n)\big).$$

Also in this case, there is a first-passage representation of the stationary r.v. V (Siegmund [339]):

$$\mathbb{P}(V \ge x) = \mathbb{P}(S_{\tau[x-K,x)} \ge x), \quad 0 \le x \le K,$$

where $\tau[x - K, x) \stackrel{\text{def}}{=} \inf\{n > 0 : S_n \notin [x - K, x)\}$. In this example, $\tau[x - K, x) < \infty$ a.s., and hence CMC simulation of $z \stackrel{\text{def}}{=} \mathbb{P}(V \ge x)$ is feasible with $Z \stackrel{\text{def}}{=} \mathbb{1}\{S_{\tau[x-K,x)} \ge x\}$.   □

Example 7.3 Let $\{V(t)\}_{t \ge 0}$ be a storage process with release rule $r(x)$ and compound Poisson input $\{A(t)\}$, i.e., $A(t) = \sum_1^{N(t)} Y_i$, where $\{N(t)\}$ is a Poisson process with intensity β and the Y_i are independent of $\{N(t)\}$ with distribution B concentrated on $(0, \infty)$,

$$V(t) = V(0) + A(t) - \int_0^t r\big(V(s)\big) \, ds.$$

Then, similarly to the two preceeding examples, the stationary distribution can be represented as a first-passage probability

$$\mathbb{P}(V \ge x) = \mathbb{P}(\tau(x) < \infty), \tag{7.1}$$

where $\tau(x) \stackrel{\text{def}}{=} \inf\{t > 0 : R(t) \le 0 \mid R(0) = x\}$ and $\{R(t)\}_{t \ge 0}$ is given by

$$R(t) \stackrel{\text{def}}{=} R(0) + \int_0^t r\big(R(s)\big) \, ds - A(t).$$

Again, Monte Carlo estimation of either side of (7.1) leads to an infinite-horizon problem, and we briefly mention a relevant importance sampling scheme in XIV.6c.   □

The examples above all fit into a common framework in which one has two processes $\{V(t)\}$, $\{R(t)\}$ with state space $[0, \infty]$, or subintervals, connected via the formula (7.1). The first such general construction is due to Siegmund [339] in a Markov-process context. Starting from $\{V(t)\}$, Siegmund constructed $\{R(t)\}$ in terms of its transition probabilities by

$$\mathbb{P}\big(R(t) \le y \,\big|\, R(0) = x\big) \;=\; \mathbb{P}\big(V(t) \ge x \,\big|\, V(0) = y\big). \qquad (7.2)$$

For this to define a transition semigroup, it is necessary and sufficient that $\{V(t)\}$ be stochastically monotone and that $\mathbb{P}\big(V(t) \ge x \,\big|\, V(0) = y\big)$ be right-continuous in y for any fixed x. Note that taking $x = y = 0$ shows that state 0 is absorbing for $\{R(t)\}$. If (7.2) holds, one then obtains (7.1) by taking $y = 0$ and letting $t \to \infty$. The processes $\{V(t)\}$ and $\{R(t)\}$ are said to be in *Siegmund duality*.

In Example 7.1 with $V = W$, one can define $R_n = x - S_n$ as long as $x - S_n > 0$; when $(-\infty, 0]$ is hit, R_n is reset to 0 and remains there forever. Example 7.2 is the same except for a further resetting to ∞ when $(K, \infty]$ is hit. Example 7.3 is also a particular case.

Asmussen & Sigman [29] gave a generalization beyond the Markov case by considering a recursive discrete-time setting in which $\{V_n\}$ is given recursively by $V_0 = y$, $V_{n+1} = f(V_n, U_n)$, where the driving sequence $\{U_n\}$ is not i.i.d. as for the Markov case, but strictly stationary, w.l.o.g. with doubly infinite time ($n \in \mathbb{Z}$). Assuming $f(v, u)$ to be nondecreasing in v for fixed u, one then defines g by letting $g(\cdot, u)$ be a (generalized) inverse of $f(\cdot, u)$ and letting $R_0 = x$, $R_{n+1} \stackrel{\mathrm{def}}{=} g(R_n, U_{-n})$. Then again (7.2) holds, and hence so does (7.1).

Steady-state simulation via (7.1) is typically elegant and (when combined with say importance sampling) efficient when it is feasible. The main limitation is that monotonicity and existence of the inverse of $f(\cdot, u)$ (which most often limits the state space to be one-dimensional) are required; some progress to get beyond this was recently obtained by Blaszczyszyn & Sigman [49], but the practical usefulness for simulation seems questionable so far. Also a general non-Markovian theory in continuous time is missing. Some further relevant references for duality are [16, Sections IX.4, XIV.3] (a general survey and references), Asmussen & Rubinstein [27] (simulation aspects) and Asmussen [14] (Markov-modulated continuous-time models).

8 Perfect Sampling

Obviously, one could avoid concerns relating to the burn-in period and bias if it were possible to generate an r.v. with the stationary distribution π, because then we could just take $X_0 = Z$ and would have a stationary Markov chain.

If such a Z can be generated, one speaks of *perfect sampling*, also going under names such as *exact sampling*, *perfect simulation*, etc. It is not a priori obvious whether perfect simulation is possible, but in fact, the answer is positive in the case of a finite Markov chain. The first algorithm was given by Asmussen, Glynn, & Thorisson [19] in 1992, but it was prohibitively inefficient in terms of computer time. We describe here some algorithms developed by Propp & Wilson [295] in 1996 and later (the area of perfect simulation remains an active research area).

Write $E = \{1, \ldots, p\}$ and denote the transition probabilities by q_{ij}, $i, j \in E$. We assume the Markov chain to be ergodic, which in the finite case just means that it be irreducible and aperiodic. It will be convenient to represent the Markov chain simulation in terms of an *updating rule*. By this we understand a random vector

$$\boldsymbol{Y} = \big(Y(1), \ldots, Y(p)\big)$$

such that $Y(i) \in E$ has distribution $\boldsymbol{q}_{i\cdot}$, i.e., $\mathbb{P}(Y(i) = j) = q_{ij}$ (note that the p components of $\boldsymbol{Y}$ are not necessarily independent; we return to this point later). From $\boldsymbol{Y}$, we construct a doubly infinite sequence $(\boldsymbol{Y}_n)_{n=0,\pm 1,\pm 2,\ldots}$ of i.i.d. random vectors distributed as $\boldsymbol{Y}$. We can then construct $\{X_n\}_{n=0,1,\ldots}$ recursively by $X_{n+1} = Y_n(X_n)$ and $X_0 = i$, where i is the initial state. More generally, we can for each $N \in \mathbb{Z}$ and each $i \in E$ define a version $\big\{X_n^N(i)\big\}_{n=N,N+1,\ldots}$ of $\{X_n\}$ starting at i at time N by

$$X_N^N(i) = i, \ X_{N+1}^N(i) = Y_N(i) = Y_N\big(X_N^N(i)\big), \ \ldots, X_{n+1}^N(i) = Y_n\big(X_n^N(i)\big).$$

Note the important point that if $N, N' \leq n$, then the updatings of $X^N(i)$ and $X^{N'}(i')$ from n to $n+1$ use the same $\boldsymbol{Y}_n$.

The *forward coupling time* is defined as

$$\tau_+ \stackrel{\text{def}}{=} \inf\{n = 1, 2, \ldots : X_n^0(1) = \ldots = X_n^0(p)\},$$

i.e., as the first time at which the Markov chains

$$\{X_n^0(1)\}_{n=0,1,\ldots}, \ \ldots, \ \{X_n^0(p)\}_{n=0,1,\ldots}$$

started at time 0 in the p different states coalesce. See Figure 8.1.

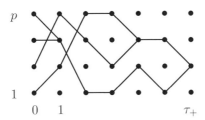

FIGURE 8.1: The forward coupling

Whether this forward coupling time is a.s. finite depends on the updating rule, i.e., the specific dependence between the p components of $\boldsymbol{Y}_n$. Call the updating rule *independent* if these p components are independent.

Proposition 8.1 *In the case of independent updating,* $\mathbb{P}(\tau_+ < \infty) = 1$.

Proof. Let $q_{ij}^{(n)} \overset{\text{def}}{=} \mathbb{P}(X_n = j \mid X_0 = i)$ be the n-step transition probabilities. Since $q_{ij}^{(n)} \to \pi_j > 0$, $n \to \infty$, we can choose first some arbitrary state j and next N such that $q_{ij}^{(N)} > \varepsilon > 0$ for all $i \in E$. Since the probability of coalescence before N is at least the probability that p independent Markov chains starting at time 0 in the p different states will all be in state j at time N, we get $\mathbb{P}(\tau_+ \leq N) \geq \varepsilon^k$. Similarly, $\mathbb{P}(\tau_+ \leq 2N \mid \tau_+ > N) \geq \varepsilon^k$, so ("geometric trial argument")

$$\mathbb{P}(\tau_+ > N) \leq 1 - \varepsilon^k, \quad \mathbb{P}(\tau_+ > 2N) \leq (1 - \varepsilon^k)^2, \quad \dots,$$

which implies that $\tau_+ < \infty$ a.s. $\qquad\square$

Rather than forward coupling, the Propp–Wilson algorithm uses *coupling from the past*. The *backward coupling time* is defined as

$$\tau \overset{\text{def}}{=} \inf\{n = 1, 2, \dots : X_0^{-n}(1) = \cdots = X_0^{-n}(p)\},$$

i.e., as the first time at which the Markov chains $X^{-n}(1), X^{-n}(p)$ started at time $-n$ in the p different states coalesce. Equivalently, coalescence means that the value set

$$\{X_0^{-n}(1), \dots, X_0^{-n}(p)\}$$

contains only one point (note that the cardinality of this set is a nonincreasing function of n). See Figure 8.2.

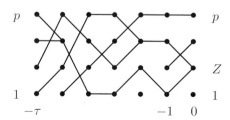

FIGURE 8.2: The backward coupling

Theorem 8.2 *Assume that the updating rule is such that $\tau_+ < \infty$ a.s. Then $\tau < \infty$ a.s. as well, $Z = X_0^{-\tau}(i)$ does not depend on i, and Z has distribution π.*

Proof. The first statement follows since $\mathbb{P}(\tau \leq k) = \mathbb{P}(\tau_+ \leq k)$ goes to 1 as $k \to \infty$. That Z does not depend on i is immediate from the definition of τ.

Now consider $X_0^{-n}(i)$ for some fixed i. On $\tau \leq n$, we have $X_0^{-n}(i) = Z$ and hence $\mathbb{P}(X_0^{-n}(i) = j) \to \mathbb{P}(Z = j)$ for all j. On the other hand, $\mathbb{P}(X_0^{-n}(i) = j) = q_{ij}^n \to \pi_j$. Hence $\mathbb{P}(Z = j) = \pi_j$ as desired. $\qquad\square$

Remark 8.3 The forward coupling time enters only as a tool to show that the backward coupling time is finite. It is definitely not correct that $X^0_{\tau_+}(i)$ has the stationary distribution! $\qquad\square$

Corollary 8.4 (GENERAL MARKOV CHAIN, INDEPENDENT UPDATING) *In the case of independent updating, one has $\tau < \infty$ a.s., $Z = X_0^{-\tau}(i)$ does not depend on i, and Z has distribution π.*

Now assume that there is defined some partial order $\preceq$ on $\{1,\ldots,p\}$ such that 1 is a minimal element and p a maximal one, $1 \preceq i \preceq p$ for all $i = 1,\ldots,p$.

Recall that $X = \{X_n\}$ is called *stochastically monotone* if $i \preceq j$ implies that $X_1^0(i) \preceq X_1^0(j)$ in stochastic order. In terms of transition probabilities, for any ℓ,

$$\sum_{k:\,\ell \preceq k} q_{ik} \le \sum_{k:\,\ell \preceq k} q_{jk} \quad \text{if } i \preceq j.$$

Example 8.5 A first simple example of a monotonic Markov chain is a random walk reflected at the barriers 0 and p, cf. Example 7.2,

$$X_{n+1} = \min\big(p, \max(0, X_n + B_n)\big)$$

where $B_1, B_2, \ldots$ are i.i.d. on $\mathbb{Z}$. Such chains show up in many finite buffer queuing problems. A particular case is *Moran's model for the dam*, where X_n is the content of a water reservoir at time n and $B_n = V_n - m$, where V_n is the amount of water flowing into the reservoir at time n, and m the maximal amount of water that can be released. $\qquad\square$

Example 8.6 In many applications in mathematical physics and image analysis, the state space E is the set of all $-1, 1$ configurations on a finite lattice, say $\{1,\ldots,N\}^2$. Thus, the number of states is 2^{N^2}. The order is defined componentwise so that if we identify the configuration of all -1's with state 1 and the configuration of all 1's with state $p = 2^{N^2}$, we have $1 \preceq i \preceq p$ for all configurations i. $\qquad\square$

Under such monotonicity assumptions, a variant of the Propp–Wilson algorithm is often more efficient. It is defined by *monotone updating*, requiring

$$Y(i) \preceq Y(j) \quad \text{if } i \preceq j.$$

This implies $X_n^N(i) \preceq X_n^N(j)$ for all N and all $n \ge N$, in particular

$$X_n^N(1) \preceq X_n^N(i) \preceq X_n^N(p) \tag{8.1}$$

for all i and all $n \ge N$. For example, in Example 8.5 the natural monotone updating rule is $Y(i) = \min(p, \max(0, i + B))$ with the same B for all i (in the case of independent updating, one would need to take the B's to be independent for different i). We define

$$\tau^m \stackrel{\text{def}}{=} \inf\{n = 1, 2, \ldots : X_0^{-n}(1) = X_0^{-n}(p)\}.$$

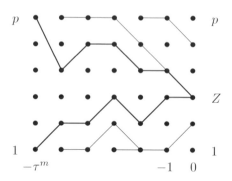

FIGURE 8.3: The monotone coupling

Corollary 8.7 (STOCHASTICALLY MONOTONE MARKOV CHAIN, MONO-
TONE UPDATING) *In the case of monotone updating, $\tau^m = \tau < \infty$ a.s.,
$Z = X_0^{-\tau^m}(i)$ does not depend on i, and Z has distribution π.*

Proof. That $\tau_+ < \infty$ a.s. follows since $\tau(p,1) \stackrel{\text{def}}{=} \inf\left\{n : X_n^0(p) = 1\right\}$ is
finite by recurrence and $X_{\tau(p,1)}^0(i) = 1$ for all i by (8.1).

Clearly, by definition $\tau^m \leq \tau$, where as above, τ is the backward coupling
time. On the other hand, $X_0^{-\tau^m}(i) = X_0^{-\tau^m}(1) = X_0^{-\tau^m}(p)$ for all i by (8.1). $\square$

Remark 8.8 In the monotone case, Propp & Wilson suggest to take
alternatively

$$\widetilde{\tau}^m \stackrel{\text{def}}{=} \inf\left\{n = 1, 2, 4, 8, \ldots : X_0^{-n}(1) = X_0^{-n}(p)\right\},$$

$\widetilde{Z} = X_0^{-\widetilde{\tau}^m}(1) = X_0^{-\widetilde{\tau}^m}(p)$. That $\widetilde{\tau}^m < \infty$ follows since $\widetilde{\tau}^m \leq 2^k$ when
$\tau^m = \tau \leq 2^k$, and $\mathbb{P}(\widetilde{Z} = j) = \pi_j$ then follows exactly as above. For the
advantages of using monotone updating and $\widetilde{Z}$, see the original Propp–
Wilson paper [295]. $\square$

Wilson's forward algorithm [365] is an algorithm developed for the
purpose of requiring less storage than the backward algorithms. In the
monotone case, it is as follows:

(i) Run two coupled versions $\{X_n^{(-)}\}$, $\{X_n^{(+)}\}$ starting from $X_0^{(-)} = 1$,
$X_0^{(+)} = p$ in forward time until they meet. Let t be the next multiple
of m and $X_t^* = X_t^{(+)} = X_t^{(-)}$.

(ii) Simulate $\{X_n^*\}_{n=t,\ldots,t+m}$, $\{\widetilde{X}_n^{(-)}\}_{n=t,\ldots,t+m}$, $\{\widetilde{X}_n^{(+)}\}_{n=t,\ldots,t+m}$ starting from $\widetilde{X}_n^{(-)} = 1$, $\widetilde{X}_n^{(+)} = p$.

(iii) If $\widetilde{X}_{t+m}^{(+)} = \widetilde{X}_{t+m}^{(+)}$, return X_t^* (not X_{t+m}^*!). Otherwise, let $t = t + m$
and return to 2.

For tutorials on perfect sampling, see Kendall [214] and references in [214, p. 96]. Some further interesting papers in the area are Fill [117], Foss & Tweedie [121], Møller [264] and Propp & Wilson [296]. An extensive bibliography can be found at the web site [$\mathbf{w}^3$.24].

Exercises

8.1 (A) Perform perfect sampling of the Moran dam (Example 8.5) for the case in which V is geometric with mean 2 and $m = 3$, and p is variable. Use both independent updating and monotone updating, and compare the two methods in terms of the size of values of p for which you are able to produce Z within reasonable time.

8.2 (TP) Let $\{Y_n\}$ be a Markov chain and $X_n = f(Y_n)$. Assume that it is possible to perform perfect sampling from $\{Y_n\}$, say T_e units of computer time are needed on average to produce one copy Y^* of the stationary r.v. For estimating $f(Y^*)$, we start $\{Y_n\}$ with $Y_0 = Y^*$, run the chain up to time $m - 1$, and let $Z = Z_m = \sum_0^{m-1} Y_n/m$.

If one updating of $\{Y_n\}$ requires T units of computer time and the stationary covariance function of $\{X_n\}$ has the form $\rho_k = z^k$ with $0 < z < 1$, how would you choose m to minimize the variance for a given simulation budget c? Give numerical examples for different values of z, T_e, T, say $T_e = 10T, 100T, 1000T$.

Chapter V
Variance-Reduction Methods

The aim of variance reduction is to produce an alternative estimator $\hat{z}_{\mathrm{VR}}$ of a number $z = \mathbb{E}Z$ having, one hopes, much smaller variance than the crude Monte Carlo (CMC) estimator $\hat{z}_{\mathrm{CMC}}$. The study of such methods is a classical area in simulation, and the literature is considerable.

Variance reduction is typically most readily available in well-structured problems. Also, variance reduction typically involves a fair amount of both theoretical study of the problem in question and added programming effort. For this reason, variance reduction is most often worthwhile only if it is substantial. Assume, for example, that a sophisticated method reduces the variance by 25% compared to the CMC method, i.e., $\sigma_{\mathrm{VR}}^2 = 0.75\sigma_{\mathrm{CMC}}^2$ in obvious notation, and consider the numbers R_{CMC}, R_{VR} of replications to obtain a given precision ε (say in terms of half-width of the 95% confidence interval). Then

$$\varepsilon = \frac{1.96\sigma_{\mathrm{CMC}}}{\sqrt{R_{\mathrm{CMC}}}} = \frac{1.96\sigma_{\mathrm{VR}}}{\sqrt{R_{\mathrm{VR}}}}, \quad R_{\mathrm{VR}} = \frac{\sigma_{\mathrm{VR}}^2}{\sigma_{\mathrm{CMC}}^2}R_{\mathrm{CMC}} = 0.75\,R_{\mathrm{CMC}},$$

so that at best (assuming that the expected CPU times $T_{\mathrm{CMC}}, T_{\mathrm{VR}}$ for one replication are about equal), one can reduce the computer time by only 25%. This is in most cases of no relevance compared to the additional effort to develop and implement the variance reduction method. If $T_{\mathrm{VR}} > T_{\mathrm{CMC}}/0.75$, as may easily be the case, there is no gain at all.

Perhaps the most commonly used methods are importance sampling, control variates, and stratification, for which there is a large number of examples in which the variance reduction has turned out to be considerable. However, variance reduction is a favorite topic of both textbooks and the research literature, and many other interesting and clever ideas have been developed. We aim here for a broad survey. The examples we give are mainly at a toy level, but we will meet more substantial examples later in the text.

1   Importance Sampling

1a   Basic Theory and Examples

Suppose that we want to compute an expectation of the form $z = \mathbb{E}Z$. Importance sampling obtains its variance reduction by modifying the sampling distribution $\mathbb{P}$ so that most of the sampling is done in that part of the state space that contributes the most to z (and is therefore most "important"). Specifically, if we choose a sampling (or importance) distribution $\widetilde{\mathbb{P}}$ for which there exists a density (or Radon–Nikodym derivative) L such that

$$\mathbb{1}\{Z(\omega) \neq 0\}\,\mathbb{P}(\mathrm{d}\omega) \;=\; \mathbb{1}\{Z(\omega) \neq 0\}L(\omega)\,\widetilde{\mathbb{P}}(\mathrm{d}\omega)\,, \qquad (1.1)$$

in the sense of equality of measures, then

$$z \;=\; \mathbb{E}Z \;=\; \widetilde{\mathbb{E}}[ZL]\,,$$

where $\widetilde{\mathbb{E}}$ is the expectation associated with $\widetilde{\mathbb{P}}$. Output analysis is performed precisely as for the crude Monte Carlo method: generate R i.i.d. replicates $Z_1 L_1, \ldots, Z_R L_R$ from $\widetilde{\mathbb{P}}$, estimate by z by the average, and use the empirical variance to form a confidence interval.

Example 1.1 For a simple example, let $\mathbb{P}$ be the distribution of n i.i.d. r.v.'s $X_1, \ldots, X_n$ with a common density $f(x)$ and assume $Z = g(X_1, \ldots, X_n)$. We can then again let $X_1, \ldots, X_n$ be i.i.d. under $\widetilde{\mathbb{P}}$ but with the density changed to say $\widetilde{f}(x)$. If f and $\widetilde{f}$ have the same support, then $\mathbb{P}(\mathrm{d}\omega) = L(\omega)\,\widetilde{\mathbb{P}}(\mathrm{d}\omega)$, where

$$L \;=\; \prod_{i=1}^{n} \frac{f(X_i)}{\widetilde{f}(X_i)}\,,$$

and so certainly (1.1) holds. $\qquad\qquad\qquad\qquad\qquad\qquad\qquad\qquad\square$

Variance reduction may or may not be obtained: it depends on the choice of the importance measure $\widetilde{\mathbb{P}}$, and the problem is to make an efficient choice. There is an easily described optimal importance distribution:

Theorem 1.2 *Let* $\mathbb{P}^*$ *be defined by*

$$\frac{\mathrm{d}\mathbb{P}^*}{\mathrm{d}\mathbb{P}} \;=\; \frac{|Z|}{\mathbb{E}|Z|}\,, \quad \text{i.e.,} \quad \mathbb{P}^*(\mathrm{d}\omega) \;=\; \frac{|Z|}{\mathbb{E}|Z|}\,\mathbb{P}(\mathrm{d}\omega)$$

or $L^* = \mathbb{E}|Z|/|Z|$. Then the importance sampling estimator ZL^* under $\mathbb{P}^*$ has smaller variance than the estimator ZL under any other $\widetilde{\mathbb{P}}$. If in particular $Z \geq 0$ $\mathbb{P}$-a.s., then the $\mathbb{P}^*$-variance of ZL^* is 0.

Proof. The stated optimality follows immediately from the Cauchy–Schwartz inequality: we have

$$\mathbb{E}^*\left[(ZL^*)^2\right] = \left(\mathbb{E}|Z|\right)^2 = \left(\widetilde{\mathbb{E}}\big[|Z|L|\big]\right)^2 \leq \widetilde{\mathbb{E}}\big[(ZL)^2\big],$$

so that the estimator under $\mathbb{P}^*$ has the smaller second moment and hence the smaller variance. If $Z \geq 0$ $\mathbb{P}$-a.s., then

$$\mathbb{V}ar^*(ZL^*) = \mathbb{E}^*\left[(ZL^*)^2\right] - \left(\mathbb{E}^*(ZL)\right)^2 = \left(\mathbb{E}Z\right)^2 - z^2 = 0. \qquad \square$$

The optimal choice $\widetilde{\mathbb{P}} = \mathbb{P}^*$ can, however, never be implemented in practice, since if say $Z \geq 0$, the evaluation of the estimator involves knowledge of the unknown z. Nevertheless, Theorem 1.2 suggests that large variance reduction can be achieved by sampling outcomes $\omega \in \Omega$ in rough proportion to $|Z(\omega)|$.

Example 1.3 (COMPUTING PROBABILITIES) If $Z = \mathbb{P}(A)$, then

$$\mathbb{P}^*(\mathrm{d}\omega) = \frac{\mathbb{1}\{\omega \in A\}}{\mathbb{P}(A)}\,\mathbb{P}(\mathrm{d}\omega),$$

so that $\mathbb{P}^*(\cdot) = \mathbb{P}(\cdot\,|\,A)$. Thus, when computing probabilities, we wish to use a sampling distribution $\widetilde{\mathbb{P}}$ that resembles as closely as possible the conditional distribution of $\mathbb{P}$ given A. This insight underlies many of the most succesful "rare event" importance sampling algorithms in Chapter VI.
$\square$

Example 1.4 Consider Monte Carlo integration of $g : [0,1]^d \to \mathbb{R}$, i.e., we wish to compute

$$z \stackrel{\text{def}}{=} \int_{[0,1]^d} g(\boldsymbol{u})\,\mathrm{d}\boldsymbol{u} = \mathbb{E}g(\boldsymbol{U}),$$

where $\boldsymbol{U}$ follows the uniform distribution $\mathbb{P}(\mathrm{d}\boldsymbol{u}) = \mathrm{d}\boldsymbol{u}$ over the hypercube $(0,1)^d$. In this case,

$$\mathbb{P}^*(\mathrm{d}\boldsymbol{u}) = \frac{|g(\boldsymbol{u})|}{z}\,\mathrm{d}\boldsymbol{x}.$$

The key here is therefore to create an importance distribution that samples more frequently from the part of $[0,1]^d$ in which $|g|$ is largest (and hence most "important"). While creating such a sampling distribution is straightforward when $d = 1$ (where, of course, conventional numerical integration generally is more efficient), the challenges of creating an efficient $\widetilde{\mathbb{P}}$ can be substantial for large values of d.
$\square$

Example 1.5 (COMPUTING GAUSSIAN PROBABILITIES) Suppose that we want to compute $z = \mathbb{P}(\boldsymbol{X} \in A)$, where $\boldsymbol{X}$ is a d-dimensional Gaussian random vector with mean $\boldsymbol{0}$ and covariance matrix $\boldsymbol{C}$. If A contains the mean $\boldsymbol{0}$, the event $\boldsymbol{X} \in A$ is typically not rare, and the use of importance sampling is generally unnecessary. If, on the other hand, $\boldsymbol{0} \notin A$ and z is small, one may wish to consider the use of importance sampling. The measure $\mathbb{P}^*$ suggests sampling in proportion to

$$\varphi_{\boldsymbol{C}}(\boldsymbol{x})\mathbb{1}\{\boldsymbol{x} \in A\}\,\mathrm{d}\boldsymbol{x}\,,$$

where $\varphi_{\boldsymbol{C}}(\cdot)$ is the density of $\boldsymbol{X}$. Given the rapid decay of $\varphi_{\boldsymbol{C}}(\boldsymbol{x})$ for large $\boldsymbol{x}$, most of the mass of $\mathbb{P}^*$ is typically located at the maximizer $\boldsymbol{x}^*$ of $\varphi_{\boldsymbol{C}}(\cdot)$ over A (which we assume exists uniquely). This suggests using a sampling distribution $\widetilde{\mathbb{P}}$ that concentrates most of its mass near $\boldsymbol{x}^*$, that makes L easily computable, and from which variates can efficiently be generated. One such distribution is the Gaussian distribution having mean $\boldsymbol{x}^*$ and covariance matrix $\boldsymbol{C}$, in which case

$$L \;=\; \exp\Big\{-\boldsymbol{x}^{*\top}\boldsymbol{C}^{-1}\boldsymbol{X} + \frac{1}{2}\boldsymbol{x}^{*\top}\boldsymbol{C}^{-1}\boldsymbol{x}^*\Big\}\,.$$

When $A \subseteq \prod_1^d[x_i, \infty)$, the resulting variance reduction can often be substantial. $\qquad\square$

Example 1.6 Consider the problem of evaluating $\mathbb{P}(X > K)$ or $\mathbb{E}[X - K]^+$, where X is a r.v. and K a constant; thus, $Z = \mathbb{1}\{X > K\}$ or $Z = [X - K]^+$. For example, X is the potential loss of an insurance company or investment fund in a certain period, or, as in option pricing, the price of an underlying asset at maturity. Suppose further that K substantially exceeds the typical range of X (the "out-of-the-money" case in option pricing) and that X has a distribution F that can naturally be viewed as a member of a certain parametric class $(F_\theta)_{\theta \in \Theta}$, say $F = F_{\theta_0}$. A simple way to implement the idea of making large values of Z more likely is to use an importance sampling in which θ_0 is replaced by a $\widetilde{\theta}$ with the property that K is in the center of $F_{\widetilde{\theta}}$. For example, this may be achieved by choosing $\widetilde{\theta}$ to satisfy $\mathbb{E}_{\widetilde{\theta}}X = K$; we discuss this choice in more detail in VI.2c. In the option pricing example, a natural alternative is $\mathbb{E}_{\widetilde{\theta}}\log X = \log K$.

In Asian options, X is the (discretely or continuously sampled) average of a geometric Brownian motion $S(0)e^{W(t)}$, i.e., W is a Brownian motion with drift say θ and variance σ^2. The equation $\mathbb{E}_{\widetilde{\theta}}X = K$ is somewhat complicated, though certainly solvable numerically, but a naive way to make large values of Z more likely is to replace θ by a larger $\widetilde{\theta}$. $\qquad\square$

1b   Exponential Change of Measure

The approach followed in Example 1.5 can easily be generalized. If $\boldsymbol{X}$ is a random vector for which $\mathbb{E}\exp\{\boldsymbol{\theta}^\top\boldsymbol{X}\} < \infty$, then the "exponentially tilted measure"

$$\mathbb{P}_{\boldsymbol{\theta}}(\boldsymbol{X} \in \mathrm{d}x) \overset{\text{def}}{=} \exp\{\boldsymbol{\theta}^{\mathsf{T}}\boldsymbol{x} - \kappa(\boldsymbol{\theta})\}\,\mathbb{P}(\boldsymbol{X} \in \mathrm{d}x) \qquad (1.2)$$

(where $\kappa(\boldsymbol{\theta}) \overset{\text{def}}{=} \log \mathbb{E}\exp\{\boldsymbol{\theta}^{\mathsf{T}}\boldsymbol{X}\}$) is a probability distribution that in many cases has the same parametric form as that of $\boldsymbol{X}$, so that variates can easily be generated. The family of all distributions of this form is called the *exponential family* generated by (the distribution of) $\boldsymbol{X}$.

One-dimensional examples are the following ($f(x)$ is the density of X in the continuous case, $p(k)$ the probability mass function in the discrete case):

(i) the exponential distribution $f(x) = \lambda e^{-\lambda x}$, where $f_\theta(x) = \lambda_\theta e^{-\lambda_\theta x}$ where $\lambda_\theta = \lambda - \theta$, $-\infty < \theta < \lambda$. More generally, for the Gamma density $f(x) = \lambda^\alpha x^{\alpha-1}e^{-\lambda x}/\Gamma(\alpha)$, one gets $f_\theta(x) = \lambda_\theta^\alpha x^{\alpha-1}e^{-\lambda_\theta x}/\Gamma(\alpha)$, where again $\lambda_\theta = \lambda - \theta$;

(ii) the binomial (N, α) distribution $p(k) = \binom{N}{k}\alpha^k(1-\alpha)^{N-k}$, $k = 0, \ldots, N$, where p_θ is binomial (N, α_θ) with $\alpha_\theta = \alpha e^\theta/(1 - \alpha + \alpha e^\theta)$, $\theta \in \mathbb{R}$;

(iii) the Poisson distribution $p(k) = e^{-\mu}\mu^k/k!$, where p_θ is again Poisson with $\mu_\theta = \mu e^\theta$, $\theta \in \mathbb{R}$;

(iv) The $\mathcal{N}(\mu, \sigma^2)$ distribution, where $f_\theta(x)$ is the $\mathcal{N}(\mu + \theta\sigma^2, \sigma^2)$ density.

A standard multivariate example is

(v) the $\mathcal{N}(\boldsymbol{\mu}, \boldsymbol{C})$ distribution, where the exponentially tilted measure is $\mathcal{N}(\boldsymbol{\mu} + \boldsymbol{C}\boldsymbol{\theta}, \boldsymbol{C})$.

Example 1.7 (I.I.D. SAMPLING FROM EXPONENTIAL FAMILIES) As in Example 1.1, let $X_1, \ldots, X_n$ be i.i.d. with common density $f(x)$ and suppose that the importance distribution preserves the i.i.d. property but changes $f(x)$ to $\widetilde{f}(x) = f_\theta(x) = e^{\theta x - \kappa(\theta)}f(x)$. Then the likelihood ratio takes a particularly simple form,

$$L = \prod_{i=1}^{n} \frac{f(X_i)}{\widetilde{f}(X_i)} = e^{-\theta S_n + n\kappa(\theta)},$$

where $S_n \overset{\text{def}}{=} X_1 + \cdots + X_n$. $\qquad\qquad\square$

Exponential tilting is often used in problems involving light-tailed r.v.'s (meaning that $\kappa(\theta)$ exists for all sufficiently small $\theta > 0$) to make, for instance, large values of $S_n = X_1 + \cdots + X_n$ more likely. If we aim for values of x_0 or larger, a common choice is to choose θ such that $\mathbb{E}_\theta S_n = x_0$; this θ is often referred to as a *saddle point* (analytically, we have $\mathbb{E}_\theta S_n = n\kappa'(\theta)$). In fact, in VI.2c we will show certain optimality properties of exponential change of measure in this setting. It should be stressed, however, that exponential change of measure implemented via the saddle-point method often is efficient in situations not incorporated in this i.i.d. sums setting. For detailed efficiency analysis in some simple examples of exponential change of measure, see VI.1.1–1.4.

For some distributions, the exponentially tilted distribution does not belong to the same parametric family as f; an example is the Pareto distribution with $f(x) = \alpha/(1+x)^\alpha$, $x > 0$, where $f_\theta(x)$ is well defined for $\theta < 0$ but is not a standard distribution. In such examples, r.v. generation may not always be straightforward.

1c   Stochastic Process Examples

Example 1.8 (BROWNIAN MOTION) Let $\{W(t)\}_{0 \le t \le T}$ be Brownian motion with drift μ and variance constant σ^2 on an interval $[0, T]$. Consider the likelihood $d\widetilde{\mathbb{P}}/dP = e^{\theta W(T) - \kappa_T(\theta)}$ where

$$\kappa_T(\theta) \overset{\text{def}}{=} \log \mathbb{E}e^{\theta W(T)} = T\mu\theta + T\theta^2\sigma^2/2\,.$$

Then under $\widetilde{\mathbb{P}}$, $\{W(t)\}$ is Brownian motion with drift $\mu + \theta\sigma^2$ and the same the variance σ^2. Indeed, the discussion in Section 1b shows that for any n the increments $W(iT/n) - W((i-1)T/n)$, $i = 1, \dots, n$, are i.i.d. and $\mathcal{N}((\mu + \theta\sigma^2)T/n, \sigma^2 T/n)$ as should be, and this is sufficient for the Brownian property (the continuity of sample paths follows by absolute continuity).

The quadratic variation $\langle W, W \rangle_T$ of the Brownian motion in $[0, T]$ is $T\sigma^2$, which implies that the set $A(\sigma^2)$ given by

$$A(\sigma^2) \overset{\text{def}}{=} \left\{ \frac{1}{T} \sum_{i=1}^{2^k} \left[W(iT/2^k) - W((i-1)T/2^k) \right]^2 \overset{\text{a.s.}}{\to} \sigma^2 \right\}$$

has $\mathbb{P}$-probability 1. Therefore, if $\widetilde{\mathbb{P}}$ is a distribution under which $\{W(t)\}$ is Brownian motion with some arbitrary drift $\widetilde{\mu}$ and variance $\widetilde{\sigma}^2 \ne \sigma^2$, $\widetilde{\mathbb{P}}$ is concentrated on $A(\widetilde{\sigma}^2)$, so that $\widetilde{\mathbb{P}}[A(\sigma^2)] = 0$, which together with $\mathbb{P}[A(\sigma^2)] = 1$ excludes the absolute continuity. Therefore, importance sampling using a changed variance is not possible in the idealized situation in which one could generate a whole trajectory and not just discrete skeletons. □

Example 1.9 (STOPPED SEQUENCES) As an extension of Example 1.1, let $X_1, X_2, \dots$ be i.i.d. with common density $f(x)$ and assume $Z = g(X_1, \dots, X_\tau)\mathbb{1}\{\tau < \infty\}$, where τ is a stopping time adapted to the X_i. The formulas of Example 1.1 can then easily be generalized to

$$z = \mathbb{E}Z = \widetilde{\mathbb{E}}\big[g(X_1, \dots, X_\tau)L_\tau; \tau < \infty\big] = \widetilde{\mathbb{E}}[ZL_\tau]\,, \qquad (1.3)$$

where $L_n \stackrel{\text{def}}{=} \prod_{i=1}^{n} f(X_i)/\widetilde{f}(X_i)$. To see this, assume w.l.o.g. $Z \geq 0$. We then get

$$
\begin{aligned}
\mathbb{E}Z &= \mathbb{E}[Z; \tau < \infty] = \sum_{n=1}^{\infty} \mathbb{E}[Z; \tau = n] = \sum_{n=1}^{\infty} \widetilde{\mathbb{E}}[ZL_n; \tau = n] \\
&= \widetilde{\mathbb{E}}ZL_\tau \mathbb{1}\{\tau < \infty\} = \widetilde{\mathbb{E}}[ZL_\tau].
\end{aligned}
$$

$\square$

Example 1.10 (MARKOV CHAINS IN DISCRETE TIME) Suppose that we wish to compute $z \stackrel{\text{def}}{=} \mathbb{E}[g(X_0, X_1, \ldots, X_\tau); \tau < \infty]$, where τ is a stopping time adapted to $\{X_n\}_{n \in \mathbb{N}}$. If $\{X_n\}$ evolves as an E-valued Markov chain (possibly time-inhomogeneous) under $\mathbb{P}$, it is natural to use importance measures preserving the Markov property (so as to guarantee that the paths can be efficiently simulated under $\widetilde{\mathbb{P}}$). Assume that the importance measure satisfies

$$
\begin{aligned}
\mathbb{P}(X_0 \in dy) &= p_0(y)\widetilde{\mathbb{P}}(X_0 \in dy), \\
\mathbb{P}(X_n \in dy \mid X_{n-1} = x) &= p_n(x, y)\widetilde{\mathbb{P}}(X_n \in dy \mid X_{n-1} = x)
\end{aligned}
$$

for $n \geq 1$. It follows that

$$
z = \widetilde{\mathbb{E}}[g(X_0, X_1, \ldots, X_\tau)L_\tau; \tau < \infty],
$$

where L_τ is the likelihood ratio

$$
L_\tau \stackrel{\text{def}}{=} p_0(X_0) \prod_{i=1}^{\tau} p_n(X_{i-1}, X_i).
$$

Note that L_n can be recursively computed as a function of n. Of course, only importance measures with $\widetilde{\mathbb{P}}(\tau < \infty) = 1$ are practicable, and then $z = \widetilde{\mathbb{E}}[g(X_0, X_1, \ldots, X_\tau)L_\tau]$.

In particular, if $\mathbb{P}$, $\widetilde{\mathbb{P}}$ correspond to discrete time-homogeneous Markov chains with transition probabilities $p_{ij}, \widetilde{p}_{ij}$ and the same initial distribution, then $L_n = \prod_1^n p_{X_{i-1}X_i}/\widetilde{p}_{X_{i-1}X_i}$. $\square$

Example 1.11 (DISCRETE MARKOV PROCESSES IN CONTINUOUS TIME) Let $\{X(t)\}_{t \geq 0}$ be a stochastic process taking values in a discrete state space E and evolving under $\mathbb{P}$ as a nonexplosive time-homogeneous Markov chain with intensity matrix $\boldsymbol{A} = (A(x, y))_{x,y \in E}$. Let $J(t)$ be the number of jumps of $\{X(t)\}$ in $[0, t]$ and $T_1, T_2, \ldots$ the consecutive jump epochs. If we choose an importance measure $\widetilde{\mathbb{P}}$ under which $\{X(t)\}$ continues to be time-homogeneous Markov, with intensity matrix $\widetilde{\boldsymbol{A}} = (\widetilde{A}(x, y))_{x,y \in E}$ satisfying

$$
\mathbb{P}(X(0) \in dy) = p_0(y)\widetilde{\mathbb{P}}(X(0) \in dy), \quad A(x, y) = a(x, y)\widetilde{A}(x, y),
$$

then

$$z = \mathbb{E}\big[g\big((X(s))_{0 \le s \le \tau}\big); \tau < \infty\big] = \widetilde{\mathbb{E}}\big[g\big((X(s))_{0 \le s \le \tau}\big)L_\tau; \tau < \infty\big],$$

where

$$L_\tau = p(X(0)) \prod_{i=1}^{J(\tau)} a\big(X(T_{i-1}), X(T_i)\big)$$

$$\times \exp\Big\{ -\int_0^\tau \widetilde{A}\big(X(s), X(s)\big)\big[1 - a\big(X(s), X(s)\big)\big]\,\mathrm{d}s \Big\},$$

provided that τ is a stopping time adapted to $\{X(t)\}$. □

Example 1.12 (GENERALIZED SEMI-MARKOV PROCESSES) Let $\{S(t)\}_{t \ge 0}$ be a nonexplosive S-valued process that evolves according to a GSMP under $\mathbb{P}$, cf. II.6. As in our earlier examples, it is natural to choose $\widetilde{\mathbb{P}}$ preserving the GSMP structure. Specifically, suppose that the probability $\widetilde{p}(s'; s, e^*)$ governing state transitions from s to s' initiated by the event e^* (under $\widetilde{\mathbb{P}}$) is positive whenever $p(s'; s, e^*)$ is positive. In addition, suppose that the distributions $\widetilde{F}(\cdot; s', e', s, e)$ used by $\widetilde{\mathbb{P}}$ to stochastically set clock e' in state s' after a transition from s triggered by event e satisfy

$$F\big(\mathrm{d}t; s', e', s, e\big) = f\big(\mathrm{d}t; s', e', s, e\big)\widetilde{F}\big(\mathrm{d}t; s', e', s, e\big).$$

Let $J(t)$ be the number of transitions of $\{S(t)\}$ over $[0, t]$, and let $T_1, T_2, \dots$ be the corresponding transition epochs. Then

$$z \overset{\text{def}}{=} \mathbb{E}\big[g\big(S(u)\big)_{0 \le u \le t}\big] = \widetilde{\mathbb{E}}\big[g\big(S(u)\big)_{0 \le u \le t}L_t\big],$$

where

$$L_t = \prod_{i=1}^{J(t)} \frac{p\big(S(T_i); S(T_i-), e^*(T_i-)\big)}{\widetilde{p}\big(S(T_i); S(T_i-), e^*(T_i-)\big)}$$

$$\times \prod_{i=1}^{J(t)} \prod_{e' \in N_i} f\big(C_{e'}(T_i); S(T_i), e', S(T_i-), e^*(T_i-)\big),$$

where $N_i \overset{\text{def}}{=} N\big(S(T_i); S(T_i-), e^*(T_i-)\big)$. Hence even in the setting of such general discrete-event systems, the likelihood ratio for complex importance distributions can easily be computed. □

Example 1.13 (COMPOUND POISSON PROCESSES) Let $X(t) = Y_1 + \cdots + Y_{N(t)}$, where $\{N(t)\}, Y_1, Y_2, \dots$ are independent, $\{N(t)\}$ is Poisson(λ) with epochs $T_1, T_1 + T_2, \dots$, and the Y_i have common density $f(y)$. If we take $\widetilde{\mathbb{P}}$ to again make $\{X(t)\}$ compound Poisson with parameters $\widetilde{\lambda}, \widetilde{f}$, then

$$L_t = \frac{\mathrm{e}^{-\lambda R(t)}}{\mathrm{e}^{-\widetilde{\lambda} R(t)}} \prod_{i=1}^{N(t)} \frac{f(Y_i)}{\widetilde{f}(Y_i)} \cdot \frac{\lambda \mathrm{e}^{-\lambda T_i}}{\widetilde{\lambda} \mathrm{e}^{-\widetilde{\lambda} T_i}} = \Big(\frac{\lambda}{\widetilde{\lambda}}\Big)^{N(t)} \mathrm{e}^{-(\lambda - \widetilde{\lambda})t} \prod_{i=1}^{N(t)} \frac{f(Y_i)}{\widetilde{f}(Y_i)},$$

where $R(t) = t - T_1 - \cdots - T_{N(t)}$ is the time from the last event to t.   □

1d Remarks

Remark 1.14 In choosing the importance distribution in a parametric class $(\mathbb{P}_\theta)_{\theta \in \Theta}$ (say all multivariate normals with a fixed covariance matrix but varying mean θ as in Examples 1.5, 1.6), the picture is often the following. A certain θ^* is optimal in the sense of minimizing the variance. In moving in a wrong direction (away from θ^* instead of towards θ^*), say by overdoing the importance sampling in the sense of moving beyond θ^*, the variance blows up and may even become infinite, so that the usual normal confidence intervals do not make sense (note that this problem is not automatically discovered by a simulation run, since the empirical variance will always be finite!). Even within the range where the variance is finite but large, the normal approximation to $\hat{z}$ will be poor, which is reflected in poor coverage properties of the confidence intervals. For example, in rare event simulation as studied in Chapter VI, the estimates will have a highly skewed distribution if the importance sampling is overdone, so that they are typically on the low side (even if they are unbiased in the statistical sense), and the variance estimates will be on the low side as well, so that the confidence interval will end up being below the correct value of z.

To illustrate some of these effects, consider the (highly unrealistic!) example of estimating $z = \int_0^\infty x e^{-x}\, dx$ via the CMC estimator $Z = X(1)$, where $X(\lambda)$ is an exponential r.v. with rate λ. Doing importance sampling by changing 1 to some other λ, the estimator becomes

$$Z(\lambda) = X(\lambda)\frac{e^{-X(\lambda)}}{\lambda e^{-\lambda X(\lambda)}} = X(\lambda)e^{-(1-\lambda)X(\lambda)}/\lambda.$$

Here

$$\mathbb{E}Z(\lambda)^2 = \int_0^\infty x^2 e^{-2(1-\lambda)}/\lambda^2 \cdot \lambda e^{-\lambda x}\, dx = \begin{cases} \dfrac{2}{\lambda(2-\lambda)^3}, & 0 < \lambda < 2, \\ \infty, & \lambda \geq 2, \end{cases}$$

so $\mathbb{V}\mathrm{ar}\, Z(\lambda) = \infty$ when $\lambda > 2$.

Intuitively, taking $\lambda > 1$ is moving in the wrong direction, and indeed one also finds the optimal λ to be $\lambda^* = 1/2$ (the maximizer of $\lambda(2-\lambda)^3$); the corresponding variance is $\mathbb{V}\mathrm{ar}\, Z(\lambda^*) = 2/\lambda^*(2-\lambda^*)^3 - 1 = 5/27$, to be compared with $\mathbb{V}\mathrm{ar}\, Z(1) = 1$ for the CMC estimator. When moving beyond λ^* to smaller λ's, there is no longer variance reduction when $\lambda(2-\lambda)^3 < 1$, i.e., when $\lambda < 0.161$, and as $\lambda \downarrow 0$, $\mathbb{V}\mathrm{ar}\, Z(\lambda) \to \infty$.

Similarly,

$$\mathbb{E}\big(Z(\lambda)^3 - z\big)^3 = \mathbb{E}Z(\lambda)^3 - 3z\mathbb{E}Z(\lambda)^2 + 3z^2\mathbb{E}Z(\lambda) - z^3$$

$$= \frac{6}{\lambda^2(3-2\lambda)^4} - \frac{6}{\lambda(2-\lambda)^3} + 3 - 1.$$

Thus the skewness $\mathbb{E}(Z(\lambda)^3 - z)^3 / \mathbb{V}ar^{3/2} Z(\lambda)$ goes to ∞ as $\lambda \downarrow 0$ or $\lambda \uparrow 3/2$, indicating a degradation of the CLT for $\widehat{z}(\lambda)$ (recall that the first-order Edgeworth correction in the CLT for a normalized average is proportional to the skewness). $\qquad\square$

Remark 1.15 Consider a CMC simulation with $Z > 0$, using i.i.d. replicates $Z_1, \ldots, Z_R$. Then $\omega_r \stackrel{\text{def}}{=} Z_r / (Z_1 + \cdots + Z_R)$ is the fraction with which replication r contributes to the overall estimate $\widehat{z}$. From the standard fact that

$$\frac{1}{R} \leq \sum_{r=1}^{R} \omega_r^2 \leq 1\,,$$

with the lower bound obtained if and only if all $\omega_r = 1/R$ and the upper if and only if one ω_r equals 1, one may take a value of $\sum \omega_r^2$ that is close to 1 as an indication that a few Z_r contain the main contribution to $\widehat{z}$, which could motivate trying importance sampling.

Similarly, for a given importance sampling scheme one may take a value close to 1 of

$$\sum_{r=1}^{R} \frac{L_r^2 Z_r^2}{\left(\sum_1^R L_s^2 Z_s^2\right)^2}$$

as a warning that the importance sampling has not been successful in terms of sampling evenly in the important area.

See further the discussion in Evans & Swartz [114, pp. 170–171]. $\qquad\square$

Example 1.16 In Chapter XIII, we will study the problem of obtaining information on a distribution Π^* whose density w.r.t. some μ, say Lebesgue measure or counting measure, is known only up to a constant. That is, $d\Pi^*/d\mu = \pi(x)/\overline{\pi}$, where $\overline{\pi} \stackrel{\text{def}}{=} \int \pi \, d\mu$ is unknown but $\pi(x)$ is known.

If we just want a functional of the form $z = \int g \, d\Pi^* = \int g\pi \, d\mu/\overline{\pi}$, there is an easy solution using importance sampling: assume that it is easy to sample from a distribution with a density $\widetilde{f}(x)$ w.r.t. μ. Then generate R replicates $X_1, \ldots, X_R$ from $\widetilde{f}(x)$ and let

$$\overline{z} = \frac{g(X_1)\pi(X_1)/\widetilde{f}(X_1) + \cdots + g(X_R)\pi(X_R)/\widetilde{f}(X_R)}{\pi(X_1)/\widetilde{f}(X_1) + \cdots + \pi(X_R)/\widetilde{f}(X_R)}\,,$$

noting that the average of the numerator estimates $\int g\pi \, d\mu$ and the average of the denominator $\int \pi \, d\mu = \overline{\pi}$.

Note that in this setting, the main goal of importance sampling is not variance reduction but to obtain information on a distribution that is not easily simulated itself (cf. also Example I.5.2). Examples and further discussion will be given in XIV.6e, XIV.2. $\qquad\square$

Remark 1.17 One sometimes hears the claim that importance sampling is inefficient in high-dimensional spaces because the variance of the likelihood ratio blows up.

There is certainly some truth to this. Consider just i.i.d. sampling with R replications, where the nominal density f and the importance density $\widetilde{f}$ do not depend on R. The likelihood ratio $L = L_R$ in Example 1.1 can then be written as $Y_1 \cdots Y_n$, where $Y_k \overset{\text{def}}{=} f(X_k)/\widetilde{f}(X_k)$. Here

$$\delta \overset{\text{def}}{=} \widetilde{\mathbb{E}} Y_k^2 > (\widetilde{\mathbb{E}} Y_k)^2 = 1^2 = 1 \, .$$

Therefore $\widetilde{\mathbb{E}} L_R^2 = \delta^R$ grows exponentially fast in R and hence so does $\widetilde{\mathbb{V}ar}\, L_R$, since $\mathbb{E} L_R = 1$.

The same problem occurs in importance sampling for long realizations of a Markov chain, as will be shown in XIV.5.5.

Consider, however, a single realization of standard Brownian motion $\{B(t)\}$ in the interval $[0,1]$ and an importance distribution that corresponds to changing the drift from 0 to μ. High dimensionality then occurs if the sample path is generated at high resolution by sampling $B(0), B(1/n), \ldots, B(1)$ for some large n. However, according to Example 1.8, the likelihood ratio $L_n = e^{-\mu B(1) + \mu^2/2}$ does not depend on n, and so of course neither does $\mathbb{V}ar\, L_n$!   □

Exercises

1.1 (TP) Consider Monte Carlo integration with $g \geq 0$ and assume that $g \leq ch$ for some $c > 1$ and some density h. The *hit-or-miss* estimator of $z \overset{\text{def}}{=} \int_0^1 g(u)\, du$ is $c\mathbb{1}\{Uch(Y) \leq g(Y)\}$, where U, Y are generated independent as uniform$(0,1)$, respectively from the density h. Show that its expectation is $z = \int g$ as desired, but that the variance is always at least the variance of the importance sampling estimator using sampling from h.

1.2 (TP) Let F be $\mathcal{N}(\mu, \sigma^2)$ and consider a likelihood ratio $(\mathrm{d}\widetilde{F}/\mathrm{d}F)(x)$ proportional to $e^{ax - bx^2}$. Identify $\widetilde{F}$.

1.3 (TP) In the counting problem in I.5.3, show that $\mathbb{V}ar\, Z$ is minimized by taking $p(\cdot)$ as uniform on S.

1.4 (A) In Example 1.5, take $d = 2$, $\boldsymbol{A} = \{(x,y) : x \geq a,\ y \geq a\}$. and

$$\boldsymbol{C} = \begin{pmatrix} 4 & -1 \\ -1 & 4 \end{pmatrix}.$$

For $a = 1, 3, 10$:

(i) Try first to give simulation estimates of $z = \mathbb{P}(\boldsymbol{X} \in \boldsymbol{A})$ and associated 95% confidence intervals using the CMC method.

(ii) Find next the point $\boldsymbol{b} \in \boldsymbol{A}$ that maximizes the $\mathcal{N}(\boldsymbol{0}, \boldsymbol{C})$ density and repeat (i), with the CMC method replaced by importance sampling, where the importance distribution is $\mathcal{N}(\boldsymbol{b}, \boldsymbol{C})$.

(iii) In (ii), experiment with importance distributions of the form $\mathcal{N}(\boldsymbol{b}, \delta\boldsymbol{C})$ for different $\delta \in \mathbb{R}$.

1.5 (TP) Verify the claims (i)–(v) in Section 1b.

1.6 (TP) Consider as in Exercise I.6.4 the Monte Carlo evaluation of the level z of the test for $\theta = \theta_0$ in the binomial distribution, which rejects for large values of $-2 \log Q$, where Q is the likelihood ratio. An obvious idea is to perform importance sampling, where one simulates from $\widetilde{\theta}$ rather than θ_0. When doing so, it was found that this did not provide any variance reduction, but that in contrast, simulation from θ_0 was close to being variance minimal. Can you find an explanation for this? And can you come up with an idea for how to choose an importance distribution that does give variance reduction?

1.7 (A) Consider a European call option with a maturity of $T = 3$ years, strike price K, underlying asset price process $\{S(t)\}_{0 \le t \le T}$ with $S(0) = 100$, and risk-free interest rate 4%. It is assumed that $\{S(t)\}_{0 \le t \le T}$ evolves like geometric Brownian motion with stochastic volatility $\sigma(t)$, such that $\{\sigma(t)\}$ is Markov with two states $\sigma = 0.25$ (the baseline volatility) and 0.75, and switching intensities $\lambda_+ = 1$, $\lambda_- = 3$ for transitions $0.25 \to 0.75$, respectively $0.75 \to 0.25$. It is easy to see that a risk-neutral measure can be described by

$$\mathrm{d}X(t) = \left(r - \sigma(t)^2/2\right)\mathrm{d}t + \sigma(t)\,\mathrm{d}B(t),$$

where $X(t) \overset{\text{def}}{=} \log S(t) - \log S(0)$ and B is standard Brownian motion..
Taking $K = 150$, do importance sampling using a change of drift $0 = \mu \to \mu^*$ for $\{B(t)\}$, choosing μ^* to satisfy

$$S(0) \exp\left\{ (r - \sigma^2/2)T + \mathbb{E}_{\mu^*} B(T) \right\} \approx K$$

(cf. Example 1.6; the implementation here simply ignores the period with the higher volatility).
What do you expect from the same idea when $K = 100$ or 50?

1.8 (A) Consider a European call basket option with payout $e^{-rT}\left[S(T) - K\right]^+$ and $S(t) = S_1(t) + \cdots + S_{10}(t)$, where the log-returns of the 10 spot prices are geometric Brownian motions and we for simplicity assume independence and that the yearly volatilities σ_i all equal $\sigma = 0.25$. The initial spot prices are $6, \ldots, 15$, we take $T = 2$ years, $r = 4\%$, and $K = 300$. Since $S(0) = 105$, we are thus in the "out-of-the-money" case. The assignment is to illustrate the efficiency of importance sampling by comparing the half-width of the confidence interval for the price Π to that of the crude Monte Carlo method. The importance distribution (this is only one possibility) is obtained by adding the same μ to the drifts $r - \sigma_i^2$ under the risk-neutral measure, with μ determined from pivotal runs such that $\widetilde{\mathbb{E}}S(T) \approx K$.

1.9 (A) Consider the Strauss model for a point process on $[0,4]^2$ with $\lambda = 1$, $r = 0.5$; cf. I.5.2. Use Example 1.16 to give histograms of the distribution of the number $m(x)$ of points for different η.

1.10 (TP) Let $X = \{X_n\}_{n \in \mathbb{N}}$ be a Markov chain with finite state space E and transition matrix $\boldsymbol{P} = (p_{xy})_{x,y \in E}$. Let $B \subset E$ be a proper subset such that all states in B communicate and that $p_{xy} > 0$ for some $x \in B$ and some $y \notin B$, and define $T \overset{\text{def}}{=} \inf\{n : X_n \notin B\}$. We are interested in estimating $z \overset{\text{def}}{=} \mathbb{P}_x(T > n)$, $x \in B$, by simulation for large n. More precisely, we will suggest an estimator $Z = Z_n$ that has the asymptotically best-possible order of variance (see further Chapter VI).

To this end, let $\boldsymbol{P}^B \overset{\text{def}}{=} (p_{xy})_{x,y \in B}$. By the Perron–Frobenius theorem, $\boldsymbol{P}^B$ has an eigenvalue $0 < \theta < 1$ with associated right eigenvector $\boldsymbol{h} = \big(h(y)\big)_{y \in B}$ with strictly positive entries $h(y)$, i.e., $\boldsymbol{P}^B \boldsymbol{h} = \theta\boldsymbol{h}$. Define $\boldsymbol{Q} \stackrel{\text{def}}{=} (q_{xy})_{x,y \in B}$ by

$$q_{xy} \stackrel{\text{def}}{=} \frac{1}{\theta h(x)} q_{xy} h(y), \quad x, y \in B.$$

(i) Show that $\boldsymbol{Q}$ is a transition matrix on B.
(ii) Suppose we simulate X on B according to $\boldsymbol{Q}$. Compute the likelihood ratio for X on $[0, t]$.
(iii) Use (ii) and the existence of a limiting stationary distribution for $\boldsymbol{Q}$ to show that $\mathbb{P}(T > n) \sim C\theta^n$ for some constant $0 < C < \infty$.
(iv) If Z is the importance sampling estimator for $z = \mathbb{P}_x(T > n)$ given by (ii) and Z^* any other unbiased estimator, show that

$$\limsup_{n \to \infty} \frac{\mathbb{E}Z^2}{\mathbb{E}Z^{*2}} < \infty.$$

1.11 (TP) Find the density of an inhomogeneous Poisson process with intensity $\lambda(t)$ on $[0, T]$ w.r.t. the standard Poisson process at unit rate. Generalize to multidimensions.

2   Control Variates

The idea is to look for an r.v. W that has a strong correlation (positive or negative) with Z and a *known* mean w, generate $(Z_1, W_1), \dots, (Z_R, W_R)$ rather than $Z_1, \dots, Z_R$, and combine the empirical means $\widehat{z}, \widehat{w}$ to an estimator with lower variance than the CMC estimator $\widehat{z}$ of $z = \mathbb{E}Z$.

The naive method is to choose some arbitrary constant α and consider the estimator $\widehat{z} + \alpha(\widehat{w} - w)$. The point is that since w is known, we are free just to add a term $\alpha(\widehat{w} - w)$ with mean zero to the CMC estimator $\widehat{z}$, so that unbiasedness is preserved. The variance is

$$\sigma_Z^2 + \alpha^2 \sigma_W^2 + 2\alpha \sigma_{ZW}^2, \tag{2.1}$$

where

$$\sigma_Z^2 \stackrel{\text{def}}{=} \mathbb{V}\mathrm{ar}\, Z, \quad \sigma_W^2 \stackrel{\text{def}}{=} \mathbb{V}\mathrm{ar}\, W, \quad \sigma_{ZW}^2 \stackrel{\text{def}}{=} \mathbb{C}\mathrm{ov}(Z, W).$$

In general, nothing can be said about how (2.1) compares to the variance σ_Z^2 of the CMC estimator $\widehat{z}$ (though sometimes a naive choice such as $\alpha = 1$ works to produce a lower variance). However, it is easily seen that (2.1) is minimized for $\alpha = -\sigma_{ZW}^2/\sigma_W^2$, and that the minimum value is

$$\sigma_Z^2(1 - \rho^2), \quad \text{where} \quad \rho \stackrel{\text{def}}{=} \mathbb{C}\mathrm{orr}(Z, W) = \frac{\sigma_{ZW}^2}{\sqrt{\sigma_Z^2 \sigma_W^2}}. \tag{2.2}$$

One then simply estimates the optimal α by replacing σ_{ZW}^2, σ_W^2 by their empirical values,

$$\widehat{\alpha} \stackrel{\text{def}}{=} -\frac{s_{ZW}^2}{s_W^2},$$

where

$$s_Z^2 \overset{\text{def}}{=} s^2 \overset{\text{def}}{=} \frac{1}{R-1} \sum_{r=1}^{R} (Z_r - \widehat{z})^2, \quad s_W^2 \overset{\text{def}}{=} \frac{1}{R-1} \sum_{r=1}^{R} (W_r - \widehat{w})^2,$$

$$s_{ZW}^2 \overset{\text{def}}{=} \frac{1}{R-1} \sum_{r=1}^{R} (Z_r - \widehat{z})(W_r - \widehat{w}),$$

and uses the estimator $\widehat{z}_{\text{CV}} = \widehat{z} + \widehat{\alpha}(\widehat{w} - w)$, which has the same asymptotic properties as $\widehat{z} + \alpha(\widehat{w} - w)$; in particular, the asymptotic variance is $\sigma_Z^2(1 - \rho^2)/R$, and a confidence interval is constructed by replacing σ_Z^2, ρ^2 by their empirical values $s_Z^2, s_{ZW}^4/s_Z^2 s_W^2$.

The procedure reduces the variance by a factor $1 - \rho^2$. Thus, one needs to look for a control variate W with $|\rho|$ as close to 1 as possible. The exact value of ρ will be difficult to asses a priori, so that in practice one would just try to make W and Z as dependent as possible (in some vague sense). It is, however, an appealing feature that even if one is not very succesful, the resulting variance is never increased.

There is also an interesting relation to standard regression analysis. In fact, the calculation of $\widehat{z}_{\text{CV}}$ amounts to using a regression of Z on W, fitting a regression line by least squares, and calculating the level of the line at the known value w of $\mathbb{E}W$; see Figure 2.1. This is seen as follows: The assumption underlying the regression is

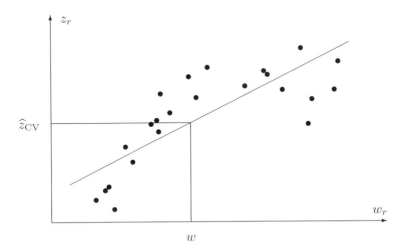

FIGURE 2.1

$$\mathbb{E}Z_i = m' + \beta w_i = m + \beta(w_i - \widehat{w}) \qquad (2.3)$$

$(m \overset{\text{def}}{=} m' + \beta \widehat{w})$, with least squares estimates

$$\widehat{m} = \widehat{z}, \quad \widehat{\beta} = \frac{\sum_1^R (Z_i - \widehat{z})(w_i - \widehat{w})}{\sum_1^R (w_i - \widehat{w})^2} = -\widehat{\alpha}.$$

Thus, the level of the fitted regression line at the known mean w of the control is $\widehat{m} + \widehat{\beta}(w - \widehat{w})$, which (replacing the w_i by the W_i) is the same as $\widehat{z}_{\text{CV}}$.

For this reason, often the term *regression-adjusted control variates* is used. The similarity is, however, formal: regression analysis via least squares is based on the assumption of linear dependence (and preferably normal errors), whereas nothing like this is needed for regression-adjusted control variates (one may, however view the method as inference in the limiting bivariate normal distribution of $(\widehat{z}, \widehat{w})$). The literature pays quite a lot of attention to control variates without regression adjustment (i.e., α is assigned some arbitrary value), but to our mind, it is difficult to imagine situations in which one would prefer this to regression adjustment.

Figure 2.1 also gives a graphical illustration of why control variates provide a variance reduction: without controls, the variance of the estimator of z is given by the variation in the vertical direction, that is, around the horizontal line with level $\widehat{z}$, whereas with controls it is given by the variation around the regression line, which obviously has one degree of freedom more.

Example 2.1 Consider the estimator $Z = 4\mathbb{1}\{U_1^2 + U_2^2 \leq 1\}$ of π. Clearly, the indicator $W \overset{\text{def}}{=} W_1 \overset{\text{def}}{=} \mathbb{1}\{U_1 + U_2 \leq 1\}$ of the triangle A in Figure 2.2 is positively correlated with Z and has the known mean $1/2$, so it is a candidate for a control.

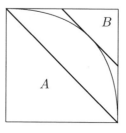

FIGURE 2.2

A simulation run gave $1 - \rho^2 = 0.727$, i.e., some modest variance reduction. Similarly, the indicator $W \overset{\text{def}}{=} W_2 \overset{\text{def}}{=} \mathbb{1}\{U_1 + U_2 \leq \sqrt{2}\}$ of the triangle B is negatively correlated with Z with mean $(2 - \sqrt{2})^2/2$, and simulation gave $1 - \rho^2 = 0.242$, i.e., a more substantial variance reduction, as should also be expected from Figure 2.2. Note that it is unimportant whether to use $\mathbb{1}_B$ or $\mathbb{1}_{B^c}$ as control—for any control W, using $a + bW$ as alternative control produces exactly the same estimates.

In some cases, a clever choice of controls can push ρ^2 much closer to 1 than in these examples. For example, in XIV.13c we will see an example in which $\rho = 0.999$! $\square$

Example 2.2 Consider Monte Carlo integration for calculating $z \stackrel{\text{def}}{=} \int_0^1 g(x)\,dx$. A suitable control for $Z \stackrel{\text{def}}{=} g(U)$ is then $W = f(U)$, where f is close to g (to get ρ close to 1) and $w = \mathbb{E}f(U) = \int_0^1 f(x)\,dx$ is analytically available. $\square$

Example 2.3 A famous example of control variates occurs in Asian options, where the key step in estimating the price is evaluating the expected value of $\left[S(0)A - K \right]^+$, where $A \stackrel{\text{def}}{=} \sum_1^p e^{B(iT/p)}/p$ is the average of a discretely sampled geometric Brownian motion $\{B(t)\}$, with drift say μ and variance σ^2 ($S(0) > 0, K, T$ are constants). The idea is that whereas the distribution of A is intractable, such is not the case for the geometric average

$$
A^* \stackrel{\text{def}}{=} \left(\prod_{i=1}^p e^{B(iT/p)} \right)^{1/p} = \prod_{i=1}^p e^{(p-i+1)Y_i/p},
$$

where $Y_i \stackrel{\text{def}}{=} B(iT/p) - B((i-1)T/p)$. Namely, since the Y_i are i.i.d. $\mathcal{N}(\mu T/p, \sigma^2 T/p)$, we have that $\log A^*$ is normal with mean and variance

$$
\theta \stackrel{\text{def}}{=} \frac{\mu T}{p^2} \sum_{i=1}^p (p-i+1), \quad \text{respectively} \quad \omega^2 \stackrel{\text{def}}{=} \frac{\sigma^2 T}{p^3} \sum_{i=1}^p (p-i+1)^2
$$

(θ, ω^2 can be reduced but we omit the details). Thus, we can take $W \stackrel{\text{def}}{=} \left[S(0)A^* - K \right]^+$ as control variate, since the expectation

$$
\int_{\log(K/S(0))}^{\infty} (s_0 e^z - K) \frac{1}{\sqrt{2\pi\omega^2}} e^{-(z-\theta)^2/2\omega^2}\,dz
$$

is explicitly available by the Black–Scholes formula I.(3.4). $\square$

An extension of the control variate method is *multiple controls*, where the single control W above is replaced by a (row) vector $\boldsymbol{W} \stackrel{\text{def}}{=} (W_1 \ldots W_p)$, with the means $w_1, \ldots, w_p$ explicitly available. Denote by $\boldsymbol{W}_r = (W_{r1} \ldots W_{rp})$ the copy of $\boldsymbol{W}$ generated together with Z_r and let $\widehat{\boldsymbol{W}} = (\widehat{W}_1 \ldots \widehat{W}_p)$ be the average over $r = 1, \ldots, R$. The multiple-control estimator is then

$$
\widehat{z} + \sum_{i=1}^p \alpha_i(\widehat{W}_i - w_i) = \widehat{z} + (\widehat{\boldsymbol{W}} - \boldsymbol{w})\boldsymbol{\alpha} \tag{2.4}
$$

(representing the α_i as a column vector $\boldsymbol{\alpha}$). Writing the covariance matrix of the $(Z_r, \boldsymbol{W}_r)$ as

$$\begin{pmatrix} \sigma^2 & \boldsymbol{\Sigma}_{Z\boldsymbol{W}} \\ \boldsymbol{\Sigma}_{\boldsymbol{W}Z} & \boldsymbol{\Sigma}_{\boldsymbol{W}\boldsymbol{W}} \end{pmatrix},$$

the variance of (2.4) is

$$\frac{1}{R}\left(\sigma_Z^2 + \boldsymbol{\alpha}^{\mathsf{T}}\boldsymbol{\Sigma}_{\boldsymbol{W}\boldsymbol{W}}\boldsymbol{\alpha} + 2\boldsymbol{\Sigma}_{Z\boldsymbol{W}}\boldsymbol{\alpha}\right) = \frac{1}{R}\left(\sigma_Z^2 + \sum_{i,j=1}^{p}\alpha_i\alpha_j\sigma_{W_iW_j} + 2\sum_{i=1}^{p}\alpha_i\sigma_{ZW_i}\right). \tag{2.5}$$

The minimization w.r.t. $\boldsymbol{\alpha}$ is straightforward: equating the partial derivatives w.r.t. the α_i to 0 yields a system of linear equations with solution $\widehat{\boldsymbol{\alpha}} = -\boldsymbol{\Sigma}_{\boldsymbol{W}\boldsymbol{W}}^{-1}\boldsymbol{\Sigma}_{\boldsymbol{W}Z}$, and (2.5) then becomes

$$\frac{1}{R}\left(\sigma_Z^2 - \boldsymbol{\Sigma}_{Z\boldsymbol{W}}\boldsymbol{\Sigma}_{\boldsymbol{W}\boldsymbol{W}}^{-1}\boldsymbol{\Sigma}_{\boldsymbol{W}Z}\right) = \frac{1}{R}(1 - \rho_{Z\boldsymbol{W}}^2)\sigma_Z^2,$$

where $\rho_{Z\boldsymbol{W}}^2 \stackrel{\text{def}}{=} \boldsymbol{\Sigma}_{Z\boldsymbol{W}}\boldsymbol{\Sigma}_{\boldsymbol{W}\boldsymbol{W}}^{-1}\boldsymbol{\Sigma}_{\boldsymbol{W}Z}/\sigma_Z^2$ is the *multiple squared correlation coefficient* between Z and $\boldsymbol{W}$, commonly interpreted as the fraction of the variance of Z that can be explained by linear dependence with the W_i.

In practice, the unknowns $\boldsymbol{\Sigma}_{Z\boldsymbol{W}}, \boldsymbol{\Sigma}_{\boldsymbol{W}\boldsymbol{W}}$ have to be replaced by estimates, that is, the empirical values

$$\boldsymbol{S}_{Z\boldsymbol{W}} \stackrel{\text{def}}{=} \frac{1}{R-1}\sum_{r=1}^{R}(Z_r - \widehat{z})(\boldsymbol{W}_r - \widehat{\boldsymbol{W}}),$$

$$\boldsymbol{S}_{\boldsymbol{W}\boldsymbol{W}} \stackrel{\text{def}}{=} \frac{1}{R-1}\sum_{r=1}^{R}(\boldsymbol{W}_r - \widehat{\boldsymbol{W}})^{\mathsf{T}}(\boldsymbol{W}_r - \widehat{\boldsymbol{W}}),$$

so that the multiple-control estimator and its variance estimate are

$$\widehat{z} - \boldsymbol{S}_{\boldsymbol{W}\boldsymbol{W}}^{-1}\boldsymbol{S}_{\boldsymbol{W}Z}, \quad \text{respectively} \quad \frac{1}{R}(s^2 - \boldsymbol{S}_{Z\boldsymbol{W}}\boldsymbol{S}_{\boldsymbol{W}\boldsymbol{W}}^{-1}\boldsymbol{S}_{\boldsymbol{W}Z}).$$

Again, the calculations of the method are formally similar to regression, this time the multiple regression model

$$\mathbb{E}Z_r = z' + \sum_{i=1}^{p}\beta_i w_{ri} = z + \sum_{i=1}^{p}\beta_i(w_{ri} - \overline{w}_i) = z + \boldsymbol{T}\boldsymbol{\beta}, \tag{2.6}$$

where $\boldsymbol{T}$ has rith element $w_{ri} - \widehat{w}_i$ (here $\widehat{w}_i \stackrel{\text{def}}{=} \sum_1^R w_{ri}/R$). The least squares estimator of $\boldsymbol{\beta} \stackrel{\text{def}}{=} (\beta_1 \ldots \beta_p)^{\mathsf{T}}$ is

$$\widehat{\boldsymbol{\beta}} = (\boldsymbol{T}^{\mathsf{T}}\boldsymbol{T})^{-1}\boldsymbol{T}^{\mathsf{T}}\boldsymbol{Z} = \boldsymbol{S}_{\boldsymbol{W}\boldsymbol{W}}^{-1}\boldsymbol{S}_{\boldsymbol{W}Z},$$

and the residual variance is

$$\frac{1}{R-p-1}\left\|\boldsymbol{Z} - \widehat{z} - \boldsymbol{T}\widehat{\boldsymbol{\beta}}\right\|^2 = \frac{1}{R-p-1}\sum_{r=1}^{R}\left(Z_r - \widehat{z} - \sum_{i=1}^{p}\widehat{\beta}_i(w_{ri} - \widehat{w}_i)\right)^2.$$

Example 2.4 Consider the same estimator $Z \overset{\text{def}}{=} 4\mathbb{1}\{U_1^2 + U_2^2 \leq 1\}$ of π as in Example 2.1 and the controls

$$W_1 \overset{\text{def}}{=} \mathbb{1}\{U_1 + U_2 \leq 1\}, \quad W_2 \overset{\text{def}}{=} \mathbb{1}\{U_1 + U_2 \leq \sqrt{2}\}.$$

We add here a third,

$$W_3 \overset{\text{def}}{=} (U_1 + U_2 - 1)\mathbb{1}\{1 < U_1 + U_2 < \sqrt{2}\};$$

since $V = U_1 + U_2$ has density $2 - v$ on $(1, 2)$, the mean is

$$w_3 = \int_1^{\sqrt{2}} (v - 1)(2 - v) \, dv = \frac{\left(\sqrt{2} - 1\right)^2}{2} - \frac{\left(\sqrt{2} - 1\right)^3}{3}.$$

The motivation for introducing W_3 is that $\mathbb{E}[Z \mid V]$ decreases from 1 to 0 as V varies from 1 to $\sqrt{2}$, and hence a first approximation is a linear function. Running a simulation with all seven possible subsets of the three controls gave the following values of $1 - \rho^2$:

1	2	3	1, 2	1, 3	2, 3	1, 2, 3
0.727	0.242	0.999	0.222	0.620	0.181	0.175

It is seen that W_3 alone is useless as control, but when added to either W_1 or W_2 produces a notable variance reduction. Considering the full set of all three controls gives little compared to the combination W_2, W_3, while combining W_1 and W_2 gives little compared to using W_2 alone. $\qquad\square$

Remark 2.5 If Z has a considerable functional dependence on a single control W, but this dependence is clearly nonlinear as in Figure 2.1, an idea (similar to polynomial regression) is to use the $W_i \overset{\text{def}}{=} W^i$ for some small $i > 1$, say 2 or 3, as additional multiple controls (assuming of course that the means are computable).

The extension to multidimensions is similar. Say we improve two controls W_1, W_2 by adding $W_3 \overset{\text{def}}{=} W_1^2, W_4 \overset{\text{def}}{=} W_2^2, W_5 \overset{\text{def}}{=} W_1 W_2$. $\qquad\square$

Exercises

2.1 (A) By invoking $|U_1 - U_2|$ and/or its powers as control(s) and possible further ideas of your own, experiment further with variance reduction in the simulation of π in Example 2.4.

2.2 (A) Consider a European call option in the same stochastic volatility setting as in Exercise 1.7 and with the same $r, T, \sigma, S(0)$.
Give simulation estimates of the option price for $K = 50, 100, 150$ using the control variates

$$S(T), \; Y(T), \; Y(T)^2, \; e^{Y(T)}, \; [S(0)e^{Y(T)} - K]^+,$$

where $Y(t)$ is given by $Y(0) = 0$,

$$dY(t) = (r - \sigma^2/2) \, dt + \sigma \, dB(t)$$

(with the same driving BM B). Use both single and multiple controls and report on the explanatory power of the control variates in the form of the multiple correlation coefficient corresponding to various subsets of the controls.

3 Antithetic Sampling

Here one generates the replicates $Z_1, \ldots, Z_R$ of Z not as i.i.d., but as pairwise dependent and as negatively correlated as possible. That is, one takes $R = 2M$ and generates M i.i.d. random pairs

$$(Z_1, Z_2), (Z_3, Z_4), \ldots, (Z_{M-1}, Z_M),$$

such that the marginal distribution of Z_r is the same (as for the CMC method) for all r (even and odd) but Z_{2j-1} and Z_{2j} may be dependent. The estimator is $\widehat{z}_{\mathrm{Anth}} \overset{\text{def}}{=} (Z_1 + \cdots + Z_R)/R$ with variance

$$\frac{1}{R}\sigma_{\mathrm{Anth}}^2 = \frac{1}{M}\mathbb{V}ar\left(\frac{Z_1 + Z_2}{2}\right) = \frac{1}{4M}(\sigma_{\mathrm{CMC}}^2 + \sigma_{\mathrm{CMC}}^2 + 2\sigma_{\mathrm{CMC}}^2\rho)$$

$$= \frac{1}{R}\sigma_{\mathrm{CMC}}^2(1 + \rho),$$

where $\rho = \mathbb{C}orr(Z_1, Z_2)$. Thus, ρ should be negative for obtaining variance reduction, and preferably as close to -1 as possible for the method to be efficient.

Example 3.1 In Monte Carlo integration (considering dimension $d = 1$ for simplicity), a standard choice is to take $Z_1 \overset{\text{def}}{=} g(U)$, $Z_2 \overset{\text{def}}{=} g(1 - U)$ when estimating $z = \int_0^1 g$. If g is monotone, Chebyshev's covariance inequality (see A7) then yields $\rho \leq 0$.

If $g(x) = x$, one has $Z_1 + Z_2 = 1$, so the antithetic estimator has zero variance. For $g(x) = x^2$, one gets

$$\rho = \frac{\mathbb{E}\left[U^2(1-U)^2\right] - (\mathbb{E}U^2)^2}{\mathbb{V}ar\,U^2} = \frac{1/5 + 1/3 - 1/2 - 1/3^2}{1/5 - 1/3^2} = -\frac{7}{8}.$$

In general, for $g(x) = x^n$,

$$|\rho| \leq \frac{\mathbb{E}\left[U^n(1-U)^n\right] + (\mathbb{E}U^n)^2}{\mathbb{V}ar\,U^n} \leq \frac{1/4^n + 1/(n+1)^2}{1/(2n+1) - 1/(n+1)^2} \sim \frac{2}{n},$$

so that the variance reduction vanishes as $n \to \infty$. □

We know of no realistic example in which the variance reduction obtained by antithetic sampling is dramatic. This may in part be understood from the following example:

Example 3.2 For Z_1, Z_2 having joint c.d.f. $F(z_1, z_2)$ and the same marginal c.d.f. $F(z)$, one has

$$\mathbb{E}[Z_1 Z_2] = \int_0^\infty \int_0^\infty \left[1 - F(z_1) - F(z_1) + F(z_1, z_2)\right] dz_1 dz_2. \qquad (3.1)$$

The r.h.s. and therefore $\rho = \mathbb{C}orr(X, Y)$ are minimized when $Z_1 = F^{\leftarrow}(U)$, $Z_2 = F^{\leftarrow}(1 - U)$, where U is uniform$(0, 1)$, cf. the characterization of the Fréchet–Hoeffding lower bound in II.3.17. For example, if X, Y have standard exponential marginals, then

$$\rho = \mathbb{C}orr(Z_1, Z_2) = \mathbb{C}ov(Z_1, Z_2) \geq \mathbb{E}[\log U \log(1 - U)] - 1 = -0.645,$$

which shows that the variance can be reduced at most by a factor of 0.355 in estimating $z = \mathbb{E}Z_1$ by antithetic sampling. □

For some nonstandard discussion of antithetic sampling, see Evans & Swartz [114].

Exercises

3.1 (TP) Verify (3.1).

4 Conditional Monte Carlo

Here $Z = Z_{\mathrm{CMC}}$ is replaced by $Z_{\mathrm{Cond}} \overset{\mathrm{def}}{=} \mathbb{E}[Z_{\mathrm{CMC}} \,|\, W]$ for some r.v. W (more generally, one could consider $\mathbb{E}[Z_{\mathrm{CMC}} \,|\, \mathscr{G}]$ for some σ-field $\mathscr{G}$). Clearly, $\mathbb{E}Z_{\mathrm{Cond}} = \mathbb{E}Z_{\mathrm{CMC}} = z$. Since

$$\begin{aligned}
\sigma_{\mathrm{CMC}}^2 &= \mathbb{V}ar(Z_{\mathrm{CMC}}) = \mathbb{V}ar\big(\mathbb{E}[Z_{\mathrm{CMC}} \,|\, W]\big) + \mathbb{E}\big(\mathbb{V}ar[Z_{\mathrm{CMC}} \,|\, W]\big) \\
&= \sigma_{\mathrm{Cond}}^2 + \mathbb{E}\big(\mathbb{V}ar[Z_{\mathrm{CMC}} \,|\, W]\big) \geq \sigma_{\mathrm{Cond}}^2,
\end{aligned}$$

conditional Monte Carlo always provides variance reduction, which is appealing. The difficulty is to find W such that the conditional expectation is computable.

Example 4.1 Consider again as in the previous sections the example of estimation of $z = \pi$ via $Z_{\mathrm{CMC}} = 4\mathbb{1}\{U_1^2 + U_2^2 < 1\}$, with U_1, U_2 independent and uniform$(0, 1)$. We can then take

$$Z_{\mathrm{Cond}} = \mathbb{E}[Z_{\mathrm{CMC}} \,|\, U_1] = 4\mathbb{P}(U_2^2 < 1 - U_1^2 \,|\, U_1) = 4\sqrt{1 - U_1^2}.$$

A simulation run gave an estimate of 3.38 of the variance reduction factor $\sigma_{\mathrm{CMC}}^2 / \sigma_{\mathrm{Cond}}^2$. □

Example 4.2 Let $z \overset{\mathrm{def}}{=} \mathbb{P}(X_1 + X_2 > x)$, where X_1, X_2 are independent with distribution F (F is known, one can simulate from F, but the convolution F^{*2} is not available). Then $Z_{\mathrm{CMC}} = \mathbb{1}\{X_1 + X_2 > x\}$, and taking $W \overset{\mathrm{def}}{=} X_1$, we get $Z_{\mathrm{Cond}} = \overline{F}(x - X_1)$.

Related algorithms are presented in VI.3a and in fact sometimes provide a dramatic variance reduction.   □

Example 4.3 Conditional Monte Carlo ideas can be applicable to density estimation as well. As a simple example, let $X_1, X_2, \ldots$ be i.i.d. with density $f(x)$, so that $S_n \stackrel{\text{def}}{=} X_1 + \cdots + X_n$ has density $f^{*n}(x) = \mathbb{P}(S_n \in \mathrm{d}x)$. The conditional Monte Carlo estimator of $f^{*n}(x)$ given S_{n-1} is then $f(x - S_{n-1})$. The advantages to the standard kernel-smoothing estimate using simulated values of S_n (cf. III.6) rather than S_{n-1} are, for example, that the arbitrariness in the choice of the bandwidth h is avoided and that one avoids such problems that the estimate of $f^{*n}(x)$ comes out positive for negative x even if the X_i are nonnegative.

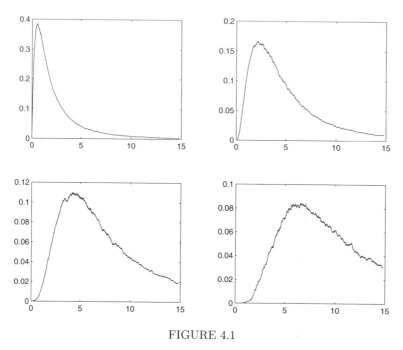

FIGURE 4.1

As an example, we took $f(x)$ as the Pareto density $\alpha/(1 + x)^{\alpha+1}$ with $\alpha = 3/2$. Here $f^{*n}(x)$ is not explicitly available for $n > 1$. Figure 4.1 gives the conditional Monte Carlo estimates given by the average of $R = 10{,}000$ replications of $f(x - S_{n-1})$ in the interval $0 < x < 15$ for $n = 2, 4, 6, 8$ [the small amount of nonsmoothness can be explained by the discontinuity of $F(x)$ at $x = 0$, and that the effect increases with n is a scaling phenomenon because less and less of the mass is in $0 < x < 15$].   □

Exercises

4.1 (A) A bank has a portfolio of $N = 100$ loans to N companies and wants to evaluate its credit risk. Given that company n defaults, the loss for the bank is a

$\mathcal{N}(\mu, \sigma^2)$ r.v. X_n where $\mu = 3$, $\sigma^2 = 1$. Defaults are dependent and described by indicators $D_1, \ldots, D_N$ and a background r.v. P (measuring, say, general economic conditions), such that $D_1, \ldots, D_N$ are i.i.d. Bernoulli(p) given $P = p$, and P has a Beta$(1, 19)$ distribution, that is, density $(1 - p)^{18}/19$, $0 < p < 1$. Estimate $\mathbb{P}(L > x)$, where $L = \sum_1^N D_n X_n$ is the loss, using both CMC and conditional Monte Carlo, where the conditioning is on $\sum_1^N D_n$. For x, take

$$x \;=\; 3\mathbb{E}L \;=\; 3N\,\mathbb{E}P\,\mathbb{E}X \;=\; 3 \cdot 100 \cdot 0.05 \cdot 3 \;=\; 45.$$

4.2 (TP) In Exercise 4.1, identify saddle-point exponential change of measure given $P = p$. That is, the conditional likelihood is $\mathrm{d}\mathbb{P}^*/\mathrm{d}\mathbb{P} = e^{\theta L}/\varphi(\theta|p)$, where $\varphi(\theta|p) \stackrel{\text{def}}{=} \mathbb{E}\big[e^{\theta L} \,\big|\, P = p\big]$ and $\theta = \theta(p)$ satisfies $\mathbb{E}_\theta[L \,|\, P = p] = x$.

4.3 (A) Consider the density proportional to $\big(1 - \cos(2\pi x)\big)/(1 + x)^{\alpha+1}$. Give a conditional Monte Carlo estimate of the renewal density $\sum_{n=1}^\infty \mathbb{P}(S_n \in \mathrm{d}x)$ in the interval $0 < x < 8$. Cf. also I.5.21.

5 Splitting

Assume we want to estimate $z = \mathbb{E}Z$, where Z has the form $Z = \varphi(X, Y)$ for independent r.v.'s X, Y. Assume further that the expected times a, b to generate X, respectively Y, satisfy $a \gg b$ but that Z is more influenced by Y than by X (examples follow below). It is then tempting to reuse the more difficultly sampled values $X_1, \ldots, X_R$ of X by sampling several (say S) values Y_{rs} for each X_r, such that the overall estimator takes the form

$$\frac{1}{RS} \sum_{r=1}^R \sum_{s=1}^S \varphi(X_r, Y_{rs}) \,;$$

this procedure is called *splitting* and has a long history in physics.

Splitting may be seen as CMC using the estimator

$$\widetilde{Z} \stackrel{\text{def}}{=} \frac{1}{S} \sum_{s=1}^S \varphi(X, Y_s) \,.$$

This estimator is in turn closely related to conditional Monte Carlo, since it is a Monte Carlo estimate of $\mathbb{E}\big[Z \,|\, X\big]$.

The question is how to choose S and whether a substantial reduction of the computational effort can be attained. To analyze this, we have to compare the variance per unit time of $\widetilde{Z}$ with the corresponding quantity for the CMC estimator Z, which leads to comparing $\widetilde{e} \stackrel{\text{def}}{=} (a + Sb)\mathbb{V}ar\widetilde{Z}$ and $e \stackrel{\text{def}}{=} (a + b)\mathbb{V}arZ$; cf. III.10. Assume w.l.o.g. that $\mathbb{V}ar\,Z = 1$ and let

$$\rho \stackrel{\text{def}}{=} \mathbb{C}orr\big(\varphi(X_r, Y_{rs_1}), \varphi(X_r, Y_{rs_2})\big), \quad s_1 \neq s_2 \,.$$

Then

$$\mathbb{V}ar\, \widetilde{Z} \;=\; \frac{1}{S^2}\big[S + 2S(S-1)\rho\big], \tag{5.1}$$

$$\widetilde{e} \;=\; \Big(\frac{1}{S} + \frac{2(S-1)\rho}{S}\Big)(a + Sb). \tag{5.2}$$

Consider first $S = 2$. Then $\widetilde{e} < e$ if $(1/2 + \rho)(a + 2b) < a + b$, which is easily seen to mean $\rho < a/(2a + 4b)$. Since a/b is large, this shows in particular that an efficiency gain cannot be expected when $\rho > 1/2$. In general, for the efficiency gain to be considerable, S clearly must be large. Viewing S as a continuous variable, the approximation $(S^{-1} + 2\rho)(a + SB)$ for $\widetilde{e}$ above is then minimized for $S = \sqrt{a/(2\rho b)}$ (in practice, this value has to be rounded), leading to the optimal efficiency

$$\widetilde{e}^{*} \;=\; \big(\sqrt{2\rho b/a} + 2\rho\big)\big(a + \sqrt{ab/(2\rho)}\,\big) \;=\; a\big(\sqrt{b/a} + \sqrt{2\rho}\,\big)^{2},$$

to be compared with $e = a+b$. This shows that if both b/a and ρ are small, $\widetilde{e}/e$ is of order $(b/a) \wedge 2\rho$ (that ρ being small is a reasonable assumption follows from Z being more influenced by Y than by X).

Example 5.1 A restaurant has peak hour 12:30–1:30 p.m. where the average arrival rate exceeds the average service rate. It wants to estimate the expected number z of lost customers from 12:00 to 2:00. We can then take X as the number of customers who are present at 12:00, supplemented with their residual service times, and Y as the point process of arrivals in $[12, 14]$ marked with the service times. ☐

Example 5.2 Assume that we want to estimate the expectation of a functional $Z = Z(B)$ of standard Brownian motion B in $[0, 1]$ using $2^{10} = 1024$ grid points. The naive method generates B by summing up 1024 i.i.d. normals with the appropriate variances. For splitting, we could take Y as the values at a coarser grid, $Y = \big(B(i/2^k)\big)_{i=1,\dots,2^k}$ for some small k, say $k = 2, 3$, or 4. Then X is the 2^k Brownian bridges of length 2^{-k} required to get the whole path by replacing the linear interpolation between the 2^k points at the coarse grid by a Brownian bridge (recall that a Brownian bridge of length $T \le 1$ is distributed as $\{B(T) - tB(T)\}_{0 \le t \le T}$). Thus (up to a proportionality factor) $a = 2^k$ and $b = 2^k \cdot 2^{10-k} = 1024$. Note that for small k, $a + b$ is for all practical purposes equal to the number 1024 required for the naive method. For example, for $k = 3$, we get $a/b = 128$; the optimal choice of k depends of course on how ρ depends on k. ☐

Exercises

5.1 (TP) Assume that Z is generated from r.v.'s X, Y and again that the times to generate are a, b with $a \gg b$. The simulator judges that only the Y taking values in a certain set E contribute significantly to z. He therefore does not necessarily generate X if $Y \notin E$ but only if a coin toss comes out with heads; say the probability is p.

Write up an unbiased estimator for z and give an efficiency analysis similar to the one in the text for splitting, discussing in particular the choice of p (you will need to invoke $\mathbb{P}(Y \in E)$).
The procedure is known as *Russian roulette*.

6   Common Random Numbers

This is a method for comparing the means z', z'' of two r.v.'s Z', Z'' that depend on the same random input in a way that is in some sense similar. More specifically, assume that we want to estimate the difference $z \overset{\text{def}}{=} z' - z''$ between the expectations z', z'' of two r.v.'s Z', Z'' of the form $Z' = Z'(U_1, \ldots, U_N)$, $Z'' = Z''(U_1, \ldots, U_N)$ for a fixed number N of i.i.d. uniforms, and that similarity means that Z', Z'' have approximately the same variance σ^2 and a relatively high positive correlation ρ when evaluated for the same $U_1, \ldots, U_N$. The method would then create R replicates of

$$ Z \overset{\text{def}}{=} Z'(U_1, \ldots, U_N) - Z''(U_1, \ldots, U_N) $$

and take the average $\widehat{z}$ as an estimate of $z = z' - z''$, resulting in a variance of approximately $2\sigma^2(1 - \rho)/R$. This is to be compared with independent sampling in which one would generate $R/2$ replicates of Z' and $R/2$ independent replicates of Z'', resulting in an approximate variance of $2\sigma^2/R$. Thus, variance reduction is indeed obtained under the stated assumption that ρ be positive.

We will see a convincing application of common random numbers in VII.2a in connection with derivative estimation using finite differences. A restriction of the method is that similarity is most often lost if N is random, for example if the r.v. generation involves rejection.

The method of common random numbers is also often used more generally for functions $z(\theta) = \mathbb{E}Z(U_1, \ldots, U_N; \theta)$. While it is less clear what variance reduction means here, the method has the advantage of producing a smoother (in θ) estimate of the curve $z(\theta)$ than the more jagged one produced by independent sampling.

Example 6.1 Outside the $Z' = Z'(U_1, \ldots, U_N)$, $Z'' = Z''(U_1, \ldots, U_N)$ setting, another application of common random numbers is the comparison of the performance of systems governed by the same input. As a simple example, we took three symmetric α-stable Lévy processes (cf. XII.1.1) in the time interval $[0, 1000]$ and plotted sample paths in which the increments are generated using inversion of common uniforms $U_1, \ldots, U_{1000}$; to get a reasonably easy inversion, we took $\alpha = 2$ (Brownian motion), $\alpha = 1$ (the Cauchy process), and $\alpha = 1/2$ (the symmetric inverse Gaussian process). The scale was chosen so as to make the processes comparable on the same plot, which was achieved by choosing the c.d.f. of the increments to be

$$\Phi(x), \quad \arctan(x/50)/\pi + 1/2, \quad \text{respectively } 1 - \Phi\big(1/\sqrt{x/500,000}\big)$$

for $x > 0$ (by symmetry, $F(x) = 1 - F(-x)$ for $x < 0$). The results are

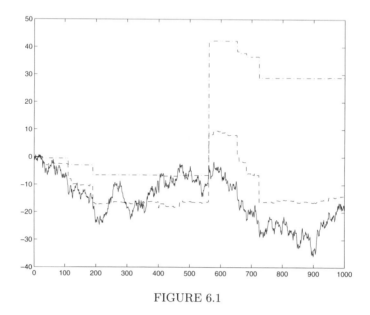

FIGURE 6.1

given in Figure 6.1 with $\alpha = 2$ corresponding to the solid graph, $\alpha = 1$ to the dashed line and $\alpha = 1/2$ to the dash-dotted one. The use of common random numbers serves to show how the α-stable process becomes more and more dominated by large jumps (a U_k close to 0 or 1) as α decreases. □

7   Stratification

Here the sample space Ω is divided into S regions $\Omega_1, \ldots, \Omega_S$, called *strata*, and the aim is to eliminate as much of the variation within strata as possible.

The strata are often obtained by subdividing the range of one or more of the most important r.v.'s driving the simulation. For a simple example, consider Monte Carlo integration of a function $g : [0,1] \to \mathbb{R}$ of a single variable as estimated by $Z = g(U)$. Here an obvious choice is $\Omega_s = \{\omega : (s-1)/S \le U(\omega) < s/S\}$. See further Example 7.1 below.

Let Z_s be an r.v. having the distribution of Z conditioned on Ω_s, i.e.,

$$\mathbb{P}(Z_s \in A) = \frac{1}{\mathbb{P}(\Omega_s)} \mathbb{P}\big(\{Z \in A\} \cap \Omega_s\big).$$

Write $p_s \overset{\text{def}}{=} \mathbb{P}(\Omega_s)$, $z_s \overset{\text{def}}{=} \mathbb{E}Z_s$. By the law of total probability, $\mathbb{E}Z = p_1 z_1 + \cdots + p_S z_S$. So we can divide the total number R of replicates into $R_1, \ldots, R_S$, for each s simulate R_s replicates of Z_s, estimate z_s by the empirical average $\widehat{z}_s$, and estimate z by

$$\widehat{z}_{\text{Str}} \overset{\text{def}}{=} p_1 \widehat{z}_1 + \cdots + p_S \widehat{z}_S$$

(in order that this scheme be feasible, it is of course essential that the p_s be known and that the Z_s can be generated).

Let $\sigma_s^2 \overset{\text{def}}{=} \mathbb{V}ar Z_s$,

$$\sigma_{\text{Str}}^2 \overset{\text{def}}{=} R \mathbb{V}ar \, \widehat{z}_{\text{Str}} = R \sum_{s=1}^{S} p_s^2 \frac{\sigma_s^2}{R_s} = \sum_{s=1}^{S} p_s^2 \frac{\sigma_s^2}{R_s/R}. \tag{7.1}$$

Then it is clear that

$$\frac{\sqrt{R}}{\sigma_{\text{Str}}} \left(\widehat{z}_{\text{Str}} - z \right) \ \to \ \mathcal{N}(0,1) \tag{7.2}$$

as $R \to \infty$ in such a way that the R_s/R have nonzero limits. The obvious estimator $\widehat{\sigma}_{\text{Str}}^2$ for σ_{Str}^2 is obtained by replacing σ_s^2 with the empirical variance $\widehat{\sigma}_s^2$ within the R_s replicates in strata s, and so the 95% confidence interval to be used in practice is $\widehat{z}_{\text{Str}} \pm 1.96 \, \widehat{\sigma}_{\text{Str}}/\sqrt{R}$, where

$$\widehat{\sigma}_{\text{Str}}^2 \overset{\text{def}}{=} \sum_{s=1}^{S} p_s^2 \frac{\widehat{\sigma}_s^2}{R_s/R}.$$

It remains to discuss when variance reduction is obtained and what is the best allocation of the R_s (and possibly the strata).

A frequently made choice is *proportional allocation*, where $R_s/R = p_s$ (ignoring rounding issues here and in the following). Subject to this choice,

$$\sigma_{\text{Str}}^2 = \sum_{s=1}^{S} p_s \sigma_s^2. \tag{7.3}$$

This can be interpreted as $\mathbb{E}[\mathbb{V}ar(Z_J | J)]$, where J is a $\{1, \ldots, S\}$ valued r.v. with $\mathbb{P}(J = s) = p_s$. But clearly $Z_J \overset{\mathscr{D}}{=} Z$, so that by a standard variance decomposition we obtain

$$\sigma^2 = \mathbb{V}ar(Z) = \mathbb{V}ar(Z_J) \geq \mathbb{E}[\mathbb{V}ar(Z_J | J)] = \sigma_{\text{Str}}^2,$$

and it follows that *stratification with proportional allocation always produces variance reduction*.

Example 7.1 Consider as above Monte Carlo integration of $g : [0,1] \to \mathbb{R}$ using proportional allocation and the strata

$$\Omega_s \overset{\text{def}}{=} \{ \omega : u_{s-1}/S \leq U(\omega) < u_s \} ,$$

where $u_s \stackrel{\text{def}}{=} s/S$. Here $p_s = 1/S$ and

$$
\begin{aligned}
\sigma_s^2 &= \operatorname{Var}\big[g(U) \,|\, u_{s-1} \le U < u_s\big] = \operatorname{Var} g\big(u_{s-1} + U/S\big) \\
&\stackrel{\text{def}}{=} \operatorname{Var}\Big[g(u_{s-1}) + \frac{h(s,S,U)}{S}\Big] = \operatorname{Var}\Big[\frac{h(s,S,U)}{S}\Big],
\end{aligned}
$$

where, provided g is smooth enough, the $h(s,S,\cdot)$ are bounded uniformly in s and S by $c \stackrel{\text{def}}{=} \|g'\|_\infty$. Thus $\sigma_s^2 \le c^2/S^2$ and

$$
\sigma_{\mathrm{Str}}^2 = \sum_{s=1}^{S} S^{-1} \mathrm{O}\big(S^{-2}\big) = \mathrm{O}\big(S^{-2}\big),
$$

so that for large S, stratification with proportional allocation reduces the variance with a factor of S^{-2}.

For further elaboration of these ideas, see Remark 7.4 below.   □

Remark 7.2 The smoothness assumption in Example 7.1 is essential, and without it, one cannot in general expect the rate S^{-2} of variance reduction in Example 7.1. As an example, consider the estimator $Z = 4\mathbb{1}\{U_1^2 + U_2^2 < 1\}$ of π and the $S = S_0^2$ strata

$$
\Omega_{ij} \stackrel{\text{def}}{=} \{(i-1)/S_0 < U_1 < i/S_0 \,,\; (j-1)/S_0 < U_2 < j/S_0\}\,;
$$

cf. Figure 7.1, where $S_0 = 16$, $S = 256$.

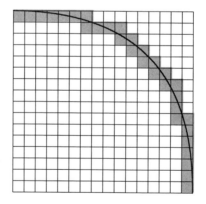

FIGURE 7.1

In Figure 7.1, σ_s^2 is nonzero only on the gray-shaded strata (on the strata to the SW, $Z_s \equiv 1$, so $\sigma_s^2 = 0$, and on the strata to the NE, $Z_s \equiv 0$, so here also $\sigma_s^2 = 0$). The number of gray-shaded strata is of order S_0, and within each we have binomial sampling with a probability z_s, which is typically not very close to 0 or 1. Therefore in (7.3) we have $p_s = S_0^{-2}$, whereas σ_s^2 is nonzero on $\mathrm{O}(S_0)$ strata and then typically not very close to 0, which leads to $\sigma_{\mathrm{Str}}^2 \approx S_0 \cdot S_0^{-2} = S^{-1/2}$. Thus for large S, stratification with proportional allocation reduces the variance with a factor of $S^{-1/2}$ compared to the S^{-2} of Example 7.1.   □

More ambitiously than just considering proportional allocation, let us consider the optimal (in the sense of variance minimization) allocation. This means minimizing σ_{Str}^2 (viewed as function of $R_1, \ldots, R_S$, cf. (7.1)) subject to $R_1 + \cdots + R_S = R$. Introducing a Lagrangian multiplier λ, we get

$$\frac{p_s^2 \sigma_s^2}{R_s^2} = -\lambda/R, \quad s = 1, \ldots, S.$$

Since the r.h.s. is independent of s, such is the l.h.s., which implies that the optimal allocation is

$$R_s^* = \frac{p_s \sigma_s R}{\sum_1^S p_t \sigma_t}.$$

The practical implementation of this choice meets the difficulty that the σ_s^2 are typically unknown; a possibility is to estimate them by a pilot run or use an adaptive scheme.

A variant of proportional allocation is *poststratification*, whereby the allocation on strata is not determined in advance but by the simulation, in such a way that the probability of strata s is p_s. Formally, let the simulation generate i.i.d. r.v.'s (Σ_r, Z_r), $r = 1, \ldots, R$, such that $\mathbb{P}(\Sigma_r = s) = p_r$ and the conditional distribution of Z_r given $\Sigma_r = s$ is the distribution of Z_s, and let R_s be the number of r with $\Sigma_r = s$. We then estimate z_s by the empirical average

$$\widehat{z}_s = \frac{V_s}{R_s}, \quad \text{where} \quad V_s \stackrel{\text{def}}{=} \sum_{r: \Sigma_r = s} Z_r$$

($\widehat{z}_s = 0$ if $R_s = 0$) and z by $\widehat{z}_{\text{PostStr}} \stackrel{\text{def}}{=} p_1 \widehat{z}_1 + \cdots + p_S \widehat{z}_S$. That is, the estimator is the same as for a deterministic allocation of the R_r, and further, *a CLT with the same variance constant (7.3) as for proportional allocation holds*. To see this, just note that since the V_r are conditionally independent given $R_1, \ldots, R_S$, (7.2) continues to hold given $R_1, \ldots, R_S$. Since the limit is independent of the conditioning, it is also an unconditional limit, and in the variance constant, one can then replace R_s/R by its limit p_s.

Despite the identical large-sample properties, it is often argued that proportional allocation is preferable to poststratification, except for situations in which it is not straightforward to generate the Z_s but one nevertheless wants to take advantage of the knowledge of the p_s.

Remark 7.3 *Latin hypercube sampling* is a sampling scheme for R d-dimensional r.v.'s $\boldsymbol{V}_1, \ldots, \boldsymbol{V}_R \in [0,1]^d$ that is somewhat related to stratification. The scheme has the property that the $\boldsymbol{V}_r$ all are uniform on $[0,1]^d$ (but dependent), and that in the matrix

$$V \stackrel{\text{def}}{=} (V_1 \ldots V_R) = \begin{pmatrix} V_{11} & \cdots & V_{1R} \\ \vdots & & \vdots \\ V_{d1} & \cdots & V_{dR} \end{pmatrix},$$

each row is stratified according to the strata

$$[0,1/R), [1/R, 2/R), \ldots, [(R-1)/R, 1].$$

That is, each of the interval $(0,1)$'s R strata is represented exactly once in each row, and further, the representations in different rows are random. This is achieved by generating d random permutations $\pi_1, \ldots, \pi_d$ of $\{1, \ldots, R\}$, $\pi_j \stackrel{\text{def}}{=} (\pi_j(1) \ldots \pi_j(R))$, and dR i.i.d. uniform$(0,1)$ r.v.'s u_{jr}, and letting

$$V_{jr} \stackrel{\text{def}}{=} \frac{1}{R}(\pi_j(r) - 1 + U_{jr}).$$

The dependence between the V_r is of course due to the fact that V_{jr} cannot be in the same stratum as $V_{jr'}$ when $r \neq r'$.

The scheme was introduced by McKay et al. [251], and a recent asymptotic study is in Loh [242].   □

Remark 7.4 The ideas of Example 7.1 can be extended to provide an example of a Monte Carlo estimator with a supercanonical rate (faster than $R^{-1/2}$). Consider again Monte Carlo integration of $g : [0,1] \to \mathbb{R}$ but now using the estimator

$$\frac{1}{R} \sum_{r=1}^{R} f((r - 1 + U_r)/R) \tag{7.4}$$

with $U_1, \ldots, U_R$ i.i.d. uniform(0,1). Then, under appropriate smoothness assumptions, (7.4) has an asymptotic variance of $O(R^{-3})$ and hence a rate of $O(R^{-3/2})$. To see this, just note that as in Example 7.1, the variance of each term under the sum sign is bounded by c^2/R^2, and therefore the variance of (7.4) is $O(R^{-2} \cdot R \cdot R^{-2})) = O(R^{-3})$.

A further, somewhat related, example of a supercanonical rate is the estimator of $\int_0^1 g$ given by Yakowitz et al. [369],

$$\sum_{r=-1}^{R} (U_{(r+1)} - U_{(r)}) \frac{g(U_{(r+1)}) + g(U_{(r)})}{2},$$

where $U_{(1)} < \cdots < U_{(R)}$ are the order statistics of R uniforms and $U_{(-1)} \stackrel{\text{def}}{=} 0$, $U_{(R+1)} \stackrel{\text{def}}{=} 1$. This estimator even has rate $O(R^{-2})$.   □

Exercises

7.1 (TP) Suggest some variance reduction methods for evaluating

$$\int_0^\infty (x + 0.02x^2) \exp\left\{0.1\sqrt{1 + \cos x} - x\right\} \mathrm{d}x$$

by Monte Carlo integration.

Exercises X.2.1–3 also contain stratification as an essential part.

8   Indirect Estimation

In some cases, some parts of the expectation z of Z can be evaluated analytically. In others, z may be related to the (known or unknown) expectations of some further r.v.'s. We give a few examples of how such facts may be exploited for variance reduction, without attempting a general discussion (to this end, see Glynn & Whitt [155], who coined the term *indirect estimation*).

Example 8.1 (CALL–PUT PARITY) For a European option, we have under the lognormal model in the usual notation that the call and put prices z_{call}, z_{put} are the expectations of

$$Z_{\mathrm{call}} \stackrel{\mathrm{def}}{=} \mathrm{e}^{-rT}\left[S(T) - K\right]^+, \quad \text{respectively}\ \ Z_{\mathrm{put}} \stackrel{\mathrm{def}}{=} \mathrm{e}^{-rT}\left[K - S(T)\right]^+,$$

under the risk-neutral measure $\mathbb{P}^*$. The definition of $\mathbb{P}^*$ (the martingale property of $\{\mathrm{e}^{-rt}S(t)\}$) implies $\mathbb{E}^*\mathrm{e}^{-rT}S(T) = S(0)$. Thus $a = a^+ - (-a)^+$ gives the estimator $S(0) - \mathrm{e}^{-rT}K + Z_{\mathrm{put}}$ as an (indirect) alternative to Z_{call} for estimating the call price z_{call}.

The following table gives a comparison of the two estimators in terms of their 95% confidence intervals (we took $S(0) = 100$, $r = 5\%$, $\sigma = 0.25$, $T = 3$ and considered different strike prices K; all confidence intervals are based on the same $R = 100{,}000$ normal random numbers, whereas the figure for z_{call} is computed from the Black–Scholes formula):

K	z_{call}	direct	indirect
25	78.48	78.63±0.28	78.48±0.00
50	57.24	57.40±0.28	57.25±0.01
75	38.24	38.39±0.26	38.25±0.04
100	23.84	23.98±0.23	23.83±0.09
125	14.25	14.35±0.19	14.20±0.14
150	8.33	8.39±0.15	8.25±0.18
175	4.83	4.86±0.12	4.72±0.21
200	2.80	2.81±0.09	2.66±0.23

It is seen that the precision of the direct estimator is increasing in K and that of the indirect estimator decreasing, indicating that each of the estimators has its range of superiority (small K, i.e., the in-the-money case, for the indirect one and vice versa for the direct one).   □

Example 8.2 (LITTLE'S FORMULA) The relation $L = \lambda W$, where L is the mean queue length, W the mean sojourn time of a customer, and λ the arrival rate, is one of the most celebrated formulas of queuing theory, for example because of the great generality in which it holds. Here L, λ, W can be defined either as the steady-state quantities or as limits as $T \to \infty$ of the averages

$$\widehat{L} \stackrel{\text{def}}{=} \frac{1}{T} \int_0^T L(t)\, \mathrm{d}t, \quad \widehat{W} \stackrel{\text{def}}{=} \frac{1}{N(T)} \sum_{n=1}^{N(T)} W_n, \quad \widehat{\lambda} \stackrel{\text{def}}{=} \frac{N(T)}{T}, \qquad (8.1)$$

where $N(T)$ is the number of arrivals in $[0, T]$, $L(t)$ the number in system at time t, and W_n the sojourn time of customer n.

In a simulation context, natural direct estimators for L, W are $\widehat{L}, \widehat{W}$. Indirect alternatives are $\widehat{\lambda}\widehat{W}$, respectively $\widehat{L}/\widehat{\lambda}$, and, assuming that λ is known (as is often the case), $\lambda\widehat{W}$, respectively $\widehat{L}/\lambda$.

Here $\widehat{\lambda}\widehat{W}$ differs from $\widehat{L}$ only by R_T, the residual sojourn times of customers in system at time T or time $t = 0$. Typically, $\widehat{\lambda}\widehat{W}$ and $\widehat{L}$ obey a CLT with variances of order $\sqrt{T}$, and R_T will converge in distribution by regenerative-process arguments. Thus

$$\sqrt{T}\left(\widehat{\lambda}\widehat{W} - \widehat{L}\right) \stackrel{\mathbb{P}}{\to} 0 \,,$$

which shows that $\widehat{\lambda}\widehat{W}$, $\widehat{L}$ have exactly the same asymptotic properties. A similar argument shows that indirect estimation via $\widehat{L}/\widehat{\lambda}$ cannot present an improvement upon $\widehat{W}$.

Now suppose that λ is known. In many situations such as the GI/G/s queue, one expects a positive dependence between the three estimators in (8.1), so that an association argument (see A7) gives

$$\mathbb{E}\big[\widehat{\lambda}\widehat{W}\big]^2 \geq \mathbb{E}\widehat{\lambda}^2 \cdot \mathbb{E}\widehat{W}^2 \sim \lambda^2 \cdot \mathbb{E}\widehat{W}^2,$$

$$\mathbb{E}\big[\widehat{L}/\widehat{\lambda}\big]^2 \leq \mathbb{E}\widehat{L}^2 \cdot \mathbb{E}\widehat{\lambda}^{-2} \sim \mathbb{E}\widehat{L}^2/\lambda^2.$$

Thus one expects the indirect estimator $\lambda\widehat{W}$ for L to perform better than the either of $\widehat{\lambda}\widehat{W}$ or $\widehat{L}$, but the direct estimator for W to be better than the indirect one. However, the difference is not always that big, as demonstrated by Glynn & Whitt [155] to whom we refer for further discussion.   □

Example 8.3 Let $T_1, T_2, \ldots$ be i.i.d. and nonnegative, and let $Z \stackrel{\text{def}}{=}$ $\sup \{n : S_n \leq t\}$ be the number of renewals up to time t, where $S_n \stackrel{\text{def}}{=}$ $T_1 + \cdots + T_n$ ($z = \mathbb{E}Z$ is then the renewal function at t). Letting $\tau \stackrel{\text{def}}{=} \inf \{n : S_n > t\}$, we have $Z = \tau - 1$. By Wald's identity,

$$\mathbb{E}S_\tau = \mu\mathbb{E}\tau = \mu(z + 1).$$

But we can write $S_\tau = t + \xi$, where $\xi \overset{\text{def}}{=} S_\tau - t$ is the overshoot. This yields

$$z = \frac{t + \mathbb{E}\xi}{\mu} - 1 \,,$$

and an alternative estimator is

$$\widetilde{Z} \overset{\text{def}}{=} \frac{t + \xi}{\mu} - 1.$$

For example, if the T_i are standard exponential and $t = 50$, then Z is Poisson(50), so that $\mathbb{V}ar Z = 50$. In contrast, since ξ is again standard exponential, $\mathbb{V}ar \widetilde{Z} = 1$.   □

For a further example of indirect estimation, see the Minh–Sorli algorithm in XIV.13.

Chapter VI
Rare-Event Simulation

1 Efficiency Issues

In this chapter, we consider the problem of estimating $z = \mathbb{P}(A)$ when z is small, say of the order 10^{-3} or less; i.e., A is a *rare event*. Examples occur in telecommunications ($z =$ bit-loss rate, probability of buffer overflow), reliability ($z =$ the probability of failure before t), insurance risk ($z =$ the ruin probability), etc. Some general references in the area are Asmussen & Rubinstein [27], Heidelberger [178], and Juneja & Shahabuddin [204].

In the crude Monte Carlo (CMC) method, $Z = \mathbb{1}_A$ so we have Bernoulli sampling and hence a variance of $\sigma_Z^2 = z(1 - z)$. This goes of course to zero as $z \downarrow 0$, i.e., we have a small absolute error σ_Z for small z. However, this is not the most important observation to be made from $\sigma_Z^2 \sim z$. In fact, rather than the absolute error it is the relative error σ_Z/z that is the relevant performance measure, and this is high:

$$\frac{\sigma_Z}{z} = \frac{\sqrt{z(1-z)}}{z} \sim \frac{1}{\sqrt{z}} \to \infty, \quad z \downarrow 0.$$

To motivate focusing on the relative error, suppose, for example, that we obtain a point estimate $\widehat{z}$ of order 10^{-5} and a confidence interval of half-width 10^{-4}. This confidence interval may look narrow, but it does not help to tell whether z is of the magnitude 10^{-4}, 10^{-5}, or even much smaller. Another way to illustrate the problem is in terms of the sample size R needed to acquire a given relative precision, say 10%, in terms of the half-width of the 95% confidence interval. This leads to the equation $1.96\,\sigma_Z/(z\sqrt{R}) = 0.1$, i.e.,

$$R = \frac{100 \cdot 1.96^2 z(1-z)}{z^2} \sim \frac{100 \cdot 1.96^2}{z}, \qquad (1.1)$$

which increases like z^{-1} as $z \downarrow 0$. Thus, if z is small, large sample sizes are required, and when we get to probabilities of the order $z \approx 10^{-9}$, which occurs in many telecommunications applications, CMC simulation is not only inefficient but in fact impossible.

For a formal setup permitting the discussion of such efficiency concepts, let $\{A(x)\}$ be a family of rare events, where $x \in (0, \infty)$ or $x \in \mathbb{N}$, assume that $z(x) \stackrel{\text{def}}{=} \mathbb{P}A(x) \to 0$ as $x \to \infty$, and for each x let $Z(x)$ be an unbiased estimator for $z(x)$, i.e., $\mathbb{E}Z(x) = z(x)$. An algorithm is defined as a family $\{Z(x)\}$ of such r.v.'s.

The best performance that has been observed in realistic rare-event settings is *bounded relative error* as $x \to \infty$, meaning

$$\limsup_{x \to \infty} \frac{\mathbb{V}ar\, Z(x)}{z(x)^2} < \infty. \qquad (1.2)$$

In particular, such an algorithm will have the feature that R as computed in (1.1), with $z(1-z)$ replaced by $\mathbb{V}ar\, Z(x)$, remains bounded as $x \to \infty$.

An efficiency concept slightly weaker than (1.2) is *logarithmic efficiency*: $\mathbb{V}ar\big(Z(x)\big) \to 0$ so quickly that

$$\limsup_{x \to \infty} \frac{\mathbb{V}ar\, Z(x)}{z(x)^{2-\varepsilon}} = 0 \qquad (1.3)$$

for all $\varepsilon > 0$, or, equivalently, that

$$\liminf_{x \to \infty} \frac{\big|\log \mathbb{V}ar\, Z(x)\big|}{\big|\log z(x)^2\big|} \geq 1. \qquad (1.4)$$

Note that (1.3) is slightly weaker than bounded relative error. For example, if $z(x) \sim Ce^{-\gamma x}$, it allows $\mathbb{V}ar\, Z(x)$ to decrease like $x^p e^{-2\gamma x}$ or even $e^{-2\gamma x + \beta\sqrt{x}}$. The reasons for working with logarithmic efficiency rather than bounded relative error are the following: the difference is minor from a practical point of view; in some main examples, logarithmically efficient estimators exist, whereas estimators with bounded relative error do not (or have at least not yet been discovered); and logarithmic efficiency is often much easier to verify than bounded relative error.

Asymptotics of the form (1.4), (1.3) are commonly referred to as *logarithmic asymptotics* and are the framework of most of large-deviations theory (see Section 6 and references there). In practice, requirements (1.2)–(1.4) are most often verified by replacing $\mathbb{V}ar\, Z(x)$ by the upper bound $\mathbb{E}Z(x)^2$.

In accordance with discussions of run lengths in III.10, it would have been more logical to replace $\mathbb{V}ar\, Z(x)$ by $T(x)\,\mathbb{V}ar\, Z(x)$ in (1.2), (1.3), where $T(x)$ is the expected CPU time to generate one $Z(x)$, and also to include a comparison with the run length of the CMC method. One can check that in the examples we discuss, $T(x)$ grows so slowly with x that this makes no practical difference.

It should be noted that the rare event setting $\mathbb{P}A(x) \to 0$ is of course just a mathematical formalism, and a given rare event A can be embedded in such a setting in more than one way. For example, if $A = \{S_n > nx\}$ for a sum of i.i.d. r.v.'s, A may be rare because of LLN effects (n is large and x exceeds the mean), but also because x is so large that nx is far out in the tail. Thus, the limits $x \to \infty$ and $n \to \infty$ may both be relevant.

Much of the work on rare-events simulation is focused on importance sampling as a potential (though not the only) way to design efficient algorithms; in fact, many of our main examples in Sections 2 and 3 employ this method. The optimal change of measure $\widetilde{\mathbb{P}}$ (as discussed generally for importance sampling in V.1.2, V.1.3) is given by

$$\widetilde{\mathbb{P}}(B) = \mathbb{E}\left[\frac{Z}{z}; B\right] = \frac{1}{z}\mathbb{P}(AB) = \mathbb{P}(B \mid A).$$

That is, *the optimal $\widetilde{\mathbb{P}}$ is the conditional distribution given A.* However, just the same problem as for importance sampling in general comes up: it is usually not practicable to simulate from $\mathbb{P}(\cdot \mid A)$, and we cannot compute the likelihood ratio since z is unknown. Again, we may try to make $\widetilde{\mathbb{P}}$ look as much like $\mathbb{P}(\cdot \mid A)$ as possible; for example, the importance sampling algorithms in Section 2a and 2c, which are perhaps the two most classical examples in the area, may be seen in this light.

Our discussion here is mainly focused on algorithms that have bounded relative error or are logarithmically efficient. In complicated practical situations, it is usually impossible a priori to assert which algorithms have these properties, and one will try simpler ideas, say just perform importance sampling by changing some distribution within its parametric class; bounded relative error or logarithmical efficiency then may or may not be achieved. The following (toy) examples illustrate these points.

Example 1.1 Let N be geometric on $\{1, 2, \ldots\}$ with success parameter π, i.e. $\mathbb{P}(N = n) = \pi(1 - \pi)^{n-1}$. Consider

$$z = \mathbb{P}(N \le m) = \sum_{n=1}^{m} \pi(1 - \pi)^{n-1} = 1 - (1 - \pi)^m,$$

where $\pi = \pi(x)$ and $m = m(x)$ both depend on a parameter x such that $z = z(x) \to 0$ as $x \to \infty$. This means that $(1 - \pi)^m \to 1$, i.e., $m\pi \to 0$, which implies $z \sim m\pi$.

For simulating z, we consider importance sampling such that N is instead simulated from $\widetilde{\mathbb{P}}$ corresponding to a geometric distribution with success parameter $\widetilde{\pi}$. How should $\widetilde{\pi}$ be chosen?

The estimator (simulated from $\widetilde{\mathbb{P}}$) is

$$Z \stackrel{\text{def}}{=} \mathbb{1}\{N \le m\} \frac{\pi(1-\pi)^{N-1}}{\widetilde{\pi}(1-\widetilde{\pi})^{N-1}}.$$

Thus

$$
\begin{aligned}
\widetilde{\mathbb{E}}Z^2 &= \frac{\pi^2}{\widetilde{\pi}^2}\widetilde{\mathbb{E}}\left[\left(\frac{1-\pi}{1-\widetilde{\pi}}\right)^{2(N-1)}; N \le m\right] = \frac{\pi^2}{\widetilde{\pi}}\sum_{n=1}^{m}\frac{(1-\pi)^{2(n-1)}}{(1-\widetilde{\pi})^{n-1}} \\
&= \frac{\pi^2}{\widetilde{\pi}}\frac{(1-\pi)^{2m}/(1-\widetilde{\pi})^m - 1}{(1-\pi)^2/(1-\widetilde{\pi}) - 1}.
\end{aligned}
\tag{1.5}
$$

An obvious candidate for $\widetilde{\pi}$ is obtained by the saddle-point argument, equating $\widetilde{\mathbb{E}}N$ to m. Since $\mathbb{E}N = 1/\pi$, this means that $\widetilde{\pi} = 1/m$. From $\pi \sim z/m$ and (1.5), we then get

$$\widetilde{\mathbb{E}}Z^2 \sim \frac{z^2/m^2}{1/m}\frac{1/e^{-1} - 1}{1/(1-1/m) - 1} \sim z^2/m\frac{1/e^{-1} - 1}{1/m} = z^2(e-1),$$

so that the estimator has bounded relative error.

More generally, bounded relative error holds also by taking $\widetilde{\pi} = c/m$, since then $\widetilde{\mathbb{E}}Z^2 \sim z^2(e^c - 1)/c^2$. The optimal $c = c^*$ is obtained by minimizing $(e^c - 1)/c^2$, which yields $c^* = 1.59$. The corresponding variance-reduction factor $(e-1)c^{*2}/(e^{c^*} - 1) = 1.12$, is, however, minimal.   □

The next two examples, as well as many examples later on, deal with sums, and the following simple lemma will often be useful:

Lemma 1.2 *Let* $z \stackrel{\text{def}}{=} z(x) \stackrel{\text{def}}{=} \mathbb{P}(S_n > x)$, *where* $S_n \stackrel{\text{def}}{=} X_1 + \cdots + X_n$ *is a sum of a fixed number n of i.i.d. r.v.'s $X_1, \ldots, X_n$ with density $f(x)$. Consider importance sampling, where $X_1, \ldots, X_n$ are simulated as i.i.d. with density $\widetilde{f}(x)$; let L be the likelihood ratio and $Z \stackrel{\text{def}}{=} L\mathbb{1}\{S_n > x\}$. Then* $\widetilde{\mathbb{E}}Z^2 = c_\#^{-n}\mathbb{P}^\#(S_n > x)$, *where* $(c_\#)^{-1} \stackrel{\text{def}}{=} \int f^2/\widetilde{f}$, $f^\# \stackrel{\text{def}}{=} c_\# f^2/\widetilde{f}$.

Proof.

$$
\begin{aligned}
\widetilde{\mathbb{E}}Z^2 &= \widetilde{\mathbb{E}}\left[\left(\prod_{i=1}^{n}\frac{f(X_i)}{\widetilde{f}(X_i)}\right)^2; S_n > x\right] \\
&= \int \cdots \int_{x_1+\cdots+x_n>x}\frac{f(x_1)^2}{\widetilde{f}(x_1)^2}\cdots\frac{f(x_n)^2}{\widetilde{f}(x_n)^2}\widetilde{f}(x_1)\cdots\widetilde{f}(x_n)\,\mathrm{d}x_1\cdots\mathrm{d}x_n \\
&= c_\#^{-n}\int\cdots\int_{x_1+\cdots+x_n>x}f^\#(x_1)\cdots f^\#(x_n)\,\mathrm{d}x_1\cdots\mathrm{d}x_n \\
&= c_\#^{-n}\mathbb{P}^\#(S_n > x).
\end{aligned}
$$

□

Example 1.3 Consider the setting of Lemma 1.2 with $f(y) = e^{-y}$ standard exponential. Since S_n is Gamma(α, λ) with $\lambda = 1$, $\alpha = n$, we have

$$z(x) \sim \frac{x^{n-1}}{(n-1)!}e^{-x}, \quad x \to \infty. \tag{1.6}$$

Consider the importance density $\tilde{f}(x) = \lambda e^{-\lambda x}$. Then

$$\frac{1}{c_\#} = \lambda^{-1}\int_0^\infty e^{-(2-\lambda)y}\,dy = \frac{1}{\lambda(2-\lambda)}, \quad f^\#(y) = \lambda^\# e^{-\lambda^\# y},$$

where $\lambda^\# = 2 - \lambda$. It seems reasonable to choose λ such that $\widetilde{\mathbb{E}}S_n = n/\lambda$ is of order x, so we write tentatively $\lambda = c/x$. We get

$$\begin{aligned}
\widetilde{\mathbb{E}}Z^2 &= c_\#^{-n}\mathbb{P}^\#(S_n > x) = c_\#^{-n}\int_x^\infty \frac{(2-\lambda)^n y^{n-1}}{(n-1)!}e^{(2-\lambda)y}\,dy \\
&= c_\#^{-n}\int_{(2-\lambda)x}^\infty \frac{z^{n-1}}{(n-1)!}e^{-z}\,dz \sim c_\#^{-n}\frac{((2-\lambda)x)^{n-1}}{(n-1)!}e^{-(2-\lambda)x} \\
&= \frac{e^c x^{2n-1}}{(2-\lambda)c^n(n-1)!}e^{-2x} \sim \frac{e^c x^{2n-1}}{2c^n(n-1)!}e^{-2x},
\end{aligned}$$

which together with (1.6) shows that we have logarithmic efficiency but not bounded relative error (then the power of x should have been $2n-2$). It is also seen that the asymptotically optimal c is the minimizer n of e^c/c^n that corresponds to $\widetilde{\mathbb{E}}S_n \sim x$. $\qquad\square$

Example 1.4 Consider again standard exponential sums, but assume now that $\widetilde{\mathbb{P}}$ corresponds to i.i.d. sampling from the Gamma(α, λ) density with $\lambda = 1$. Then

$$\frac{1}{c_\#} = \int_0^\infty \frac{e^{-2y}}{y^{\alpha-1}e^{-y}/\Gamma(\alpha)}\,dy = \begin{cases} \Gamma(\alpha)\Gamma(2-\alpha) & 0 < \alpha < 2, \\ \infty & \alpha \geq 2. \end{cases}$$

This is discouraging, since for the importance sampling algorithm to have good properties, one expects $\alpha = \alpha(x) \to \infty$, but we see that we have infinite variance for $\alpha \geq 2$. For $\alpha \in (0,2)$, $\Gamma(\alpha)$ is bounded below, say by $\delta > 0$, so that

$$\begin{aligned}
\widetilde{\mathbb{E}}Z^2 &= \Gamma(\alpha)^n\Gamma(2-\alpha)^n\mathbb{P}^\#(S_n > x) \\
&= \Gamma(\alpha)^n\Gamma(2-\alpha)^{n-1}\int_x^\infty y^{1-\alpha}e^{-y}\,dy \\
&\geq \delta^{2n-1}\int_x^\infty y^{-1}e^{-y}\,dy \sim \delta^{2n-1}\frac{1}{x}e^{-x},
\end{aligned}$$

which shows that no choice of α can lead to logarithmic efficiency (for this, it should have been e^{-2x} instead of e^{-x}). $\qquad\square$

Exercises

1.1 (TP) Show that $\lambda = \lambda(x) = o(1)$ is necessary and sufficient for logarithmic efficiency in Example 1.3 whenever $\lambda x \to \infty$. Find similarly the necessary and sufficient condition when $\lambda x \to 0$.

2 Examples of Efficient Algorithms: Light Tails

In this and the next section, we give examples of algorithms meeting the efficiency criteria discussed in Section 1. One will note that they all deal with extremely simple problems. In more complex situations, one should not expect to be able to find rare-event estimators that are, say, logarithmically efficient. Rather, the ideas behind algorithms like the ones we study will then provide guidelines on how to proceed to obtain some substantial variance reduction without necessarily meeting the efficiency criteria in full.

We assume in this section that the relevant tails are light, i.e., that they decay at an exponential rate or faster. More precisely, $X_1, X_2, \ldots$ will be i.i.d. with distribution F with m.g.f. $\widehat{F}[s] \overset{\text{def}}{=} \mathbb{E}e^{sX} = \int e^{sx} F(\mathrm{d}x)$, and we will need the finiteness of $\widehat{F}[s]$ for certain $s > 0$ depending on the context.

Exponential change of measure (ECM) is defined as in V.1b by

$$F_\theta(\mathrm{d}x) \overset{\text{def}}{=} \frac{e^{\theta x}}{\widehat{F}[\theta]} F(\mathrm{d}x) = e^{\theta x - \kappa(\theta)} F(\mathrm{d}x),$$

where $\kappa(\theta) \overset{\text{def}}{=} \log \widehat{F}[\theta]$ is the c.g.f., and will be the main ingredient in the examples to be presented. The likelihood ratio $(\mathrm{d}\mathbb{P}/\mathrm{d}\mathbb{P}_\theta)_n$ for $X_1, \ldots, X_n$ is

$$L_{n,\theta} \overset{\text{def}}{=} \prod_{k=1}^{n} \frac{\widehat{F}[\theta]}{e^{\theta X_k}} = e^{-\theta S_n} \widehat{F}[\theta]^n = e^{-\theta S_n + n\kappa(\theta)},$$

where $S_n \overset{\text{def}}{=} X_1 + \cdots + X_n$. Thus if Y_n is $\sigma(X_1, \ldots, X_n)$-measurable, we have

$$\mathbb{E}Y_n = \mathbb{E}_\theta[Y_n L_{n,\theta}] = \mathbb{E}_\theta[Y_n e^{-\theta S_n + n\kappa(\theta)}]. \tag{2.1}$$

A convenient condition ensuring the finiteness of the $\widehat{F}[s]$ that will be needed is that $\widehat{F}$ be *steep*. For the definition, let $\theta_{\max} \overset{\text{def}}{=} \sup\{\theta : \widehat{F}[\theta] < \infty\}$ (for light-tailed distibutions, $0 < \theta_{\max} \leq \infty$). Then steepness means $\widehat{F}[\theta] \uparrow \infty$ as $\theta \uparrow \theta_{\max}$.

An illustration of the m.g.f. $\widehat{F}[\theta]$ and the c.g.f. $\kappa(\theta)$ of a distribution F with negative mean is given in Figure 2.1. Here γ_0 is the solution of $\widehat{F}'[\theta] = \kappa'(\theta) = 0$ or, equivalently, the minimizer of $\widehat{F}[\theta]$ and/or $\kappa(\theta)$, and γ is the nonzero solution of $\kappa(\gamma) = 0$, or, equivalently, of $\widehat{F}[\gamma] = 1$. The particular role of γ and γ_0 will become clear later on. When performing

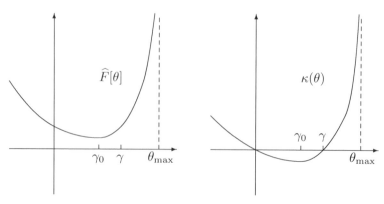

FIGURE 2.1

ECM, it is crucial to understand that the changed drift is

$$\mu_\theta \overset{\text{def}}{=} \mathbb{E}_\theta X = \mathbb{E}\left[\frac{Xe^{\theta X}}{\widehat{F}[\theta]}\right] = \frac{\widehat{F}'[\theta]}{\widehat{F}[\theta]} = \kappa'(\theta). \qquad (2.2)$$

Thus $\mu_\theta < 0$ when $\theta < \gamma_0$, $\mu_\theta > 0$ when $\theta > \gamma_0$, and $\mu_{\gamma_0} = 0$.

2a Siegmund's Algorithm

Assume that F is not concentrated on $(-\infty, 0]$ and that $\mathbb{E}X < 0$. We consider the problem of estimating

$$z(x) \overset{\text{def}}{=} \mathbb{P}\big(\tau(x) < \infty\big), \quad \text{where} \quad \tau(x) \overset{\text{def}}{=} \inf\{n : S_n > x\},$$

when x is large and hence $z(x)$ small. This problem has many applications: GI/G/1 waiting times (cf. I.(1.7)), ruin probabilities (cf. I.5.15), and sequential tests (cf. Exercise 2.4).

When implementing importance sampling via exponential change of measure, (2.1) takes the form

$$\begin{aligned}
z(x) &= \mathbb{P}\big(\tau(x) < \infty\big) = \mathbb{E}_\theta\left[L_{\tau(x),\theta}; \tau(x) < \infty\right] \\
&= \mathbb{E}_\theta\left[e^{-S_{\tau(x)} + \tau(x)\kappa(\theta)}; \tau(x) < \infty\right]. \qquad (2.3)
\end{aligned}$$

In choosing θ, the first step is to ensure that $\mathbb{P}_\theta\big(\tau(x) < \infty\big) = 1$, i.e., $\mathbb{E}_\theta X \geq 0$. By (2.2), this means that $\theta \geq \gamma_0$, where γ_0 is as in Figure 2.1. For such a θ, (2.3) becomes

$$z(x) = \mathbb{P}\big(\tau(x) < \infty\big) = \mathbb{E}_\theta L_{\tau(x),\theta} = e^{-\theta S_{\tau(x)} + \tau(x)\kappa(\theta)}. \qquad (2.4)$$

Thus, we may perform the simulation by the Monte Carlo method with $Z(x) = L_{\tau(x),\theta}$.

The crucial fact is now that typically the value γ of θ^1 is optimal. We will assume existence, i.e., that $\widehat{F}[\gamma] = 1$ has a solution $\gamma > 0$, and further that $\widehat{F}'[\gamma] < \infty$; in view of $\mathbb{E}X < 0$ and convexity, this basically says only that enough exponential moments exist, cf. Figure 2.1, and steepness is certainly a sufficient condition. For this special case, (2.4) becomes

$$z(x) = \mathbb{P}\big(\tau(x) < \infty\big) = \mathbb{E}_\gamma e^{-\gamma S_{\tau(x)}} = e^{-\gamma x}\mathbb{E}_\gamma e^{-\gamma \xi(x)}, \qquad (2.5)$$

where $\xi(x) \stackrel{\text{def}}{=} S_{\tau(x)} - x$ is the overshoot. Indeed we shall prove the following result:

Theorem 2.1 *The algorithm given by $Z(x) = e^{-\gamma x}e^{-\gamma \xi(x)}$ [simulated from $\mathbb{P}_\gamma$] has bounded relative error.*

Before giving the proof, we present two simple examples indicating what the change of measure looks like:

Example 2.2 Assume that F is $\mathscr{N}(-\mu, 1)$, where $\mu > 0$. Then $\widehat{F}[s] = \exp\{-\mu s + s^2/2\}$, so that γ solves $0 = -\mu\gamma + \gamma^2/2$, which in view of $\gamma > 0$ implies $\gamma = 2\mu$. We then get

$$\widehat{F}_\gamma[s] = \widehat{F}[s + \gamma] = \exp\{\mu s + s^2/2\},$$

which shows that F_γ is $\mathscr{N}(\mu, 1)$. □

Example 2.3 Assume that $X = U - T$ is the independent difference between two exponential r.v.'s with rates δ, respectively β, where $\beta < \delta$. This corresponds to the M/M/1 queue with arrival rate β and service rate δ. Then $\widehat{F}[\gamma] = 1$ means that

$$1 = \mathbb{E}e^{\gamma U}\mathbb{E}e^{-\gamma T} = \frac{\delta}{\delta - \gamma}\frac{\beta}{\beta + \gamma},$$

which has the positive solution $\gamma = \delta - \beta$. We then get

$$\widehat{F}_\gamma[s] = \widehat{F}[s + \gamma] = \frac{\beta}{\beta - s}\frac{\delta}{\delta + s},$$

which shows that F_γ is the distribution of the independent difference between two exponential r.v.'s with respective rates β and δ. That is, the changed measure corresponds to the M/M/1 queue with arrival rate δ and service rate β (the rates are switched). □

Proof of Theorem 2.1. The process $\{\xi(x)\}_{x \geq 0}$ is regenerative (cf. IV.6b; regeneration occurs at each partial maximum of $\{S_n\}$). Thus (assuming F to be aperiodic in the lattice case and nonlattice otherwise) $\xi(x) \stackrel{\mathscr{D}}{\to} \xi(\infty)$ and

$$\mathbb{E}_\gamma e^{-\gamma \xi(x)} \to C \stackrel{\text{def}}{=} \mathbb{E}_\gamma e^{-\gamma \xi(\infty)}, \qquad x \to \infty.$$

---

[1]This number γ is commonly referred to as the *Cramér root*; in insurance risk, γ goes under the name of the *adjustment coefficient*.

It follows that

$$z(x) = \mathbb{P}\big(\tau(x) < \infty\big) \sim Ce^{-\gamma x}, \tag{2.6}$$

a celebrated result going back to Cramér in 1930 and commonly referred to as the *Cramér-Lundberg approximation*.

The calculations are almost the same as for $\mathbb{V}ar_\gamma Z(x)$. Recalling that $Z = e^{-\gamma x}e^{-\gamma \xi(x)}$, we get

$$\mathbb{E}_\gamma Z^2 = e^{-2\gamma x}\mathbb{E}_\gamma e^{-2\gamma \xi(x)} \sim C_1 e^{-2\gamma x},$$

where $C_1 \stackrel{\text{def}}{=} \mathbb{E}_\gamma e^{-2\gamma \xi(\infty)}$. By Jensen's inequality, $C_1 > C^2$, and hence

$$\mathbb{V}ar_\gamma Z(x) \sim C_1 e^{-2\gamma x} - \big(Ce^{-\gamma x}\big)^2 \sim C_2 e^{-2\gamma x}, \tag{2.7}$$

where $C_2 \stackrel{\text{def}}{=} C_1 - C^2 > 0$. The relative error is thus

$$\frac{\sqrt{\mathbb{V}ar_\gamma Z}}{z(x)} \sim \frac{C_2^{1/2}e^{-\gamma x}}{Ce^{-\gamma x}} = C_3$$

($C_3 \stackrel{\text{def}}{=} C_2^{1/2}/C$), which does not increase with x, completing the proof of Theorem 2.1. □

Note that in the Siegmund algorithm, the expected time $T(x)$ to generate one replication satisfies $T(x) \sim \mathbb{E}_\gamma \tau(x) = \mathrm{O}(x)$.

2b   Uniqueness of the Change of Measure in Siegmund's Algorithm

Consider as above an importance sampling algorithm for estimating $z(x) = \mathbb{P}\big(\tau(x) < \infty\big)$ for a random walk with negative drift $\mu = \mu_F$, with the extension that we allow an arbitrary candidate $\widetilde{F}$ for the changed distribution of the X_k. That is, we simulate $X_1, X_2, \ldots$ from $\widetilde{F}$ and use the estimator

$$Z(x) \stackrel{\text{def}}{=} L_{\tau(x)}(F \mid \widetilde{F}) = \frac{\mathrm{d}F}{\mathrm{d}\widetilde{F}}(X_1) \cdots \frac{\mathrm{d}F}{\mathrm{d}\widetilde{F}}(X_{\tau(x)}), \tag{2.8}$$

where $\mathrm{d}F/\mathrm{d}\widetilde{F}$ means Radon–Nikodym derivative (e.g., if F and $\widetilde{F}$ both have densities f, $\widetilde{f}$ w.r.t. Lebesgue measure, then $(\mathrm{d}F/\mathrm{d}\widetilde{F})(x) = f(x)/\widetilde{f}(x)$). Note that we must impose two conditions on $\widetilde{F}$: that $\mathrm{d}F/\mathrm{d}\widetilde{F}$ exist and that $\widetilde{f}$ have positive mean $\widetilde{\mu}$ (otherwise, the simulation does not terminate in finite time).

Theorem 2.4 *The importance sampling algorithm* (2.8) *is logarithmically efficient if and only if* $\widetilde{F} = F_\gamma$.

Proof (Asmussen & Rubinstein [27]). Sufficiency [even with the stronger conclusion of bounded relative error] is contained in Theorem 2.1, so we assume that the IS distribution is $\widetilde{F} \neq F_\gamma$.

By the chain rule for Radon–Nikodym derivatives,

$$\widetilde{\mathbb{E}}Z(x)^2 \;=\; \widetilde{\mathbb{E}}L^2_{\tau(x)}(F\,|\,\widetilde{F}) \;=\; \widetilde{\mathbb{E}}\left[L^2_{\tau(x)}(F\,|\,F_\gamma)L^2_{\tau(x)}(F_\gamma\,|\,\widetilde{F})\right]$$

$$=\; \mathbb{E}_\gamma\left[L^2_{\tau(x)}(F\,|\,F_\gamma)L_{\tau(x)}(F_\gamma\,|\,\widetilde{F})\right] \;=\; \mathbb{E}_\gamma \exp\{K_1 + \cdots + K_{\tau(x)}\},$$

where

$$K_i \;\stackrel{\text{def}}{=}\; \log\left(\frac{\mathrm{d}F_\gamma}{\mathrm{d}\widetilde{F}}(X_i)\left(\frac{\mathrm{d}F}{\mathrm{d}F_\gamma}(X_i)\right)^2\right) \;=\; -\log\frac{\mathrm{d}\widetilde{F}}{\mathrm{d}F_\gamma}(X_i) - 2\gamma X_i.$$

Here

$$\mathbb{E}_\gamma K_i \;=\; \varepsilon' - 2\gamma\mathbb{E}_\gamma X_i \;=\; \varepsilon' - 2\gamma\mu_\gamma,$$

where $\mu_\gamma = \mu_{F_\gamma} > 0$ and

$$\varepsilon' \;\stackrel{\text{def}}{=}\; -\mathbb{E}_\gamma \log\frac{\mathrm{d}\widetilde{F}}{\mathrm{d}F_\gamma}(X_i) \;>\; 0$$

by the information inequality A.1. Since $K_1, K_2, \ldots$ are i.i.d., Jensen's inequality and Wald's identity yield

$$\widetilde{\mathbb{E}}Z(x)^2 \;\geq\; \exp\{\mathbb{E}_\gamma(K_1 + \cdots + K_{\tau(x)})\} \;=\; \exp\{\mathbb{E}_\gamma\tau(x)(\varepsilon' - 2\gamma\mu_\gamma)\}.$$

Since $\mathbb{E}_\gamma\tau(x)/x \to 1/\mu_\gamma$, it thus follows (using (2.6)) that for $0 < \varepsilon'' < \varepsilon'$, $0 < \varepsilon < \varepsilon''/\gamma\mu_\gamma$,

$$\liminf_{x\to\infty} \frac{\widetilde{\mathbb{E}}Z(x)^2}{z(x)^{2-\varepsilon}} \;=\; \liminf_{x\to\infty} \frac{\widetilde{\mathbb{E}}Z(x)^2}{C^{2-\varepsilon}\mathrm{e}^{-2\gamma x + \varepsilon\gamma x}}$$

$$\geq\; \liminf_{x\to\infty} \frac{\mathrm{e}^{x(\varepsilon''/\mu_\gamma - 2\gamma)}}{C^{2-\varepsilon}\mathrm{e}^{-2\gamma x + \varepsilon\gamma x}} \;=\; \infty,$$

which completes the proof. $\qquad\square$

2c   Efficient Simulation of $\mathbb{P}\big(S_n > n(\mu + \varepsilon)\big)$

Consider again a random walk $S_n \stackrel{\text{def}}{=} X_1 + \cdots + X_n$, where $X_1, X_2, \ldots$ are i.i.d. with common distribution F with mean μ (the sign of μ is unimportant). The rare event in question is now $A(n) \stackrel{\text{def}}{=} \{S_n > n(\mu + \varepsilon)\}$, where $\varepsilon > 0$ [thus the rare-event index is n and not x as before, and is discrete in this example]. That $z(n) = \mathbb{P}A(n) \to 0$ as $n \to \infty$, and hence that the event $A(n)$ is rare indeed, is immediate from the LLN.

We shall again employ exponential change of measure so that

$$Z(n) \;=\; \mathrm{e}^{-\theta S_n + n\kappa(\theta)}\mathbb{1}\{S_n > n(\mu + \varepsilon)\}\,.$$

The relevant choice of θ turns out to be given by the saddle-point method:

$$\mathbb{E}_\theta X = \kappa'(\theta) = \mu + \varepsilon\,, \tag{2.9}$$

which in particular implies $\theta > 0$ (since κ' is strictly increasing according to the strict convexity of κ) and $I > 0$, where $I \overset{\text{def}}{=} \theta(\mu + \varepsilon) - \kappa(\theta)$. Cf. Figure 2.2, where $\mu > 0$.

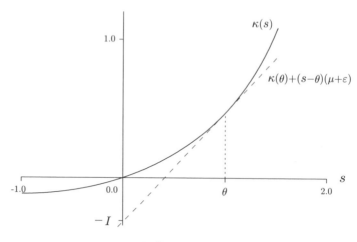

FIGURE 2.2

Theorem 2.5 *The exponential change of measure given by* (2.9) *is logarithmically efficient, and is the only importance distribution* $\widetilde{F}$ *with this property.*

As for Siegmund's algorithm, the proof is a small variant of the standard estimates for obtaining the asymptotics of $z(n)$ itself. In this case, these are given in [16, pp. 355 ff.], but parts are reproduced below (Lemmas 2.6, 2.8) for the sake of self-containedness.

Lemma 2.6 (CHERNOFF BOUND) $z(n) \leq e^{-nI}$.

Proof. Using the basic likelihood ratio identity and $\theta > 0$, we get

$$
\begin{aligned}
z(n) &= \mathbb{P}\,A(n) = \mathbb{E}_\theta\big[L_{n;\theta};\,A(n)\big] = \mathbb{E}_\theta\big[e^{-\theta S_n + n\kappa(\theta)};\,S_n > n(\mu + \varepsilon)\big] \\
&= e^{-nI}\mathbb{E}_\theta\big[e^{-\theta(S_n - n(\mu + \varepsilon))};\,S_n > n(\mu + \varepsilon)\big] \leq e^{-nI}. \qquad (2.10)
\end{aligned}
$$

$\square$

Lemma 2.7 $\mathrm{Var}_\theta\, Z(n) \leq e^{-2nI}$.

Proof. As in (2.10),

$$
\begin{aligned}
\mathbb{E}_\theta Z(n)^2 &= \mathbb{E}_\theta\big[e^{-2\theta S_n + 2n\kappa(\theta)};\,S_n > n(\mu + \varepsilon)\big] \\
&= e^{-2nI}\mathbb{E}_\theta\big[e^{-2\theta(S_n - n(\mu + \varepsilon))};\,S_n > n(\mu + \varepsilon)\big] \leq e^{-2nI}.
\end{aligned}
$$

$\square$

Lemma 2.8 $\displaystyle\liminf_{n\to\infty}\big(e^{nI + \theta\sqrt{n}}z(n)\big) > 0$.

Proof. Since

$$\frac{S_n - n(\mu + \varepsilon)}{\sqrt{n}} \xrightarrow{\mathscr{D}} \mathscr{N}(0, \sigma_\theta^2)$$

in $\mathbb{P}_\theta$-distribution ($\sigma_\theta^2 \stackrel{def}{=} \kappa''(\theta) > 0$), we have

$$\lim_{n \to \infty} \mathbb{P}_\theta \left(\frac{S_n - n(\mu + \varepsilon)}{\sqrt{n}} \in (0, 1) \right) = \Phi\left(\frac{1}{\sigma_\theta}\right) - \Phi(0) \stackrel{def}{=} c > 0.$$

Hence by (2.10),

$$\liminf_{n \to \infty} \left(e^{nI + \theta\sqrt{n}} z(n) \right)$$

$$\geq \quad \liminf_{n \to \infty} e^{\theta\sqrt{n}} \mathbb{E}_\theta \left[e^{-\theta(S_n - n(\mu + \varepsilon))}; \frac{S_n - n(\mu + \varepsilon)}{\sqrt{n}} \in (0, 1) \right]$$

$$\geq \quad \liminf_{n \to \infty} e^{\theta\sqrt{n}} e^{-\theta\sqrt{n}} \mathbb{P}_\theta \left(\frac{S_n - n(\mu + \varepsilon)}{\sqrt{n}} \in (0, 1) \right)$$

$$= \quad c > 0.$$

$\square$

The first part of Theorem 2.5 now follows by combining Lemmas 2.7 and 2.8. The second (uniqueness of the importance sampling distribution) can be proved by similar arguments as for Siegmund's algorithm. $\square$

See Bucklew, Ney, & Sadowsky [63] for further discussion and extension of the algorithm.

Remark 2.9 As sharpening of Lemmas 2.6–2.8, one can prove one of the basic results in the theory of saddle-point approximations, that subject to some smoothness assumptions,

$$z(n) \sim \frac{e^{-nI}}{\theta\kappa''(\theta)\sqrt{2\pi n}};$$

see, e.g., Petrov [289] and Jensen [196] for this and sharper versions. $\square$

2d   Compound Poisson Sums

Consider as in I.5.14 an r.v. C of the form $C \stackrel{def}{=} V_1 + \cdots + V_N$, where $V_1, V_2, \ldots$ are i.i.d. and nonnegative, with distribution say F, and N is an independent Poisson(λ) r.v. As explained in I.5.14, problems from insurance risk and queues call for the evaluation of the rare-event probability $\mathbb{P}A(x)$, where $A(x) \stackrel{def}{=} \{C > x\}$ and x is large.

By a variant of the analysis of Section 2c, we shall show that exponential change of measure is again logarithmically efficient provided F is light-tailed and satisfies some regularity conditions. To this end, we need first the c.g.f. $\varphi(\alpha) \stackrel{def}{=} \log \mathbb{E}e^{\alpha C}$ of C, which by a straightforward conditioning on N comes out as $\lambda(\widehat{F}[\alpha] - 1)$, where $\widehat{F}[\alpha] \overset{\text{def}}{=} \mathbb{E}e^{\alpha V}$ is the m.g.f. of F. Under exponential change of measure, the c.g.f. of C is therefore

$$\varphi_\theta(\alpha) \;=\; \varphi(\alpha + \theta) - \varphi(\theta) \;=\; \lambda\big(\widehat{F}[\alpha + \theta] - \widehat{F}[\theta]\big) \;=\; \lambda_\theta\big(\widehat{F_\theta}[\alpha] - 1\big),$$

where $\lambda_\theta \overset{\text{def}}{=} \lambda\widehat{F}[\theta]$, $F_\theta(\mathrm{d}x) \overset{\text{def}}{=} e^{\theta x} F(\mathrm{d}x)/\widehat{F}[\theta]$. In this notation, the classical Esscher approximation for $z(x)$ (Jensen [196] and references there) is

$$z(x) \;=\; \mathbb{P}(C > x) \;\sim\; \frac{e^{-\theta x + \varphi(\theta)}}{\theta\sqrt{2\pi\,\lambda\,\widehat{F}''[\theta]}}, \qquad x \to \infty, \tag{2.11}$$

where $\theta \overset{\text{def}}{=} \theta(x)$ is again determined by the saddle-point argument as the solution of $\mathbb{E}_\theta C = x$, i.e., by $x = \varphi'(\theta) = \lambda\widehat{F}'[\theta]$.

The conditions for (2.11) require some regularity of the density $f(x)$ of F. In particular, either of the following conditions is sufficient:

A. f is Gamma-like, i.e., bounded with

$$f(x) \;\sim\; c_1 x^{\alpha - 1} e^{-\delta x}, \qquad x \to \infty; \tag{2.12}$$

B. f is log-concave, or, more generally, $f(x) = q(x)e^{-h(x)}$, where $q(x)$ is bounded away from 0 and ∞ and $h(x)$ is convex on an interval of the form $[x_0, x^*)$, where $x^* \overset{\text{def}}{=} \sup\{x : f(x) > 0\}$. Furthermore, $\int_0^\infty f(x)^\zeta \, \mathrm{d}x < \infty$ for some $\zeta \in (1, 2)$.

For example, **A** covers the exponential distribution, phase-type distributions, and the inverse Gaussian distribution; **B** covers distributions with finite support or with a density not too far from e^{-x^α} with $\alpha > 1$. The role of these assumptions is mainly to verify a (local) CLT in $\mathbb{P}_\theta$-distribution for C as $\theta \uparrow \theta^* \overset{\text{def}}{=} \sup\{\theta : \widehat{F}[\theta] < \infty\}$, with variance constant $\kappa''(\theta) = \lambda\widehat{F}''[\theta]$ and mean x; for details, see [196]. One then gets (2.11) as follows:

$$
\begin{aligned}
z(x) &= \mathbb{E}_\theta\left[L(0 \mid \theta); C > x\right] = e^{-\theta x + \varphi(\theta)}\mathbb{E}_\theta\left[e^{-\theta(S - x)}; C > x\right] \\
&\sim e^{-\theta x + \varphi(\theta)} \int_0^\infty e^{-\theta\sqrt{\lambda\widehat{F}''[\theta]}\,y} \frac{1}{\sqrt{2\pi}}e^{-y^2/2} \, \mathrm{d}y \\
&= \frac{e^{-\theta x + \varphi(\theta)}}{\theta\sqrt{2\pi\lambda\widehat{F}''[\theta]}} \int_0^\infty e^{-z}e^{-z^2/(2\theta^2\lambda\widehat{F}''[\theta])} \, \mathrm{d}z \\
&\sim \frac{e^{-\theta x + \varphi(\theta)}}{\theta\sqrt{2\pi\lambda\widehat{F}''[\theta]}} \int_0^\infty e^{-z} \, \mathrm{d}z = \frac{e^{-\theta x + \varphi(\theta)}}{\theta\sqrt{2\pi\lambda\widehat{F}''[\theta]}}.
\end{aligned}
$$

Theorem 2.10 *Assume that either of* **A, B** *holds. Then the estimator* $Z(x) \overset{\text{def}}{=} e^{-\theta S + \kappa(\theta)}\mathbb{1}\{C > x\}$ *for* $z(x)$ *is logarithmically efficient.*

Proof. The outline of the proof of (2.11) is made precise in, e.g., [196]. In just the same way as there, one can rigorously verify the heuristics

$$\mathbb{E}_\theta \left[e^{-2\theta(C-x)}; C > x) \right] \approx \int_0^\infty e^{-2\theta\sqrt{\lambda \widehat{F}''[\theta]} y} \frac{1}{\sqrt{2\pi}} e^{-y^2/2} \, dy$$

$$= \frac{1}{2\theta\sqrt{2\pi \lambda \widehat{F}''[\theta]}} \int_0^\infty e^{-z} e^{-z^2/(8\theta^2 \lambda \widehat{F}''[\theta])} \, dz$$

$$\sim \frac{1}{2\theta\sqrt{2\pi \lambda \widehat{F}''[\theta]}} \int_0^\infty e^{-z} \, dz = \frac{1}{2\theta\sqrt{2\pi \lambda \widehat{F}''[\theta]}}.$$

Thus we get

$$\mathbb{E}_\theta Z(x)^2 = e^{-2\theta x + 2\varphi(\theta)} \mathbb{E}_\theta \left[e^{-2\theta(C-x)}; C > x) \right] \sim \frac{e^{-2\theta x + 2\varphi(\theta)}}{2\theta\sqrt{2\pi \lambda \widehat{F}''[\theta]}}.$$

Comparing this expression with (2.11) shows that all that remains to be verified is

$$\log(\theta\sqrt{\widehat{F}''[\theta]}) = o(\theta x - \varphi(\theta)). \tag{2.13}$$

In case **A**, $\theta_{max} = \delta$, and explicit calculus easily yields

$$\widehat{F}[\theta] \sim \frac{c_1}{(\delta - \theta)^\alpha}, \quad \widehat{F}'[\theta] \sim \frac{c_2}{(\delta - \theta)^{\alpha+1}}, \quad \widehat{F}''[\theta] \sim \frac{c_3}{(\delta - \theta)^{\alpha+2}}, \quad \theta \uparrow \theta_{max}.$$

Since $\theta = \theta(x) \uparrow \theta_{max} = \delta < \infty$ as $x \to \infty$, we get

$$\theta x - \varphi(\theta) = \theta\lambda\widehat{F}'[\theta] - \lambda(\widehat{F}[\theta] - 1) \sim c_4\widehat{F}'[\theta] \sim \frac{c_5}{(\delta - \theta)^{\alpha+1}},$$

from which (2.13) immediately follows, .

In case **B** with $\theta_{max} = \infty$, we write

$$\theta x - \varphi(\theta) = \theta\varphi'(\theta) - \int_0^\theta \varphi'(s) \, ds = \int_0^\theta s\varphi''(s) \, ds$$

$$\geq \theta \cdot \int_0^\theta \frac{s}{\theta} \, ds \cdot \int_0^\theta \frac{\varphi''(s)}{\theta} \, ds = \frac{\theta}{2}(\varphi'(\theta) - \varphi'(0)),$$

where we have used the inequality $\mathbb{E}[f(X)g(X)] \geq \mathbb{E}f(X)\,\mathbb{E}g(X)$ for increasing functions (cf. A7); in the present setting, $f(s) = s$, $g(s) = \varphi'(s)$ and X is uniform on $(0, \theta)$). By equation (4.11) of [196], $\widehat{F}''[\theta]$ is of the order of magnitude $\varphi'(\theta)^2\varphi(\theta)$, which is $O(\varphi'(\theta)^3)$. Since obviously $\log(\theta\varphi'(\theta)^{3/2}) = o(\theta\varphi'(\theta))$, (2.13) follows. □

Exercises

2.1 (TP) Show that the m.g.f. of the inverse Gaussian distribution fails to be steep.

2.2 (TP) Verify the formula $\varphi(\theta) = \lambda(\widehat{F}[\theta] - 1)$ in Section 2d.

2.3 (TP) Consider as in Section 2d a compound sum $C = V_1 + \cdots + V_N$, but assume now that N is geometric rather than Poisson, $\mathbb{P}(N = n) = (1 - \rho)\rho^n$ (cf. the Pollaczek–Khinchine formula in queuing and insurance risk). Choose $\gamma > 0$ such that $\rho \int_0^\infty e^{\gamma x} F(\mathrm{d}x) = 1$ and consider the importance distribution $\widetilde{F}(\mathrm{d}x) \overset{\text{def}}{=} \rho e^{\gamma x} F(\mathrm{d}x)$. Let further $\tau(x) \overset{\text{def}}{=} \inf\{n : V_1 + \cdots + V_n > x\}$ and $\xi(x) \overset{\text{def}}{=} V_1 + \cdots + V_{\tau(x)} - x$. Show that the estimator $Z(x) \overset{\text{def}}{=} e^{-\xi(x)}$ is unbiased for $z(x) \overset{\text{def}}{=} \mathbb{P}(S > x)$ and has bounded relative error.

2.4 (TP) Let $Y_1, Y_2, \ldots$ be i.i.d. with common density $f(y)$, and consider the problem of testing $H_0 : f = f_0$ versus the alternative $H_1 : f = f_1$. For a given fixed n, the usual likelihood ratio test rejects if the log likelihood $\ell_n \overset{\text{def}}{=} \sum_1^n X_k$ (where $X_k \overset{\text{def}}{=} \log[f_1(Y_k)/f_0(Y_k)]$) is large, say $\ell_n > b_n'$. The sequential test (Siegmund [340]) is formed by fixing $a < 0 < b$ and continuing observation until time $\tau \overset{\text{def}}{=} \inf\{n : \ell_n \notin [a, b]\}$. One rejects if $S_\tau > b$ and accepts if $\ell_\tau < a$. The level α is the probability of rejecting a true null hypothesis, i.e., $\mathbb{P}_0(\ell_\tau > b)$. Show that subject to H_0, $\{S_n\}$ is a random walk with negative drift, and that exponential change of measure with γ chosen as in the Siegmund algorithm gives relative bounded error in the limit $b \to \infty$ with a fixed when estimating α.

2.5 (A) Consider as in Section 2d a compound Poisson sum with $\lambda = 500$ and $f(x) = xe^{-x}$ (this random sum occurs, for example, in insurance risk as the total claims amount in a given year, where N is the number of claims and V_i the size of the ith claim). Compare empirically the efficiency of the technique of Section 2d with those in which instead you change only λ, respectively f.

2.6 (A) Consider the M/D/1 queue with reneging (customers being discouraged by long lines). That is, a customer arriving when $Q(t) = n$ customers are in the system leaves without joining the line w.p. p_n, where p_n is assumed to have a limit $p \in (0, 1)$ as $n \to \infty$. Your assignment is to compute the steady-state probability $\mathbb{P}(Q \geq N)$ of at least N customers in the system for some large N by regenerative simulation combined with importance sampling. More precisely, let λ denote the arrival rate and let services have length d. Then in each cycle, change the arrival rate to the $\widetilde{\lambda}$ you would use in Siegmund's algorithm if the arrival rate were $(1 - p)\lambda$ but only until $Q(t)$ reaches level N, then switch off the importance sampling.

A main difficulty is to figure out how the naive estimate

$$\int_0^C \mathbb{1}\{Q(t)) \geq N\}\, \mathrm{d}t$$

of

$$\mathbb{E}\left[\int_0^C \mathbb{1}\{Q(t)) \geq N\}\, \mathrm{d}t\right] = \mathbb{E}C \cdot \mathbb{P}(Q \geq N)$$

should be modified by the importance sampling! We recommend that you compute the mean length $\mathbb{E}C$ of the busy cycle C by independent sampling without importance sampling.

Parameters: $\lambda = 1/2$, $d = 1$, $p_n = (1 - 1/2^n)/2$, $N = 6$.

3 Examples of Efficient Algorithms: Heavy Tails

That a distribution F has heavy (right) tails has been given various meanings. We follow here the tradition of considering subexponential distributions, which (for the case in which F is concentrated on $[0, \infty)$) means that

$$\frac{\overline{F}^{*2}(x)}{\overline{F}(x)} \to 2 \ \text{ or equivalently } \ \frac{\mathbb{P}(X_1 + X_2 > x)}{\mathbb{P}(X_1 > x)} \to 2, \ x \to \infty, \quad (3.1)$$

where X_1, X_2 are i.i.d. with distribution F. This can in fact be shown to be equivalent to the apparently stronger

$$\frac{\overline{F}^{*n}(x)}{\overline{F}(x)} \to n \ \text{ or equivalently } \ \frac{\mathbb{P}(X_1 + \cdots + X_n > x)}{\mathbb{P}(X_1 > x)} \to n \quad (3.2)$$

for all n. Practitioners may safely restrict attention to the following main examples:

Regular variation, $\overline{F}(x) = L(x)/x^\alpha$, where $\alpha > 0$ and L is slowly varying, i.e., satisfies $L(tx)/L(x) \to 1$ for any fixed $t > 0$.

The **lognormal distribution**, where $X_1 = e^Y$ with $Y \sim \mathcal{N}(\mu, \sigma^2)$.

The **Weibull distribution** with $\overline{F}(x) = e^{-cx^\beta}$ for some $0 < \beta < 1$.

Probably the most important example among these is regular variation. Here the most prominent example is the Pareto distribution, with tail $1/(1+x)^\alpha$ and density $f(x) = \alpha/(1+x)^{\alpha+1}$, $0 < x < \infty$, or (after a change of location and scale) $f(x) = \alpha b^\alpha/(b+x-a)^{\alpha+1}$, but also α-stable distributions with $0 < \alpha < 2$ and others are included.

The general theory of subexponential distributions is surveyed in, for example, Embrechts et al. [108] and Asmussen [15, Section IX.1]. Heavy-tail modeling has had a boom in the last decade in particular in queuing theory and its telecommunications applications and in insurance risk[2] (it is also related to long-range dependence, to which we return in XI.6). See, for example, Adler et al. [5], Resnick [298], and references there. We discuss statistical and modeling aspects in Section 4.

Rare-event simulation with heavy tails is far less developed than with light tails. In fact, Asmussen, Binswanger, & Højgaard [18] present a number of counterexamples showing that the main ideas from the light-tailed case do not carry over to heavy tails, so that efficient algorithms have to look different. We will discuss here only a very simple problem, to simulate $z(x) = \mathbb{P}(S_n > x)$, where $S_n = X_1 + \cdots + X_n$ with $X_1, X_2, \ldots$ i.i.d. and non-

---

[2]In mathematical finance, heavy tails usually means tails heavier than the normal, say NIG tails that are light-tailed in the framework of this chapter. The problems in the area are usually not rare-event problems anyway, with the possible exception of some barrier options and out-of-the-money options.

negative with common subexponential distribution F. Thus $z(x) \sim n\overline{F}(x)$ by (3.2), but it should be noted that there are many reported cases in which this approximation is very inaccurate, so that simulation may be realistic. Making $n = N$ random, the $\mathbb{P}(S_n > x)$ setting incorporates cases such as the M/G/1 waiting-time tail (N geometric, cf. the Pollaczeck–Khinchine formula I.(5.3)) or accumulated claims (say N Poisson) in insurance, but we will not discuss this here and refer to the original articles cited below. It should be noted than in contrast to the algorithms of Section 2c, which also deal with $\mathbb{P}(S_n > x)$, we study here the limit $x \to \infty$ rather than $n \to \infty$ (this is mainly for convenience in order to fit into the subexponential framework).

3a   Conditional Monte Carlo Algorithms

The CMC estimator is $Z_1(x) \overset{\text{def}}{=} \mathbb{1}\{S_n > x\}$, and so a conditional Monte Carlo estimator has the form $\mathbb{P}(S_n > x \,|\, \mathscr{F})$, where $\mathscr{F} \subset \sigma(X_1, \ldots, X_n)$. We recall from V.4 that this always gives variance reduction. The problem is to find an $\mathscr{F}$ for which this is substantial, hopefully so much that the estimator is logarithmically efficient or even has bounded relative error.

The first and obvious idea (cf. V.4.2) is to condition on $X_1, \ldots, X_{n-1}$, which leads to

$$Z_2(x) \overset{\text{def}}{=} \mathbb{P}(S_n > x \,|\, X_1, \ldots, X_{n-1}) = \overline{F}(x - S_{n-1}).$$

Thus, we generate only $X_1, \ldots, X_{n-1}$. As a conditional Monte Carlo estimator, $Z_2(x)$ has a smaller variance than $Z_1(x)$. However, asymptotically it presents no improvement: the variance is of the same order of magnitude $\overline{F}(x)$. To see this, just note that

$$\mathbb{E}Z_2(x)^2 \geq \mathbb{E}\big[\overline{F}(x - S_{n-1}); X_1 > x\big] = \mathbb{P}(X_1 > x) = \overline{F}(x)$$

(here we used that by positivity of the X_i, $S_{n-1} > x$ when $X_1 > x$, and that $\overline{F}(y) = 1$, $y < 0$).

The reason that this algorithm does not work well is that the probability of one single X_i to become large is too big. Asmussen & Binswanger [17] suggested that one could circumvent this problem by discarding the largest of the X_i and considering only the remaining ones. For the simulation, we thus generate $X_1, \ldots, X_n$, form the order statistics

$$X_{(1)} < X_{(2)} < \cdots < X_{(n)},$$

throw away the largest one $X_{(n)}$, and let

$$\begin{aligned} Z_3(x) &\overset{\text{def}}{=} \mathbb{P}\big(S_n > x \,\big|\, X_{(1)}, X_{(2)}, \ldots, X_{(n-1)}\big) \\ &= \frac{\overline{F}\big((x - S_{(n-1)}) \vee X_{(n-1)}\big)}{\overline{F}(X_{(n-1)})}, \end{aligned}$$

where $S_{(n-1)} \stackrel{\text{def}}{=} X_{(1)} + X_{(2)} + \cdots + X_{(n-1)}$ (to avoid multiple-match problems, we assume the existence of a density $f(x)$ here and in the following). To check the formula for the conditional probability, note first that

$$\mathbb{P}\big(X_{(n)} > x \,\big|\, X_{(1)}, X_{(2)}, \ldots, X_{(n-1)}\big) = \frac{\overline{F}\big(X_{(n-1)} \vee x\big)}{\overline{F}(X_{(n-1)})}.$$

We then get

$$
\begin{aligned}
\mathbb{P}\big(S_n > x \,\big|\, X_{(1)}, X_{(2)}, \ldots, X_{(n-1)}\big) \\
= \quad & \mathbb{P}\big(X_{(n)} + S_{(n-1)} > x \,\big|\, X_{(1)}, X_{(2)}, \ldots, X_{(n-1)}\big) \\
= \quad & \mathbb{P}\big(X_{(n)} > x - S_{(n-1)} \,\big|\, X_{(1)}, X_{(2)}, \ldots, X_{(n-1)}\big) \\
= \quad & \frac{\overline{F}\big((x - S_{(n-1)}) \vee X_{(n-1)}\big)}{\overline{F}(X_{(n-1)})}.
\end{aligned}
$$

Theorem 3.1 *In the regularly varying case, the estimator $\{Z_3(x)\}$ is logarithmically efficient.*

This appears to be the first example of a logarithmically efficient algorithm in the heavy-tailed case. The idea of the proof is to bound the density $f_{X_{(n-1)}}(y)$ of the r.v. $X_{(n-1)}$ using

$$f_{X_{(n-1)}}(y) = n(n-1)F^{n-2}(y)\overline{F}(y)f(y) \le c\overline{F}(y)f(y), \qquad (3.1)$$

and evaluate the second moment of $Z_3(x)$ separately over the regions $X_{(n-1)} \le x/n$, $x/n < X_{(n-1)} \le x/2$, and $X_{(n-1)} > x/2$. We will, however, not give the details (but see Exercise 3a.1), since another equally simple conditional Monte Carlo developed later by Asmussen & Kroese [23] performs better. The idea there is to partition according to which X_i is the largest, i.e., for which i one has $M_n \stackrel{\text{def}}{=} X_{(n)} = X_i$, and condition on the X_j with $j \ne i$. Since clearly by symmetry $z(x) = n\mathbb{P}(S_n > x, M_n = X_n)$, this gives the estimator

$$
\begin{aligned}
Z_4(x) \;\stackrel{\text{def}}{=}\; & n\,\mathbb{P}\big(S_n > x, M_n = X_n \mid X_1, \ldots, X_{n-1}\big) \\
= \;& n\,\overline{F}\big(M_{n-1} \vee (x - S_{n-1})\big). \qquad (3.2)
\end{aligned}
$$

Theorem 3.2 *The estimator $Z_4(x)$ has bounded relative error in the regularly varying case, and is logarithmically efficient in the Weibull case provided $\beta < \bar{\beta} = \log(3/2)/\log 2 = 0.585$.*

Proof. We consider only the regularly varying case. If $M_{n-1} \le x/n$, then $S_{n-1} \le (n-1)x/n$ and therefore always $M_{n-1} \vee (x - S_{n-1}) \ge x/n$. Therefore

$$
\begin{aligned}
\frac{\mathbb{E}Z_4(x)^2}{\overline{F}(x)^2} \;&\le\; n^2 \frac{\overline{F}(x/n)^2}{\overline{F}(x)^2} \;=\; n^2 \frac{L(x/n)^2/(x/n)^{2\alpha}}{L(x)^2/x^{2\alpha}} \\
&= \; n^{2+2\alpha} \frac{L(x/n)^2}{L(x)^2} \;\sim\; n^{2+2\alpha}.
\end{aligned}
$$

Noting that $z(x) \sim n\overline{F}(x)$ by subexponentiality completes the proof. □

In the case of a random $n = N$, it is suggested in [23] suggested that either N be used as a control variate or that N be stratified, and a substantial variance reduction was obtained. A theoretical support for the control-variate approach was provided by Hartinger & Kortschak [176], who showed that in fact, in this setting the relative error goes to 0 as $x \to \infty$.

3b   Importance Sampling Algorithms

We first consider an importance sampling scheme in which the density f of $X_1, \ldots, X_n$ is changed to $\widetilde{f}$.

Asmussen, Binswanger, & Højgaard [18] considered the case in which $\widetilde{f}$ does not depend on x and proved the following:

Proposition 3.3 *Let $F_{\#}$ be the distribution with density proportional to $f^2/\widetilde{f}$. Assume that $F_{\#}$ is subexponential and satisfies*

$$\liminf_{x\to\infty} \frac{|\log \overline{F}_{\#}(x)|}{2|\log \overline{F}(x)|} \geq 1.$$

Then the importance sampling algorithm given by $\widetilde{f}$ is logarithmically efficient.

Proof. Let $c_{\#}^{-1} \stackrel{\text{def}}{=} \int_0^\infty f^2/\widetilde{f}$. Then by Lemma 1.2, the second moment of the estimator is

$$\mathbb{E}^* \left[\left(\prod_{i=1}^n \frac{f(X_i)}{\widetilde{f}(X_i)} \right)^2 ; S_n > x \right] = c_{\#}^{-n} \mathbb{P}^{\#}(S_n > x) \sim c_{\#}^{-n} n\overline{F}_{\#}(x),$$

where the last step used the subexponentiality of $F_{\#}$. Since $z(x) \sim n\overline{F}(x)$ and $c_{\#}$ does not depend on x, the second assumption on $F_{\#}$ therefore immediately implies that (1.4) holds. □

Example 3.4 If F is regularly varying with $\alpha > 0$, one can take the tail of $\widetilde{F}$ as, e.g., $1/\log(e+x)$ (then $\widetilde{F}$ is a regularly varying distribution with $\alpha = 0$). Indeed, then $\widetilde{f}(x) = L_1(x)/x$ and Karamata's theorem $\int_x^\infty L(y)/y^\alpha \, dy \sim L(x)/(\alpha-1)x^{\alpha-1}$ for a slowly varying function L (see Feller [116]) implies

$$\overline{F}_{\#}(x) = c_1 \int_x^\infty \frac{L(x)^2/x^{2\alpha+2}}{L_1(x)/x} \, dx \sim c_2 \frac{L_2(x)}{x^{2\alpha}},$$

where L_1 and $L_2 \stackrel{\text{def}}{=} L^2/L_1$ are slowly varying. □

Proposition 3.3 is, however, one of the notorius reminders that a limit theorem does not always tell the truth on how an algorithm performs for a given set of parameters: all numerical experience shows poor performance. Juneja & Shahabuddin [203] developed more efficient importance sampling algorithms by allowing $\widetilde{f}$ to depend on x. More precisely, they suggested that the tail of F be changed to $c_1 \overline{F}(x)^{\theta(x)}$ on $[x_0, \infty)$ and that the density $c_2 f(x)$ be used on $(0, x_0)$, where $\theta(x) \to 0$ and c_1, c_2 have to be chosen in a certain way. We will not give the details but only present a simplied version of the algorithm in the Pareto case, where $\widetilde{f}$ again is Pareto where $\widetilde{\alpha} \stackrel{\text{def}}{=} \widetilde{\alpha}(x) \stackrel{\text{def}}{=} \alpha\theta(x) \to 0$ (the regularly varying case is an easy extension).

Example 3.5 Let $f(x) = \alpha(1 + x)^{-\alpha-1}$ and $\widetilde{f}(x) = \widetilde{\alpha}(1 + x)^{-\widetilde{\alpha}-1}$, where $\widetilde{\alpha} \to 0$. By Lemma 1.2, the second moment of the estimator is $c_\#^{-n} \mathbb{P}_{2\alpha-\widetilde{\alpha}}(S_n > x)$, where

$$c_\#^{-1} = \int_0^\infty \frac{\alpha^2/(1+x)^{2\alpha+2}}{\widetilde{\alpha}/(1+x)^{\widetilde{\alpha}+1}} \, dx = \frac{\alpha^2}{\widetilde{\alpha}(2\alpha - \widetilde{\alpha})} \sim \frac{\alpha}{2\widetilde{\alpha}}. \qquad (3.3)$$

Bounding $\mathbb{P}_{2\alpha-\widetilde{\alpha}}(S_n > x)$ above and below by

$$\mathbb{P}_{2\alpha-\varepsilon}(S_n > x) \sim \frac{n}{x^{2\alpha-\varepsilon}}, \quad \text{respectively} \quad \mathbb{P}_{2\alpha+\varepsilon}(S_n > x) \sim \frac{n}{x^{2\alpha+\varepsilon}},$$

and letting $\varepsilon \downarrow 0$ gives easily that the algorithm is logarithmically efficient provided $\widetilde{\alpha} = \widetilde{\alpha}(x)$ satisfies $\log \widetilde{\alpha}/\log x \to 0$.

What is, more precisely, the best choice of $\widetilde{\alpha}$? Equation (3.3) suggests (at least at the heuristical level) that the variance minimizer is the minimizer $\widetilde{\alpha}^* = n/\log x$ of the r.h.s. of

$$\widetilde{\mathbb{E}}Z^2 \sim c_\#^{-n} \mathbb{P}_{2\alpha-\widetilde{\alpha}}(S_n > x) \sim \left(\frac{\alpha}{2\widetilde{\alpha}}\right)^n \frac{1}{x^{2\alpha-\widetilde{\alpha}}}.$$

Indeed, Juneja & Shahabuddin [203] suggested that one take $\widetilde{\alpha} = b/\log x$ for some unspecified b. We will give an alternative argument for this choice in Section 8. See in particular Example 8.2, which also leads to the identification $b = n$. ☐

We finally present an algorithm that uses the same symmetry argument as in (3.2), but now combined with importance sampling rather than conditioning. Changing only the density of X_n, this means that we simulate $X_1, \ldots, X_n$ as independent with density f for $X_1, \ldots, X_{n-1}$ and $\widetilde{f}$ for X_n. Thus, the estimator is

$$n \frac{f(X_n)}{\widetilde{f}(X_n)} \mathbb{1}\{S_n > x, M_n = X_n\}. \qquad (3.4)$$

This was proposed in Asmussen & Kroese [23], who also demonstrated numerically that the algorithm is more efficient than any of the importance sampling algorithms above (but that still the conditional Monte Carlo estimator (3.2) is better) and showed the expected logarithmic efficiency. We omit the proof, which is somewhat more complicated than the ones above.

Exercises

3.1 (TP) Prove Theorem 3.1 for $n = 2$.
3.2 (TP) Prove Theorem 3.2 for $n = 2$ in the Weibull case.

3.3 (TP) Prove that if F is Weibull, then any regularly varying F^* can be used in Proposition 3.3.

4 Tail Estimation

The fact that the order of rare-event probabilities usually depends crucially on the tail and that the asymptotics as well as the efficient simulation algorithms are very different in light- and heavy-tailed regimes poses the problem of which distributional tail to employ. Of course, this is a general statistical problem but definitely something that the simulator needs to take seriously. We give here a brief introduction and refer to Embrechts, Klüppelberg, & Mikosch [108] and de Haan & Ferreira [84] for more detailed and broader expositions (one should note, however, that the area is rapidly expanding).

We will consider the problem of fitting a distribution F with particular emphasis on the tail to a set of data $X_1, \ldots, X_n \geq 0$ assumed to be i.i.d. with common distribution F. As usual, $X_{(1)}, \ldots, X_{(n)}$ denotes the order statistics.

Inference on $\overline{F}(x)$ beyond $x = X_{(n)}$ is of course extrapolation of the data, and in a given situation, it will far from always be obvious that this makes sense. However, some extrapolation seems inevitable in rare-event simulation. In particular, trace-driven simulation (r.v. generation from the empirical distribution F_n^*) as discussed in III.9 is not feasible in rare-event settings: because F_n^* has the finite upper bound $X_{(n)}$, the method will invariably underestimate the tail and in particular always postulate that it is light.

4a The Mean Excess Plot

A first question is to decide whether to use a light- or a heavy-tailed model. The approach most widely used is based on the *mean excess function*, defined as the function

$$e(x) \stackrel{\text{def}}{=} \mathbb{E}\big[X - x \mid X > x\big] = \frac{1}{\overline{F}(x)} \int_x^\infty \overline{F}(y) \, dy$$

of $x > 0$ (in insurance mathematics, the term *stop-loss transform* is common).

The reason that the mean excess function $e(x)$ is useful is that it typically asymptotically behaves quite differently for light and heavy tails. Namely, for a subexponential heavy-tailed distribution one has $e(x) \to \infty$, whereas with light tails it will typically hold that $\limsup e(x) < \infty$; say a sufficient condition is

$$\overline{F}(x) \sim \ell(x) e^{-\alpha x} \tag{4.1}$$

for some $\alpha > 0$ and some $\ell(x)$ such that $\ell(\log x)$ is slowly varying (e.g., $\ell(x) = x^\gamma$ with $-\infty < \gamma < \infty$).

The mean excess test proceeds by plotting the empirical version

$$e_n(x) = \frac{1}{\#j : X_j > x} \sum_{j: X_j > x} (X_j - x)$$

of $e(x)$, usually only at the (say) K largest X_j. That is, the plot consists of the pairs formed by $X_{(n-k)}$ and

$$\frac{1}{k} \sum_{\ell=n-k+1}^{n} \left(X_{(\ell)} - X_{(n-k)}\right),$$

where $k = 1, \ldots, K$. If the plot shows a clear increase to ∞ except possibly at very small k, one takes this as indication that F is heavy-tailed, otherwise one settles for a light-tailed model.

Example 4.1 Figure 4.1 contains the mean excesses of simulated data with $n = 1,000$ from six different distributions. Each row is generated from i.i.d. r.v.'s $Y_1, Y_2, \ldots$ such that $X = Y_1$ in the left column and $X = Y_1 + Y_2 + Y_3$ in the right. In row 1, Y is Pareto with $\alpha = 3/2$; in row 2, Y is Weibull with $\beta = 1/2$; and in row 3, Y is exponential; the scale is chosen such that $\mathbb{E}Y = 3$ in all cases.

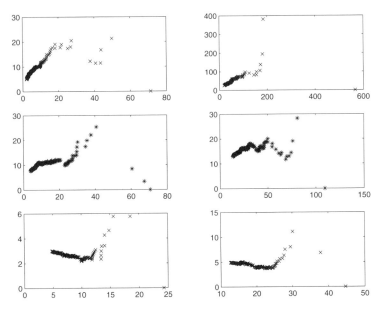

FIGURE 4.1

To our mind, the story told by these pictures is not all that clear, so one conclusion is certainly that using the mean excess plot is not entirely straightforward.   □

4b The Hill Estimator

We now assume that F is either regularly varying, $\overline{F}(x) = L(x)/x^\alpha$, or light-tailed satisfying (4.1). The problem is to estimate α.

Even with L or ℓ completely specified, the maximum likelihood estimator (MLE) is not adequate in this connection, because maximum likelihood will try to adjust α so that the fit is good in the center of the distribution, without caring too much about the tail, where there are fewer observations. The Hill estimator is the most commonly used (though not the only) estimator designed specifically to take this into account.

To explain the idea, consider first the setting of (4.1). If we ignore fluctuations in $\ell(x)$ by replacing $\ell(x)$ by a constant, the $X_j - x$ with $X_j > x$ are i.i.d. exponential(α). Since the standard MLE of α in the (unshifted) exponential distribution is $n/(X_1 + \cdots + X_n)$, the MLE α based on these selected X_j alone is

$$\frac{\#j : X_j > x}{\sum_{j: X_j > x}(X_j - x)}.$$

The *Hill plot* is this quantity plotted as function of x or the number $\#j :$ $X_j > x$ of observations used. As for the mean excess plot, one usually plots only at the (say) k largest j or the k largest X_j. That is, one plots

$$\frac{k}{\sum_{\ell=n-k+1}^{n}\left(X_{(\ell)} - X_{(n-k)}\right)}. \tag{4.2}$$

as function of either k or $X_{(n-k)}$. The *Hill estimator* $\alpha_{n,k}^{H}$ is (4.2) evaulated at some specified k. However, most often one checks graphically whether the Hill plot looks reasonably constant in a suitable range and takes a typical value from there as the estimate of α.

The regularly varying case can be treated by entirely the same method, or one may remark that it is 1-to-1 correspondance with (4.1) because X has tail $L(x)/x^\alpha$ if and only if $\log X$ has tail (4.1). Therefore, the Hill estimator in the regularly varying case is

$$\frac{k}{\sum_{\ell=n-k+1}^{n}\left(\log X_{(\ell)} - \log X_{(n-k)}\right)}. \tag{4.3}$$

It can be proved that if $k = k(n) \to \infty$ but $k/n \to 0$, then weak consistency $\alpha_{n,k}^{H} \xrightarrow{\mathbb{P}} \alpha$ holds. No conditions on L are needed for this. One might think that the next step would be the estimation of the slowly varying function L, but this is in general considered impossible among statisticians. In fact, we will see below that there are already difficulties enough with $\alpha_{n,k}^{H}$ itself.

Example 4.2 Figure 4.2 contains the Hill plot (4.3) of simulated data (now with $n = 10{,}000$) from the same six distributions as in Example 4.1 and the number $k = 10, \ldots, 2{,}000$ of order statistics used on the horizontal axis.

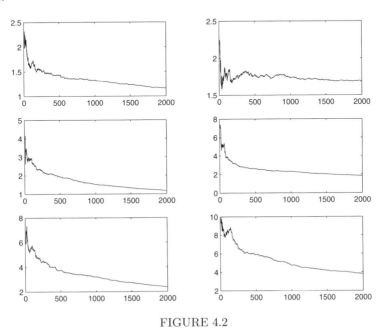

FIGURE 4.2

Of course, only the first row, Pareto(3/2), is meaningful, since the distributions in the remaining rows are not regularly varying. Nevertheless, the appearance of the second row of plots, Weibull, is so close to the first that it is hard to assert from this alone that the distribution is not regularly varying (the evidence from the mean excess plot in Figure 4.1 is not that conclusive either). The same holds, though maybe in a somewhat weaker form, for the exponential case in the third row.

The first row also clearly demonstrates the difficulty in choosing k. Maybe one would settle for a value between 50 and 500 in the left panel, giving an estimate of α between 1.6 and 1.4. The implications for the correct estimation of $\overline{F}(x)$ down to 10^{-5} (recall that we have 10^4 observations) are illustrated in the following table, where the first row gives the theoretical values of the 99%, 99.9%, and 99.99% quantiles of $F_{1.5}$:

	20.5	99.0	463	2153
1.6	0.007	0.0006	0.00005	0.000005
1.5	0.010	0.0010	0.00010	0.000010
1.4	0.014	0.0016	0.00018	0.000022

There is also a CLT $k^{1/2}\left(\alpha_{n,k}^{H} - \alpha\right) \to \mathcal{N}\left(0, \alpha^2\right)$. For this, however, stronger conditions on $k = k(n)$ and L are needed. In particular, the correct choice of k requires delicate estimates of L. That L can present a major complication has also been observed in the many "Hill horror plots" in the literature; cf. Exercise 4.3.   □

4c Fitting the Whole Distribution

Even if one believes that one has an acceptable estimate of the tail index α of a regularly varying distribution F, there are still some steps before one arrives at the distribution $\widehat{F}$ to be used in the simulation. Figure 4.3 contains histograms of simulated observations from the two distributions in the first row of Figure 4.1, which have the same α. In the left panel it may be acceptable to take $\widehat{F}$ to be Pareto, say with $\alpha = \alpha_{n,k}^{H}$. Even so, the MLE would probably provide a somewhat better fit in the part of the support given in the figure.

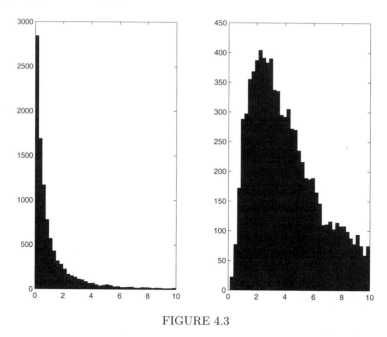

FIGURE 4.3

In the right panel, the Pareto obviously does not have the right shape, and neither do any of the standard regularly varying distributions. A solution is to take $\widehat{F}$ as a mixture of the MLE within a parametric (possibly light-tailed) class, say the Gammas, and a Pareto with support $[x_0, \infty)$ and $\alpha = \alpha_{n,k}^{H}$. However, although this is probably widely done in practice, the criteria for how to calibrate the parameters appear to be arbitrary, and we have seen little systematic discussion of this practically important point.

Exercises

4.1 (TP) Verify that $e(x)$ stays bounded when (4.1) holds.
4.2 (TP) Show that

$$e(x) \sim \frac{x}{\alpha - 1}, \quad e(x) \sim \frac{\sigma^2 x}{\log x}, \quad \text{respectively } e(x) \sim \frac{x^{1-\beta}}{\beta}$$

in the three main examples of subexponentiality listed at the beginning of Section 3. You will need Karamata's theorem, cf. Example 3.4.
4.3 (A) Simulate observations from the distribution with tail $\overline{F}(x) = (1+e)/(x+e)\log(x+e)$, $x > 0$, and draw a Hill plot. To check your program, first do the same with $\overline{F}(x) = (1+e)/(x+e)$ instead and check that your results do not look too different from the first row in Figure 4.2.
4.4 (A) In the GARCH model II.(4.1), σ_n^2 can be shown to be heavy-tailed in the limiting stationary situation. Thus one would expect this to show up in a mean-excess plot of the first R values in a nonstationary situation. Give some numerical examples to illustrate this point.
4.5 (A) In Exercise 4.4, it can be shown more precisely (Kesten [215]) that in stationarity, σ_n^2 is regularly varying with index α given as the solution of

$$\mathbb{E}\left(\alpha_1 Z_0^2 + \beta\right)^\alpha = 1.$$

Compare this with a Hill plot of $\sigma_1^2, \ldots, \sigma_R^2$, using some arbitrary initial condition.

5 Conditioned Limit Theorems

The optimal (zero-variance) change of measure for importance sampling is the conditional distribution $\mathbb{P}^{(x)}(\cdot) \overset{\text{def}}{=} \mathbb{P}(\cdot|A(x))$ given $A(x)$. Therefore an obvious way to look for a good importance distribution is to try to find a simple asymptotic description of $\mathbb{P}^{(x)}(\cdot)$ and to simulate using this asymptotic description.

For the random-walk setting in Section 2a, where $A(x) = \{\tau(x) < \infty\}$, it turns out that an asymptotic description of $\mathbb{P}^{(x)}(\cdot)$ is available. The results state roughly that up to $\tau(x)$, the random walk behaves as though it changed increment distribution from F to F_γ, which is precisely the type of behavior needed to infer (at least heuristically) the optimality of γ. A variety of precise statements supporting this informal description were given by Asmussen [9]. For example:

Proposition 5.1 *Let $\{B(x)\}$ be any sequence of events with $B(x) \in \mathscr{F}_{\tau(x)}$, $B(x) \subseteq \{\tau(x) < \infty\}$, $\mathbb{P}_\gamma(B(x)) \to 1$, $x \to \infty$. Then $\mathbb{P}^{(x)}(B(x)) \to 1$ as well.*

Proof. From our basic likelihood ratio identities (e.g., V.1.9) and the Cramér–Lundberg approximation (2.6), stating that $\mathbb{P}(\tau(x) < \infty) \sim$

$Ce^{-\gamma x}$, we get

$$\mathbb{P}^{(x)}(B^c(x)) = \frac{\mathbb{P}(B^c(x); \tau(x) < \infty)}{\mathbb{P}(\tau(x) < \infty)} = \frac{\mathbb{E}_\gamma[L_{\tau(x);\gamma}; B^c(x)]}{\mathbb{P}(\tau(x) < \infty)}$$

$$\leq \frac{e^{-\gamma x}\mathbb{P}_\gamma(B^c(x))}{\mathbb{P}(\tau(x) < \infty)} \sim \frac{\mathbb{P}_\gamma(B^c(x))}{C} \to 0.$$

□

As a main example, consider the one-dimensional empirical distribution of the X_i. Define

$$F_n^*(y) \stackrel{\text{def}}{=} \frac{1}{n}\sum_{i=1}^{n} \mathbb{1}\{X_i \leq y\}.$$

Corollary 5.2 *As $x \to \infty$, $\mathbb{P}^{(x)}(\|F_{\tau(x)}^* - F_\gamma\| > \varepsilon) \to 0$, where $\|\cdot\|$ denotes the supremum norm.*

Proof. By the Glivenko–Cantelli theorem, $\|F_n^* - F_\gamma\| \to 0$ $\mathbb{P}_\gamma$-a.s. as $n \to \infty$. Hence also $\|F_{\tau(x)}^* - F_\gamma\| \to 0$, and we can take $B(x) \stackrel{\text{def}}{=} \{\|F_{\tau(x)}^* - F_\gamma\| > \varepsilon\}$.

□

The results of [9] are in fact somewhat more general by allowing inference also on the dependency structure in the conditional limit. For example, it is straightforward to show that

$$\frac{1}{\tau(x)}\sum_{i=1}^{\tau(x)} \mathbb{1}\{X_i \leq y_1, \ldots, X_{i+k-1} \leq y_k\} \to F_\gamma(y_1)\cdots F_\gamma(y_k)$$

in $\mathbb{P}(\cdot \mid \tau(x) < \infty)$-probability for any fixed k.. Perhaps, the most convincing indication that the X_i are asymptotically conditionally independent is the fact that variance constants arising in conditional approximations by Brownian motion and Brownian bridge are the same as in the unconditional F_γ-random walk. See [9] for more details.

In the setting $A(x) = \{X_1 + \cdots + X_n > n(\mu + \varepsilon)\}$ of Section 2c, the conditioned limit theorem supporting that the X_i are i.i.d. with distribution F_θ given $A(x)$ (where $\theta \stackrel{\text{def}}{=} \theta(x)$ satisfies $\mathbb{E}_\theta X_1 = \mu + \varepsilon$) is a classical result from statistical physics going under the name of *Boltzmann's law* (but note that it is not easy to give a reference from the probability literature; cf. [16, p. 376]).

For heavy tails, the picture is entirely different, and often a rare event occurs as a consequence of one r.v. being large. The typical example is exceedance of a random walk $S_n = X_1 + \cdots + X_n$ of a high level x, where in the subexponential case one of the X_i exceeds x and the remaining X_j basically behave "normally."

As an example, we took exceedance of a random walk $S_n = X_1 + \cdots + X_n$ of level x before returning to 0. The increments correspond to a D/G/1 queue with traffic intensity 0.8 and service-time density $e^{-x/2}/2$ in the first panel, $1.5/(1+x)^{2.5}$ in the last two. In the first two panels of Figure 5.1, x is chosen such that the corresponding rare-event probability is of order 0.0002; more precisely, $x = 35$ and $x = 650$. The figure gives the sample paths out of a total of 25,000 in which x was exceeded. The last panel has the same Pareto service times, but now $x = 35$ as in the first panel, and a total of 1,000 runs. One sees clearly the dominant role of the big jumps in the heavy-tailed case (two last panels) as compared with the gradual buildup in the first panel (light tails).

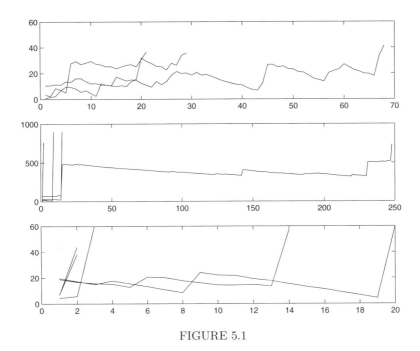

FIGURE 5.1

There are exceptions to the "one big jump" heuristics, such as many-server queues and on–off models in which several of the driving r.v.'s may need to be large in order for a buildup to occur. See further [16, Notes to Section X.9].

Different views on conditioned limit theorems are discussed in Sections 6, 7.

5a   A Counterexample

It is important to point out that examples have begun to appear in the literature that clearly show that using an asymptotic description of $\mathbb{P}^{(x)}(\cdot)$ as an importance distribution for computing a rare event probability has its limitations. We give one of them, taken from Glasserman & Wang [140].

As in Section 2c, we consider exceedances in the LLN but this time two-sided,

$$A(n) \stackrel{\text{def}}{=} \{S_n > n\varepsilon \text{ or } S_n < -n\varepsilon'\}$$

(taking $\mu = 0$). We choose ε' such that

$$\frac{\mathbb{P}(S_n > n\varepsilon)}{\mathbb{P}(S_n < -n\varepsilon')} \to \infty, \tag{5.1}$$

so that

$$\mathbb{P}^{(n)}(\cdot) \sim \mathbb{P}(\cdot \mid S_n > n\varepsilon).$$

Thus, Boltzmann's law suggests that one use the same exponential change of measure as in Section 2c and the estimator

$$Z(n) \stackrel{\text{def}}{=} e^{-\theta S_n + n\kappa(\theta)} \mathbb{1}\{S_n > n\varepsilon \text{ or } S_n < -n\varepsilon'\}.$$

However, we will see that then the contribution to $\mathbb{V}ar_\theta Z(n)$ from the event $\{S_n < -n\varepsilon'\}$ blows up the variance.

Fix ε, and define θ, θ' as the solutions of $\kappa'(\theta) = \varepsilon$, respectively $\kappa'(\theta') = -\varepsilon'$, and let $I = \theta\varepsilon - \kappa(\theta)$, $I' = -\theta'\varepsilon' - \kappa(\theta')$. We choose ε' such that $I' > I$ (then (5.1) holds by Lemmas 2.6, 2.7) and that

$$\delta' = \theta(\varepsilon + \varepsilon') + I - I' > 0, \tag{5.2}$$

which can be obtained by first choosing ε' such that $I' = I$ and next replacing ε' by a slightly larger value to get $I' > I$ without violating (5.2).

Proposition 5.3 *If* $0 < \delta < \delta'$, *then* $\liminf\limits_{n\to\infty} \dfrac{\mathbb{V}ar_\theta Z(n)}{z(n)^2 e^{\delta n}} = \infty$.

Proof.

$$\begin{aligned}
\mathbb{E}_\theta Z(n)^2 &\geq \mathbb{E}_\theta\left[L_{n,\theta}^2; S_n < -n\varepsilon'\right] = \mathbb{E}\left[L_{n,\theta}; S_n < -n\varepsilon'\right] \\
&= \mathbb{E}\left[e^{-\theta S_n + n\kappa(\theta)}; S_n < -n\varepsilon'\right] \geq e^{n\theta\varepsilon' + n\kappa(\theta)}\mathbb{P}(S_n < -n\varepsilon') \\
&\geq c_1 e^{n\theta\varepsilon' + n\kappa(\theta)} e^{-nI' - \theta'\sqrt{n}} \\
&= c_1 \exp\left\{n\left[\theta(\varepsilon + \varepsilon') - I - I'\right] - \theta'\sqrt{n}\right\},
\end{aligned}$$

using Lemma 2.8 for the last inequality. Hence by Lemma 2.6,

$$\begin{aligned}
\liminf_{n\to\infty} \frac{\mathbb{V}ar_\theta Z(n)}{z(n)^2 e^{\delta n}} \\
\geq c_1 \liminf_{n\to\infty} \exp\left\{n\left[\theta(\varepsilon + \varepsilon') - I - I'\right] - \theta'\sqrt{n} + 2nI - n\delta\right\} \\
= c_1 \liminf_{n\to\infty} \exp\left\{n\left[\delta' - \delta\right] - \theta'\sqrt{n}\right\} = \infty.
\end{aligned}$$

$\square$

Exercises

5.1 (TP) In Example 1.3, $\mathbb{P}(\cdot \mid S_n > x)$ can be found explicitly. Do it.

6   Large-Deviations or Optimal-Path Approach

The large-deviations (LD) approach to optimal exponential change of measure has several variants. We give one involving the concept of the *optimal path* or *most likely path*, which may be seen as an alternative means of computing an asymptotic description of the conditional distribution given the rare event.

We will work in the setting of discrete random walks or continuous-time Lévy processes, and write $\kappa(\theta) \overset{\text{def}}{=} \log \widehat{F}[\theta]$ in the random-walk case, $\kappa(\theta) \overset{\text{def}}{=} \log \mathbb{E}e^{\theta X_1}$ in the Lévy-process case. We first introduce the function

$$I(y) \overset{\text{def}}{=} \sup_{\theta \in \Theta}\big(\theta y - \kappa(\theta)\big), \quad y \in \mathscr{Y} \overset{\text{def}}{=} \{\kappa'(\theta) : \theta \in \Theta\},$$

which in the literature goes under names such as the *LD rate function*, the *Legendre transform*, the *Legendre–Fenchel transform*, the *Cramér transform* (sometimes the sign is reversed). For simplicity, we assume that the interval Θ is open. For $y \in \mathscr{Y}$, we define $\theta(y)$ (the saddle-point of y) by $\kappa'\big(\theta(y)\big) = y$, so that $I(y) = \theta(y)y - \kappa\big(\theta(y)\big)$. Note that we have already encountered $I(y)$ once, in Section 2c, where $I = I(\mu + \varepsilon)$. It is not too hard to show that $I(\cdot)$ is non-negative, convex, and attains its minimum 0 for $y = \kappa'(0) = \mu$.

One of the main themes of LD theory is to give estimates of the probability that a random walk (or some more general process) follows an atypical path. In the random-walk setting, this means that $S^{(n)}(\cdot)$ follows a path different from the one $\varphi_0(t) \overset{\text{def}}{=} \mu t$ given by the LLN, where $S^{(n)}(t) \overset{\text{def}}{=} S_{\lfloor nt \rfloor}/n$, $0 \le t \le 1$ (here $\lfloor \cdot \rfloor$ = integer part). The relevant LD result, known as *Mogulskii's theorem* (Theorem 5.1 in Dembo & Zeitouni [87]), states that under appropriate regularity conditions,

$$\mathbb{P}(S^{(n)}(\cdot) \in \mathscr{S}) \approx \exp\left\{-n \inf_{\varphi \in \mathscr{S}} \int_0^1 I\big(\varphi'(t)\big)\,\mathrm{d}t\right\} \qquad (6.1)$$

for suitable subsets $\mathscr{S}$ of continuous paths with $\varphi_0 \notin \mathscr{S}$. In many examples, there is a single path φ^* for which the minimum is attained, and this is the *optimal path*.

Note that $S^{(n)}$ is a description of the conditioned process in "law of large numbers scale" (often also called the "fluid scale"). Of course, such a fluid-scale description characterizes only the "coarse scale" behavior of the process under the conditioning.

6a   The Siegmund Algorithm

The crucial fact for optimal exponential change of measure in the setting of the Siegmund algorithm is given in the following lemma:

Lemma 6.1 $\displaystyle\min_{0<y<\infty} \frac{I(y)}{y}$ *is attained for* $y^* = \kappa'(\gamma)$.

Proof. Obviously, $I(y)/y \to \infty$ as $y \downarrow 0$ and (cf. (6.2) below) $I(y)/y$ is non-decreasing for large y, so that the minimum is attained. By straightforward differentiation, we get

$$
\begin{aligned}
I'(y) &= \theta'(y)y + \theta(y) - \theta'(y)\kappa'(\theta(y)) = \theta(y), \\
\frac{\mathrm{d}}{\mathrm{d}y}\frac{I(y)}{y} &= \frac{yI'(y) - I(y)}{y^2} = \frac{y\theta(y) - \theta(y)y + \kappa(\theta(y))}{y^2} \\
&= \frac{\kappa(\theta(y))}{y^2}.
\end{aligned}
\tag{6.2}
$$

Setting the last expression equal to 0 yields $\theta(y^*) = \gamma$ (since we look at minimum values for $y > 0$ only, $\theta(y^*) = 0$ is excluded), from which we immediately get $y^* = \kappa'(\gamma)$. □

In order to understand how the random walk reaches the high level x, we perform the optimization in (6.1) not only over φ but also over n. We then write x in the form $x = ny$ and let $\mathscr{S}$ be the set of continuous functions on $[0, 1]$ with $\varphi(0) = 0$, $\varphi(1) = y$. Then (6.1) takes the form

$$
\mathbb{P}(S_n \sim x) \sim \exp\left\{-x\frac{1}{y}\inf_{\varphi\in\mathscr{S}}\int_0^1 I(\varphi'(t))\,\mathrm{d}t\right\}.
\tag{6.3}
$$

By Jensen's inequality and the convexity of I,

$$
\int_0^1 I(\varphi'(t))\,\mathrm{d}t \geq I\left(\int_0^1 \varphi'(t)\,\mathrm{d}t\right) = I(y),
$$

with equality if and only if $\varphi(t) = ty$. Hence for fixed n,

$$
\mathbb{P}(S_n \sim x) \sim \exp\left\{-x\frac{I(y)}{y}\right\}.
$$

Viewing x as fixed and taking into the account that minimizing over n is the same as minimizing over y, we obtain by Lemma 6.1 that the minimizer is $y^* = \kappa'(\gamma)$. In conclusion, if x is large, the most likely way in which the random walk can cross level x is by crossing at time

$$
\tau(x) = n = \frac{x}{y} = \frac{x}{\kappa'(\gamma)}
$$

and by moving linearly at rate $y = \kappa'(\gamma)$ up to that time. But *this is precisely the same way as that in which the random walk with increment distribution F_γ crosses level x*, which motivates the conclusion that the conditional distribution given the rare event $\tau(x) < \infty$ is that of the random walk with increment distribution F_γ.

6b   *Further Examples*

Example 6.2 For a digital barrier option, the problem arises (after some rewriting) of estimating

$$z = \mathbb{P}\big(\underline{W}(T) \le -a, W(T) \ge b\big), \quad \text{where } \underline{W}(T) \overset{\text{def}}{=} \inf_{t \le T} W(t)$$

and W is $\mathrm{BM}(\mu, \sigma^2)$. If a, b are not too close to 0, this is a rare-event problem regardless the sign of μ. It seems reasonable that the optimal path φ^* should be piecewise linear as in Figure 6.1.

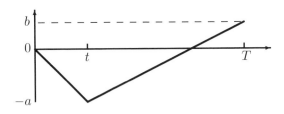

FIGURE 6.1

Thus we are left with a one-dimensional minimization problem, looking for the infimum over t of

$$
\begin{aligned}
2 \int_0^T I\big(\varphi'(t)\big)\, \mathrm{d}t &= \int_0^t (-a/t - \mu)^2\, \mathrm{d}s + \int_t^T (c/(T-t) - \mu)^2\, \mathrm{d}s \\
&= t(a/t + \mu)^2 + (T-t)(c/(T-t) - \mu)^2
\end{aligned}
$$

where $c = b + a$, we have taken $\sigma^2 = 1$, and we have used that $I(y) = (y - \mu)^2/2$ for $\mathrm{BM}(\mu, 1)$. Elementary calculus shows that the minimum is attained at $t = aT/(a+c)$. This means that the optimal path is linear with slope $-\mu^*$ on $[0, t]$ and slope μ^* on $(t, T]$ where $\mu^* = (a + 2b)/T$. See further the discussion in Glasserman [133, pp. 264 ff].   □

6c   *Discussion*

Notice that the LD argument is somewhat more heuristic than those in the preceding subsections. Also, in simple settings, LD results are typically not the strongest possible, involving only logarithmic asymptotics.

However, the motivation for the LD approach is its generality and the fact that the mathematical state of the area is very advanced, providing a considerable body of theory to draw upon. The philosophy is that once it has been understood how to paraphrase the optimality properties of optimal exponential change of measure (ECM) for simple systems in LD language, the generality of LD theory will allow one to find suitable importance distributions also in more complicated settings.

At least in the present authors' opinion, the success in implementing this program has been slightly more moderate than sometimes claimed. As we see it, the alternative approaches to optimal ECM (in the sense of how to derive it and how to study its optimality properties) in simple models are both simpler and more precise, and lead to stronger conclusions. Actually, one can argue for many specific cases, say for the GI/G/1 queue, that the use of LD theory often overshoots the goal. Also, as noted in Section 5a, counterexamples to the traditional LD approach are notable and suggest that care must be taken in applying LD to rare-event simulation. Section 7 below offers additional insight into this issue, and presents an alternative view of the use of LD in the rare-event setting.

Bucklew [62] gives an introduction to the LD point of view in simulation. There is an abundance of research papers, and we mention here just Atar & Dupuis [30], Bucklew et al. [63], Cottrell et al. [76], Dupuis & Ellis [101], Dupuis & Wang [102], Lehtonen & Nyrhinen [236], [237], Sadowsky [324], [325], and Sadowsky & Bucklew [327]. See also the discussion in XIV.6d and references there. For LD theory in general, we refer to Bucklew [61], Dembo & Zeitouni [87], den Hollander [184], and Shwartz & Weiss [337].

Exercises

6.1 (TP) Compute $I(y)$ when F is exponential, Poisson, or any other light-tailed distribution you may choose.

6.2 (TP) What is the most likely path for the Brownian bridge in $[0,1]$ with maximum $m > 0$?

7   Markov Chains and the h-Transform

As noted above, the LD theory provides a description of the conditioned process in fluid scale. A finer-scale description can be computed for the class of "exit probability" problems for Markov chains. In particular, let $X = \{X_n\}_{n \in \mathbb{N}}$ be Markov with state space E and let R be a subset of E. Suppose that we want to compute the exit probability $h(x) \stackrel{\text{def}}{=} \mathbb{P}_x(X_\tau \in A, \ \tau < \infty)$, where $\tau \stackrel{\text{def}}{=} \inf \{n \geq 0 : X_n \in R\}$ is the first exit time from R^c and $A \subseteq R$.

Many rare-event simulation problems can be formulated as exit calculations. For example, if X is a random walk with $X_0 = 0$:

(a) If X is a random walk with $X_0 = 0$ and we are interested in the probability z of ever exceeding level $\ell > 0$, taking $R = A = [\ell, \infty)$ yields $z = h(0)$.

(b) If instead we ask for the probability z that the random walk exceeds ℓ before attaining negative values, taking $A = [\ell, \infty)$, $R = (-\infty, 0) \cup A$ again yields $z = h(0)$.

(c) For computing tail probabilities of the form $\mathbb{P}(S_n > x)$ for a sum $S_n = Y_1 + \cdots + Y_n$ of r.v.'s, X is the space-time chain $\{(k, S_k)\}_{k \in \mathbb{N}}$ and $R = A = \{(n, y) : y > x\}$.

Conditioning on X_1 shows immediately that $h(x) = \mathbb{E}_x h(X_1)$, that is, h is *harmonic*. Therefore

$$P^h(x, dy) \stackrel{\text{def}}{=} P(x, dy) \frac{h(y)}{h(x)} \tag{7.1}$$

is a transition kernel. With $\mathbb{P}_x^h$ the probability measure governing the distribution of a Markov chain X with $X_0 = x$ and transition kernel P^h, we then have

$$\mathbb{E}_x^h f(X_1) \; = \; \frac{1}{h(x)} \mathbb{E}_x \big[f(X_1) h(X_1) \big] . \tag{7.2}$$

The following result establishes that under the zero-variance conditional distribution corresponding to an exit problem, X evolves according to the Markov transition kernel P^h.

Theorem 7.1 *For all $x \in E$, we have $\mathbb{P}_x^h (X_\tau \in A, \tau < \infty) = 1$. Further, the conditional $\mathbb{P}_x$-distribution of $(X_0, \ldots, X_\tau)$ given $\{X_\tau \in A, \tau < \infty\}$ and the $\mathbb{P}_x^h$-distribution of $(X_0, \ldots, X_\tau)$ are the same.*

Proof. For notational simplicity, we will assume that E is countable, and in terms of the given transition probabilities $p(x, y)$, (7.2) then means

$$p^h(x, y) \; = \; \frac{p(x, y)h(y)}{h(x)} . \tag{7.3}$$

We have

$$
\begin{aligned}
\mathbb{P}_x^h (X_1 = x_1, \ldots, X_n = x_n) \; &= \; p^h(x, x_1) p^h(x_1, x_2) \cdots p^h(x_{n-1}, x_n) \\
&= \; \frac{h(x_n)}{h(x)} p(x, x_1) p(x_1, x_2) \cdots p(x_{n-1}, x_n) \\
&= \; \frac{h(x_n)}{h(x)} \mathbb{P}_x (X_1 = x_1, \ldots, X_n = x_n).
\end{aligned}
$$

Summing over $x_1, \ldots, x_{n-1} \notin R$, $x_n \in A$, and using $h(x_n) = 1$ for $x_n \in A$ gives

$$
\begin{aligned}
\mathbb{P}_x^h (S_\tau \in A, \tau = n) \; &= \; \mathbb{P}_x^h (X_1 \notin R, \ldots, X_{n-1} \notin R, X_n \in A) \\
&= \; \frac{1}{h(x)} \mathbb{P}_x (X_1 \notin R, \ldots, X_{n-1} \notin R, X_n \in A) \\
&= \; \frac{1}{h(x)} \mathbb{P}_x (S_\tau \in A, \tau = n) .
\end{aligned}
$$

Summing next over n and using the definition of $h(x)$ gives $\mathbb{P}_x^h(S_\tau \in A, \tau < \infty) = 1$. Similarly, for any infinite sequence $x_1, x_2, \ldots$,

$$
\begin{aligned}
& \mathbb{P}_x\left(X_1 = x_1, \ldots, X_\tau = x_\tau \mid S_\tau \in A,\, \tau < \infty\right) \\
&= \frac{1}{h(x)} \sum_{n=0}^\infty \mathbb{P}_x\left(X_1 = x_1, \ldots, X_n = x_n, \tau = n\right) \\
&= \frac{1}{h(x)} \sum_{n=0}^\infty \frac{h(x)}{h(x_n)} \mathbb{P}_x^h\left(X_1 = x_1, \ldots, X_n = x_n, \tau = n\right) \\
&= \sum_{n=0}^\infty \mathbb{P}_x^h\left(X_1 = x_1, \ldots, X_n = x_n, \tau = n\right) \mathbb{1}\{x_n \in R\} \\
&= \mathbb{P}_x^h\left(X_1 = x_1, \ldots, X_\tau = x_\tau\right).
\end{aligned}
$$

$\square$

The transition kernel P^h is the so-called h-*transform* of P; see Doob [94] where h could be any harmonic function; an early reference noting the simulation relevance is Glynn & Iglehart [153]. It describes the fine-scale behavior of X under the conditioning. Of course, h is the object to be computed and is therefore unknown to the simulator. On the other hand, in some practical settings, we have available analytical approximations to $h(\cdot)$ or at least an intuitive sense of how to approximate h. Let v be an approximation to h, and consider simulating X according to the transition function

$$
\widetilde{P}(x, dy) \stackrel{\text{def}}{=} P(x, dy) \frac{v(y)}{w(x)}, \tag{7.4}
$$

where $w(\cdot)$ is the normalizing constant given by

$$
w(x) \stackrel{\text{def}}{=} \int_{R^c \cup A} P(x, dy)\, v(y).
$$

In building an importance sampling algorithm based on $\widetilde{P}$, one obvious question that arises is the degree of variance reduction obtained by simulating under $\widetilde{P}$ rather than P. A more subtle question is how to efficiently generate transitions of X under $\widetilde{P}$ rather than P. In particular, the question of efficient variate generation under $\widetilde{P}(x, \cdot)$ is of fundamental interest.

Example 7.2 Consider problem (a) above where h(x) is the probability that a random walk X starting from x ever exceeds ℓ. As has been seen earlier in this chapter, the easily implemented exponential twisting Siegmund algorithm is then the obvious candidate for computing $h(x)$ for $x < \ell$.

The Cramér–Lundberg approximation (2.6) suggests setting $v(x) = Ce^{-\gamma(\ell-x)}$ for $x < \ell$ and putting $v(x) = 1$ for $x \geq \ell$. With this choice of the approximation v, the normalization $w(\cdot)$ appears explicitly and the algorithm no longer corresponds to the Siegmund algorithm. On the other hand, if one sets $v(x) = Ce^{-\gamma(\ell-x)}$ over the entire real line, this degrades the quality of the approximation (at least for $x \geq \ell$), but the resulting algorithm now corresponds to the exponential twisting suggested by Siegmund. This example illustrates the basic principle that a good choice of v should reflect not only the quality of the approximation to h but also the degree of numerical difficulty associated with computing the normalization $w(\cdot)$ and the related efficiency of the variate generation scheme associated with generating transitions under $\widetilde{\mathbb{P}}$.

The example also demonstrates that in general, a key issue in a literal implementation of (7.4) is to compute the normalization $w(\cdot)$. $\qquad\square$

Example 7.3 In computing $h(n, nx)$, where $h(k, z) \stackrel{\text{def}}{=} \mathbb{P}(S_k > z)$, the zero-variance transition function for the random walk is

$$P^h\left(Y_{k+1} \in dy \mid S_k\right) = \mathbb{P}(Y_{k+1} \in dy)\frac{h(n-k-1, nx-y-S_k)}{h(n-k, nx-S_k)}.$$

Observe that if k and S_k are small, then (cf. Remark 2.9)

$$\frac{h(n-k-1, nx-y-S_k)}{h(n-k, nx-S_k)} \approx \exp\{\theta(x)y - \kappa(\theta(x))\}, \qquad (7.5)$$

where $\theta(x)$ satisfies $\kappa'(\theta(x)) = x$. In other words, the approximation yields the exponential twisting algorithm of Section 2c (which was proved there to be logarithmically efficient). More generally, if $S_k \approx kx$ (regardless of the magnitude of k), the approximation (7.5) is a good one. Note that $S_k \approx kx$ holds, with high probability, under the exponentially twisted path using twisting parameter $\theta(x)$.

Suppose, however, that a random fluctuation throws the random walk off the fluid path on which $S_k \approx kx$. In such a case, the approximation

$$\frac{h(n-k-1, nx-y-S_k)}{h(n-k, nx-S_k)} \approx \exp\left\{\theta\left(\frac{nx-S_k}{n-k}\right)y - \kappa\left(\theta\left(\frac{nx-S_k}{n-k}\right)\right)\right\}$$

is better. This importance measure no longer produces increments with a state-independent distribution. Instead, the particular exponential twist that is used depends on the level of progress that the random walk has made by time k in exceeding nx at time n. This state-dependent change of measure leads to an importance sampling algorithm having bounded relative error (rather than being logarithmically efficient) as $n \to \infty$; see Blanchet & Glynn [46]. While this second algorithm differs from the first on the fine scale of individual transitions, its fluid-scale behavior is identical to that of the first algorithm. $\qquad\square$

Example 7.4 Consider the counterexample of Section 5a ($S_n > n\varepsilon$ or $S_n < -n\varepsilon'$). The state-dependent exponential twisting idea of Example 7.3 can again be applied here. Under this algorithm, the importance distribution will attempt to push the random walk to exit through $n\varepsilon'$ if S_k is unusually small, unlike the state-independent increment distribution used in Section 5a. Of course, under fluid scaling, both the state-dependent and state-independent implememtations behave identically. But the state-dependent change of measure just described yields logarithmic efficiency, see Dupuis & Wang [103] and Blanchet, Liu, & Glynn [48], whereas the state-independent algorithm has infinite variance.   □

Example 7.5 Consider again as in Example 7.2 the level-crossing problem for a random walk, but assume now that the right tail of the increment is heavy-tailed. In this setting, it is easily shown that for a fixed k and x,

$$\mathbb{P}_x\big(Y_i \in dy_i,\ i = 1,\ldots,k \,\big|\, \tau < \infty\big) \;=\; \prod_{i=1}^{k} \mathbb{P}(Y_i \in dy_i)$$

as $\ell \to \infty$. This result just expresses the fact that the great majority of increments behave "normally" under the conditioning. So, the finite-dimensional distributions of the conditional process provide very little information about the appropriate importance distribution to use here.

On the other hand, it is known (e.g., [16, Section X.9]) that under suitable regularity conditions,

$$h(x) \;\sim\; \frac{1}{|\mathbb{E}Y_1|}\int_{\ell-x}^{\infty}\mathbb{P}(Y_1 > y)\,dy$$

as $\ell \to \infty$. For example, this asymptotic holds when the right tail of Y_1 is regularly varying. This result suggests generating transitions of the random walk in a state-dependent fashion, so that

$$\widetilde{\mathbb{P}}\big(Y_{k+1} \in dy \,\big|\, S_k, \tau > k\big) \;=\; \frac{v(S_k + y)}{v(S_k)}\mathbb{P}(Y_1 \in dy)\,, \qquad (7.6)$$

where $v(x) \stackrel{\text{def}}{=} \int_x^{\infty}\mathbb{P}(Y_1 > y)\,dy$. This transition distribution can be used when S_k is far from the exit set $R = [\ell, \infty)$ (so that the approximation $v(S_k + y) \approx h(S_k + y)$ is good). When S_k lies in the "boundary layer" $[\ell - x_0, \ell)$, in which the approximation is poor, the increments are generated according to the original increment distribution. When such an importance sampling distribution is used, the algorithm achieves bounded relative error (under suitable regularity conditions on $\overline{F}(y)$; see Blanchet & Glynn [47]).

One key implementation issue here is the question of how to efficiently generate transitions of the process under the kernel (7.6). This can be done, for example, via a rejection algorithm when $\overline{F}(y)$ is regularly varying, but no completely general effective algorithm is currently known for implementing this.   □

In our view, the h-transform perspective serves to unify much of the rare-event simulation literature. Note, for example, that the $S_n > n\varepsilon$ or $S_n < -n\varepsilon'$ counterexample of Glasserman & Wang [140] (reproduced in Section 5a) presents a situation in which the most straightforward importance sampling algorithm that is consistent with the fluid-scale LD path (i.e., one with state-independent increments) fails to have finite variance.

However, there are infinitely many such changes of measure having the consistency property. The state-dependent change of measure suggested by the h-transform analysis (see Example 7.4) is one such importance distribution that is consistent with the fluid-scale LD analysis, and that provides the desired level of variance reduction. More generally, one can expect that for problems in which the LD fluid scale path can be computed, there exists a change of measure, consistent with the fluid-scale path, that achieves logarithmic efficiency (the counterexample of [140] shows that not all such importance distributions achieve a variance reduction).

In addition, the h-transform perspective provides a unified means of developing algorithms for both light-tailed and heavy-tailed problems. Of course, a practical obstacle is the requirement that the simulator develop a good approximation to the function h. This can be an enormous challenge in practical problems.

Some promising recent developments on state-dependent importance sampling involving ideas exposed in this section are given by Dupuis & Wang [104], [105]. Nevertheless, in the view of the authors, the efficient implementation of rare-event simulation in the context of large and/or unstructured stochastic models remains an open issue deserving of significant additional research attention. One problem that is common to both the traditional importance sampling approach and the state-dependent extension treated in this section is the need to develop good approximations for rare-event probabilities in such complex (and thereby also more realistc and practically challenging!) models. In particular, when implemented via LD theory this leads to variational problems that do not have an explicit solution. The application of the methods of this and the preceding section can therefore be expected to require a preliminary numerical computation that either implicitly or explicitly solves some version of the variational problem or implements a sampling-based learning algorithm to build a good importance distribution. Only for relatively simple models should one expect to find an explicit closed-form optimal importance distribution.

8 Adaptive Importance Sampling via the Cross-Entropy Method

Let X be an r.v. (possibly multidimensional) with density $f(x)$. The *cross-entropy* or *Kullback–Leibler distance* between f and a different density g is then defined as

$$\mathbb{E}_g \log \frac{g(X)}{f(X)} = \int g(x) \log g(x)\, \mathrm{d}x - \int g(x) \log f(x)\, \mathrm{d}x. \qquad (8.1)$$

A lower bound follows from Jensen's inequality:

$$\mathbb{E}_g \log \frac{g(X)}{f(X)} = -\mathbb{E}_g \log \frac{f(X)}{g(X)} \geq -\log \mathbb{E}_g \frac{f(X)}{g(X)} = \log 1 = 0,$$

with equality if and only if $f = g$. This motivates viewing cross-entropy as a measure of closeness, such that the f in a given classs of densities that is closest to g is the one with minimal cross-entropy or equivalently with the highest value of

$$\int g(x) \log f(x) \, \mathrm{d}x = \mathbb{E}_g \log f(X).$$

For the following, note that $\mathbb{E}_G \log f(X)$ also makes sense if the distribution G does not necessarily possess a density g.

The implication for importance sampling for a rare event probability $\mathbb{P}(A)$ is that if we look for an importance density in a certain parametric class $\{f(x;\theta)\}_{\theta \in \Theta}$, the one $F(x, \widetilde{\theta})$ maximizing $\mathbb{E}\big[\log f(x;\theta) \,\big|\, A\big]$ should be the closest within the class to the conditional density given the rare event A. The cross-entropy method implements the maximization in two ways: either by using theory to insert an asymptotic form of $\mathbb{P}(\cdot \,|\, A)$ to compute a closed-form expression for $\widetilde{\theta}$, or by an adaptive algorithm providing in its final step r.v.'s that are approximately sampled from the conditional distribution given A.

In both cases, it is a crucial feature of the method that the maximization most often can be carried out analytically. This may be understood by the relation to maximum likelihood from statistics: if $V_1, \ldots, V_n$ are observations from an unknown density $f(x;\theta)$, the MLE $\widehat{\theta}$ is obtained by maximizing the log likelihood

$$\sum_{i=1}^{n} \log f(V_i;\theta) = n \, \mathbb{E}_{F_n^*} \log f(V;\theta),$$

where F_n^* is the empirical distribution giving mass $1/n$ to each of the V_i. Not only do we have a maximization problem of similar form, but since $\widehat{\theta}$ is most often a simple explicit functional of F_n^*, the particular expression for $\widehat{\theta}$ most often suggests that a similar explicit formula for the maximizer also is valid in the cross-entropy setting. This contrasts with methods based on variance minimization, which most often must be carried out empirically.

The cross-entropy method was introduced by Rubinstein [316]. A tutorial is given by de Boer et al. [83] and a comprehensive treatment by Rubinstein & Kroese [318]. An extensive bibliography can be found at the web site [w^3.21].

We start with two simple examples using the asymptotics of $\mathbb{P}(\cdot \,|\, A)$.

Example 8.1 Let $A \overset{\mathrm{def}}{=} \{X_1 + \cdots + X_n > n\varepsilon\}$ where $X_1, \ldots, X_n$ are i.i.d. $\mathcal{N}(0, 1)$, and consider the $\mathcal{N}(\theta, 1)$ densities as candidates for the importance density $\widetilde{f}$. The MLE is

$$\widehat{\theta} = \frac{1}{n}\sum_{i=1}^{n} V_i = \int_{-\infty}^{\infty} u\, F_n^*(\mathrm{d}u)\,.$$

By Boltzmann's law (cf. Section 5), the X_i are approximately governed by the $\mathcal{N}(\varepsilon, 1)$ density. Thus the θ suggested by the cross-entropy argument is

$$\widetilde{\theta} = \int_{-\infty}^{\infty} u\,\frac{1}{\sqrt{2\pi}}\mathrm{e}^{-(u-\varepsilon)^2/2}\,\mathrm{d}u = \varepsilon\,,$$

which we know from the study of the saddle-point method in Section 2c to be asymptotically efficient as $n \to \infty$.   □

Example 8.2 (ASMUSSEN, KROESE, & RUBINSTEIN [22]) Let $X_1, \ldots, X_n$ be i.i.d. Pareto with $\overline{F}(x) = (1+x)^{-\alpha}$, that is, having density $f(x) = \alpha(1+x)^{-\alpha-1}$. We look for $\widetilde{F}$ as another distribution of this form, with parameter $\widetilde{\alpha}$, say, when the rare event is $A \stackrel{\mathrm{def}}{=} \{X_1 + \cdots + X_n > x\}$ for some large x.

First, we need to compute the MLE $\widehat{\alpha}$. The log likelihood is $n\log\alpha - (\alpha+1)\sum_1^n \log(1+V_i)$, which in a straightforward way yields

$$\widehat{\alpha} = \frac{n}{\sum_1^n \log(1+V_i)} = \left(\int_0^\infty \log(1+y)F_n^*(\mathrm{d}y)\right)^{-1}.$$

Subexponential theory predicts that A occurs by $X_i > x$ for one i, whereas the rest do not change their distribution much. The conditional distribution of X given $X > x$ has density

$$\frac{\alpha(1+x)^\alpha}{(1+y)^{\alpha+1}}, \quad y > x.$$

Thus, we take $\widetilde{\alpha} = 1/J_x$, where

$$J_x \stackrel{\mathrm{def}}{=} \int_0^\infty \log(1+y)\left(\frac{n-1}{n}\frac{\alpha}{(1+y)^{\alpha+1}} + \frac{1}{n}\frac{\alpha(1+x)^\alpha}{(1+y)^{\alpha+1}}\mathbb{1}_{y>x}\right)\mathrm{d}y$$

$$= \frac{n-1}{n\,\alpha} + \frac{1}{n}\left(\log(1+x) + \frac{1}{\alpha}\right) = \frac{\log(1+x)}{n} + \frac{1}{\alpha}\,.$$

It follows that for large x,

$$\widetilde{\alpha} \approx \widetilde{\alpha}_0 \stackrel{\mathrm{def}}{=} \frac{n}{\log(1+x)}. \tag{8.2}$$

This is to be compared with the suggestion of Juneja & Shahabuddin [203] to take $\widetilde{\alpha} = b/\log x$, with b unspecified but arbitrary, and with the suggestion of Asmussen, Binswanger, & Højgaard [18] to take $\widetilde{F}$ with asymptotic tail $1/\log x$, which is even heavier and may be considered a particular instance of the boundary case $b = 0$. It can be proved that (8.2) gives logarithmic efficiency. In fact, this is the case whenever $\widetilde{\alpha} = b/\log x$ for some

b, and numerical evidence in [22] shows that $b = n$ is also close to being variance minimal.   □

We next turn to the adaptive cross-entropy algorithm. Here it is assumed that the rare event has the form $A = \{h(X) \geq \eta\}$ for some large η and some (typically high-dimensional) r.v. X. We assume the given density $f(x) = f(x; \theta_0)$ to be a member of a parametric family $\{f(x; \theta)\}_{\theta \in \Theta}$ and choose the changed density $\widetilde{f}(x)$ in importance sampling within the family, i.e., $\widetilde{f}(x) = f(x; \widetilde{\theta})$ for some $\widetilde{\theta} \in \Theta$. The adaptive cross-entropy algorithm is an automatic search for a good $\widetilde{\theta}$. In each step, importance sampling is performed using a trial θ^*, and a certain proportion δ of the X_i with the highest h-values is used for the updating of θ^*:

(i) $\theta^* \longleftarrow \theta$ for some arbitrary θ.

(ii) Generate a sample $X_1, \ldots, X_N$ from $f(x; \theta^*)$.

(iii) Determine η' such that δN of the $h(X_i)$ are larger than η'.

(iv) If $\eta' < \eta$, let $N' \overset{\text{def}}{=} \delta N$ and let $X'_1, \ldots, X'_{N'}$ be the X-values with $h(X_i) \geq \eta'$.
     Otherwise, let N' be the number of the X_i such that $X_i \geq \eta$ and let $X'_1, \ldots, X'_{N'}$ be the corresponding X'_i.

(v) Choose $\widehat{\theta}$ as maximizer of $\displaystyle\sum_{i=1}^{N'} \frac{f(X'_i; \theta_0)}{f(X'_i; \theta^*)} \log f(X'_i; \theta)$.

(vi) If $\eta' < \eta$, let $\theta^* \longleftarrow \widehat{\theta}$ and return to (ii).

(vii) Else $\widetilde{\theta} \longleftarrow \theta^*$.

(viii) Run a full importance sampling experiment with R replications and sampling from $f(x; \widetilde{\theta})$.

To understand why the algorithm works, we exploit again the relation to maximum likelihood. The updated value of θ^* is the same as the MLE for the (artificial) set of data in which (ignoring rounding issues) $f(X'_i; \theta_0)/f(X'_i; \theta^*)$ is the number of observations taking the value X'_i. Since the X'_i are selected with frequencies proportional to $f(x; \theta^*)\mathbb{1}\{h(x) \geq \eta'\}$, this means that the data have a frequency of x-values that is the same as for the density $f(x; \theta_0)$ conditioned to the set $h(x) \geq \eta'$. Thus, when $\eta' = \eta$, we can expect $\widetilde{\theta}$ to be close to the $\theta_\#$ that minimizes the entropy distance to the θ_0-conditional distribution of X given $h(X) > \eta$.

The θ in step (i) needs to be chosen such that the number of X-values with $h(X) \geq \eta'$ is nonzero for the given N.

In the implementation, N must be much smaller than the number R of replications that will eventually be used in step 8 (if say $R = 1,000,000 = 10^6$, reasonable values may be $N = 1,000 = 10^3$ or maybe $N = 10,000 = $

10^4). The choice of δ requires some comment. Choosing δ small will speed up the convergence from the initial value θ to the final one $\widetilde{\theta}$, but may also require a larger N. In fact, if δ is too small, the algorithm may converge only after a huge number of steps or not at all, as is illustrated by the following example:

Example 8.3 Let $f(x;\theta) = \theta e^{-\theta x}$ and $h(x) = x$. If z is large, the desired $\widetilde{\theta}$ should therefore be small. Consider the Markov chain $\theta_0 \overset{\text{def}}{=} \theta, \theta_1 \overset{\text{def}}{=} \theta_1^*, \theta_2 \overset{\text{def}}{=} \theta_2^*, \dots$. In the extreme case $\delta = 1$, the updating of θ^* to θ_{n+1} from θ_n simply means that θ_{n+1} is the MLE $n/(X_1 + \cdots + X_n)$. Hence by Jensen's inequality,

$$\mathbb{E}\left[\theta_{n+1} \mid \theta_n\right] > \frac{n}{n\,\mathbb{E}_{\theta_n} X} = \theta_n, \tag{8.3}$$

and the same inequality will then hold also for certain $\delta < 1$. The unfortunate conclusion of (8.3) is that θ_n may eventually go to ∞ without ever reaching the desired small level. □

In many situations, $\delta \in (1\%, 10\%)$ has been found to work well. If in a given situation η' gets stuck below the desired level z, the obvious device is to decrease δ.

Example 8.4 We will illustrate a simple case of the adaptive cross-entropy method, importance sampling within the family of bivariate normals, by a military application. A target is located at $(0,0)$, and a sensitive civilian object is in the rectangle with SW corner $(4, -0.25)$ and NE corner $(5, 0.25)$ (the unit is 100 m). A bomb aimed at the target falls at $X = (U, V)$, where U, V are independent standard normals. What is the probability of major damage to the civilian object, as defined by (U, V) being closer than 100 m to the object? It is certainly small, since the object is several standard deviations away from the origin, so importance sampling may be needed. To fit this in the general setup, let $h(x)$ denote the distance from $x = (u, v)$ to the object, so we want $\mathbb{P}\big(h(U, V) \leq 1\big)$ (doing an obvious reversal of $\geq$ and $\leq$).

A simple importance sampling scheme is changing the mean of U only, say to θ' (it is apparent that no gain can be expected by making the mean of V nonzero). A first naive guesss is that $\theta' = 4.5$ is the best choice, but more reflection shows that the conditional distribution of U given $h(U, V) \leq 1$ is left-skewed, so that a smaller value is probably better.

Let $\varphi_{\theta, \sigma^2}$ denote the density of $\mathcal{N}(\theta, \sigma^2)$. The maximization step of the algorithm is then to maximize $\sum_1^{N'} w_i \log \varphi_{\theta,1}(U_i)$ w.r.t. θ, where $w_i \overset{\text{def}}{=} \varphi_{0,1}(U_i')/\varphi_{\theta^*,1}(U_i')$. Interpreting again w_i as the number of observations equal to U_i' and recalling that the MLE for θ in the normal distribution is just the empirical mean, the maximizer is $\widehat{\theta} = \sum w_i U_i'/\sum w_i$.

To check the dependence on the choice of parameters, the algorithm was run 100 times for 8 sets of parameters, namely $N = 500$ and $5{,}000$, and $\delta =$

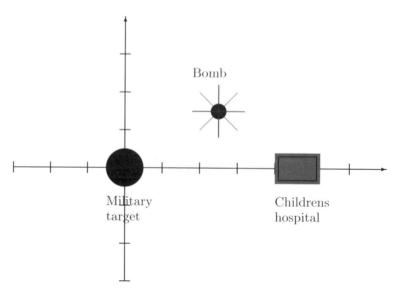

FIGURE 8.1

$25\%, 10\%, 5\%, 1\%$. In all cases (note the independence of δ!), the estimates for $\widetilde{\theta}$ came out with a mean of 3.34, with a standard deviation of order 0.03 for $N = 500$ and 0.01 for $N = 5000$. The number of iterations required for reaching $\eta' \leq \eta = 1$ depended only on δ (no random fluctuations were observed) and were $4, 3, 2, 2$ for the values under consideration.

The algorithm was run also for $\delta = 50\%$, but $\eta' \leq \eta = 1$ was not attained, at least within a reasonable time. In contrast, the algorithm appeared to get stuck at around $\eta' = 1.2$, corresponding to θ^*-values of order 3.1.

Performing the importance sampling with the selected $\widetilde{\theta} = 3.34$ and $R = 500{,}000$ replications gave the confidence interval $(7.95 \pm 0.05) \times 10^{-4}$. To check that 3.34 is a good value, we repeated with other choices of $\widetilde{\theta}$ and observed the following half-widths (in units of 10^{-4}) of the confidence interval [all intervals overlapped, so that the point estimates appeared trustworthy within the range considered]:

$\widetilde{\theta}$	0	1	2	3	3.34	3.66	4	5	6
	0.76	0.19	0.08	0.05	0.05	0.05	0.06	0.13	0.43

The impression is that there is a minimum in the range $\widetilde{\theta} \in (3, 4)$ providing a substantial improvement on crude Monte Carlo simulation, but that the minimum is quite flat. The precision of the cross-entropy algorithm for picking a good $\widetilde{\theta}$ seems more than adequate.

We next looked at the possibility of changing the standard deviations of U, V to $\widetilde{\sigma}_U, \widetilde{\sigma}_V$. The algorithm was run once more with $N = 5{,}000$, $\delta = 2\%$.

The maximization step consists now in maximizing

$$\sum_{i=1}^{N'} w_i \big(\log \varphi_{\theta,\sigma_U}(U_i') + \log \varphi_{0,\sigma_V}(V_i')\big)$$

w.r.t. $\theta, \sigma_U, \sigma_V$, where

$$w_i \stackrel{\text{def}}{=} \frac{\varphi_{0,1}(U_i')\varphi_{0,1}(V_i')}{\varphi_{\theta',\sigma_U^{*2}}(U_i')\varphi_{0,\sigma_V^{*2}}(V_i')}.$$

Again, the maximization relates to a standard statistical problem in the normal distribution, namely the estimation of the mean of the first component and the standard deviations in an independent bivariate setting with the mean of the second component fixed at 0. This immediately suggests the correct maximizer:

$$\widehat{\theta} = \frac{\sum w_i U_i'}{\sum w_i},$$

$$\widehat{\sigma}_U^{*\,2} = \frac{\sum w_i U_i'^2}{\sum w_i} - \widehat{\theta}^2, \quad \widehat{\sigma}_V^{*\,2} = \frac{\sum w_i V_i'^2}{\sum w_i}.$$

Using this, the CE algorithm ended up with $\widetilde{\theta} = 3.36$ (almost unchanged), and standard deviations changed in the smaller direction (as perhaps was to be expected), $\widetilde{\sigma}_1 = 0.30$, $\widetilde{\sigma}_2 = 0.49$. The importance sampling with $N = 500{,}000$ gave the confidence interval $(7.95 \pm 0.03) \times 10^{-4}$ for $\mathbb{P}(A)$; that is, some further variance reduction was obtained. $\qquad\square$

Exercises

8.1 (TP) Find the λ suggested by the CE method in Example 1.3 (you will need your solution to Exercise 5.1).

8.2 (A) Consider the M/M/∞ queue with arrival rate $\lambda = 1$ and service rate $\mu = 3$. That is, the embedded Markov chain on $\mathbb{N}$ goes up one step from state n w.p. $p_n = \lambda/(\lambda + n\mu)$, and down one step from $n > 0$ w.p. $q_n = n\mu/(\lambda + n\mu)$. Use the adaptive CE algorithm to estimate the probability that the maximal number in system within a busy cycle is $N = 7$ or more. Compare also the $\widetilde{p}_n$ used in the final step to the values q_n that would correspond to using the change of measure in the Siegmund algorithm locally at each n (see further XIV.6c).

9   Multilevel Splitting

Multilevel splitting is a set of ideas for estimating the probability $z(x)$ that a Markov chain $\{X_n\}$ hits a rare set $B(x)$ in a regenerative cycle. The method is basically a multistage version of splitting used for variance reduction in V.5, and there are many variants of the method around. Some main early references are Kahn & Harris [206], Hammersley & Handscomb [173], and Bayes [39], and some more recent ones are Villén-Altamirano

[357], [358] (RESTART), Glasserman et al. [137], [138], [139], and Haraszti & Townsend [174].

The method uses a decomposition of the state space E into subsets $E_0, \ldots, E_m$ with $E_m = B(x)$; see Figure 9.1.

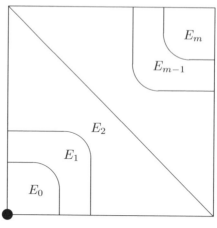

FIGURE 9.1

Think of $\mathbf{0} = (0,0)$ as the regeneration point ($\mathbf{0} \in E_0$) and that the event of hitting E_m occurs through succesive hits of the E_k. That is, define

$$\tau_k \stackrel{\text{def}}{=} \inf\{n : X_n \in E_k \cup E_{k+1} \cup \cdots \cup E_m\}.$$

We then assume

$$\mathbb{P}_i(X_{\tau_k} \in E_k \mid \tau_k < \infty) = 1, \quad i \in E_{k-1}, \tag{9.1}$$

which implies

$$z(x) = p_1 p_2 \cdots p_m, \tag{9.2}$$

where $z(x) \stackrel{\text{def}}{=} \mathbb{P}_{\mathbf{0}}(\tau_m < C)$ with $C \stackrel{\text{def}}{=} \inf\{n > 0 : X_n = \mathbf{0}\}$ the regenerative cycle and $p_k \stackrel{\text{def}}{=} \mathbb{P}_{\mathbf{0}}(\tau_k < \infty \mid \tau_{k-1} < \infty)$, $p_1 \stackrel{\text{def}}{=} \mathbb{P}_{\mathbf{0}}(\tau_1 < \infty)$.

For the simulation, we generate R_1 sample paths $\{X_n^{(r)}\}$ starting from $X_0 = \mathbf{0}$; out of these, $N_1 \stackrel{\text{def}}{=} \sum_1^{R_1} \mathbb{1}\{\tau_1^{(r)} < C^{(r)}\}$ will hit E_1 before returning to $\mathbf{0}$, and we have the obvious estimate $\hat{p}_1 = N_1/R_1$. For the N_1 successes, we let ν_1 denote the empirical distribution of the entrance state in E_1,

$$\nu_1(j) \stackrel{\text{def}}{=} \frac{1}{N_1} \sum_{r=1}^{R_1} \mathbb{1}\{X_{\tau_1^{(r)}} = j, \tau_1^{(r)} < C^{(r)}\}.$$

We then generate R_2 copies of $\{X_n\}$ with initial distribution ν_1, and let N_2 be the number of copies entering E_2 before returning to $\mathbf{0}$, ν_2 the corresponding empirical distribution of the successful hits of E_2, $\widehat{p}_2 \stackrel{\text{def}}{=} N_2/R_2$, and so on. The estimator is $\widehat{z} \stackrel{\text{def}}{=} \widehat{z}(x) \stackrel{\text{def}}{=} \widehat{p}_1 \cdots \widehat{p}_m$.

The generation of a copy of $\{X_n\}$ with initial distribution ν_1 (and similarly in the following steps) can be performed in more than one way. One approach is to choose the initial value X_0 by randomly sampling from the successful hits of E_1. Another is to use each such hit, say at $x \in E_1$, to generate a new number R_1' of paths starting from x. Thus $R_2 = R_1' N_1$; from this, the term *splitting*. Irrespective of the choice, we have the following result:

Proposition 9.1 $\widehat{z}$ *is an unbiased estimator of z.*

Proof. Let

$$\mathscr{F} \stackrel{\text{def}}{=} \sigma\Big(X_n^{(r)} : r = 1, \ldots, R_1,\ n = 0, \ldots, \tau_1^{(r)}\Big), \quad q_2(y) \stackrel{\text{def}}{=} \mathbb{P}_y(\tau_2 < C),$$

$y \in E_1$. Then for $m = 2$,

$$\mathbb{E}\big[\widehat{p}_2 \,\big|\, \mathscr{F}\big] \;=\; \int_{E_1} q_2(y)\,\nu_1(dy) \;=\; \frac{1}{N_1} \sum_{r=1}^{R_1} \mathbb{1}\{\tau_1^{(r)} < C^{(r)}\} q_2\big(X_{\tau_1^{(r)}}\big),$$

$$\mathbb{E}\widehat{z} \;=\; \mathbb{E}\big[\widehat{p}_1 \widehat{p}_2\big] \;=\; \mathbb{E}\big[\widehat{p}_1 \mathbb{E}(\widehat{p}_2 \,|\, \mathscr{F})\big] \;=\; \frac{N_1}{R_1} \mathbb{E}\big(\widehat{p}_2 \,|\, \mathscr{F}\big)$$

$$\;=\; \frac{1}{R_1} \mathbb{E} \sum_{r=1}^{R_1} \mathbb{1}\{\tau_1^{(r)} < C^{(r)}\} q_2\big(X_{\tau_1^{(r)}}\big) \;=\; p_1 p_2 \;=\; z.$$

The case $m > 2$ follows similarly by induction. $\qquad\square$

With the right choice of m and the E_k, the splitting algorithm can be highly efficient. We illustrate this via a simple example, a birth–death chain on $E = \{0, \ldots, x\}$ with transition matrix

$$
\begin{pmatrix}
b_0 & a_0 & 0 & 0 & \cdots & 0 & 0 & 0 \\
b_1 & 0 & a_1 & 0 & & 0 & 0 & 0 \\
0 & b_2 & 0 & a_2 & & 0 & 0 & 0 \\
\vdots & & & & \ddots & & & \vdots \\
0 & 0 & 0 & 0 & & b_{x-1} & 0 & a_{x-1} \\
0 & 0 & 0 & 0 & \cdots & 0 & b_x & a_x
\end{pmatrix}
$$

$(a_i + b_i = 1)$. The rare set is $B(x) \stackrel{\text{def}}{=} \{x\}$ and thus

$$x \;=\; z(x) \;=\; \mathbb{P}_0\big(X_n = x \text{ for some } n < C\big).$$

The level sets are $E_k \stackrel{\text{def}}{=} \{x_k, x_k + 1, \ldots, x_{k+1} - 1\}$, where $0 = x_0 < x_1 < \cdots < x_{m-1} < x_m = x$.

Proposition 9.2 *Take $R_k = w_k R$ and consider the limit $R \to \infty$ with x, m, the w_k and the level sets $E_0, \ldots, E_m$ fixed. Then $\widehat{z} \stackrel{\text{def}}{=} \widehat{z}_R$ satisfies a*

CLT $R^{1/2}(\widehat{z}_R - z) \xrightarrow{\mathscr{D}} \mathscr{N}(0, \sigma^2)$, *where*

$$\sigma^2 \stackrel{\text{def}}{=} \sigma^2(x, m, w_1, \ldots, w_m, p_1, \ldots, p_m) \stackrel{\text{def}}{=} z^2 \sum_{k=1}^{m} \frac{1 - p_k}{w_k p_k}.$$

In particular, if for each x it is possible to choose the level sets such that $m \approx -\log z/c$, $p_1 \approx p_2 \approx \cdots \approx p_m$, then taking $w_k = 1/m$ one has $\sigma^2 \sim z^2 |\log z|^2 (e^c - 1)/c^2$, $x \to \infty$.

Proof. Note that ν_k is concentrated at x_k and thus the $\widehat{p}_k$ are simply obtained by independent binomial sampling, so that

$$\widehat{p}_k \stackrel{\mathscr{D}}{\approx} p_k + \frac{\left(p_k(1 - p_k)\right)^{1/2}}{R_k^{1/2}} Y_k = p_k + \frac{\left(p_k(1 - p_k)\right)^{1/2}}{w_k^{1/2} R^{1/2}} Y_k$$

for independent $\mathscr{N}(0, 1)$ r.v.'s $Y_1, \ldots, Y_m$. Therefore

$$\widehat{z}_R \stackrel{\mathscr{D}}{\approx} \prod_{k=1}^{m} \left[p_k + \frac{\left(p_k(1 - p_k)\right)^{1/2}}{w_k^{1/2} R^{1/2}} Y_k \right]$$

$$= z + \sum_{k=1}^{m} \frac{z}{p_k} \frac{\left(p_k(1 - p_k)\right)^{1/2}}{w_k^{1/2} R^{1/2}} Y_k + \mathrm{O}(R^{-1}).$$

This gives immediately the CLT (for a more formal proof, appeal to the delta method in III.3). With the parameter choices in the last part of the theorem, we then obtain $p_k \approx z^{1/m} \approx e^{-c}$ for all k and

$$\sigma^2 \approx z^2 \cdot m \cdot \frac{1}{1/m} \frac{1 - e^{-c}}{e^{-c}} \approx z^2 |\log z|^2 (e^c - 1)/c^2. \qquad \square$$

Remark 9.3 The last part of the theorem is obviously somewhat reminiscent of logarithmic efficency, though it does not fit formally into the definition of Section 1.

The specific birth–death assumption is essentially needed only to infer that we have independent binomial sampling. In general, one expects this to hold asymptotically as $R \to \infty$, so the general picture should remain unchanged. Of course, an aspect we have not discussed is the time needed to generate one of the Bernoulli r.v.'s needed for p_k, which may vary greatly among the k and among different models

In practice, one can at best hope to get close to the given order $z^2 |\log z|^2$ of σ^2, first, because it will seldom be apparent how to choose the level sets to make the p_k approximately equal, and second, because the p_k are not continuous variables in a discrete model. Thus, one should think of Proposition 9.2 as a guideline only.

For asymptotic studies of multilevel splitting, we refer in particular to Garvels & Kroese [125], Glasserman et al. [139], and Villén-Altamirano [358]. $\qquad \square$

Exercises

9.1 (A) A particle moves on the quarter

$$\{(x,y) : x \geq 0, \, y \geq 0, \, x^2 + y^2 \leq 25\}$$

of the disk with radius 5, such that the movements in the x- and y-directions are independent copies of the Ornstein–Uhlenbeck process

$$dZ(t) = -Z(t)dt + dB(t),$$

where $\{B(t)\}$ is standard Brownian motion. The starting point is $(1,1)$.
Use multilevel splitting to compute the probability that the particle hits the quarter-circle

$$\{(x,y) : x > 0, \, y > 0, \, x^2 + y^2 = 25\}$$

before either the x- or the y-axis. You are not supposed to try to fix the level sets in an optimal way!

9.2 (A) Consider a two-node Jackson network as in Figure 9.2.

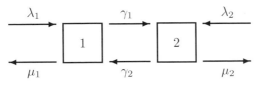

FIGURE 9.2

That is, λ_i is the service rate at node i, μ_i the rate of a customer being served and leaving the network, and γ_i the rate of a customer being served and going to the other node. Use multilevel splitting[3] to compute the probability that the total number in system during a busy cycle exceeds $N = 12$, taking $\lambda_1 = 1$, $\lambda_2 = 1/2$, $\mu_1 = 4$, $\mu_2 = 2$, $\gamma_1 = 1/2$, $\gamma_2 = 2$.

9.3 (A) In Exercise I.6.1, find improved estimates of the probabilities for the two weakest teams $12, 14$ to win the soccer tournament.

---

[3]Note that it is currently not known for all parameter combinations how to perform efficient exponential change of measure; see the references at the end of XIV.6d.

Chapter VII
Derivative Estimation

We consider a stochastic system and assume (highly simplified) that its performance can be summarized as a single real number z, which typically can be expressed as $\mathbb{E}Z$ for some r.v. Z. For example, in a PERT network Z can be the (random) length of the maximal path; in a data network Z can be the steady-state delay of a packet or z the probability that a packet is transmitted without errors; in an insurance risk model z can be the probability $\mathbb{P}(C > x)$ that the claims C within one year exceed a large value x (typically, $C = \sum_1^N V_i$ where N is the number of claims and $V_1, V_2, \ldots$ the claim sizes); in option pricing Z is the pay-out of the option and $\mathbb{E}$ the risk neutral expectation; and so on.

In all these examples, z will typically depend on a number of parameters. For example, in the insurance risk model, N could be Poisson with rate parameter λ with the claims having a distribution F_θ depending on a parameter θ. One might then be interested not only in the performance z but also in its (partial) derivatives $(\partial/\partial\lambda)z$, $(\partial/\partial\theta)z$, ... w.r.t. the parameters $\lambda, \theta, \ldots$ Here $(\partial/\partial\lambda)z$ is called the *sensitivity* of z w.r.t. λ (and similarly for $(\partial/\partial\theta)z$, etc.), and the vector

$$\nabla z \stackrel{\text{def}}{=} \left(\frac{\partial}{\partial\lambda}z \quad \frac{\partial}{\partial\theta}z \quad \ldots \right)$$

is the *gradient*. The problem we address in this chapter is how to estimate such sensitivities by simulation.

There are numerous reasons for being interested in the sensitivities. In particular:

(i) For identifying the most important system parameters; see further Remark 0.4.

(ii) To assess the effect of a small change of a parameter.

(iii) To produce confidence intervals for z if some parameters are estimated. For example, if β is a Poisson parameter estimated via the empirical rate $\overline{\beta} = N(T)/T$ of Poisson events in $[0, T]$, then $\overline{\beta}$ is asymptotically normal $N(\beta, \beta/T)$ as $T \to \infty$. So, if $z(\beta), (\partial/\partial\beta)z$ are analytically available as functions of β, the delta method establishes that

$$z(\overline{\beta}) \pm 1.96 \frac{\overline{\beta}^{1/2} |(\partial/\partial\beta)z|}{\sqrt{T}}$$

is an asymptotic 95% confidence interval for $z(\beta)$. More generally, if $z(\beta)$ needs to be evaluated by simulation, the confidence interval is

$$\widehat{z}(\overline{\beta}) \pm 1.96 \sqrt{\frac{\overline{\beta}\widehat{z}'(\beta)^2}{T} + \frac{\widehat{\sigma}^2}{R}} \,,$$

where $\widehat{z}(\overline{\beta})$ is a CMC estimator of $z(\overline{\beta})$ based on R replications and with associated variance estimator $\widehat{\sigma}^2$, with $\widehat{z}'(\beta)$ an estimator of the sensitivity. For additional details, see Heidelberger & Towsley [180] and Rubinstein & Shapiro [315, pp. 96–100].

(iv) In stochastic optimization, where we want to find the maximum (or minimum) of $z = z(\theta)$ w.r.t. some parameter θ, most algorithms require estimates of the gradient (see also Chapter VIII).

(v) In an empirical root-finding problem in which we want to find the θ^* satisfying $z(\theta^*) = 0$ for some function $z(\theta)$ representable as $\mathbb{E}u(X; \theta)$, the obvious estimator is the solution $\widehat{\theta}$ of

$$\frac{1}{R} \sum_{r=1}^{R} u(X_r; \widehat{\theta}) = 0.$$

When $u(X; \cdot)$ is smooth, the delta method again implies that the confidence interval for $\widehat{\theta}$ requires an estimate of $(\partial/\partial\theta)z$, cf. III.4 and Exercise 1.3.

(vi) Finally, there are examples in which the sensitivities are of intrinsic interest, for example, in financial mathematics, where the sensitivities go under the name of the *Greeks*. In particular, if the price of an option is $z = \Pi$ and $\theta = S(0)$ is the initial asset price, $(\partial/\partial S(0))\Pi$ is called the *delta* and $(\partial^2/\partial S(0)^2)\Pi$ the *gamma*; the delta and the gamma play a major role in the most commonly implemented hedging strategies; cf. Exercise I.6.3 and [45]. In the GBM Black–Scholes model with volatility σ, $(\partial/\partial\sigma)\Pi$ is the *vega* and so on. In addition to hedging, the calculation of Greeks may sometimes also be needed to meet requirements set by regulators of the financial industry.

Many important applications of derivative estimation require the whole gradient and not just its individual components. For example, stochastic optimization is often performed w.r.t. several parameters and not just a single one. A portfolio of stocks, options, etc., may consist of d assets, where d may well run into the hundreds or thousands. Thus, delta hedging requires a gradient with thousands of components, and delta–gamma hedging a Hessian (the matrix of second derivatives) with millions of components. From a practical point of view, this presents a considerable challenge. However, from the methodological point of view, there is little difference between the one-dimensional and the multidimensional cases: gradients and Hessians are computed componentwise. Therefore, to simplify notation, we will focus on a one-dimensional setting in this chapter in which the parameter in question is θ, defined on a real interval Θ. The corresponding performance is denoted by $z(\theta)$ and its derivative by $z'(\theta)$, and we use the notation $Z(\theta)$, $D(\theta)$ for Monte Carlo estimators of $z(\theta)$, respectively $z'(\theta)$.

We will present the three general standard methods:[1] finite differences (FDs), infinitesimal perturbation analysis (IPA), and the likelihood ratio (LR) method for derivative estimation. An important difference is that for finite differences and IPA, the dependence on θ is in the realizations of the relevant r.v.'s (sample paths in a stochastic process setting), whereas for the LR method, it is in the underlying measures. Further, IPA and the LR method produce unbiased estimates, while finite differences not.

Remark 0.4 In (i), the naive and often well working definition of "most important system parameters" is those having sensitivities with the largest absolute values. An inconsistency is, however, that $z'(\theta)$ depends on the scale of θ (for example, if θ is a Poisson rate, on whether the time unit is years or seconds). This may motivate considering $\mathrm{d}z(\theta)/\mathrm{d}\log\theta = \theta z'(\theta)$ instead of $z'(\theta)$.

When studying several system performances $z_1(\theta), z_2(\theta), \ldots$, similar scale problems may arise which could motivate looking at $\mathrm{d}\log z(\theta)/\mathrm{d}\theta = z'(\theta)/z(\theta)$ or $\mathrm{d}\log z(\theta)/\mathrm{d}\log\theta = \theta z'(\theta)/z(\theta)$. For an example, see Exercise 1.4.   □

---

[1]For special problems, different methods may exist, and in particular we mention an approach based on Malliavin calculus that has been developed for estimating the Greeks in mathematical finance; see Fournié et al. [120] and Kohatsu-Riga & Montero [221] and references there.

1   Finite Differences

Assume that for each θ we are in a position to generate an r.v. $Z(\theta)$ with expectation $z(\theta)$. We want a simulation estimate of $z'(\theta)$.

The starting point for the method of FDs is the formulas

$$f'(\theta) \;=\; \lim_{h \downarrow 0} \frac{f(\theta + h) - f(\theta)}{h} \;=\; \lim_{h \downarrow 0} \frac{f(\theta + h/2) - f(\theta - h/2)}{h} \qquad (1.1)$$

for the derivative of a deterministic function $f(\theta)$. In the context of simulation, this suggests performing a CMC experiment using either

$$\widetilde{D}(\theta) \;\overset{\text{def}}{=}\; \frac{Z(\theta + h) - Z(\theta)}{h} \quad \text{or} \quad D(\theta) \;\overset{\text{def}}{=}\; \frac{Z(\theta + h/2) - Z(\theta - h/2)}{h} \qquad (1.2)$$

as derivative estimator for some small h, where as a first naive attempt we take $Z(\theta + h), Z(\theta)$ as independent for $\widetilde{D}(\theta)$ (the *forward difference* estimator) and $Z(\theta + h/2), Z(\theta - h/2)$ independent for $D(\theta)$ (the *central difference* estimator).

A first important observation is that the second formula in (1.1) is preferable for numerical differentiation because

$$\frac{f(\theta + h) - f(\theta)}{h} \;=\; f'(\theta) + \frac{h}{2} f''(\theta) + \mathrm{O}(h^2),$$

$$\frac{f(\theta + h/2) - f(\theta - h/2)}{h} \;=\; f'(\theta) + \frac{h^2}{24} f'''(\theta) + \mathrm{O}(h^4),$$

as follows by straightforward Taylor expansions. These expressions establish, in the simulation context, that the bias of the central difference estimator $D(\theta)$ is an order of magnitude lower than that of the forward difference estimator $\widetilde{D}(\theta)$, so that obviously $D(\theta)$ should be preferred.[2] Thus only $D(\theta)$ is considered in the following.

The choice of h remains to be determined. If the number of replications is R and $\widehat{z}'(\theta) = \widehat{z}'_R(\theta)$ (the corresponding average of the $D(\theta)$), it seems clear that $h = h_R$ must go to zero as $R \to \infty$ to reduce bias. On the other hand, taking a smaller h increases variance so there is a trade-off. The answer is that h should be of order $R^{-1/6}$:

Proposition 1.1 *Consider the central difference estimator with independent sampling of $Z(\theta + h/2), Z(\theta - h/2)$. Then the root-mean-square error $\left[\mathbb{E}\big(\widehat{z}'(\theta) - z'(\theta)\big)^2 \right]^{1/2}$ is asymptotically minimized by letting*

$$h \;\overset{\text{def}}{=}\; h_R \;\overset{\text{def}}{=}\; \frac{1}{R^{1/6}} \frac{\left[576 \, \mathbb{V}ar(Z(\theta)) \right]^{1/6}}{\left| z'''(\theta) \right|^{1/3}}.$$

---

[2]The possible exception is a high-dimensioal setting in which the parameter vector is $\boldsymbol{\theta} = (\theta_1 \ldots \theta_d)$. Then the evaluation of the estimator for the full gradient vector requires roughly only half as many evaluations of $Z(\boldsymbol{\theta})$ in using the forward difference estimator as in using the central one.

For this choice, the root-mean-square error $\left[\mathbb{E}\left(\widehat{z}'(\theta)-z'(\theta)\right)^2\right]^{1/2}$ *is of order* $R^{-1/3}$.

Proof. Clearly, $\mathrm{Var}\left(Z(\theta\pm h/2)\right) \sim \mathrm{Var}\,Z(\theta)$, so $\mathrm{Var}\left(D(\theta)\right) \sim 2\mathrm{Var}\,Z(\theta)/h^2$ and

$$\begin{aligned}\mathbb{E}\left(\widehat{z}'(\theta) - z'(\theta)\right)^2 &= \mathrm{Var}\left(\widehat{z}'(\theta)\right) + \left(\mathbb{E}\widehat{z}'(\theta) - z'(\theta)\right)^2 \\ &\sim \frac{2}{Rh^2}\mathrm{Var}\left(Z(\theta)\right) + \left(\frac{h^2}{24}z'''(\theta)\right)^2\end{aligned}$$

Now, the function $a/x + bx^2$ of x is minimized for $x = (a/2b)^{1/3}$. Letting

$$x = h^2, \quad a = 2\mathrm{Var}\,Z(\theta)/R, \quad b = \left(z'''(\theta)/24\right)^2,$$

the result follows. $\square$

Remark 1.2 Of course, Proposition 1.1 is, in some sense, rather academic. In particular, the constant going with $R^{-1/6}$ for the optimal h_R is essentially unknown (because when $z'(\theta)$ is unknown, $z'''(\theta)$ is even more so). One important feature to note is, however, the rate $R^{-1/3}$ for the root-mean-square error, which is subcanonical compared to the canonical one $R^{-1/2}$ for the Monte Carlo method. Furthermore, a similar computation (Exercise 1.1) shows that the forward difference estimator converges at an optimal root-mean-square error rate of $R^{-1/4}$, which is achieved when h is of order $R^{-1/4}$. This shows that the lower bias of the central difference estimator permits one to use a larger difference increment h. This, in turn, reduces the variance of the difference quotient (given the presence of h in the denominator). $\square$

Remark 1.3 The discussion above rests on the minimization of mean square error. More generally, one may also consider minimization of the p norm error $(p > 0)$, as defined by $\left[\mathbb{E}\left|\widehat{z}'(\theta) - z'(\theta)\right|^p\right]^{1/p}$. The conclusion turns out to be identical to that of Proposition 1.1. In particular, the best possible choice for h is to let it be of order $R^{-1/6}$, in which case the convergence rate of $\widehat{z}'(\theta)$ to $z'(\theta)$ in pth mean is of order $R^{-1/3}$. However, the choice of the optimal constant in front of $R^{-1/6}$ depends on p; see Exercise 1.2. $\square$

As will be seen later, FDs can perform substantially better when one is using common random numbers, and the rate of convergence can sometimes be as good as $R^{-1/2}$ (which is the best one can expect and is obtained, for example, for the more sophisticated methods of IPA and likelihood ratios to be discussed in Sections 2, 3); see Section 2a for more discussion of common random numbers for computing derivatives.

Remark 1.4 A close relative to the method of FDs is a kernel estimator. Suppose, for example, that simulations have been performed at the points $\theta = 0, 1/R, 2/R, \ldots, 1$ and that a derivative estimator at $\theta_0 \in (0, 1)$ is desired. One potential estimator of $z(\theta)$ is

$$\frac{1}{Rh} \sum_{r=0}^{R} Z(r/R) k\big((\theta - r/R)/h\big),$$

where k is a kernel (typically a smooth and symmetric probability density) and $h > 0$ a smoothing constant. The corresponding derivative estimator at θ_0 is therefore

$$D_k(\theta_0) \stackrel{\text{def}}{=} \frac{1}{(R+1)h^2} \sum_{r=0}^{R} Z(r/R) k'\big((\theta_0 - r/R)/h\big).$$

It follows that for fixed h and $R \to \infty$,

$$\mathbb{E}D_k(\theta_0) = \frac{1}{(R+1)h^2} \sum_{r=0}^{R} z(r/R) k'\big((\theta_0 - r/R)/h\big)$$

$$\to \frac{1}{h^2} \int_0^1 z(\theta) k'\big((\theta_0 - \theta)/h\big) \, d\theta$$

$$= \frac{1}{h}\big[z(0)k(\theta_0/h) - z(1)k((\theta_0 - 1)/h)\big] + \frac{1}{h}\int_0^1 z'(\theta)k\big((\theta_0 - \theta)/h\big)\, d\theta.$$

Here the first term converges to 0 as $h \downarrow 0$ provided $k(x) = o(|x|)$ as $|x| \to \infty$. Noting that $k\big((\theta_0 - \theta)/h\big)/h$ is the density of $\theta_0 - hW$, where W has density k, it follows that $\mathbb{E}D_k(\theta_0)$ converges to $z'(\theta_0)$ as $h \downarrow 0$. For an appropriate choice of h as function of R, one can therefore obtain an estimator of $z'(\theta_0)$ (with a convergence rate significantly slower than $R^{-1/2}$). While this shares many of the same disadvantages as exhibited by FD estimators, it has the advantage that the R simulations can be used to estimate $z'(\cdot)$ globally, in contrast to at a single point as for the FD estimator. See also Kohatsu-Higa & Montero [221]. $\qquad\square$

In some settings (e.g., optimization or delta–gamma hedging), higher-order derivatives may be of interest. A starting point for estimating $z''(\theta)$ is the second central difference $\big(z(\theta + h) - 2z(\theta) + z(\theta - h)\big)/h^2$, suggesting that we can estimate $z''(\theta)$ by drawing i.i.d. replicates of the r.v.

$$\frac{Z(\theta + h) - 2Z(\theta) + Z(\theta - h)}{h^2}.$$

When $Z(\theta + h)$, $Z(\theta)$ and $Z(\theta - h)$ are generated independently, this yields an optimal rate of convergence of order $R^{-1/4}$ (which is obtained when h is of order $R^{-1/8}$).

1a Optimal Rates for Finite Differences

It turns out that one can construct finite difference estimators for $z'(\theta)$ with convergence rates arbitrarily close to the canonical rate $R^{-1/2}$. A starting point for deriving such estimators is to write the Taylor series for z as a formal power series in the differentiation operator $\mathscr{D}$, so that

$$z(\theta + h) \;=\; \sum_{j=0}^{\infty} z^{(j)}(\theta) \frac{h^j}{j!} \;=\; \sum_{j=0}^{\infty} \Big(\frac{(h\mathscr{D})^j}{j!} \Big) z(\theta)\,.$$

Let $\mathscr{T}_h$ be the shift operator defined by $\mathscr{T}_h z(\theta) \overset{\text{def}}{=} z(\theta + h)$. Hence, we can formally write

$$\mathscr{T}_h \;=\; \exp\{h\mathscr{D}\}\,, \tag{1.3}$$

so that $h\mathscr{D} = \log(\mathscr{T}_h)$. Formally expanding the logarithm as a power series around the identity operator $\mathscr{I}$, we arrive at

$$\mathscr{D} \;=\; \frac{1}{h} \sum_{j=1}^{\infty} \frac{(\mathscr{T} - \mathscr{I})^j}{j!} (-1)^{j+1}\,. \tag{1.4}$$

Note that $\mathscr{T}_h^k = \mathscr{T}_{kh}$. Hence, a simulation-based first-order approximation is given by $\widetilde{D}(\theta) \overset{\text{def}}{=} \big(Z(\theta + h) - Z(\theta) \big)/h$, whereas a second-order approximation (obtained by using the first two terms in the power series expansion (1.4)) is

$$\frac{-Z(\theta + 2h) + 4Z(\theta + h) - 3Z(\theta)}{2h}\,.$$

This second order approximation can easily be shown to have bias of order h^2 (just as for the central difference estimator) and hence possesses a convergence rate of $R^{-1/3}$. Use of the simulation-based kth-order approximation based on (1.4) yields a bias of order h^k and an optimal convergence rate of order $R^{-k/(2(k+1))}$ (obtained when h is of order $R^{-1/(2k+2)}$). Hence, FD estimators can attain rates arbitrarily close to the canonical rate $R^{-1/2}$.

Note that an alternative approach to constructing such low-bias FD estimators is to use (1.3) to formally write

$$\sinh(h\mathscr{D}) \overset{\text{def}}{=} \big(\exp\{h\mathscr{D}\} - \exp\{-h\mathscr{D}\} \big)/2 \;=\; (\mathscr{T}_h - \mathscr{T}_{-h})/2\,,$$

and to express $h\mathscr{D}$ in terms of the power series in $(\mathscr{T}_h - \mathscr{T}_{-h})/2$ obtained from the inverse function to $\sinh(\cdot)$. The first-order approximation obtained from this alternative approach is then our central difference estimator $D(\theta)$.

Despite the improved convergence rates associated with these estimators, they are rarely used in practice. To construct an estimator having convergence rate $R^{-k/(2k+1)}$, one needs to start with an FD approximation having bias of order h^k. This means that the derivatives of order $2, \ldots, k$ must be explicitly estimated so that their influence can be subtracted off. Hence, such an estimator must implicitly involve an FD approximation to the kth derivative, thereby requiring estimation of $z(\cdot)$ at $k + 1$ points. Thus the computational effort per replication increases linearly in k. Furthermore, such high-order estimators tend to generate weights $w_{j,k}$ on the

$Z(\theta+jh)$ that grow larger in absolute value as k increases, with corresponding alternating signs on the $w_{j,k}$ that serve to cancel most of the effect (in expectation) on $w_{j,k}Z(\theta + jh)$. However, the presence of these large $w_{j,k}$ may influence the variance of the FD estimator substantially (the alternating signs do not help). Furthermore, the alternating signs can produce numerical difficulties in dealing with floating-point implementations of the sum. For all these reasons, it is rare that estimators of higher order than the central difference estimator are used.

Exercises

1.1 (TP) Derive the analogue of Proposition 1.1 for the forward difference estimator.

1.2 (TP) Develop the details of Remark 1.3 and other similar estimates stated without proof in the text.

1.3 (A) An insurance company has capital w and (increasing) utility function $u(\cdot)$. The utility premium Π_0 for a risk $X \geq 0$ is defined as the smallest Π not decreasing the expected utility, that is, the solution of $z(\Pi_0) = 0$, where $z(\Pi) = \mathbb{E}u(w+\Pi-X) - u(w)$. Assume that $u(y) = 1 - e^{-y}$, $w = 4$, and that X is Pareto with density $\alpha/(1+x)^{\alpha+1}$, where $\alpha = 3/2$. Give a simulation estimate of Π_0 and an associated confidence interval, using the ideas of III.4.

1.4 (A+) We will consider an example of bonus systems in car insurance, for which we refer to Kaas et al. [205] for further details. The portfolio consists of D drivers and is divided into K bonus classes, such that the premium H_k in class $k = 1, \ldots, K$ depends on the class, and that the class of a driver is dynamically adjusted according to his claim experience (a driver entering the portfolio is assigned class k_0 for some fixed k_0). For a particular driver d, we denote by X_n his bonus class in year n and M_n his number of claims. Many systems then determine the bonus class X_{n+1} in year n on basis of X_n and M_n alone. That is, $X_{n+1} = \kappa(k, m)$ when $X_n = k$, $M_n = m$.

A standard assumption is this setting is that driver d generates a Poisson(λ_d) number of claims each year. The ideal premium $H^*(\lambda)$ that drivers with Poisson parameter λ pay per year is therefore of the form $c\lambda$, whereas in reality their expected premium (averaging over the total time T spent in the portfolio and a large number of drivers) is

$$H(\lambda) \stackrel{\text{def}}{=} \frac{1}{\mathbb{E}T}\mathbb{E}\sum_{n=0}^{T-1} H_{X_n(\lambda)}, \tag{1.5}$$

where $\{X_n(\lambda)\}$ is the Markov chain of bonus classes of a driver with Poisson parameter λ. As a measure of the deviation of the function $H(\lambda)$ of λ from $H^*(\lambda)$, it has been suggested to consider the *Loiramanta efficiency*

$$e(\lambda) \stackrel{\text{def}}{=} \frac{\mathrm{d}\log H(\lambda)}{\mathrm{d}\log \lambda} = \frac{\lambda H'(\lambda)}{H(\lambda)}$$

(cf. Remark 0.4). In particular, as explained in [205], $e(\lambda) \leq 1$ for all λ means that the bonus system favors the bad risks (drivers with a small λ pay too much).

(a) Explain why $e(\lambda) \approx 1$ is a desirable property that ensures the bonus system to be reasonably fair.

(b) Give plots of $H(\lambda)$ and of $e(\lambda)$, simulating at selected values of $\lambda \in (0,3)$ and using linear interpolation in between (use FDs for $H'(\lambda)$). Assume that T is geometric with mean 8, i.e., $\mathbb{P}(T = n) = (1 - \rho)\rho^{n-1}$, $n = 1, 2, \ldots$, where $\rho = 7/8$, and that the bonus system is given by the rules of a Dutch bonus system in use around 1982, which had $K = 14$, $k_0 = 2$, and the remaining parameters determined by the following table:

k	H_k	$\kappa(k,0)$	$\kappa(k,1)$	$\kappa(k,2)$	$\kappa(k,3+)$
14	30.0	14	9	5	1
13	32.5	14	8	4	1
12	35.0	13	8	4	1
11	37.5	12	7	3	1
10	40.0	11	7	3	1
9	45.0	10	6	2	1
8	50.0	9	5	1	1
7	55.0	8	4	1	1
6	60.0	7	3	1	1
5	70.0	6	2	1	1
4	80.0	5	1	1	1
3	90.0	4	1	1	1
2	100.0	3	1	1	1
1	120.0	2	1	1	1

(c) Redo (b) with $\mathbb{E}T = 4, 16, 32$.

1.5 (A) Redo Exercise 1.4 by using kernel estimates instead of FDs; see Remark 1.4.

2   Infinitesimal Perturbation Analysis

The idea of this method is *sample path differentiation*: we write $z = z(\theta)$ as $\mathbb{E}Z(\theta)$ for some r.v. $Z(\theta)$ depending on θ, and estimate $z'(\theta)$ via the r.v.

$$D \stackrel{\text{def}}{=} D(\theta) \stackrel{\text{def}}{=} \frac{\mathrm{d}}{\mathrm{d}\theta} Z(\theta) = Z'(\theta),$$

evaluated at $\theta = \theta_0$, where θ_0 is the parameter of the given system (assuming, of course, that the derivative exists w.p. 1). Thus we simulate R i.i.d. copies $D_1, \ldots, D_R$ of D and use the estimator

$$\widehat{z}'(\theta) \stackrel{\text{def}}{=} \frac{D_1 + \cdots + D_R}{R}, \tag{2.1}$$

with the obvious confidence interval based on the empirical variance of the D_r. The method goes under the name of *infinitesimal perturbation analysis* (IPA) and goes back to (at least) Ho & Cao [183] and Suri & Zazanis [348]. A standard textbook reference is Glasserman [131], stressing the GSMP setup, but there is also an extended exposition in Glasserman [133].

What is needed for consistency of $\widehat{z}'(\theta)$ as $R \to \infty$ is $\mathbb{E}D(\theta) = z'(\theta)$, that is,

$$\frac{\mathrm{d}}{\mathrm{d}\theta}\big[\mathbb{E}Z(\theta)\big] = \mathbb{E}\Big[\frac{\mathrm{d}}{\mathrm{d}\theta}Z(\theta)\Big]. \tag{2.2}$$

Of course, when (2.2) holds, averaging i.i.d. copies of $D = Z'(\theta)$ yields an unbiased estimator that converges to $z'(\theta)$ at the canonical rate $R^{-1/2}$.

We illustrate the approach via two extremely simple examples, where (2.2) holds in the first but not the second:

Example 2.1 This is a simplified version of what is needed for a PERT net. Assume that $Z = \max(X_1, X_2)$ with X_1, X_2 independent where θ is a scale parameter for X_2 (thus, the given system corresponds to $\theta_0 = 1$). That is,

$$Z(\theta) = \max(X_1, \theta X_2) = \begin{cases} X_1 & X_1 > \theta X_2, \\ \theta X_2 & X_1 < \theta X_2. \end{cases}$$

It follows that

$$D(\theta) = \begin{cases} 0 & X_1 > \theta X_2, \\ X_2 & X_1 < \theta X_2, \end{cases}$$

so that $D = D(\theta_0) = X_2 \mathbb{1}\{X_1 < X_2\}$. The check of (2.2) goes as follows: with F_1, F_2 the c.d.f.'s and f_1, f_2 the densities, we get

$$
\begin{aligned}
z(\theta) &= \mathbb{E}\max(X_1, \theta X_2) = \int_0^\infty \mathbb{P}\big(\max(X_1, \theta X_2) > x\big)\,\mathrm{d}x \\
&= \int_0^\infty \big(1 - F_1(x)F_2(x/\theta)\big)\,\mathrm{d}x.
\end{aligned}
$$

Differentiation under the integral sign (see Example 2.4 for a formal justification!) yields

$$
\begin{aligned}
z'(\theta) &= \int_0^\infty F_1(x)\,\frac{x}{\theta^2}\,f_2(x/\theta)\,\mathrm{d}x \\
&\overset{\theta=\theta_0=1}{=} \int_0^\infty \mathbb{P}(X_1 < x)\,x\,f_2(x)\,\mathrm{d}x = \mathbb{E}\big[X_2;\, X_1 < X_2\big] = \mathbb{E}D(\theta_0).
\end{aligned}
$$

$\square$

In general, the assumption (2.2) appears to be rather innocent, requiring differentiation and integration to be interchangeable. However, the following example shows that one has to be careful:

Example 2.2 Consider $\max(X_1, X_2)$ as in Example 2.1, but assume now that the relevant performance is $z = \mathbb{P}(X_2 > X_1)$. Then $Z(\theta) = \mathbb{1}\{\theta X_2 > X_1\}$ (see Figure 2.1), which is differentiable with derivative 0 except at $\theta = X_1/X_2$.

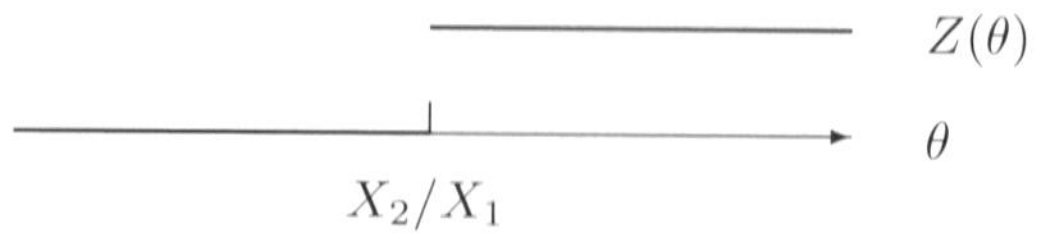

FIGURE 2.1

Since $\mathbb{P}(X_1 = X_2) = 0$, it follows that

$$D = \left.\frac{\mathrm{d}}{\mathrm{d}\theta}Z(\theta)\right|_{\theta=\theta_0=1} = 0 \text{ a.s.},$$

so that $\mathbb{E}D = 0$, which in general does not equal $z'(\theta_0)$.   □

The same phenomenon occurs often when discrete r.v.'s are involved (say the Poisson r.v. N in the insurance risk example). Just think of $Z(\theta) = Z\big(B_1(\theta),\ldots,B_p(\theta)\big)$ being a function of p i.i.d. Bernoulli(θ) r.v.'s generated by p i.i.d. uniforms as $B_1(\theta) = \mathbb{1}\{U_1 < \theta\},\ldots,B_p(\theta) = \mathbb{1}\{U_p < \theta\}$. Then $Z(\theta)$ is piecewise constant and hence $Z'(\theta) = 0$ on any θ-interval falling between two of the U_i.

The standard condition for the validity of IPA is the following:

Proposition 2.3 *Assume that $Z(\theta)$ is a.s. differentiable at θ_0 and that a.s. $Z(\theta)$ satisfies the Lipschitz condition*

$$\big|Z(\theta_1) - Z(\theta_2)\big| \leq |\theta_1 - \theta_2|M \tag{2.3}$$

for θ_1,θ_2 in a nonrandom neighborhood of θ_0, where $\mathbb{E}M < \infty$. Then (2.2) holds at $\theta = \theta_0$.

Proof. Just note that

$$z'(\theta_0) = \lim_{h\to 0}\frac{z(\theta_0 + h) - z(\theta_0)}{h} = \lim_{h\to 0}\mathbb{E}\left[\frac{Z(\theta_0 + h) - Z(\theta_0)}{h}\right]. \tag{2.4}$$

Here $[\cdots]$ is bounded by M and goes to $Z'(\theta_0) = D$, so by dominated convergence the r.h.s. of (2.4) equals $\mathbb{E}D$.   □

Example 2.4 In Example 2.1, the inequality $\big|\max(a,b_1) - \max(a,b_2)\big| \leq |b_1 - b_2|$ yields

$$\big|Z(\theta_1) - Z(\theta_2)\big| \leq |\theta_1 - \theta_2|\,|X_2|,$$

so we can just take $M = |X_2|$. In Example 2.2, consider for simplicity the case in which X_1/X_2 has a density bounded below by $A > 0$ in the interval $[1, 1+\varepsilon]$. If $1 < \theta < 1 + \varepsilon$, then

$$Z(\theta) - Z(\theta_0) = \mathbb{1}\{\theta > X_1/X_2\} - \mathbb{1}\{1 > X_1/X_2\} = \mathbb{1}\{1 \leq X_1/X_2 < \theta\}$$

equals 1 with probability at least $A(\theta - 1)$. Hence a possible M in (2.3) must satisfy $\mathbb{P}\big(M \geq 1/(\theta - 1)\big)$ w.p. at least $A(\theta - 1)$ for all $\theta \in (1, 1+\varepsilon)$, i.e., $\mathbb{P}(M \geq x) \geq A/x$ for all $x \in (\varepsilon^{-1}, \infty)$, so that

$$\mathbb{E}M \;\geq\; \int_{1/(1+\varepsilon)}^{\infty} \mathbb{P}(M > x)\,\mathrm{d}x \;\geq\; A\int_{1/(1+\varepsilon)}^{\infty} x^{-1}\,\mathrm{d}x \;=\; \infty\,,$$

which violates the integrability condition of Proposition 2.3. $\square$

Example 2.5 Consider $Z(\theta) = [\theta Y - K]^+ = \max(0, \theta Y - K)$ for $\theta > 0$ and some r.v. $Y > 0$. This corresponds to the payout function of a European call option. The Lipschitz property of the max used in Example 2.4 implies immediately that (2.3) holds with $M = Y$, so IPA is valid for estimating the delta; the corresponding estimator is $D = Z'(\theta) = Y\mathbb{1}\{\theta Y > K\}$. However, since $Z''(\theta)$ exists and is zero except at $\theta = K/Y$, we have in general that $z''(\theta) \neq \mathbb{E}Z''(\theta)$, so that *IPA is not valid for estimating the gamma*. $\square$

Remark 2.6 A general rule of thumb is that the derivative interchange (2.2) required to justify IPA is valid when $Z(\cdot)$ is a.s. continuous and is differentiable except (possibly) at finitely many random points. IPA tends to fail when $Z(\cdot)$ has discontinuities. $\square$

Example 2.7 (THE DELTA VIA SDE'S) The delta of an European option with underlying asset price process $\{X(t)\}_{0 \leq t \leq T} \stackrel{\text{def}}{=} \{X(t,x)\}_{0 \leq t \leq T}$, where $x = X(0)$, is the derivative of the price $z(x) \stackrel{\text{def}}{=} \mathbb{E}\big[f\big(X(T,x)\big)\big]$ w.r.t. $x = X(0)$. A common model is then that $\{X(t,x)\}$ satisfies a time-homogeneous SDE (see further Chapter X) with diffusion coefficients a, b, i.e.,

$$\mathrm{d}X(t) \;=\; a\big(X(t)\big)\,\mathrm{d}t + b\big(X(t)\big)\,\mathrm{d}B(t)\,,$$

$$X(t,x) \;=\; x + \int_0^t a\big(X(s,x)\big)\,\mathrm{d}s + \int_0^t b\big(X(s,x)\big)\,\mathrm{d}B(s)\,,$$

where B is standard Brownian motion. Pathwise differentiation w.r.t. x yields

$$Y(t,x) \stackrel{\text{def}}{=} \frac{\partial}{\partial x}X(t,x)$$

$$= 1 + \int_0^t a'\big(X(s,x)\big)Y(s,x)\,\mathrm{d}s + \int_0^t b'\big(X(s,x)\big)Y(s,x)\,\mathrm{d}B(s)\,.$$

This suggests that if f is smooth, then by the chain rule,

$$z'(x_0) \;=\; \mathbb{E}\big[f'\big(X(T)\big)Y(T)\big]\,, \tag{2.5}$$

where (X, Y) satisfies the coupled pair of SDEs

$$\mathrm{d}X(t) \;=\; a\big(X(t)\big)\,\mathrm{d}t + b\big(X(t)\big)\,\mathrm{d}B(t)\,,$$

$$\mathrm{d}Y(t) \;=\; a'\big(X(t)\big)Y(t)\,\mathrm{d}t + b'\big(X(t)\big)Y(t)\,\mathrm{d}B(t)\,,$$

subject to $X(0) = x_0$, $Y(0) = 1$. Similar ideas are capable of successfully computing derivatives in the more general setting in which the parameter of interest also influences the drift and variance coefficients (i.e., $a(x) = a(x, \theta)$ and $b(x) = b(x, \theta)$). □

2a Finite Differences, IPA, and Common Random Numbers

A common theme of the IPA approach is that the expectation $z(\theta)$ can be expressed in the form $\mathbb{E}Z(\theta)$, where the $Z(\theta)$ are defined on a common probability space (i.e., use a common stream of random numbers). This suggests a close relation between IPA and common random numbers, which we will next discuss.

A derivative is a limit of a finite difference quotient. We have already discussed in V.6 the fact that common random numbers generally provide a variance reduction when one computes the difference of two expectations. We pursue this point in greater detail here.

Suppose that we represent $z(\theta)$ as $\mathbb{E}Z(\theta)$, where the $Z(\theta)$ are simulated using random numbers that are the same across the range of θ's. Assume that we can synchronize our simulation across θ so effectively that $Z(\cdot)$ is a.s. differentiable at θ. In this setting, $z'(\theta)$ can potentially be estimated via sample means based on $Z'(\theta)$. This, of course, is precisely IPA as discussed above.

If IPA is valid, then

$$\frac{Z(\theta + h) - Z(\theta)}{h} \overset{\text{a.s.}}{\to} z'(\theta),$$

and the expectations also converge. If the coupling underlying the use of our common random numbers is well enough behaved that the expectations converge, then one typically has also that variances converge, implying that

$$\mathbb{V}ar\left[Z(\theta + h) - Z(\theta)\right] \sim h^2 \mathbb{V}ar\, Z'(\theta) \qquad (2.6)$$

as $h \downarrow 0$. Hence, it is typically the case that when IPA is valid, $\mathbb{V}ar\left[Z(\theta + h) - Z(\theta)\right]$ is of order h^2 (conversely, this rate implies that IPA is automatically valid). In this setting,

$$\frac{Z(\theta + h) - Z(\theta)}{h} \approx Z'(\theta) \qquad (2.7)$$

both pathwise and in expectation. In contrast to our discussion of FDs in Section 1, the variance of our common random number FD quotient in (2.7) does not blow up as $h \downarrow 0$. This means that there is no variance penalty in this setting for letting h be small, so that the bias-variance trade-off issues of Section 1 are irrelevant. Hence, an FD common-random-number estimator of $z'(\theta)$ can potentially achieve the canonical convergence rate of $R^{-1/2}$. In particular, if one chooses h small enough that the bias be of order $R^{-1/2}$ (i.e., $h = O(R^{-1/2})$ for the FD common-random-number estimator), the convergence rate is canonical; see also L'Ecuyer & Perron [235]. Of course, in this setting, one possible choice of h is $h = 0$, i.e., the IPA estimator. We are unaware of any practical situations in which under these circumstances one would prefer the common-random-number FD estimator to the one based on IPA.

As we have indicated above, there are important applications for which one can expect to find oneself in the setting (2.6). However, the more typical situation, in the presence of common random numbers, is that (2.6) fails to hold. Unlike the use of independent realizations of $Z(\theta)$ and $Z(\theta+h)$, we expect that $Z(\theta+h) \overset{\text{a.s.}}{\to} Z(\theta)$ as $h \downarrow 0$ when common random numbers are used. Hence, in contrast to the independence setting of Section 1, we always expect that $\mathbb{V}ar\left[Z(\theta+h) - Z(\theta)\right] \to 0$ when common random numbers are used. The only question is how fast the rate is.

As noted above, the variance converges to zero at rate h^2 when IPA is valid. When IPA fails to hold, our experience is that the rate typically is h. While we know of no general theorem supporting this assertion, it can be verified in all of our examples of IPA nonvalidity and holds in many other such settings. Arguments similar to those of Proposition 1.1 establish that when $\mathbb{V}ar\left[Z(\theta+h) - Z(\theta)\right]$ is of order h, then the optimal rate of convergence of the forward difference common-random-number estimator is $R^{-1/3}$ and is achieved when h is of order $R^{-1/3}$. For the central difference estimator, the best possible convergence rate is $R^{-2/5}$, achieved when h is of order $R^{-1/5}$.

These convergence rates makes clear that the use of common random numbers can significantly enhance the computational efficiency of derivative estimation as compared with the use of independent simulations, even in cases in which IPA ceases to be valid.

Exercises

2.1 (TP) In Example 2.2, discuss the validity of IPA in the case in which the support of X_1/X_2 does not contain 1.

2.2 (TP) In the GBM Black–Scholes model for a European option, show that IPA is valid for estimating the vega and write up the IPA estimator.

2.3 (A) Redo Exercise I.6.3 on delta hedging, with the modification that instead of the GBM Black–Scholes model the log asset price is assumed to follow an NIG Lévy process (see XII.1.4) having the parameters fitted by Rydberg [322] to the Dresdner Bank stock, i.e.,

$$(\alpha, \beta, \mu, \delta) = (68.28, 1.81, 0, 0.01).$$

As risk-neutral measure $\mathbb{P}^*$, use the Esscher measure (see XII.1b and Exercise XII.6.3), and generate the necessary NIG r.v.'s by subordination as in Example XII.5.1.

2.4 (A) Consider the delta $z'(\theta)$ of a weekly sampled Asian option with price $z(\theta)$ where θ is the initial price of the underlying asset. Assume the GBM Black–Scholes model, maturity $T = 1$ years, strike price $K = 120$, $\theta_0 = 100$, $r = 5\%$ per year, and $\sigma = 0.25$.

  (i) Compute $z(\theta)$, $z'(\theta)$, and associated confidence intervals.

(ii) Improve the efficiency of your method by antithetic sampling of the Brownian increments.
(iii) If time permits, do some experiments on variance reduction by changing the drift of the Brownian motion. At least outline what the changed estimates are!

3   The Likelihood Ratio Method: Basic Theory

The most generally applicable method for obtaining unbiased gradient estimates is the likelihood ratio (LR) method, for which some key references are Glynn [144] and Rubinstein & Shapiro [315]. The key feature is that the dependence of the expectation on θ is expressed in the measure $\mathbb{P}_\theta$, i.e., we assume that

$$z(\theta) \;=\; \mathbb{E}_\theta Z \;=\; \int_\Omega Z(\omega)\,\mathbb{P}_\theta(\mathrm{d}\omega)\,.$$

This is the opposite of the situation in FDs and IPA, where it is the sample function $Z(\theta)$ that depends on θ, whereas the measure $\mathbb{P}$ does not. Most often, one easily gets from one setting to the other, as illustrated by the following example:

Example 3.1 As in Example 2.1, consider $z(\theta) = \mathbb{E}\max(X_1, \theta X_2)$, where X_1, X_2 are independent with densities f_1, f_2. Suppose that (X_1, X_2) has joint positive density $h(x_1, x_2)$ under some given $\widetilde{\mathbb{P}}$. In this case,

$$z(\theta) \;=\; \widetilde{\mathbb{E}}\big[\max(X_1, X_2)\,L(\theta)\big] \quad \text{where } \ L(\theta) \stackrel{\text{def}}{=} \frac{f_1(X_1)f_2(X_2/\theta)/\theta}{h(X_1, X_2)}\,.$$

Under modest regularity conditions on f_2 (see Proposition 3.5 below), one can interchange expectation and differentiation to get

$$z'(\theta) \;=\; \widetilde{\mathbb{E}}\left[\max(X_1, X_2)\frac{f_1(X_1)}{h(X_1, X_2)}\left(-\frac{1}{\theta^2}f_2(X_2/\theta) - \frac{X_2}{\theta^3}f_2'(X_2/\theta)\right)\right].$$

A particularly natural choice is to take $h(x_1, x_2) = f_1(x_1)f_2(x_2/\theta_0)/\theta_0$ when estimating $z'(\theta_0)$. With this choice of h, the importance distribution is equal to $\mathbb{P}_{\theta_0}$. Consequently, both $z(\theta_0)$ and $z'(\theta_0)$ can be estimated from the same simulation run using the identical sampling distribution. Furthermore, the formula for the derivative estimator simplifies to

$$\max(X_1, X_2)\left(-\frac{1}{\theta_0} - \frac{X_2}{\theta_0^2}\frac{f_2'(X_2/\theta_0)}{f_2(X_2/\theta_0)}\right)$$

$$= \; \max(X_1, X_2)\frac{\mathrm{d}}{\mathrm{d}\theta}\log\left(\frac{1}{\theta}f_2(X_2/\theta)\right)\bigg|_{\theta=\theta_0}\,.$$

The random variable

$$\frac{\mathrm{d}}{\mathrm{d}\theta} \log\left(\frac{1}{\theta} f_2(X_2/\theta)\right)\bigg|_{\theta=\theta_0}$$

is a special case of the *score function* familiar to statisticians, which will be defined more generally below. □

More generally, we apply the importance-sampling idea; cf. V.1. Specifically, suppose there exists an importance distribution $\widetilde{\mathbb{P}}$ for which

$$\mathbb{1}\{Z(\omega)\} \neq 0)\mathbb{P}_\theta(\mathrm{d}\omega) \;=\; \mathbb{1}\{Z(\omega) \neq 0\}L(\theta,\omega)\widetilde{\mathbb{P}}(\mathrm{d}\omega)$$

for some likelihood ratio $L(\theta)$. It follows that

$$z(\theta) \;=\; \widetilde{\mathbb{E}}\big[ZL(\theta)\big]\,.$$

Hence, if $L(\cdot,\omega)$ is differentiable, this suggests the basic formula of likelihood ratio derivative estimation

$$z'(\theta) \;=\; \widetilde{\mathbb{E}}\big[ZL'(\theta)\big]\,, \tag{3.1}$$

where the prime denotes differentiation w.r.t. θ. A particularly natural choice is to take $\widetilde{\mathbb{P}} = \mathbb{P}_{\theta_0}$ when estimating $z'(\theta_0)$. More specifically, assume that $\mathbb{P}_\theta$ has density f_θ w.r.t. some reference measure μ independent of θ. Then $L = f_\theta/f_{\theta_0}$, and (3.1) takes the form

$$z'(\theta_0) \;=\; \mathbb{E}_{\theta_0}\big[ZS\big]\,, \quad \text{where } S \stackrel{\mathrm{def}}{=} S(\theta_0) = \frac{f'_\theta}{f_\theta}\bigg|_{\theta=\theta_0}\,. \tag{3.2}$$

We refer to $S(\theta_0)$ as the *score function* evaluated at θ_0.

For a single r.v. X with density $f_\theta(x)$ and $\theta \in \mathbb{R}$, we write $f'_\theta(x) \stackrel{\mathrm{def}}{=} (\mathrm{d}/\mathrm{d}\theta)f_\theta(x)$ and

$$S_X(\theta) \stackrel{\mathrm{def}}{=} \frac{\mathrm{d}}{\mathrm{d}\theta} \log f_\theta(X) \;=\; f'_\theta(X)/f_\theta(X)$$

or, when no ambiguity exists, just $S_X \stackrel{\mathrm{def}}{=} S_X(\theta)$. If $X_1,\ldots,X_n$ are i.i.d. with densities $f_\theta^{(1)}(x),\ldots,f_\theta^{(n)}(x)$, then the score contained in the vector $\boldsymbol{X} \stackrel{\mathrm{def}}{=} (X_1 \ldots X_n)$ becomes

$$\begin{aligned}
S_{\boldsymbol{X}} &= \frac{\mathrm{d}}{\mathrm{d}\theta} \log\big[f_\theta^{(1)}(X_1)\cdots f_\theta^{(n)}(X_n)\big] \\
&= \frac{\mathrm{d}}{\mathrm{d}\theta} \log f_\theta^{(1)}(X_1) + \cdots + \frac{\mathrm{d}}{\mathrm{d}\theta} \log f_\theta^{(n)}(X_n) \\
&= S_{X_1} + \cdots + S_{X_n}\,, \tag{3.3}
\end{aligned}$$

which we refer to as the *additive property of the score*. A further fundamental property is given in the following proposition:

Proposition 3.2 $\mathbb{E}_{\theta_0} S = 0$.

Proof. Formal differentiation of $1 = \int f_\theta \, \mathrm{d}\mu$ under the integral sign yields

$$0 = \int f'_\theta \, \mathrm{d}\mu = \int \frac{f'_\theta}{f_\theta} f_\theta \, \mathrm{d}\mu \,.$$

Letting $\theta = \theta_0$, the r.h.s. becomes $\mathbb{E}_{\theta_0} S$. $\qquad\square$

Example 3.3 Consider as in insurance risk a compound Poisson sum $C = \sum_0^N V_i$, where N is Poisson(λ) and the V_i independent and i.i.d. with density $f(x)$. Assume we are interested in $z(\lambda) = \mathbb{P}(C > x)$ and the associated sensitivity. The LR derivative estimator is then $D = S_N \mathbb{1}\{C > x\}$, where

$$S_N = \frac{\mathrm{d}}{\mathrm{d}\lambda} \log\left[\mathrm{e}^{-\lambda} \frac{\lambda^N}{N!}\right] = \frac{N}{\lambda} - 1 \,.$$

If instead λ is fixed but $f(x) = f_\theta(x)$ depends on another parameter θ, then the likelihood ratio derivative estimator of $z'(\theta)$ is

$$\left(S_{X_1}(\theta) + \cdots + S_{X_n}(\theta)\right) \mathbb{1}\{C > x\} \,.$$

For example, if $f_\theta(x) = \theta \mathrm{e}^{-\theta x}$, then $S_{X_i}(\theta) = 1/\theta - X_i$. $\qquad\square$

Example 3.4 For the likelihood ratio derivative estimator of the vega of a European call option in the GBM Black–Scholes model, we first write the (risk-neutral) payout as $\mathrm{e}^{-rT}\left[S(0)\mathrm{e}^X - K\right]^+$ where $X \sim \mathcal{N}\left((r - \sigma^2/2)T, \sigma^2 T\right)$. The desired estimator is $\mathrm{e}^{-rT}\left[S(0)\mathrm{e}^X - K\right]^+ S_X$, where

$$\begin{aligned}
S_X &= \frac{\mathrm{d}}{\mathrm{d}\sigma} \log\left[\frac{1}{\sigma\sqrt{2\pi}} \exp\{-\left(X - (r - \sigma^2/2)T\right)^2/2\sigma^2 T\}\right] \\
&= \frac{\mathrm{d}}{\mathrm{d}\sigma}\left[\log \sigma - \frac{X^2}{2\sigma^2 T} + \frac{X(r - \sigma^2/2)}{\sigma^2} - \frac{(r - \sigma^2/2)^2 T}{2\sigma^2}\right] \\
&= -\frac{1}{\sigma} + \frac{X^2}{\sigma^3 T} - \frac{2Xr}{\sigma^3} - \frac{-4\sigma^3(r - \sigma^2/2)T - 4\sigma(r - \sigma^2/2)^2 T}{4\sigma^4} \,.
\end{aligned}$$

$\qquad\square$

The interchange of gradient and expectation required to justify (3.1) holds in great generality; only some weak regularity on $L(\theta)$ is required. Consequently, the approach works in many different application settings. A rather safe rule of thumb is that it suffices that $\widetilde{\mathbb{P}}$ be equivalent to $\mathbb{P}_\theta$ for all θ and that $L(\theta, \omega)$ not have discontinuities depending on θ. At the mathematically rigorous level, we just give the following result on interchange of integration and differentiation for the one-dimensional case:

Proposition 3.5 *Let* $\left(f_\theta(x)\right)_{\theta \in \Theta}$ *be a family of densities on* $\mathbb{R}$ *such that x-a.e.* $f_\theta(x)$ *is continuously differentiable in* θ *with derivative* $f'_\theta(x)$. *Then*

$$\frac{\mathrm{d}}{\mathrm{d}\theta} \int z(x) f_\theta(x) \, \mathrm{d}x = \int z(x) f'_\theta(x) \, \mathrm{d}x$$

for all θ in a given open subinterval Θ_0 of Θ provided there exist p, q with $1/p + 1/q = 1$ such that $z \in L_q$ and $|f'_\theta(x)| \leq M(x)$ for some $M \in L_p$ and for all $\theta \in \Theta_0$ and x-a.e.

Proof. Assume $(\theta - \varepsilon, \theta + \varepsilon) \subseteq \Theta_0$. For $|h| < \varepsilon$ we then have

$$\frac{1}{h}\left[\int z(x) f_{\theta+h}(x)\, dx \; - \; \int z(x) f_\theta(x)\, dx\right]$$

$$= \int z(x) \frac{f_{\theta+h}(x) - f_\theta(x)}{h}\, dx \; = \; \int z(x) f'_{\theta+h^*(x)}(x)\, dx$$

for some $h^*(x) \in (-h, h)$. Here the integrand has limit $z(x) f'_\theta(x)$ as $h \to 0$, and is bounded by $|z(x) M(x)|$, which is in L_1 by Hölder's inequality. Dominated convergence completes the proof. $\qquad\square$

Remark 3.6 In many applications, $f_\theta(x)$ is infinitely differentiable at θ_0. The LR method can then typically be extended to provide estimators for higher-order derivatives. In particular, in the setting of (3.1), the second-order derivative estimator takes the form $\widetilde{\mathbb{E}}[ZL^{(2)}(\theta)]$, where $L^{(2)}$ is the second derivative of the likelihood ratio. For example, in the setting of (3.3), $L_n(\theta) = \prod_1^n f_\theta(X_i)/f_{\theta_0}(X_i)$ and

$$L_n^{(2)}(\theta_0) \; = \; \sum_{i=1}^n \left[\frac{f_{\theta_0}^{(2)}(X_i)}{f_{\theta_0}(X_i)} - \frac{f_{\theta_0}^{(1)}(X_i)^2}{f_{\theta_0}(X_i)^2}\right] \; + \; \left(\sum_{j=1}^n \frac{f_{\theta_0}^{(1)}(X_i)}{f_{\theta_0}(X_i)}\right)^2.$$

An estimator based on this idea will converge at rate $R^{-1/2}$, unlike the much slower FD estimator. However, given the presence of the squared sum in the above expression for $L_n^{(2)}(\theta_0)$, this suggests that $L_n^{(2)}(\theta_0)$ generally has higher variance than does $L_n^{(1)}(\theta_0)$ for large n. Thus, as in the setting of FD approximations to higher-order derivatives, higher-order derivative estimators tend to be noisier, making computation of higher-order derivatives challenging regardless of the method applied. $\qquad\square$

One should note, however, that nonpathological examples exist in which the interchange required for LR derivative estimation fails. We return to some of these in Section 5d.

Remark 3.7 The setup can be generalized to

$$z(\theta) \; = \; \int_\Omega Z(\theta, \omega)\, \mathbb{P}_\theta(d\omega),$$

where $Z(\cdot, \omega)$ is differentiable. Differentiating under the integral sign, this suggests that

$$z'(\theta) \; = \; \widetilde{\mathbb{E}}\big[Z'(\theta)L(\theta) + Z(\theta)L'(\theta)\big].$$

However, the interchange of differentiation and expectation in this setting faces the same difficulties as for IPA. $\qquad\square$

Exercises

3.1 (A-) For the compound Poisson sum C in the insurance risk setting, assume that $\lambda = 1{,}000$ and that the V_i are lognormal with $\mu = 0$, $\sigma^2 = 1$. Let $z(\lambda) \stackrel{\text{def}}{=} \mathbb{P}_\lambda(C > 5\mathbb{E}C)$. Give an estimate of $z'(\lambda)$ and an associated confidence interval by means of the LR method.

3.2 (TP) Assume $Z = \varphi(X, Y)$ where X, Y are independent with densities $f_\theta(x)$, $g(y)$. A variant of the LR method called *weak derivatives* (e.g., Pflug [294]) then consists in writing $f'_\theta(x)$ as $c(\theta)h_\theta(x) - d(\theta)k_\theta(x)$ where h_θ, k_θ are probability densities.

  (i) Show that $c(\theta) = d(\theta)$, and that $c(\theta)\big[\varphi(X_h, Y) - \varphi(X_k, Y)\big]$ is an unbiased estimator of $z'(\theta)$ when X_h, X_k have densities h_θ, k_θ.
  (ii) If $X = N$ is Poisson(λ), show that one can take $c(\lambda) = 1$, $X_h = N + 1$, and $X_k = N$.
  (iii) What happens if $X \sim \mathcal{N}(\theta, 1)$?

3.3 (A-) Redo Exercise 3.1 by means of weak derivatives (cf. Exercise 3.2(ii)), and give a comparison of the efficiencies.

4   The Likelihood Ratio Method: Stochastic Processes

4a   Recursions, Stopping Times, and Regeneration

We consider first the discrete-time case and let $\{X_n\}_{n \in \mathbb{N}}$ be a stochastic process driven by an input sequence $\{Y_n\}$. More precisely, we let $\mathscr{F}_n \stackrel{\text{def}}{=} \sigma(X_0, Y_1, \dots, Y_n)$ and assume that $\{X_n\}$ is adapted to $\{\mathscr{F}_n\}$. A simple example of this setup is a recursion $X_{n+1} = \varphi(X_n, Y_n)$ with i.i.d. innovations Y_n, where $\{X_n\}$ becomes Markov; cf. II.4.1.

Let $\widetilde{\mathbb{P}}$ be some importance distribution and $L_n(\theta)$ a function such that $(\mathrm{d}\mathbb{P}_\theta/\mathrm{d}\widetilde{\mathbb{P}})\big|_{\mathscr{F}_n} \stackrel{\text{def}}{=} L_n(\theta)$ defines a probability measure on $\mathscr{F}_n$. Assume further that $\widetilde{\mathbb{E}}\big[L_m(\theta); A\big]$ does not depend on $m \geq n$ when $A \in \mathscr{F}_n$. Then the $\mathbb{P}_\theta\big|_{\mathscr{F}_n}$ can be extended to a probability measure $\widetilde{\mathbb{P}}$ on $\mathscr{F} \stackrel{\text{def}}{=} \sigma(\mathscr{F}_0, \mathscr{F}_1, \dots)$. It is well known that a necessary and sufficient condition on the $L_n(\theta)$ for this construction to work out is that $\{L_n(\theta)\}_{n \in \mathbb{N}}$ be a nonnegative mean one martingale; see [16, Section XIII.3].

Let $C_n \stackrel{\text{def}}{=} C_n(\theta) \stackrel{\text{def}}{=} (\mathrm{d}/\mathrm{d}\theta) \log L_n(\theta)$. Under weak regularity conditions on the $L_n(\theta)$ (cf. Proposition 3.5), we can conclude that

$$\frac{\mathrm{d}}{\mathrm{d}\theta}\mathbb{E}_\theta g(X_0, Y_0, \dots, Y_n) = \mathbb{E}_\theta\big[g(X_0, Y_0, \dots, Y_n)C_n(\theta)\big] \qquad (4.1)$$

for a large class of functions g. Thinking of g as a function of $X_0, Y_1, \dots, Y_{n+1}$ also (that is, Y_{n+1} is a dummy variable), we get

$$\mathbb{E}_\theta\big[g(X_0, Y_0, \dots, Y_n)C_n(\theta)\big] = \mathbb{E}_\theta\big[g(X_0, Y_0, \dots, Y_n)C_{n+1}(\theta)\big].$$

The truth of this for a sufficiently large class of g's implies $\mathbb{E}_\theta[C_{n+1}(\theta)|\mathscr{F}_n]$ $C_n(\theta)$, and hence the following:

Proposition 4.1 *For any θ, $\{C_n(\theta)\}_{n\in\mathbb{N}}$ is a mean-zero $\mathbb{P}_\theta$-martingale.*

This result can be seen as an extension (allowing for dependence) of the fact that the score in the i.i.d. case is a mean-zero random walk, cf. Proposition 3.2 and (3.3). For notational convenience, we often omit θ in $C_n(\theta)$ in the following.

The form in which we will use (4.1) in the following is the special case

$$\frac{\mathrm{d}}{\mathrm{d}\theta}\mathbb{E}_\theta h(X_0, X_1, \ldots, X_n) = \mathbb{E}_\theta[h(X_0, X_1, \ldots, X_n)C_n]. \quad (4.2)$$

This implies in particular that $h(X_0, \ldots, X_n)C_n$ is an unbiased estimator of $z'(\theta)$, where $z(\theta) \overset{\text{def}}{=} \mathbb{E}_\theta h(X_0, \ldots, X_n)$, as follows already directly from Section 3.

If instead $z(\theta) = \mathbb{E}_\theta h(X_0, \ldots, X_\tau)$ for some stopping time $\tau < \infty$, arguments close to those used in V.1.9 show that we can just replace n by τ to get the unbiased estimator $h(X_0, \ldots, X_\tau)C_\tau$ (here we assumed implicitly that τ has no pathwise dependence on θ).

A case of particular interest arises if $\{X_n\}$ is regenerative and τ the regeneration time. In regenerative simulation, one then needs to consider the expectation $z(\theta, f)$ of

$$z(\theta, f) \overset{\text{def}}{=} \mathbb{E}_\theta Z(f), \quad \text{where } Z(f) \overset{\text{def}}{=} \sum_{k=0}^{\tau-1} f(X_k)$$

(in particular, $\tau = Z(1)$). Specializing the above discussion then leads to the unbiased estimator $Z(f)C_\tau$ of $z'(\theta, f)$.

A modification most often providing variance reduction is available as follows. Write

$$Z(f)C_\tau = \sum_{n=0}^{\infty} f(X_n)C_\tau \mathbb{1}\{\tau > n\}. \quad (4.3)$$

The expectation is left unchanged if the nth term of (4.3) is replaced by its conditional expectation given $\mathscr{F}_n$. Since $\{\tau > n\} \in \mathscr{F}_n$ and $\{C_n\}$ is a mean zero martingale, this gives the new unbiased estimator

$$\sum_{n=0}^{\infty} f(X_n)C_n\mathbb{1}\{\tau > n\} = \sum_{n=0}^{\tau-1} f(X_n)C_n. \quad (4.4)$$

Since (4.4) is obtained from $Z(f)C_\tau$ by conditioning, one expects the variance to be smaller. This is also typically the case and therefore (4.4) is the estimator that is commonly used in practice. However, (4.4) is not a conditional Monte Carlo estimator in the strict sense, since a different conditioning is used for each term. Thus, even though the variance is reduced term by term, the picture may be different for the total variance, and in fact, examples to this effect exist. We shall give a proof that indeed variance reduction is obtained for a basic case, the cycle length corresponding to $f \equiv 1$. The two estimators is question are then τC_τ and $\sum_0^{\tau-1} C_n$.

Proposition 4.2 *Assume that τ is a stopping time w.r.t. $\{\mathscr{F}_n\}$. Then*

$$\mathbb{V}ar\left(\sum_{n=0}^{\tau-1} C_n\right) \leq \mathbb{V}ar(\tau C_\tau).$$

Proof. Writing $\tau C_\tau = \sum_{n=0}^{\tau-1} C_\tau$, we get

$$\mathbb{E}(\tau C_\tau)^2 - \mathbb{E}\left(\sum_{n=0}^{\tau-1} C_n\right)^2$$

$$= \mathbb{E}\sum_{n=0}^{\tau-1}(C_\tau^2 - C_n^2) + 2\mathbb{E}\sum_{n=0}^{\tau-1}\sum_{k=n+1}^{\tau-1}(C_\tau^2 - C_n C_k). \qquad (4.5)$$

Now $\{C_n\}$ is a martingale and $\{C_n^2\}$ a supermartingale, so

$$\mathbb{1}_{\{\tau>k\}}C_k = \mathbb{1}_{\{\tau>k\}}\mathbb{E}[C_\tau \,|\, \mathscr{F}_k], \quad \mathbb{1}_{\{\tau>n\}}C_n^2 \leq \mathbb{1}_{\{\tau>n\}}\mathbb{E}[C_\tau^2 \,|\, \mathscr{F}_n]. \quad (4.6)$$

Hence conditioning on $\mathscr{F}_k$ yields

$$\mathbb{E}\left[\mathbb{1}_{\{\tau>n\}}\sum_{k=n+1}^{\tau-1} C_n C_\tau\right] = \mathbb{E}\left[\mathbb{1}_{\{\tau>n\}}\sum_{k=n+1}^{\tau-1} C_n C_k\right],$$

so using (4.6) repeatedly, we can bound (4.5) below by

$$\mathbb{E}\sum_{n=0}^{\tau-1} 0 + 2\mathbb{E}\sum_{n=0}^{\tau-1}\sum_{k=n+1}^{\tau-1}(C_\tau^2 - C_n C_\tau)$$

$$= 2\mathbb{E}\sum_{n=0}^{\tau-1}\sum_{k=n+1}^{\tau-1} C_\tau(C_\tau - C_n) \geq 2\mathbb{E}\sum_{n=0}^{\tau-1}\sum_{k=n+1}^{\tau-1}(C_n^2 - C_n^2) = 0.$$

$\square$

Example 4.3 The variance reduction provided by (4.4) is seldom dramatic. To exemplify this, we considered the M/D/1 queue with service times $V_k \equiv 1$ and Poisson(λ) arrivals. We took τ as the number of customers served in a busy period and $z(\lambda, \tau) \equiv \mathbb{E}_\lambda \tau$, $z(\lambda, W) \equiv \mathbb{E}_\lambda \sum_0^{\tau-1} W_n$, where W_n is the waiting time of customer n. The following table gives the empirical standard deviations on the derivative estimators corresponding to $R = 1,000,000$ replications for various values of $\rho = \lambda$. Here $*$ refers to (4.3) and $**$ to (4.4):

ρ	0.1	0.2	0.3	0.4	0.5	0.6	0.7	0.8	0.9
$\widehat{\sigma}_\tau^*$	14.9	11.9	13.2	17.2	24.6	39.8	81.2	209	1.224
$\widehat{\sigma}_\tau^{**}$	12.9	9.4	9.8	12.5	17.9	29.2	59.6	156	922
$\widehat{\sigma}_W^*$	6.1	9.0	15.0	27.3	54.4	121	346	1.466	16.039
$\widehat{\sigma}_W^{**}$	5.6	7.8	12.7	22.6	44.7	99.9	281	1.182	13.285

$\square$

Finally, consider the steady-state expectation $z(\theta) = z(\theta, f)/z(\theta, 1)$. The likelihood ratio estimator of

$$z'(\theta) = \frac{z'(\theta, f)z(\theta, 1) - z(\theta, f)z'(\theta, 1)}{z(\theta, 1)^2} \qquad (4.7)$$

is then

$$\widehat{z}'(\theta) = \frac{\widehat{z}'(\theta, f)\widehat{z}(\theta, 1) - \widehat{z}(\theta, f)\widehat{z}'(\theta, 1)}{\widehat{z}(\theta, 1)^2}, \qquad (4.8)$$

where $\widehat{z}'(\theta, f), \widehat{z}'(\theta, 1)$ are replications of the estimators discussed above.

4b Markov Chains and Related Processes

Suppose that $\{X_n\}_{n\in\mathbb{N}}$ is a time-homogeneous E-valued Markov chain that evolves according to the one-step transition kernel $\big(P(\theta, x, \mathrm{d}y)\big)_{x,y\in E}$ under $\mathbb{P}_\theta$. As in V.1.10, assume that

$$P(\theta, x, \mathrm{d}y) = p(\theta, x, y)\,\widetilde{P}(x, \mathrm{d}y)$$

and (for simplicity) that $\mathbb{P}_\theta(X_0 \in \cdot)$ is independent of θ. Then by V.1.10,

$$L_n(\theta) = \prod_{k=1}^{n} p(\theta, X_{k-1}, X_k),$$

and hence

$$
\begin{aligned}
C_n(\theta) &= \frac{1}{L_n(\theta)} \frac{\mathrm{d}}{\mathrm{d}\theta} L_n(\theta) \\
&= \frac{1}{L_n(\theta)} \sum_{k=1}^{n} p'(\theta, X_{k-1}, X_k) \prod_{\substack{\ell=1,\dots,n \\ \ell \neq k}} p(\theta, X_{\ell-1}, X_\ell) \\
&= \sum_{k=1}^{n} \frac{p'(\theta, X_{k-1}, X_k)}{p(\theta, X_{k-1}, X_k)} = \sum_{k=1}^{n} \frac{\mathrm{d}}{\mathrm{d}\theta} \log p(\theta, X_{k-1}, X_k). \quad (4.9)
\end{aligned}
$$

Once (4.9) is established, we are just back to the setup of Section 4a with $Y_1, Y_2, \ldots = X_0, X_1, X_2, \ldots$. In particular, all formulas for computing derivatives of expectations of quantities such as

$$z(\theta) = \mathbb{E}_\theta g(X_0, \ldots, X_\tau) \quad \text{and} \quad z(\theta) = \mathbb{E}_\theta\big[g(X_0, \ldots, X_\tau); \tau < \infty\big]$$

with τ a stopping time apply without changes.

As above, the question of how to choose $\widetilde{P}$ arises. Again, a particularly advantageous choice is to take $\widetilde{P}(x, dy) = P(\theta_0, x, dy)$ when estimating $z'(\theta_0)$. With this choice, both $z(\theta_0)$ and $z'(\theta_0)$ can be estimated using the same simulation run. Furthermore, $L_\tau(\theta_0) = 1$ with this choice, so

$$g(X_0, \ldots, X_\tau)C_\tau(\theta_0)\mathbb{1}\{\tau < \infty\} \tag{4.10}$$

is then an unbiased estimator of $z'(\theta_0)$. Because $\mathbb{V}ar\, L_n(\theta)$ grows exponentially in n whenever $\widetilde{P}(x, dy) \neq P(\theta, x, dy)$ (see XIV.5.5), this choice for $\widetilde{P}$ in (4.9) reduces variance substantially for expectations that depend on a long-time horizon τ.

Remark 4.4 Even with the above judicious choice of $\widetilde{P}$, the variance of (4.10) usually grows quite rapidly with τ. In particular, note that $C_n(\theta_0)$ is generally a square-integrable martingale under $\mathbb{P}_{\theta_0}$, so its variance typically grows linearly in n. One means of partially compensating for this increase in variance is to exploit the zero-mean property to use $C_n(\theta_0)$ as a control variate. This suggests estimating $z'(\theta_0)$ via $(W_1(\lambda) + \cdots + W_R(\lambda))/R$, where the $W_r(\lambda)$ are i.i.d. copies of

$$\left(g(X_0, \ldots, X_\tau)\mathbb{1}\{\tau < \infty\} - \lambda\right)C_\tau(\theta_0).$$

One important application of this idea arises when $\tau = n$ and

$$g(X_0, \ldots, X_n) = \frac{1}{n+1}\sum_{k=0}^{n} f(X_k)$$

as in steady-state simulation with $z(\theta) = \pi(\theta, f)$ (here $\pi(\theta, \cdot)$ is the stationary distribution under $\mathbb{P}_\theta$); cf. Chapter IV. In this setting, the optimal control coefficient λ for large n is approximately equal to $z(\theta_0)$. □

Remark 4.5 An alternative approach to (4.7) in steady-state gradient estimation uses the equilibrium equation

$$\pi(\theta, dy) = \int_E \pi(\theta, dx)p(\theta, x, y)\,\widetilde{P}(x, dy).$$

Assuming that differentiation and integration can be interchanged, we find that

$$\pi'(\theta, dy) = \int_E \left[\pi'(\theta, dx)p(\theta, x, y) + \pi(\theta, dx)p'(\theta, x, y)\right]\widetilde{P}(x, dy).$$

Hence the signed measure $\pi'(\theta_0)$ should satisfy

$$\pi'(\theta_0)\left(I - P(\theta_0)\right) = \pi(\theta_0)P'(\theta_0), \tag{4.11}$$

where $P'(\theta_0, x, dy)$ is the signed kernel defined by

$$P'(\theta_0, x, dy) \stackrel{\text{def}}{=} p'(\theta_0, x, y)\widetilde{P}(x, dy).$$

Note that (4.11) is the measure version of Poisson's equation. Under mild regularity conditions on the Markov chain (which are automatic, say, for a finite E), the solution is

$$\pi'(\theta_0) \;=\; \pi(\theta_0)P'(\theta_0)\Big[I + \sum_{j=1}^{\infty}\big(P^j(\theta_0) - \Pi(\theta_0)\big)\Big],$$

where $P^j(\theta_0, x, \mathrm{d}y) \overset{\text{def}}{=} \mathbb{P}_{\theta_0}\big(X_j \in \mathrm{d}y \mid X_0 = x\big)$, $\Pi(\theta_0, x, \mathrm{d}y) \overset{\text{def}}{=} \pi(\theta_0, \mathrm{d}y)$. Setting $\widetilde{P}(x, \mathrm{d}y) \overset{\text{def}}{=} P(\theta_0, x, \mathrm{d}y)$, we conclude that

$$\pi'(\theta_0, f) \;=\; \sum_{j=1}^{\infty} \mathbb{E}^*_{\theta_0}\Big[\frac{\mathrm{d}}{\mathrm{d}\theta}\log p(\theta_0, X_0, X_1) f_c(X_j)\Big], \tag{4.12}$$

where $\mathbb{E}^*$ denotes the stationary expectation and we write $f_c \overset{\text{def}}{=} f - \int f \mathrm{d}\pi$. Note that (4.12) is a "cross-spectral density" (between the stationary sequences $\{(\mathrm{d}/\mathrm{d}\theta)\log p(\theta_0, X_0, X_1)\}_{j=1,2,\ldots}$ and $\{f_c(X_j)\}_{j=1,2,\ldots}$) that can be estimated using methods similar to those in IV.3. $\quad\square$

We conclude this section by briefly noting the likelihood ratio gradient estimator in some other stochastic process settings.

Example 4.6 (CONTINUOUS-TIME MARKOV PROCESSES) Let $\{X(t)\}_{t\geq 0}$ be a continuous-time Markov process on a discrete state space E and evolving according to the intensity matrix $\boldsymbol{A}(\theta) = \big(A(\theta, x, y)\big)_{x,y\in E}$ under $\mathbb{P}_\theta$. As in V.1.11, assume that

$$A(\theta, x, y) \;=\; a(\theta, x, y)\widetilde{A}(x, y).$$

If the distribution of X_0 is independent of θ, then

$$z(\theta) \;=\; \mathbb{E}_\theta\big[g\big(X(s)\big)_{0\leq s\leq \tau}\mathbb{1}\{\tau < \infty\}\big] \;=\; \widetilde{\mathbb{E}}\big[g\big(X(s)\big)_{0\leq s\leq \tau}L_\tau(\theta); \tau < \infty\big],$$

where

$$L_t(\theta) \;=\; \prod_{i=1}^{J(t)} a\big(\theta, X(T_{i-1}), X(T_i)\big)$$
$$\times \exp\Big\{-\int_0^t \widetilde{A}\big(X(s), X(s)\big)\big[1 - a\big(\theta, X(s), X(s)\big)\big]\,\mathrm{d}s\Big\}$$

(here T_i is the ith jump epoch and $J(t)$ the number of jumps before t). Consequently, if we choose the sampling distribution such that $\widetilde{\mathbb{P}} = \mathbb{P}_{\theta_0}$, then

$$z'(\theta_0)] \;=\; \mathbb{E}_{\theta_0}\big[g\big(X(s)\big)_{0\leq s\leq \tau}C_\tau(\theta_0); \tau < \infty\big],$$

where

$$C_t(\theta_0) \overset{\text{def}}{=} \sum_{i=1}^{J(t)} \frac{\text{d}}{\text{d}\theta} \log a\big(\theta, X(T_{i-1}), X(T_i)\big)$$

$$+ \int_0^t \tilde{A}\big(X(s), X(s)\big) \frac{\text{d}}{\text{d}\theta} \, a\big(\theta, X(s), X(s)\big)\big] \, \text{d}s$$

evaluated at $\theta = \theta_0$. $\qquad\qquad\qquad\qquad\qquad\qquad\qquad\qquad\qquad\qquad\qquad$ □

Example 4.7 (GENERALIZED SEMI-MARKOV PROCESSES) Here we assume that $\{S(t)\}_{t>0}$ is a GSMP (cf. II.6) in which the transition probabilities $p(s'; s, e)$ and clock-setting distributions $F(\cdot; s', e', s, e)$ depend on a parameter θ. As in V.1.12, we assume that the support of $p(\theta, \cdot; s, e)$ is independent of θ and

$$F\big(\theta, \text{d}t; s', e', s, e\big) \;=\; f\big(\theta, \text{d}t; s', e', s, e\big) F\big(\theta_0, \text{d}t; s', e', s, e\big).$$

Letting $z(\theta) \overset{\text{def}}{=} \mathbb{E}_\theta g\big(S(u)\big)_{0 \le u \le t}$, we get

$$z'(\theta_0) \;=\; \mathbb{E}_\theta \left[g\big(S(u)\big)_{0 \le u \le t} L'_t(\theta_0) \right],$$

where

$$L'_t(\theta_0) \;=\; \sum_{i=1}^{J(t)} \frac{\text{d}}{\text{d}\theta} \log p\big(\theta_0, S(T_i), S(T_i-), e^*(T_i-)\big)$$

$$+ \sum_{i=1}^{J(t)} \sum_{e' \in N_i} \frac{\text{d}}{\text{d}\theta} \log f\big(\theta_0, C_{e'}(T_i); S(T_i), e', S(T_i-), e^*(T_i-)\big).$$

with $N_i \overset{\text{def}}{=} N\big(S(T_i); S(T_i-), e^*(T_i-)\big).$ $\qquad\qquad\qquad\qquad\qquad$ □

Exercises

4.1 (TP) Compute the score function for some parametric distributions selected according to your own choice.

4.2 (TP) In Exercise 2.3, give the formula for computing the delta at time t by the likelihood ratio method.

4.3 (A) In the PERT net in I.5.6, assume that the task durations Y_i are exponential with rates λ_i, where

$$\lambda_A = 1/3, \quad \lambda_B = \lambda_C = \lambda_E = 1, \quad \lambda_D = 1/4, \quad \lambda_F = 1/2$$

(the unit is days). The project has a deadline of $\ell = 21$ days. Give estimates of the sensitivities of $\mathbb{P}(L > \ell)$ (violation of the deadline) w.r.t. the λ_i.

5   Examples and Special Methods

5a   Numerical Examples

When comparing IPA and the LR method, it appears that IPA is usually more efficient when both methods work, but that the LR method has a broader scope. We do not, however, know of either theoretical results supporting this statement or broader numerical studies (but only some scattered examples).

To illustrate this point, we shall report on a few numerical experiments of our own.

Example 5.1 Consider a simple PERT net problem in which

$$z(\theta) = \mathbb{E}\max(\theta X_1 + X_2, (1-\theta)X_3)$$

with $0 < \theta < 1$ (think of a unit resource that can be shared between edges $1, 3$ at the user's convenience). We took X_1, X_2, X_3 as Erlang(2) with means $1, 2, 3$, $\theta = 0.1, 0.2, \ldots, 0.9$, and $R = 100{,}000$ replications. The following table summarizes the experiment; "std.rat." is the ratio between the estimated standard deviations on $\widehat{z}'(\theta)$ using the LR method, respectively IPA.

θ	0.1	0.2	0.3	0.4	0.5	0.6	0.7	0.8	0.9
$\widehat{z}(\theta)$	6.6	6.3	6.0	5.7	5.6	5.5	5.5	5.6	5.8
$\widehat{z}'(\theta)$	−3.9	−3.4	−2.8	−2.0	−1.2	−0.3	0.7	1.5	1.9
std.rat.	18.4	9.1	6.4	5.4	5.1	5.7	7.5	14.1	49.7

The conclusion is that indeed IPA is substantially more efficient, though the degree of variance reduction depends strongly on θ. □

Example 5.2 Let z be the payoff of a European call option in the GBM Black–Scholes model with maturity $T = 1$ year, yearly volatility $\sigma = 0.25$, short rate $r = 4\%$, and strike price $K = 1$. The sensitivity we looked at is the delta, that is, the sensitivity $z'(\theta)$ w.r.t. the initial price θ of the underlying asset. This means that $Z(\theta) = \mathrm{e}^{-rT}[\theta\mathrm{e}^Y - 1]^+$, where $Y \sim \mathcal{N}(\mu, \sigma^2)$ with $\mu = r - \sigma^2/2$. The IPA estimator of $z'(\theta)$ is $\mathrm{e}^{-rT}\mathrm{e}^Y \mathbb{1}\{\theta\mathrm{e}^Y > 1\}$. For the LR method, we use the chain rule

$$\frac{\mathrm{d}}{\mathrm{d}\theta}\mathbb{E}_{\mu+\log\theta}[\mathrm{e}^Y - 1]^+ = \frac{1}{\theta}\frac{\mathrm{d}}{\mathrm{d}\mu}\mathbb{E}_\mu[\mathrm{e}^Y - 1]^+\Big|_{\mu=\mu+\log\theta},$$

which gives the likelihood ratio estimator as

$$\frac{1}{2\sigma^2\theta}\mathrm{e}^{-rT}[\mathrm{e}^Y - 1]^+(Y - \mu - \log\theta)$$

(the score in a $\mathcal{N}(\mu, \sigma^2)$ r.v. V is $(V - \mu)/2\sigma^2$), where Y is now generated as $\mathcal{N}(\mu + \log\theta, \sigma^2)$.

We took $\theta = 0.6, 0.7, \ldots, 1.4$ and $R = 100{,}000$ replications. The following table summarizes the experiment using the same notation as in Example 5.1:

θ	0.6	0.7	0.8	0.9	1.0	1.1	1.2	1.3	1.4
$\widehat{z}(\theta)$	0.002	0.01	0.03	0.07	0.12	0.19	0.27	0.35	0.45
$\widehat{z}'(\theta)$	0.04	0.13	0.27	0.45	0.61	0.75	0.85	0.91	0.95
std.rat.	1.58	1.67	1.83	2.07	2.45	3.00	3.77	4.76	5.88

Again IPA is more efficient, though not so marked as in Example 5.1.   □

Note that in both examples both IPA and the LR method are so straight-forward to apply that the differing computational effort is hardly an issue in the comparison.

5b Nonsmoothness: Generalities

In many settings, both IPA and LR derivative estimation can be easily applied, in which case one has the opportunity to choose the more efficient estimator (or create a new one by taking a mixture of the two estimators). We now proceeed to describe ideas that can be useful in contexts in which neither method applies directly.

In particular, consider the function $z(\theta) = \mathbb{E}Z(\theta)$, where $Z(\theta)$ is non-smooth (or even discontinuous). Such a problem appears unlikely to yield an unbiased derivative estimator. Suppose, for the sake of concreteness, that as in Examples 2.1 and 3.1, $Z = g(X_1, \theta X_2)$, where g is nonsmooth (e.g., $Z(\theta) = \mathbb{1}\{\theta X_2 \leq X_1\}$ in one dimension) and $X_1 \in \mathbb{R}^n$, $X_2 \in \mathbb{R}^m$. If X_1, X_2 possess a joint density $f(x_1, x_2)$, then

$$z(\theta) = \int_{\mathbb{R}^{n+m}} g(x_1, \theta x_2) f(x_1, x_2) \, dx_1 \, dx_2 . \qquad (5.1)$$

The nonsmoothness of g is clearly likely to cause difficulties in the interchange of integral and derivative, so it is unlikely that $z'(\theta_0) = \mathbb{E}Z'(\theta_0)$.

One idea that is sometimes useful in dealing with this problem is to recognize that integration tends to smooth discontinuties. Specifically, observe that

$$z(\theta) = \int_{\mathbb{R}^n} \int_{\mathbb{R}^m} g(x_1, \theta x_2) f_2(x_2 \mid x_1) \, dx_2 \, f_1(x_1) \, dx_1 ,$$

where $f_2(\cdot \mid x_1)$ is the conditional density of X_2 given $X_1 = x_1$ and $f_1(\cdot)$ is the (marginal) density of X_1. Note that

$$h(\theta, x_1) \stackrel{\text{def}}{=} \int_{\mathbb{R}^m} g(x_1, \theta x_2) f_1(x_2 \mid x_1) \, dx_2 = \mathbb{E}\big[Z(\theta) \, \big| \, X_1 = x_1 \big] \quad (5.2)$$

may be smooth even when g is not. This presents the possibility of estimating $z'(\theta)$ by averaging i.i.d. replicates of $h'(\theta, X_1)$, where $h'(\theta, x_1)$ is the derivative of (5.2). What is being used here is the smoothing properties of conditional expectations to derive an appropriate estimator. This method relies on our ability to find a conditioning vector X_1 (or, more generally, a σ-field $\mathscr{G}$) for which $\mathbb{E}\big[Z(\theta)\,|\,X_1\big]$ (or, more generally, $\mathbb{E}\big[Z(\theta)\,|\,\mathscr{G}\big]$) is smooth in θ and computable.

The second idea that can be useful in dealing with (5.1) can potentially be exploited when f is smooth. The idea is to "push out" the parameter from the nonsmooth g and push it into the function f, by executing an appropriate change of variables. Specifically, if we put $v \overset{\text{def}}{=} \theta x_2$, we can rewrite (5.1) as

$$z(\theta) \;=\; \int_{\mathbb{R}^{n+m}} g(v,x_1) f(x_1, v/\theta) \theta^{-n}\,\mathrm{d}x_1\,\mathrm{d}v. \tag{5.3}$$

A natural approach to representing (5.3) as an expectation is to recognize that the random vector $(X_1, \theta X_2)$ has joint density $f(x_1, v/\theta)\theta^{-n}$, and hence $z(\theta) = \mathbb{E}_\theta g(X_1, V)$ where V has density $f(x_1, v/\theta)\theta^{-n}$ under $\mathbb{P}_\theta$. We can now apply the LR derivative estimator to the above representation of $z(\theta)$ (this was exactly the idea underlying Example 3.1).

We now present these ideas and extensions in more detail.

5c   Smoothed IPA

The idea of using

$$\frac{\mathrm{d}}{\mathrm{d}\theta}\mathbb{E}\big[Z(\theta)\,|\,\mathscr{G}\big]$$

as an estimator of $z'(\theta)$ is due to Gong [160] and Gong & Ho [161] (see also the survey of Glasserman [131]), and is sometimes referred to as "smoothed IPA." Obviously, the method has some ideas in common with conditional Monte Carlo; see in particular Example V.4.3.

Example 5.3 Consider, as in Example 2.2, $Z(\theta) = \mathbb{1}\{\theta X_2 > X_1\}$. Here, $\mathbb{E}\big[Z(\theta)\,|\,X_2\big] = F_1(\theta X_2)$, where F_1 is the c.d.f. of X_1. If F_1 has a positive density f_1, then the smoothed IPA estimator for $z'(\theta)$ is $f_1(\theta X_2)X_2$.   □

The second approach we present follows Glasserman [132] and is based on the concept of a *smoothing complement* $C(\theta)$. This is an r.v. designed to cancel the discontinuities (in θ) of $Z(\theta)$, which are most often the main problem in naive IPA. Formally, we assume that

(a) $Z(\theta)$ is differentiable in θ except at a set $\Theta^\#$ (typically discrete and possibly random),

(b) $C(\theta) = 0$ when $\theta \in \Theta^\#$,

(c) $\mathbb{E}\big[C(\theta)\,|\,\mathscr{G}\big] = 1$ for some σ-field $\mathscr{G}$ such that $Z(\theta)$ is $\mathscr{G}$-measurable.

Since

$$\begin{aligned}
\mathbb{E}\big[Z(\theta)C(\theta)\big] &= \mathbb{E}\,\mathbb{E}\big[Z(\theta)C(\theta)\,|\,\mathscr{G}\big] = \mathbb{E}\big(Z(\theta)\mathbb{E}\big[C(\theta)\,|\,\mathscr{G}\big]\big) \\
&= \mathbb{E}Z(\theta)\cdot 1 \;=\; z(\theta),
\end{aligned}$$

the r.v. $Z(\theta)C(\theta)$ is unbiased for $z(\theta)$, and one can hope that the properties of $C(\theta)$ will ensure that its sample path derivative

$$D_{\mathrm{sc}} \overset{\text{def}}{=} Z'(\theta)C(\theta) + Z(\theta)C'(\theta)$$

is unbiased for $z'(\theta)$. Again, we use an example to illustrate the approach:

Example 5.4 Consider the expectation $z(\theta)$ of $Z(\theta) = h(X(\theta))$ of a function h of a discrete r.v. $X(\theta)$ that is concentrated on $\mathbb{N}$ with point probabilities $p_n(\theta)$ and generated from a uniform r.v. U by inversion, i.e.,

$$X(\theta) = n \quad \text{when} \quad P_n(\theta) \leq U < P_{n+1}(\theta),$$

where $P_0(\theta) \overset{\text{def}}{=} 0$, $P_n(\theta) \overset{\text{def}}{=} p_0(\theta) + \cdots + p_{n-1}(\theta)$. Here $Z'(\theta)$ exists (except at the $P_k(\theta)$) but equals 0, so that IPA is in trouble.

For the smoothing complement, let

$$g(u, \theta) = \begin{cases} u - P_n(\theta) & P_n(\theta) \leq u < (P_n(\theta) + P_{n+1}(\theta))/2, \\ P_{n+1}(\theta) - u & (P_n(\theta) + P_{n+1}(\theta))/2 \leq u < P_{n+1}(\theta), \end{cases}$$

be the distance from $0 \leq u \leq 1$ to the closest $P_k(\theta)$; see Figure 5.1.

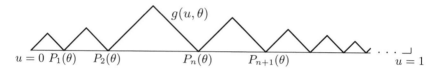

$u = 0 \quad P_1(\theta) \quad P_2(\theta) \qquad\qquad P_n(\theta) \qquad P_{n+1}(\theta) \qquad\qquad\qquad\qquad u = 1$

FIGURE 5.1

We then let

$$C(\theta) \overset{\text{def}}{=} \frac{g(U, \theta)}{\mathbb{E}\big[g(U, \theta) \,\big|\, X(0)\big]}.$$

Then requirements (a), (c) for a smoothing complement are clear, whereas for (b) we can take

$$\Theta^{\#} = \Big\{ P_0(\theta), P_1(\theta), \dots, (P_0(\theta) + P_1(\theta))/2, (P_1(\theta) + P_2(\theta))/2, \dots \Big\}.$$

Conditionally on $X(\theta) = n$, U is uniform on $[P_n(\theta), P_{n+1}(\theta))$, and inspection of the graph of $g(u, \theta)$ then shows that

$$\mathbb{E}\big[g(U, \theta) \,\big|\, X(\theta) = n\big] = \frac{P_{n+1}(\theta) - P_n(\theta)}{4} = \frac{p_n(\theta)}{4}.$$

It follows that on $\{X(\theta) = n\}$,

$$C'(\theta) = \frac{d}{d\theta}\left[\frac{4\big(U - P_n(\theta)\big)}{p_n(\theta)}\right] = -\frac{4P_n'(\theta)}{p_n(\theta)} - \frac{4p_n'(\theta)\big(U - P_n(\theta)\big)}{p_n^2(\theta)}$$

when $P_n(\theta) < U < \big(P_n(\theta) + P_{n+1}(\theta)\big)/2$, and that

$$C'(\theta) = \frac{d}{d\theta}\left[\frac{4\big(P_{n+1}(\theta) - U\big)}{p_n(\theta)}\right] = \frac{4P'_{n+1}(\theta)}{p_n(\theta)} - \frac{4p'_n(\theta)\big(P_{n+1}(\theta) - U\big)}{p_n^2(\theta)}$$

when $\big(P_n(\theta) + P_{n+1}(\theta)\big)/2 < U < P_{n+1}(\theta)$. The estimator of $z'(\theta)$ can now be computed as $D_{\mathrm{sc}} \stackrel{\text{def}}{=} Z(\theta)C'(\theta) = h\big(X(\theta)\big)C'(\theta)$ (note that $Z'(\theta) = 0$ a.s.).

For a rigorous proof that $Z(\theta)C'(\theta)$ is unbiased for $z'(\theta)$, one now either has to verify the Lipschitz condition of Proposition 2.3 or give a direct proof. We choose the latter way. Now

$$\mathbb{E}D_{\mathrm{sc}} = \sum_{n=0}^{\infty} \mathbb{E}\big[Z(\theta)C'(\theta);\, X(\theta) = n\big] = \sum_{n=0}^{\infty} h(n)\mathbb{E}\big[C'(\theta);\, X(\theta) = n\big].$$

Here the event $B \stackrel{\text{def}}{=} \{X(\theta) = n\}$ occurs w.p. $p_n(\theta)$, and each of the two subevents B_1, B_2 separating the two expressions for $C'(\theta)$ occur w.p. $p_n(\theta)/2$. Hence

$$\mathbb{E}\big[C'(\theta);\, X(\theta) = n\big]$$

$$= \frac{p_n(\theta)}{2}\left[-\frac{4P'_n(\theta)}{p_n(\theta)} + \frac{4P'_{n+1}(\theta)}{p_n(\theta)}\right.$$

$$\left. - \frac{4p'_n(\theta)\big(\mathbb{E}[U \mid B_1] - P_n(\theta)\big)}{p_n^2(\theta)} - \frac{4p'_n(\theta)\big(P_{n+1}(\theta) - \mathbb{E}[U \mid B_2]\big)}{p_n^2(\theta)}\right]$$

$$= 2\big(P'_{n+1}(\theta) - P'_n(\theta)\big)$$

$$- \frac{2p'_n(\theta)\big(P'_{n+1}(\theta) - P'_n(\theta) + \mathbb{E}[U \mid B_1] - \mathbb{E}[U \mid B_2]\big)}{p_n(\theta)}$$

$$= 2p'_n(\theta) - \frac{2p'_n(\theta)\big(p_n(\theta) - p_n(\theta)/2\big)}{p_n(\theta)} = p'_n(\theta),$$

so that

$$\mathbb{E}D_{\mathrm{sc}} = \sum_{n=0}^{\infty} h(n)p'_n(\theta) = z'(\theta).$$

Note that this problem can also be easily attacked via the LR ideas of Section 3. It leads to

$$z'(\theta) = \mathbb{E}\left[h\big(X(\theta)\big)\frac{p'\big(X(\theta)\big)}{p\big(X(\theta)\big)}\right],$$

where, of course, $[\cdot]$ is a much simpler estimator than the smoothing complement estimator. $\qquad\square$

Remark 5.5 A realistic instance of estimating $z'(\theta)$ in the setting $Z(\theta) = \mathbb{E}h\big(X(\theta)\big)$ of Example 5.4 occurs for a compound sum $A(\theta) = \sum_0^{X(\theta)} V_i$

in insurance risk, where $X(\theta)$ is Poisson(θ) and one may be interested in $z(\theta) \stackrel{\text{def}}{=} \mathbb{P}(A > x)$ (cf. I.5.14) and the sensitivity $z'(\theta)$ w.r.t. the Poisson parameter θ. Namely, if F is the common distribution of the V_i, then $h(n) = \overline{F}^{*n}(x)$.

Typically, $h(n)$ has to be estimated, which makes only a minor difference: just replace $Z(\theta) = h(X(\theta))$ in Example 5.4 by $Z(\theta) = \mathbb{1}\{V_1 + \cdots + V_{X(\theta)} > x\}$ and keep $C(\theta)$ unchanged.

It is useful to compare the complexity of the smoothed IPA procedure with the simplicity of likelihood ratio method: since the score function of a Poisson(θ) r.v. $X(\theta)$ is $X(\theta)/\theta - 1$, the likelihood ratio estimator is just

$$\left[X(\theta)/\theta - 1\right]\mathbb{1}\{V_1 + \cdots + V_{X(\theta)} > x\} \;! \qquad\qquad \square$$

The ideas of Example 5.4 can be carried over to Markov chains; cf. again [132].

5d   The Push-Out Method. Continuity Corrections

In some situations, a parameter θ important for the behavior of a stochastic system appears in a different way than through a density $f_\theta(x)$. Examples are:

(a) A PERT net in which the maximal path L is a maximum of r.v.'s of the form $Y_{i_1} + \cdots + Y_{i_N}$, where $Y_1, \ldots, Y_M$ is a given set of r.v.'s with different distributions. Here one of the Y_i may be concentrated at θ;

(b) a GI/D/1 queue with deterministic service times θ, or a D/G/1 queue with deterministic interarrival times θ;

(c) an insurance risk model in which the claims $V_1, V_2, \ldots$ are excess-of-loss reinsured at level θ, so that the claims facing the insurance company in reality are $V_1 \wedge \theta, V_2 \wedge \theta, \ldots$.

When one is trying to evaluate the sensitivity $z'(\theta)$ w.r.t. such a θ, the usual machinery of the likelihood ratio method does not apply. In many cases, IPA may work, but in others not without problems; for example, in (b) above, one may need the number of customers in a busy period for regenerative simulation, and the discreteness is then a problem for IPA.

The *push-out method*, initiated by Rubinstein [315], is an extension of the likelihood ratio method to some such situations. The idea is to rewrite the model such that θ appears in the usual way as the parameter of a density. We have in fact already seen one such (obvious) example, the option-pricing example of Section 5a, where we replaced an r.v. of the form θe^Y with $Y \sim \mathcal{N}(\mu, \sigma^2)$ by $e^{\widetilde{Y}}$ with $\widetilde{Y} \sim \mathcal{N}(\mu + \log \theta, \sigma^2)$.

In the D/G/1 example with service times $V_0, V_1, \ldots$ with density $b(x)$, $x > 0$, a similar idea would be to work with a random walk $S_n = X_1 + \cdots + X_n$ with increments distributed as $V_0 - \theta$, that is, with density $b_\theta(x) \stackrel{\text{def}}{=}$

$b(x + \theta)\mathbb{1}\{x > -\theta\}$. Say we are interested in the expected number $z(\theta) = \mathbb{E}\tau$ of customers in a busy period, where $\tau \stackrel{\text{def}}{=} \inf\{n > 0 : S_n < 0\}$. We then face the problem that the support of $b_\theta(x)$ depends on θ, which may violate the regularity conditions of Proposition 3.5. More precisely, if $b(0) > 0$, then for all $x < 0$ differentiability of $b_\theta(x)$ in θ fails at $\theta = -x$.

The implication is not that likelihood ratio derivative estimation is hopeless, but rather that we need a correction:[3]

Proposition 5.6 *Let* $\big(f_\theta(x)\big)_{\theta \in \Theta}$ *be a family of densities on* $\mathbb{R}$ *such that* $f_\theta(x)$ *is continuous and differentiable in* x *except for a discontinuity at* $x = a(\theta)$. *Write* $\Delta(\theta) \stackrel{\text{def}}{=} f_\theta\big(a(\theta)-\big) - f_\theta\big(a(\theta)+\big)$, *let* $X, X_1, X_2, \ldots$ *be i.i.d. with density* $f_\theta(x)$, *and let the score be defined the usual way as* $S_X \stackrel{\text{def}}{=} (\mathrm{d}/\mathrm{d}\theta)\log f_\theta(X)$. *Then:*

(i) *For a function* $g(x)$ *of one variable and* $Z \stackrel{\text{def}}{=} g(X)$, $z(\theta) \stackrel{\text{def}}{=} \mathbb{E}_\theta Z$, *one has*

$$z'(\theta) = a'(\theta)\Delta(\theta)g\big(a(\theta)\big) + \mathbb{E}_\theta\big[ZS_X\big].$$

(ii) *For a function* $g(x_1, \ldots, x_n)$ *of* n *variables and* $Z \stackrel{\text{def}}{=} g(X_1, \ldots, X_n)$, $z(\theta) \stackrel{\text{def}}{=} \mathbb{E}_\theta Z$, *one has*

$$z'(\theta) = a'(\theta)\Delta(\theta)\sum_{i=1}^n \mathbb{E}Z_i + \mathbb{E}_\theta\big[Z(S_{X_1} + \cdots + S_{X_n})\big],$$

where Z_i *is the r.v. obtained by replacing* X_i *in* $Z = g(X_1, \ldots, X_n)$ *by* $a(\theta)$.

Proof. In (i),

$$z(\theta) = \int_{-\infty}^{a(\theta)} g(x)f_\theta(x)\,\mathrm{d}x + \int_{a(\theta)}^{\infty} g(x)f_\theta(x)\,\mathrm{d}x.$$

Applying the integral formula (A8.1), we get

$$\begin{aligned}
z'(\theta) &= g\big(a(\theta)\big)f_\theta\big(a(\theta)-\big)a'(\theta) + \int_{-\infty}^{a(\theta)} g(x)f'_\theta(x)\,\mathrm{d}x \\
&\quad -g\big(a(\theta)\big)f_\theta\big(a(\theta)+\big)a'(\theta) + \int_{a(\theta)}^{\infty} g(x)f'_\theta(x)\,\mathrm{d}x \\
&= g\big(a(\theta)\big)\Delta(\theta)a'(\theta) + \int_{-\infty}^{\infty} g(x)f'_\theta(x)\,\mathrm{d}x \\
&= g\big(a(\theta)\big)\Delta(\theta)a'(\theta) + \mathbb{E}_\theta\big[ZS_X\big].
\end{aligned}$$

We omit the proof of (ii). □

---

[3]Not all needed regularity conditions are included in the proposition.

More general and rigorous versions of results of this type can be found in Signahl [343].

Example 5.7 Let $X_1, \ldots, X_n$ be i.i.d. with density $f(y)$ concentrated on $(0, \infty)$ with $f(0) > 0$ and consider the problem of estimating $f^{*n}(\theta)$, i.e., the sensitivity of $z(\theta) \overset{\text{def}}{=} \mathbb{P}(S_n \leq \theta)$ w.r.t. θ, where $S_n \overset{\text{def}}{=} X_1 + \cdots + X_n$. Let $X \overset{\text{def}}{=} X_1 - \theta$, $Y \overset{\text{def}}{=} X_2 + \cdots + X_n$. Thus θ has been pushed out to X, which has density $f(x + \theta)\mathbb{1}\{x > -\theta\}$, and we have $z(\theta) = \mathbb{E}h(X, Y)$ where $h(x, y) \overset{\text{def}}{=} \mathbb{1}\{x + y \leq 0\}$. We now use the following corollary of part (i) of Proposition 5.6:

$$\frac{\mathrm{d}}{\mathrm{d}\theta}\mathbb{E}_\theta h(X, Y) = a'(\theta)\Delta(\theta)\mathbb{E}h\big(a(\theta), Y\big) + \mathbb{E}_\theta\big[h(X, Y)S_X\big],$$

where Y is an independent r.v. with distribution not dependent on θ. This follows (at least at the formal level) by conditioning on $Y = y$, letting $g(x) = h(x, y)$ and integrating Y out. Here

$$S_X = \frac{f'(x + \theta)}{f(x + \theta)}, \quad a(\theta) = -\theta, \quad a'(\theta) = -1, \quad \Delta(\theta) = -f(0),$$

and so the pushed-out sensitivity estimator is

$$f(0)\mathbb{1}\{Y \leq \theta\} + \mathbb{1}\{X + Y \leq 0\}\frac{f'(X + \theta)}{f(X + \theta)}$$

$$= f(0)\mathbb{1}\{X_2 + \cdots + X_n \leq 0\} + \mathbb{1}\{S_n \leq \theta\}\frac{f'(X_1)}{f(X_1)}. \quad (5.4)$$

$\square$

Example 5.8 We will study the problem of estimating the sensitivity of $z(\theta) = \mathbb{E}_\theta\tau$, where τ is the number of customers in a busy period of the D/G/1 queue with deterministic interarrival times θ and service times $V_0, V_1, \ldots$ with density $b(x)$. Again, we push θ out by letting $X_1 = V_0 - \theta$, $X_2 = V_1 - \theta, \ldots$. Then $\tau = \inf\{n : Y_n < 0\}$, where $Y_n \overset{\text{def}}{=} X_1 + \cdots + X_n$.

We first derive an estimator of the sensitivity $z_n'(\theta)$ of $z_n(\theta) \overset{\text{def}}{=} n\mathbb{P}(\tau = n) = \mathbb{E}Z_n$, where $Z_n = nI_n$ with I_n the indicator function of the event

$$\{Y_n < 0, Y_1 \geq 0, \ldots, Y_{n-1} \geq 0\}.$$

In the following, write $Y_{i,n}$ for the random walk obtained from Y_n by replacing Y_i with $a(\theta) = -\theta$; that is, $Y_{i,n} = Y_n$ for $n < i$ and $Y_{i,n} = Y_n - X_i - \theta$ for $n \geq i$. Similarly, write $\tau_i \overset{\text{def}}{=} \inf\{n : Y_{i,n} < 0\}$ and let $I_{i,n}$ be the indicator function of the event

$$\{Y_{i,n} < 0, Y_{i,1} \geq 0, \ldots, Y_{i,n-1} \geq 0\} = \{\tau_i = n\}$$

and $Z_{i,n} \overset{\text{def}}{=} nI_{i,n}$. As in Example 5.7, $a'(\theta)\Delta(\theta) = b(0)$, and we can apply Proposition 5.6(ii) to get the pushed-out sensitivity estimator of $z'_n(\theta)$ as

$$D_n = b(0) \sum_{i=1}^{n} nI_{i,n} + nI_n(S_{X_1} + \cdots + S_{X_n}).$$

Here the last term equals $\tau I_n(S_{X_1} + \cdots + S_{X_\tau})$, and since $\sum_1^{\infty} nI_{i,n} = \tau_i$, $S_{X_i} = f'(V_i)/f(V_i)$, we get the pushed-out sensitivity estimator of $z'(\theta)$ as

$$D_{\text{po}} = \sum_{n=1}^{\infty} D_n = b(0) \sum_{i=1}^{\tau} \tau_i + \tau\left(\frac{f'(V_1)}{f(V_1)} + \cdots + \frac{f'(V_\tau)}{f(V_\tau)}\right).$$

Here we used that $X_i \geq -\theta$ implies $Y_{i,n} \leq Y_n$ for all n and hence $\tau_i \leq \tau$, $I_{i,n} = 0$ for $n > \tau$. $\square$

5e   Rare Events via Likelihood Ratios

We will work in the setting of the likelihood ratio method in Sections 3 and 4 so that the dependence on the parameter θ is in the probability measure $\mathbb{P}_\theta$. Assume that $Z = Z(\zeta) \geq 0$ is the indicator of a rare event $A(\zeta)$ depending on another parameter ζ such that $z(\zeta, \theta) \overset{\text{def}}{=} \mathbb{P}_\theta A(\zeta) \to 0$ as $\zeta \to \infty$. We shall here give a brief discussion of some the issues arising in derivative estimation in this setting and refer further to Asmussen & Rubinstein [28] and Nakayama [268].

One expects that when $z(\zeta, \theta)$ is small, then so is the derivative $z'(\zeta, \theta) \overset{\text{def}}{=} (\mathrm{d}/\mathrm{d}\theta)z(\zeta, \theta)$ w.r.t. θ. In fact, under some regularity conditions $|z'(\zeta, \theta)|$ is bounded by $z(\zeta, \theta)$ in the logarithmic sense discussed in VI.1:

Proposition 5.9 *Assume that $z'(\zeta, \theta)$ can be represented as $\mathbb{E}_\theta\big[Z(\zeta)S\big] = \mathbb{E}_\theta\big[S; A(\zeta)\big]$ for some r.v. S, say a score function. If $\mathbb{E}|S|^q < \infty$ for all $q < \infty$, then*

$$\liminf_{\zeta \to \infty} \frac{\log|z'(\zeta, \theta)|}{\log z(\zeta, \theta)} \geq 1.$$

Proof. Let $1/p + 1/q = 1$. Hölder's inequality yields

$$|z'(\zeta, \theta)| \leq \big\|\mathbb{1}_{A(\zeta)}\big\|_p \|S\|_q = z(\zeta, \theta)^{1/p}\|S\|_q.$$

The assertion follows by taking logarithms and letting first $\zeta \to \infty$ and next $p \to 1$, i.e., $q \to \infty$. $\square$

Remark 5.10 We are not aware of converses, that is, results giving conditions for $z(\zeta, \theta)$ to be (logarithmically) bounded by $|z'(\zeta, \theta)|$ and hence of the same order. In fact, this cannot be true without conditions: in the extreme, $Z(\zeta)$ may have distribution not dependent on θ, so that $z'(\zeta, \theta) = 0$.

Typically, $z(\zeta,\theta)$ and $z'(\zeta,\theta)$ will, however, have the same logarithmic asymptotics as $\zeta \to \infty$ with θ fixed. For worked-out examples, see, e.g., the ruin probability calculations in Asmussen [15, Section III.9].   □

When estimating sensitivities of rare events, we are therefore facing the same difficulties about controlling the relative error as in ordinary rare-event simulation. Whenever an efficient algorithm is available for the rare event itself, the obvious idea is to try adapting it to the sensitivity as well. We give one example and refer to [27] for further ones.

Example 5.11 Consider Siegmund's algorithm in the notation of VI.2, so that $z(\zeta,\theta)$ is the $\mathbb{P}_\theta$-probability that the maximum of a random walk $Y_n = X_1 + \cdots + X_n$ with negative drift exceeds ζ, i.e., $z(\zeta,\theta) = \mathbb{P}_\theta\big(\tau(\zeta) < \infty\big)$, where $\tau(\zeta)$ is the first passage time. Let $S_X \overset{\text{def}}{=} (\mathrm{d}/\mathrm{d}\theta) \log f_\theta(X)$ be the score in X. Then typically

$$z'(\zeta,\theta) \;=\; \mathbb{E}_\theta\big[S_{X_1} + \cdots + S_{X_{\tau(\zeta)}};\, \tau(\zeta) < \infty\big],\qquad (5.5)$$

as follows by the following formal calculation:

$$\frac{\mathrm{d}}{\mathrm{d}\theta}\mathbb{P}_\theta\big(\tau(\zeta) < \infty\big) \;=\; \frac{\mathrm{d}}{\mathrm{d}\theta}\sum_{n=1}^{\infty}\mathbb{P}_\theta\big(\tau(\zeta) = n\big) \;=\; \sum_{n=1}^{\infty}\frac{\mathrm{d}}{\mathrm{d}\theta}\mathbb{P}_\theta\big(\tau(\zeta) = n\big)$$

$$=\; \sum_{n=1}^{\infty}\mathbb{E}_\theta\big[S_{X_1} + \cdots + S_{X_n};\, \tau(\zeta) = n\big]$$

$$=\; \mathbb{E}_\theta\big[S_{X_1} + \cdots + S_{X_{\tau(\zeta)}};\, \tau(\zeta) < \infty\big].$$

What (5.5) and the analysis of VI.2 suggests is to evaluate the r.h.s. of (5.5) by the same exponential change of measure as in VI.2. That is, we solve $\mathbb{E}_\theta \mathrm{e}^{\gamma X} = 1$ for $\gamma = \gamma(\theta)$, simulate from the exponentially tilted distribution $\widetilde{F}(\mathrm{d}x) \overset{\text{def}}{=} F_\gamma(\mathrm{d}x) \overset{\text{def}}{=} \mathrm{e}^{\gamma x}F(\mathrm{d}x)$ rather than F, and use the estimator

$$D(\zeta,\theta) \overset{\text{def}}{=} \mathrm{e}^{-\gamma Y_{\tau(\zeta)}}\big[S_{X_1} + \cdots + S_{X_{\tau(\zeta)}}\big]$$

(recall that $\mathrm{e}^{-\gamma Y_{\tau(\zeta)}}$ is the likelihood ratio on $\mathscr{F}_{\tau(\zeta)}$ and that $\mathbb{P}_\gamma\big(\tau(\zeta) < \infty\big) = 1$).

We will verify that $D(\zeta,\theta)$ is in fact logarithmically efficient in the sense of VI.1. To this end, we will assume that $z'(\zeta,\theta)$ is of logarithmic order $\mathrm{e}^{-\gamma\zeta}$ (this is verified in examples in [15, Section III.9]; recall also that $z(\zeta,\theta) \sim C\mathrm{e}^{-\gamma\zeta}$ for some $C = C_\theta$). We will need also $\mathbb{E}_\theta X^2 \mathrm{e}^{\gamma X} < \infty$, which ensures that $\mathbb{E}_\gamma X^2 < \infty$ and hence ([170]) that

$$\mathbb{E}_\gamma\big(S_{X_1} + \cdots + S_{X_{\tau(\zeta)}}\big)^2 \;=\; \mathrm{O}\big(\mathbb{E}_\gamma\tau(\zeta)^2\big) \;=\; \mathrm{O}(\zeta^2).$$

From $Y_{\tau(\zeta)} \geq \zeta$ we then get $\mathbb{E}_\gamma D(\zeta,\theta)^2 = \mathrm{O}(\zeta^2\mathrm{e}^{-2\gamma\zeta})$, which suffices for logarithmic efficiency [the more precise estimates of [15, Section III.9] show in fact that at least in some cases, there is bounded relative error].   □

Exercises

5.1 (TP) In Example 5.1, give the formulas for the IPA estimator, respectively the LR estimator.

5.2 (TP) Show that by taking the conditional expectation of (5.4) w.r.t. X_1, we retrieve the conditional Monte Carlo estimator $f(\theta - X_2 - \cdots - X_n)$ of $f^{*n}(x)$ in V.4.3.

5.3 (TP) In the PERT net in I.5.6 and Exercise 2.3, assume that Y_E is deterministic and equal to θ rather than exponential. How can you estimate the sensitivity w.r.t. θ? Same question if instead $Y_C = \theta$.

Chapter VIII
Stochastic Optimization

1 Introduction

Many applied contexts require the solution of challenging optimization problems in which the objective to be (say) minimized involves some expected value. When the expectation can be computed in closed form (or, more generally, rapidly numerically evaluated), conventional numerical optimization algorithms are generally applicable and yield good solutions. On the other hand, when the expectations require evaluation via simulation, the need for simulation-based optimization becomes apparent.

Of course, all the difficulties that arise in conventional numerical optimization also manifest themselves in the setting of simulation-based optimization (as well as the new difficulties that arise due to the use of sampling-based methods). Chief among these is the fact that in the absence of convexity, it is difficult to guarantee that an iterative algorithm will converge to a global optimizer. Instead, the mathematical theory focuses on the less-ambitious goal of establishing convergence to a local optimizer. One pragmatic approach to computing a global optimizer in the nonconvex setting is to run the *iterative search algorithm* from different initial feasible points, with the goal of using the optimal among all the limit points generated by the different iterations. The same pragmatic approach is generally followed in the practical application of simulation-based optimization.

This chapter is focused on the case in which the decision variable θ is continuous. We do this largely because the theory of simulation-based optimization in the presence of discrete decision variables is, in our view, poorly understood at this time.

Some examples in which stochastic optimization may occur are given in Section 5. Some general references in the area are Kushner & Yin [224], Pflug [294], and Rubinstein & Shapiro [315].

2 Stochastic Approximation Algorithms

We consider here a rather general class of iterative algorithms that are widely used in practice for purposes of minimizing (via simulation) an objective function $z(\boldsymbol{\theta})$ over some feasible region $\boldsymbol{\Theta} \subseteq \mathbb{R}^d$.[1] We assume that $z(\cdot)$ is smooth and that the constraints are not binding at the minimizer $\boldsymbol{\theta}^*$, so that $\nabla z(\boldsymbol{\theta}^*) = \mathbf{0}$ (here ∇ as usual denotes the gradient). Stochastic approximation algorithms are iterative simulation-based algorithms that are intended to converge to a zero of $\nabla z(\boldsymbol{\theta})$.

Given that the algorithm at iteration n has produced an approximation $\boldsymbol{\theta}_n$ to $\boldsymbol{\theta}^*$, it next generates an r.v. $\boldsymbol{Y}_{n+1}$ having an expectation close to $\nabla z(\boldsymbol{\theta}_n)$. Such an r.v. may be obtained via the methods of Chapter VII. In particular, if one is able to obtain an unbiased gradient estimator (as sometimes occurs when either IPA or likelihood ratio methods are applicable), then the conditional expectation of $\boldsymbol{Y}_{n+1}$ given $\boldsymbol{\theta}_n$ is precisely $\nabla z(\boldsymbol{\theta}_n)$. On the other hand, for some applications we must satisfy ourselves with finite-difference methods, in which case $\boldsymbol{Y}_{n+1}$ has a (conditional) expectation close to $\nabla z(\boldsymbol{\theta}_n)$ but not exactly equal to it. Stochastic optimization algorithms that employ finite-difference methods are called *Kiefer–Wolfowitz algorithms*. Otherwise, they are called *Robbins–Monro algorithms* (but note that slightly different conventions occur in part of the literature).

The typical form of an unconstrained stochastic approximation algorithm is an iteration of the form

$$\boldsymbol{\theta}_{n+1} \;=\; \boldsymbol{\theta}_n - \varepsilon_n \boldsymbol{K} \boldsymbol{Y}_{n+1} \,, \tag{2.1}$$

where $\{\varepsilon_n\}_{n \geq 0}$ is a sequence of positive prespecified deterministic constants and $\boldsymbol{K}$ is a given (deterministic) square matrix. In the presence of constraints (as expressed through a feasible region $\boldsymbol{\Theta}$), we modify (2.1) to

$$\boldsymbol{\theta}_{n+1} \;=\; \Pi_{\boldsymbol{\Theta}}\big(\boldsymbol{\theta}_n - \varepsilon_n \boldsymbol{K} \boldsymbol{Y}_{n+1}\big) \,, \tag{2.2}$$

where $\Pi_{\boldsymbol{\Theta}}(\boldsymbol{x})$ is the point in $\boldsymbol{\Theta}$ closest to $\boldsymbol{x}$ (assumed to exist uniquely).

Suppose that the Y_n are uniformly bounded r.v.'s. Since

$$\boldsymbol{\theta}_n \;=\; \boldsymbol{\theta}_0 - \sum_{i=1}^n \varepsilon_{i-1} \boldsymbol{K} \boldsymbol{Y}_i \,,$$

---

[1] A more general setup is to find the root $\boldsymbol{\theta}^*$ of an equation $\varphi(\boldsymbol{\theta}) = 0$.

we must require that

$$\sum_{i=0}^{\infty} \varepsilon_i = \infty, \qquad (2.3)$$

for otherwise $\{\boldsymbol{\theta}_n - \boldsymbol{\theta}_0\}$ is a bounded sequence (and the iteration cannot reach the optimizer if the initial guess $\boldsymbol{\theta}_0$ is too far away from $\boldsymbol{\theta}^*$). On the other hand, the convergence of $\boldsymbol{\theta}_n$ to $\boldsymbol{\theta}^*$ requires that $\boldsymbol{\theta}_{n+1} - \boldsymbol{\theta}_n \to \mathbf{0}$ as $n \to \infty$. Hence, unless the variance of $\boldsymbol{Y}_{n+1}$ goes to $\mathbf{0}$, which is typically not the case, we must additionally require that

$$\varepsilon_n \downarrow 0, \quad n \to \infty. \qquad (2.4)$$

Conditions (2.3) and (2.4) are standard conditions on the ε_i that appear throughout the stochastic approximation literature. A natural choice for ε is to put $\varepsilon_n = n^{-\gamma}$ for some $0 < \gamma \le 1$.

Remark 2.1 A parameter not specified at this stage is the number of replications $m \overset{\text{def}}{=} m_n$ used in stage n to generate $\boldsymbol{Y}_{n+1}$; say $\boldsymbol{Y}_{n+1}$ is the average of m replications of any of the gradient estimators discussed in Chapter VII. This generates an additional degree of freedom in design of the algorithm that manifests itself in the convergence rate analysis to be developed in Section 3. □

To analyze the convergence of such algorithms, it is natural to apply martingale ideas that can take advantage of the recursive structure of the algorithm. To this end, let $\{\mathscr{F}_n\}_{n\ge 0}$ be the filtration in which $\mathscr{F}_n$ is the σ-algebra generated by all the r.v.'s computed in iterations $1, \ldots, n$, including $\boldsymbol{\theta}_0$ itself. We then write the algorithm (2.1) in the form

$$\boldsymbol{\theta}_{n+1} = \boldsymbol{\theta}_n - \varepsilon_n \boldsymbol{K}\nabla z(\boldsymbol{\theta}_n) - \varepsilon_n \boldsymbol{K}\boldsymbol{D}_{n+1} - \varepsilon_n \boldsymbol{K}\boldsymbol{U}_{n+1}, \qquad (2.5)$$

where

$$\boldsymbol{D}_{n+1} \overset{\text{def}}{=} \boldsymbol{Y}_{n+1} - \mathbb{E}\big[\boldsymbol{Y}_{n+1} \,\big|\, \mathscr{F}_n\big], \quad \boldsymbol{U}_{n+1} \overset{\text{def}}{=} \mathbb{E}\big[\boldsymbol{Y}_{n+1} \,\big|\, \mathscr{F}_n\big] - \nabla z(\boldsymbol{\theta}_n);$$

note that the $\boldsymbol{D}_n$ form a sequence of martingale differences.

In the Robbins–Monro setting, one can frequently choose the $\boldsymbol{Y}_n$ so that $\mathbb{E}\big[\boldsymbol{Y}_{n+1} \,\big|\, \mathscr{F}_n\big] = \nabla z(\boldsymbol{\theta}_n)$, in which case the $\boldsymbol{U}_{n+1}$ term on the r.h.s. of (2.5) vanishes. This occurs, for example, when one has the ability to generate in finite time an r.v. with expectation $\nabla z(\boldsymbol{\theta})$ for any $\boldsymbol{\theta}$. On the other hand, in the Kiefer–Wolfowitz setting, the $\boldsymbol{U}_{n+1}$ term is always present (as a consequence of the finite-difference approximation to the gradient).

Remark 2.2 Assume that $z(\cdot)$ is a steady-state performance measure, say $z(\boldsymbol{\theta})$ is the stationary $\mathbb{P}_{\boldsymbol{\theta}}$-expectation of $f(X_0)$, where $\{X_k\}$ is a Markov chain. Here one cannot generally simulate unbiased estimates in finite time (due to initial transient effects). On computing $\boldsymbol{\theta}_n$, one could generate $\boldsymbol{Y}_{n+1}$ by calculating a time-average based on initializing X_0 to a fixed (independent of n) initial condition and then simulate $\{X_k\}$ under the $\mathbb{P}_{\boldsymbol{\theta}_n}$-dynamics. Here, $\mathbb{E}[\boldsymbol{Y}_{n+1} \,|\, \mathscr{F}_n]$ is a function of $\boldsymbol{\theta}_n$ and (2.1) is said to be a stochastic approximation that is driven by "martingale difference" noise. An alternative would be to use the final state X_τ obtained by simulating the Markov chain under $\mathbb{P}_{\boldsymbol{\theta}_{n-1}}$ at iteration n as the initial state for the simulation undertaken at iteration $n+1$ to generate $\boldsymbol{Y}_{n+1}$. In this setting, $\mathbb{E}[\boldsymbol{Y}_{n+1} \,|\, \mathscr{F}_n]$ is a function of both $\boldsymbol{\theta}_n$ and X_τ, and we are dealing with a special case of what is called a stochastic optimization driven by "correlated noise." In either case, $\boldsymbol{U}_n$ appears in (2.5), and dealing with this is then an important part of any convergence analysis. $\qquad\square$

Remark 2.3 In the presence of regenerative structure of the Markov chain, one can set up the optimization algorithm so as to avoid the $\boldsymbol{U}_n$ term. To this end, note first that by VII.(4.7), a zero of $\nabla z(\cdot)$ occurs whenever

$$\varphi(\boldsymbol{\theta}) \stackrel{\text{def}}{=} z(\boldsymbol{\theta},1)\nabla z(\boldsymbol{\theta},f) - \nabla z(\boldsymbol{\theta},1)z(\boldsymbol{\theta},f) = \boldsymbol{0},$$

where $z(\boldsymbol{\theta},f) \stackrel{\text{def}}{=} \mathbb{E}_{\boldsymbol{\theta}} \sum_0^{\tau-1} f(X_n)$ with τ the regenerative cycle. Here

$$Z(\boldsymbol{\theta},f) \stackrel{\text{def}}{=} \sum_0^{\tau-1} f(X_n), \quad \nabla Z(\boldsymbol{\theta},f) \stackrel{\text{def}}{=} \sum_0^{\tau-1} f(X_n)C_n(\boldsymbol{\theta})$$

are unbiased estimators of $z(\boldsymbol{\theta},f)$, $\nabla z(\boldsymbol{\theta},f)$, where $C_n(\boldsymbol{\theta})$ is the accumulated score defined in VII.(4.9). To obtain an unbiased estimator of $\varphi(\boldsymbol{\theta})$, let

$$\boldsymbol{Y} \stackrel{\text{def}}{=} \tau_1 \nabla Z_2(\boldsymbol{\theta},f) + \tau_2 \nabla Z_1(\boldsymbol{\theta},f) - Z_1(\boldsymbol{\theta},f)\nabla Z_2(\boldsymbol{\theta},1) - Z_2(\boldsymbol{\theta},f)\nabla Z_1(\boldsymbol{\theta},1),$$

where

$$\big(\tau_1, Z_1(\boldsymbol{\theta},f), \nabla Z_1(\boldsymbol{\theta},1), \nabla Z_1(\boldsymbol{\theta},1)\big), \quad \big(\tau_2, Z_2(\boldsymbol{\theta},f), \nabla Z_2(\boldsymbol{\theta},1), \nabla Z_2(\boldsymbol{\theta},1)\big)$$

are independent copies of $\big(\tau, Z_1(\boldsymbol{\theta},f), \nabla Z(\boldsymbol{\theta},1), \nabla Z_1(\boldsymbol{\theta},1)\big)$. $\qquad\square$

3 Convergence Analysis

A key idea in the analysis of stochastic approximation algorithms is to consider the behavior of the post-n process $\{\boldsymbol{\theta}_{n+m}\}_{m\geq 0}$ on an appropriately defined time scale on which its convergence characteristics are particularly easy to understand.

Note that if $\boldsymbol{Y}_n = \nabla z(\boldsymbol{\theta}_n)$ (so that $\boldsymbol{Y}_n$ is deterministic), then the iteration (2.1) takes the form

$$\frac{\boldsymbol{\theta}_{n+1} - \boldsymbol{\theta}_n}{\varepsilon_n} = -\boldsymbol{K}\nabla z(\boldsymbol{\theta}_n).$$

This iteration resembles the dynamics traced out by the solution of an ordinary differential equation. To make the connection more precise, we define the interpolated process $\{\boldsymbol{\theta}(t)\}_{t\geq 0}$ for which $\boldsymbol{\theta}(t) = \boldsymbol{\theta}_n$ for $t_n \leq t \leq t_{n+1}$ where $t_n \overset{\text{def}}{=} \varepsilon_0 + \cdots + \varepsilon_{n-1}$. Then

$$\frac{\boldsymbol{\theta}(t_{n+1}) - \boldsymbol{\theta}(t_n)}{t_{n+1} - t_n} = -\boldsymbol{K}\nabla z\big(\boldsymbol{\theta}(t_n)\big),$$

thereby suggesting that the asymptotic behavior of (2.1) should be intimately connected to that of the deterministic dynamical system

$$\dot{\boldsymbol{\theta}}(t) = -\boldsymbol{K}\nabla z\big(\boldsymbol{\theta}(t)\big). \tag{3.1}$$

In particular, let $\boldsymbol{\theta}^n$ be the "post-iteration-n" process defined by $\boldsymbol{\theta}^n(t) = \boldsymbol{\theta}(t_n + t)$. The idea is to show that if the $\boldsymbol{\theta}_n$ satisfy (2.1) (and some additional regularity conditions), then the sequence $(\boldsymbol{\theta}^n)_{n\geq 1}$ of random processes is tight, and to establish that any weakly convergent subsequence converges to a solution of the differential equation (3.1) that lies entirely in an invariant set Λ of the dynamical system (2.1); here Λ is said to be an invariant set if for each $\boldsymbol{\theta}_0 \in \Lambda$, there exists a solution of (3.1) for which $\boldsymbol{\theta}(0) = \boldsymbol{\theta}_0$ and $\boldsymbol{\theta}(t) \in \Lambda$ for all $t \geq 0$. In the optimization context, the invariant set is typically the finite set consisting of local minimizers of $z(\cdot)$, in which case this approach permits one to conclude that the algorithm converges in probability to a local minimizer.

To implement the above style of argument, let $\boldsymbol{D}^n = \{\boldsymbol{D}^n(t)\}_{t\geq 0}$ and $\boldsymbol{U}^n = \{\boldsymbol{U}^n(t)\}_{t\geq 0}$ be the shifted and interpolated processes corresponding to the $\boldsymbol{D}_n$ and $\overline{\boldsymbol{U}}_n$ appearing in (2.5). We can then write $\boldsymbol{\theta}^n(t)$ as

$$\boldsymbol{\theta}^n(0) - \int_0^t \boldsymbol{K}\nabla z\big(\boldsymbol{\theta}^n(s)\big)\,\mathrm{d}s - \int_0^t \boldsymbol{K}\boldsymbol{D}^n(s)\,\mathrm{d}s - \int_0^t \boldsymbol{K}\boldsymbol{U}^n(s)\,\mathrm{d}s.$$

The key is to then verify the necessary tightness and weak convergence relations

$$\int_0^t \boldsymbol{D}^n(s)\,\mathrm{d}s \overset{\mathbb{P}}{\to} \boldsymbol{0}, \quad \int_0^t \boldsymbol{U}^n(s)\,\mathrm{d}s \overset{\mathbb{P}}{\to} \boldsymbol{0}$$

as $n \to \infty$. These arguments tend to be somewhat specific to the application at hand; one important point of mathematical leverage in implementing such arguments is the martingale structure of the $\boldsymbol{D}_n$.

To develop a rate-of-convergence result, suppose that the invariant set Λ for (3.1) is a single point, say $\{\boldsymbol{\theta}^*\}$. Put $\boldsymbol{\xi}_n \overset{\text{def}}{=} \beta_n^{-1/2}(\boldsymbol{\theta}_n - \boldsymbol{\theta}^*)$ for some sequence $\beta_n \downarrow 0$, and note that (2.5) implies that

$$\boldsymbol{\xi}_{n+1} = \left(\frac{\beta_n}{\beta_{n+1}}\right)^{1/2}\boldsymbol{\xi}_n + \left(\frac{\beta_n}{\beta_{n+1}}\right)^{1/2}\Big[-\varepsilon_n\beta_n^{-1/2}\boldsymbol{K}\nabla z(\boldsymbol{\theta}_n)$$

$$-\varepsilon_n\beta_n^{-1/2}\boldsymbol{K}\boldsymbol{D}_{n+1} - \varepsilon_n\beta_n^{-1/2}\boldsymbol{K}\boldsymbol{U}_{n+1}\Big]. \tag{3.2}$$

If $z(\cdot)$ is smooth at $\boldsymbol{\theta}^*$, the convergence of $\boldsymbol{\theta}_n$ to $\boldsymbol{\theta}^*$ allows us to write

$$\begin{aligned} \nabla z(\boldsymbol{\theta}_n) &= \nabla z(\boldsymbol{\theta}_n) - \nabla z(\boldsymbol{\theta}^*) \\ &= \boldsymbol{H}(\boldsymbol{\theta}^*)(\boldsymbol{\theta}_n - \boldsymbol{\theta}^*) + o\big(\|\boldsymbol{\theta}_n - \boldsymbol{\theta}^*\|\big), \end{aligned} \tag{3.3}$$

where $\boldsymbol{H}$ is the Hessian. Also, if $\beta_n = n^{-\delta}$ with $0 < \delta \le 1$, then

$$\left(\frac{\beta_n}{\beta_{n+1}}\right)^{1/2} = \begin{cases} 1 + \frac{1}{2n} + o(\beta_n) & \text{if } \delta = 1, \\ 1 + o(\beta_n) & \text{if } 0 < \delta < 1. \end{cases} \tag{3.4}$$

Note that if $\{\boldsymbol{M}^n(t)\}_{t \ge 0}$ is the shifted interpolated process for which

$$\boldsymbol{M}^n(t_{n+m} - t_n) = \sum_{j=n}^{n+m-1} \varepsilon_j \beta_j^{-1/2} \boldsymbol{K} \boldsymbol{D}_{j+1},$$

then $\boldsymbol{M}^n$ is a martingale. Because $\boldsymbol{\theta}_n$ converges to $\boldsymbol{\theta}^*$ as $n \to \infty$, one can typically establish that

$$\{\boldsymbol{M}^n(t)\} \xrightarrow{\mathscr{D}} \{(\boldsymbol{K}\boldsymbol{\Sigma}\boldsymbol{K}^{\mathsf{T}})^{1/2} \boldsymbol{B}(t)\} \tag{3.5}$$

as $n \to \infty$, where $\boldsymbol{B}$ is standard Brownian motion in $\mathbb{R}^d$ and $\boldsymbol{\Sigma}$ some suitable covariance matrix, by appealing to a suitable version of the martingale CLT. This basically comes down to showing that the predictable quadratic variation of the martingale $\boldsymbol{M}^n$ converges to that of the limit process. In other words, the kay step is to show that

$$\begin{aligned} \frac{\langle \boldsymbol{M}^n, \boldsymbol{M}^n \rangle(t_n)}{t_n} &= \boldsymbol{K} \sum_{j=0}^{n-1} \varepsilon_j^2 \beta_j^{-1} \mathbb{V}ar(\boldsymbol{D}_{j+1} \mid \mathscr{F}_j) \boldsymbol{K}^{\mathsf{T}} \Big/ \sum_{j=0}^{n-1} \varepsilon_j \\ &\to \boldsymbol{K}\boldsymbol{\Sigma}\boldsymbol{K}^{\mathsf{T}} \end{aligned} \tag{3.6}$$

as $n \to \infty$. In order to obtain the limit (3.6), one clearly needs to choose the sequence $\{\beta_n\}_{n \ge 1}$ appropriately, taking into account both the behavior of the ε_j and that of the $\mathbb{V}ar[\boldsymbol{D}_{j+1} \mid \mathscr{F}_j]$.[2]

Remark 3.1 As in Remark 2.1, assume that $\boldsymbol{Y}_{n+1}$ is the average of m replications of any of the gradient estimators $\nabla Z(\boldsymbol{\theta}_n)$ discussed in Chapter VII. Then $\boldsymbol{\theta}_n \to \boldsymbol{\theta}^*$ implies that the matrix $\mathbb{V}ar[\boldsymbol{D}_{n+1} \mid \mathscr{F}_n]$ has limit $\boldsymbol{\Sigma} \stackrel{\text{def}}{=} \mathbb{V}ar(\nabla Z(\boldsymbol{\theta}^*))/m$, from which one gets (3.5), provided the β_n are selected so that $\beta_n = \varepsilon_n$. □

Finally, one can often argue (again, in the presence of the knowledge that $\boldsymbol{\theta}_n \to \boldsymbol{\theta}^*$) that $\widetilde{\boldsymbol{U}}^n \xrightarrow{\mathscr{D}} 0$, where $\widetilde{\boldsymbol{U}}^n$ is the interpolated process for which

$$\widetilde{\boldsymbol{U}}^n(t_{n+m} - t_n) = \sum_{j=n}^{n+m-1} \varepsilon_j \beta_j^{-1/2} \boldsymbol{U}_j,$$

---

[2]Note that the $\mathbb{V}ar[\boldsymbol{D}_{j+1} \mid \mathscr{F}_j]$ may not form a bounded sequence; see the discussion below of the Kiefer–Wolfowitz algorithm.

Assuming tightness can be established, the above argument suggests that any weak limit $\boldsymbol{\xi}$ for the shifted/interpolated process $\boldsymbol{\xi}^n$ should satisfy the SDE

$$
d\boldsymbol{\xi}(t) = \begin{cases} \big(\boldsymbol{I}/2 - \boldsymbol{K}\boldsymbol{H}(\boldsymbol{\theta}^*)\big)\boldsymbol{\xi}(t)\,dt + \big(\boldsymbol{K}\boldsymbol{\Sigma}\boldsymbol{K}^{\mathsf{T}}\big)^{1/2}d\boldsymbol{B}(t) & \delta = 1, \\ -\boldsymbol{K}\boldsymbol{H}(\boldsymbol{\theta}^*)\boldsymbol{\xi}(t)\,dt + \big(\boldsymbol{K}\boldsymbol{\Sigma}\boldsymbol{K}^{\mathsf{T}}\big)^{1/2}d\boldsymbol{B}(t) & 0 < \delta < 1. \end{cases}
$$
(3.7)

Given the convergence to $\boldsymbol{\theta}^*$, we further expect that the limit should be a stationary porcess (for if $\boldsymbol{\xi}$ satisfies (3.7) and does not admit a stationary version, then $\|\boldsymbol{\xi}(t)\| \to \infty$ as $t \to \infty$). A stationary version of the vector-valued Ornstein–Uhlenbeck process (3.7) exists for $\delta = 1$ only when $\boldsymbol{I}/2 - \boldsymbol{K}\boldsymbol{H}(\boldsymbol{\theta}^*)$ has only eigenvalues with negative real parts (i.e., is a so-called *Hurwitz matrix*). Similarly, a stationary version of (3.7) exists for $\delta < 1$ only when $-\boldsymbol{K}\boldsymbol{H}(\boldsymbol{\theta}^*)$ is a Hurwitz matrix. Hence, it is standard in convergence rate analysis of stochastic approximation algorithms to assume that $\boldsymbol{I}/2 - \boldsymbol{K}\boldsymbol{H}(\boldsymbol{\theta}^*)$ and $-\boldsymbol{K}\boldsymbol{H}(\boldsymbol{\theta}^*)$ are Hurwitz matrices when $\delta = 1$ and $\delta < 1$, respectively. Such assumptions are natural, in view of the fact that $\boldsymbol{\theta}^*$ is a local attractor for the deterministic dynamical system (3.1) (i.e., $\boldsymbol{\theta}(t)$ converges to $\boldsymbol{\theta}^*$ as $t \to \infty$ for $\boldsymbol{\theta}(0)$ in a neighborhood of $\boldsymbol{\theta}^*$) only when $-\boldsymbol{K}\boldsymbol{H}(\boldsymbol{\theta}^*)$ is a Hurwitz matrix.

Remark 3.2 Consider the Robbins–Monro algorithm when driven by an unbiased gradient estimtor, so that the $\boldsymbol{U}_n$ terms vanish. If, as in Remark 3.1, the sample size m used at iteration n is independent of n, Remark 3.1 establishes that one should choose $\beta_n = \varepsilon_n = n^{-\gamma}$. Proving that $\boldsymbol{\xi}^n \overset{\mathscr{D}}{\to} \boldsymbol{\xi}$, $n \to \infty$, implies, in particular, that $\boldsymbol{\xi}^n(0) = \beta_n^{-1/2}(\boldsymbol{\theta}_n - \boldsymbol{\theta}^*) = n^{\gamma/2}(\boldsymbol{\theta}_n - \boldsymbol{\theta}^*)$ converges weakly to $\boldsymbol{\xi}(0)$, where $\boldsymbol{\xi}(0)$ has the stationary distribution of $\boldsymbol{\xi}$. Because $\boldsymbol{\xi}$ is Gaussian, $\boldsymbol{\xi}(0)$ is Gaussian. It follows from a straightforward calculation that $\mathbb{E}\boldsymbol{\xi}(0) = 0$ with covariance matrix $\boldsymbol{R} \overset{\text{def}}{=} \boldsymbol{R}(\boldsymbol{K}) \overset{\text{def}}{=} \mathbb{E}\big[\boldsymbol{\xi}(0)\boldsymbol{\xi}(0)^{\mathsf{T}}\big]$ given by

$$
\boldsymbol{R} = \int_0^\infty \exp\{-\boldsymbol{K}\boldsymbol{H}(\boldsymbol{\theta}^*)s\}\,\boldsymbol{K}\boldsymbol{\Sigma}\boldsymbol{K}^{\mathsf{T}}\exp\{-\boldsymbol{H}(\boldsymbol{\theta}^*)\boldsymbol{K}^{\mathsf{T}}s\}\,ds
$$
(3.8)

for $\gamma < 1$, so that

$$
n^{\gamma/2}(\boldsymbol{\theta}_n - \boldsymbol{\theta}^*) \overset{\mathscr{D}}{\to} \mathscr{N}(\boldsymbol{0}, \boldsymbol{R}), \quad n \to \infty.
$$

Similarly, if $\varepsilon_n = n^{-1}$, then $n^{1/2}(\boldsymbol{\theta}_n - \boldsymbol{\theta}^*) \to \mathscr{N}(0, \boldsymbol{R}_1)$, where $\boldsymbol{R}_1 \overset{\text{def}}{=} \boldsymbol{R}_1(\boldsymbol{K})$ is given by

$$
\int_0^\infty \exp\{\big(\boldsymbol{I}/2 - \boldsymbol{K}\boldsymbol{H}(\boldsymbol{\theta}^*)s\big)\}\boldsymbol{K}\boldsymbol{\Sigma}\boldsymbol{K}^{\mathsf{T}}\{\big(\boldsymbol{I}/2 - \boldsymbol{H}(\boldsymbol{\theta}^*)\boldsymbol{K}^{\mathsf{T}}\big)s\}\,ds.
$$
(3.9)

□

This analysis shows that the sequence $\varepsilon_n = n^{-1}$ maximizes the rate of convergence. As for the best possible choice of $\boldsymbol{K}$, the best choice is to set $K^* = H(\theta)^{-1}$ (assuming that the Hessian is positive definite at the minimizer θ^*), in the sense that $R_1(K) - R_1(K^*)$ is nonnegative definite for all K. Hence, the best-possible choice of the iteration (2.1) in the Robbins–Monro context is

$$\theta_{n+1} = \theta_n - \frac{H(\theta^*)^{-1}}{n} Y_{n+1}.$$

Of course, $H(\theta^*)$ (and even θ^* itself) is not known to the simulator, so that this algorithm is not practically implementable.

We next turn to the Kiefer–Wolfowitz algorithm. In this setting, Y_{n+1} is a finite-difference estimator (with difference increment c_n), either of central difference form or forward difference form. We assume that the simulation runs at the different points in the finite-difference scheme are independent (as are the simulations across the iterations) and unbiased as estimators of $z(\theta)$. As noted earlier, this is an algorithm in which the U_{n+1} are present.

Suppose that $c_n = cn^{-r}$ for some $c > 0$ and some $r > 0$. Focusing first on the central difference version of the Kiefer–Wolfowitz algorithm, recall from Chapter VII that U_n is of order $c_n^2 = c^2 n^{-2r}$. In order that the process $\widetilde{U}^n$ be tight, it follows that $\{\beta_n^{-1/2} c_n^2\}$ must be a bounded sequence, so that δ must satisfy $\delta \le 4r$.

Turning next to M^n, (3.6) requires that

$$\{\varepsilon_j \beta_j^{-1} \mathbb{V}ar[D_{j+1} \mid \mathscr{F}_j]\}$$

must be a bounded sequence. If the number of simulations performed per iteration is fixed, then $\mathbb{V}ar[D_{j+1} \mid \mathscr{F}_j]$ is of order c_j^{-2}. To compare the convergence rate for a given choice of r and γ (respecting the above inequalities), we wish to choose δ as large as possible (i.e., $\delta = 4r$). Hence, to maximize the convergence rate, for a given choice of γ, we should choose r as large as possible, namely $r = \gamma/6$. For this choice of r, establishing $\xi^n \overset{\mathscr{D}}{\to} \xi$ proves, in particular, that $n^{\gamma/3}(\theta_n - \theta^*) \overset{\mathscr{D}}{\to} \xi(0)$ as $n \to \infty$. In other words, the convergence rate of the Kiefer–Wolfowitz algorithm using central differences is of order $n^{-\gamma/3}$, with maximal convergence rate $n^{-1/3}$ achieved when $\gamma = 1$. On the other hand, when forward differences are used, δ must satisfy $\delta \le 2r$. The inequality $-\gamma + \delta + 2r \le 0$ continues to hold, leading to setting $r = \gamma/4$ and an associated convergence rate of $n^{-\gamma/4}$ which is maximized (at rate $n^{-1/4}$) when $\gamma = 1$ is used in the algorithm. So, use of central differences rather than forward differences improves the convergence rate from $n^{-1/4}$ to $n^{-1/3}$ (and use of unbiased gradient estimators, as in the Robbins-Monro algorithm, further improves the convergence rate to $n^{-1/2}$). An explicit description of the limit process ξ that arises in the Kiefer–Wolfowitz context can be found in Kushner & Yin [224, Section 10.3]. When $\delta = 4r$ and $r = \gamma/6$ (for the central difference scheme) as required for maximizing the convergence rate, the process $\widetilde{U}^n$ does not converge to zero but to a proper limit, thereby adding an additional term to the limit as described in (3.7); a similarly more complex limit (as compared to (3.7)) arises in the setting of forward differences when $\delta = 2r$ and $r = \gamma/4$.

4 Polyak–Ruppert Averaging

We note that when $\varepsilon_n = n^{-\gamma}$, the ε_n are slowly changing when n is large, in the sense that $(\varepsilon_{n+1} - \varepsilon_n)/\varepsilon_n \to 0$ as $n \to \infty$. Because the ε_n are roughly constant over long time scales, this suggests studying the stochastic approximations in the case that $\varepsilon_n = \varepsilon$ for $n \geq 1$. Assume that (2.1) is driven by "martingale difference" noise. In this case, the θ_n form a Markov chain with stationary transition probabilities. When ε is small and $-\boldsymbol{KH}(\theta^*)$ is a Hurwitz matrix, one expects the point $\boldsymbol{\theta}^*$ to be "attracting" and the chain to possess a stationary distribution with a mean close to $\boldsymbol{\theta}^*$. In such a context, it is clear that a better estimate of $\boldsymbol{\theta}^*$ will be achieved by using the average $(\boldsymbol{\theta}_1 + \cdots + \boldsymbol{\theta}_n)/n$, rather than depending solely on the final value $\boldsymbol{\theta}_n$ of the chain.

This averaging idea was independently developed by Polyak [291] and Ruppert [321]; see also Polyak & Juditsky [292]. Consider the Robbins–Monro algorithm (so that the $\boldsymbol{U}_n$ vanish and $\beta_n = \varepsilon_n$) with $\varepsilon_n = n^{-\gamma}$ and $0 < \gamma < 1$ (note that when $\gamma < 1$, the ε_n are "more constant"). Recall that the stochastic approximation is described by the limit process $\boldsymbol{\xi}$, in the sense that the shifted/interpolated process converges weakly in $D[0,\infty)$. Because we are interested in the effect of averaging, we take advantage of the functional weak convergence of $\boldsymbol{\xi}^n$ to $\boldsymbol{\xi}$ to yield

$$\frac{1}{t}\int_0^t \boldsymbol{\xi}^n(s)\,\mathrm{d}s \;\to\; \frac{1}{t}\int_0^t \boldsymbol{\xi}(s)\,\mathrm{d}s$$

for each fixed $t > 0$. Note that $\boldsymbol{\xi}^n$ over $[0,t]$ describes the algorithm (2.5) over the iterations that are numbered n through $n + \lfloor t/\varepsilon_n \rfloor$. Furthermore, the step-lengths $\varepsilon_n, \ldots, \varepsilon_{n+\lfloor t/\varepsilon_n \rfloor}$ are all roughly constant (when n is large), so that

$$\frac{\varepsilon_n}{t} \sum_{k=n}^{n+\lfloor t/\varepsilon_n \rfloor} \varepsilon_n^{-1/2}(\boldsymbol{\theta}_n - \boldsymbol{\theta}^*) \;-\; \frac{1}{t}\int_0^t \boldsymbol{\xi}^n(s)\,\mathrm{d}s \;\to\; \boldsymbol{0}$$

as $n \to \infty$, and hence

$$\frac{\varepsilon_n}{t} \sum_{k=n}^{n+\lfloor t/\varepsilon_n \rfloor} \varepsilon_n^{-1/2}(\boldsymbol{\theta}_n - \boldsymbol{\theta}^*) \;\to\; \frac{1}{t}\int_0^t \boldsymbol{\xi}(s)\,\mathrm{d}s$$

as $n \to \infty$. Since $\boldsymbol{\xi}$ is a mean-zero Gaussian process,

$$\frac{1}{t}\int_0^t \boldsymbol{\xi}(s)\,\mathrm{d}s \;\sim\; \mathcal{N}\!\left(0, \boldsymbol{\Sigma}_{\boldsymbol{\xi}}^2\right), \quad \text{where } \boldsymbol{\Sigma}_{\boldsymbol{\xi}}^2 \overset{\text{def}}{=} \frac{1}{t^2}\int_0^t\int_0^t \mathbb{E}\!\left[\boldsymbol{\xi}(u_1)\boldsymbol{\xi}(u_2)^{\mathsf{T}}\right]\mathrm{d}u_1\,\mathrm{d}u_2 .$$

The covariance function of $\boldsymbol{\xi}$ is easily computable:

$$\mathbb{E}\big[\boldsymbol{\xi}(s+u)\boldsymbol{\xi}(s)^{\mathsf{T}}\big] = \begin{cases} \exp\{-\boldsymbol{K}\boldsymbol{H}(\boldsymbol{\theta}^*)u\}\boldsymbol{R} & u \geq 0, \\ \exp\{\boldsymbol{H}(\boldsymbol{\theta}^*)\boldsymbol{K}^{\mathsf{T}}\}\boldsymbol{R} & u \leq 0, \end{cases}$$

where $\boldsymbol{R}$ is defined in (3.8). Note that

$$\int_0^t \int_0^t \mathbb{E}\big[\boldsymbol{\xi}(u_1)\boldsymbol{\xi}(u_2)^{\mathsf{T}}\big]\,\mathrm{d}u_1\,\mathrm{d}u_2$$

$$= \int_0^t \int_0^t \mathbb{E}\big[\boldsymbol{\xi}(0)\boldsymbol{\xi}(u_1-u_2)^{\mathsf{T}}\big]\,\mathrm{d}u_1\,\mathrm{d}u_2 \ \sim\ t\int_0^t \mathbb{E}\big[\boldsymbol{\xi}(0)\boldsymbol{\xi}(u)^{\mathsf{T}}\big]\,\mathrm{d}u$$

$$= t\big[(\boldsymbol{K}\boldsymbol{H}(\boldsymbol{\theta}^*))^{-1}\boldsymbol{R} + \boldsymbol{R}\big(\boldsymbol{H}(\boldsymbol{\theta}^*)\boldsymbol{K}^{\mathsf{T}}\big)^{-1}\big] \qquad (4.1)$$

as $t \to \infty$. But $\boldsymbol{R}$ satisfies the Ricatti equation

$$\boldsymbol{R}\big(\boldsymbol{H}(\boldsymbol{\theta}^*)\boldsymbol{K}^{\mathsf{T}}\big) + \big(\boldsymbol{K}\boldsymbol{H}(\boldsymbol{\theta}^*)\big)\boldsymbol{R} = \boldsymbol{K}\boldsymbol{\Sigma}\boldsymbol{K}^{\mathsf{T}}, \qquad (4.2)$$

as is easily verified by direct integration of (3.8). Premultiplying through (4.2) by $\big(\boldsymbol{K}\boldsymbol{H}(\boldsymbol{\theta}^*)\big)^{-1}$ and postmultiplying by $\big(\boldsymbol{H}(\boldsymbol{\theta}^*)\boldsymbol{K}^{\mathsf{T}}\big)^{-1}$, we conclude that

$$\big(\boldsymbol{K}\boldsymbol{H}(\boldsymbol{\theta}^*)\big)^{-1}\boldsymbol{R} + \boldsymbol{R}\big(\boldsymbol{H}(\boldsymbol{\theta}^*)\boldsymbol{K}^{\mathsf{T}}\big)^{-1}$$

$$= \big(\boldsymbol{K}\boldsymbol{H}(\boldsymbol{\theta}^*)\big)^{-1}\boldsymbol{K}\boldsymbol{\Sigma}\boldsymbol{K}^{\mathsf{T}}\big(\boldsymbol{H}(\boldsymbol{\theta}^*)\boldsymbol{K}^{\mathsf{T}}\big)^{-1} = \boldsymbol{\Lambda}_{\boldsymbol{\xi}},$$

where

$$\boldsymbol{\Lambda}_{\boldsymbol{\xi}} \overset{\mathrm{def}}{=} \boldsymbol{H}(\boldsymbol{\theta}^*)^{-1}\boldsymbol{\Sigma}\boldsymbol{H}(\boldsymbol{\theta}^*)^{-1}.$$

Thus,

$$\frac{1}{\sqrt{t}}\int_0^t \boldsymbol{\xi}(s)\,\mathrm{d}s \ \to\ \mathcal{N}(\boldsymbol{0}, \boldsymbol{\Lambda}_{\boldsymbol{\xi}}) \qquad (4.3)$$

as $t \to \infty$.

It follows that when n is large and t is large, we may conclude that

$$\Big(\frac{\varepsilon_n}{t}\Big)^{1/2}\sum_{k=n}^{n+\lfloor t/\varepsilon_n\rfloor} \varepsilon_n^{-1/2}(\boldsymbol{\theta}_n - \boldsymbol{\theta}^*) \overset{\mathscr{D}}{\approx} \mathcal{N}(\boldsymbol{0}, \boldsymbol{\Lambda}_{\boldsymbol{\xi}}). \qquad (4.4)$$

This can now be compared against the optimal rate of convergence that is achieveable for the Robbins–Monro algorithm, namely that obtained when $\varepsilon_n = n^{-1}$ and $\boldsymbol{K} = \boldsymbol{H}(\boldsymbol{\theta}^*)^{-1}$. We established earlier that with these choices, we can expect $n^{1/2}(\boldsymbol{\theta}_n - \boldsymbol{\theta}^*)$ to converge to a Gaussian distribution having mean zero and covariance matrix $\boldsymbol{\Lambda}_{\boldsymbol{\xi}}$. Hence, the covariance structure that appears when one averages over the iterates $n, n+1, \ldots, n+\lfloor t/\varepsilon_n\rfloor$ match the optimal covariance structure, at least when n and t are large. Furthermore, the standard deviation of the average depends on the "window size" $w \overset{\mathrm{def}}{=} \lfloor t/\varepsilon_n\rfloor$ through the factor $w^{-1/2}$, matching the square-root convergence rate of the optimal version of the Robbins–Monro algorithm.

This suggests attempting to use the "window" of iterates numbered 1 through n as a basis for constructing an averaging scheme that will match the optimal rate of convergence achieveable by the Robbins–Monro method. In particular, consider the estimator

$$\overline{\boldsymbol{\theta}}_n \stackrel{\text{def}}{=} \frac{1}{n} \sum_{i=1}^{n} \boldsymbol{\theta}_i \,,$$

where (as above) $\varepsilon_n = n^{-\gamma}$ with $0 < \gamma < 1$, and $\boldsymbol{K}$ is chosen arbitrarily (modulo the requirement that $-\boldsymbol{K}\boldsymbol{H}(\boldsymbol{\theta}^*)$ be a Hurwitz matrix).

Observe that when $n = k_1 + \cdots + k_m$, $\overline{\boldsymbol{\theta}}_n$ can be viewed as a weighted average of $\widehat{\boldsymbol{\theta}}_1, \ldots, \widehat{\boldsymbol{\theta}}_m$, in which each $\widehat{\boldsymbol{\theta}}_i$ is itself an average over a window of size k_i (and with weights k_i/n). By choosing k_i appropriately, (4.4) applies to each of the m "window averages" $\widehat{\boldsymbol{\theta}}_1, \ldots, \widehat{\boldsymbol{\theta}}_m$, so that

$$\widehat{\boldsymbol{\theta}}_i \stackrel{\mathscr{D}}{\approx} \boldsymbol{\theta}^* + \frac{1}{\sqrt{k_i}} \boldsymbol{V}_i \,, \quad \text{where} \quad \boldsymbol{V}_i \sim \mathscr{N}(\boldsymbol{0}, \boldsymbol{\Lambda_\xi}) \,.$$

Furthermore, because of the presence of "mixing effects" (as evidenced in the fact that the limit process $\boldsymbol{\xi}$ itself mixes), the $\boldsymbol{V}_i$ are asymptotically independent. This leads to the conclusion that

$$\overline{\boldsymbol{\theta}}_n \stackrel{\mathscr{D}}{\approx} \boldsymbol{\theta}^* + \mathscr{N}(\boldsymbol{0}, \boldsymbol{\Lambda_\xi}/\sqrt{n}) \,.$$

In other words, when one runs the Robbins–Monro algorithm with $\gamma \in (0, 1)$ and computes the average $\overline{\boldsymbol{\theta}}_n$, the average achieves the optimal rate of convergence associated with a $\gamma = 1$ implementation of Robbins–Monro, despite the fact that the optimal $\boldsymbol{K}^* = \boldsymbol{H}(\boldsymbol{\theta}^*)^{-1}$ is not known and not used by the algorithm. In addition, use of $\gamma = 1$ leads to larger step sizes, allowing to move more rapidly to the optimizer in the early iterations of the algorithm. It should also be noted that the averaging scheme provides optimal rates when $-\boldsymbol{K}\boldsymbol{H}(\boldsymbol{\theta}^*)$ is Hurwitz, while one must require $\boldsymbol{I}/2 - \boldsymbol{K}\boldsymbol{H}(\boldsymbol{\theta}^*)$ to be Hurwitz for an $\gamma = 1$ implementation of Robbins–Monro to achieve an optimal convergence rate.

The above theoretical advantages of averaging have been observed in many empirical implementations. As a consequence, averaging has become a widely used tool in the context of stochastic approximation. For a full theoretical development of this topic, see Kushner & Yin [224, Chapter 11].

Remark 4.1 Note that applying averaging for $\gamma = 1$ does not have similar beneficial effects. One way of seeing this is to note that because the covariance structure of $\boldsymbol{\xi}$ differs when $\gamma = 1$, it is no longer the case that (4.4) holds. Hence, applying averaging to the limit $\boldsymbol{\xi}$ does not lead to the optimal covariance structure achievable for Robbins–Monro when $\gamma = 1$, suggesting that averaging is not effective for $\gamma = 1$. $\Box$

5   Examples

5a   *Numerical Illustrations in a PERT Net Example*

As in VII.5.1, consider the minimization of

$$z(\theta) \;=\; \mathbb{E}\max\!\big(\theta X_1 + X_2, (1-\theta)X_3\big)$$

in a simple PERT net, where X_1, X_2, X_3 are Erlang(2) with means 1, 2, 3. The correct minimizer is $\theta^* = 0.625$.

We used the Robbins–Monro algorithm with $\varepsilon_n = cn^{-\delta}$ and the estimator $\boldsymbol{Y}$ for $z'(\theta)$ evaluated by averaging the IPA estimator (cf. again VII.5.1) over m replications. All runs used $\theta_0 = 0.4$ and $a = 0.1$.

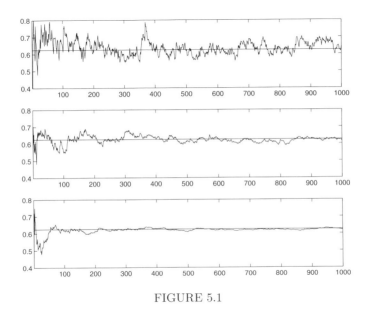

FIGURE 5.1

In Figure 5.1, we took $m = 10$ and $\delta = 0.4$ in the top panel, $\delta = 0.6$ in the middle one and $\delta = 0.8$ in the bottom one. Figure 5.2 contains the same runs, but the plotted quantity is the Polyak–Ruppert average $\sum_1^k \theta_j/k$, $k = 1, \ldots, n$.

In Figure 5.3, we took different values $m = 10, 100, 1{,}000$ in the three panels and adjusted the run length n by keeping nm fixed; the computational effort is then roughly the same, provided generation of the derivative estimates consumes the main part of the computer time.

Whereas these figures suggest certain obvious conclusions on optimal choice of parameters, one should be aware that we are dealing with an unrealistically simple example, and that practical experience shows that tuning of the parameters in general has substantial influence on the performance of the Robbins–Monro procedure and has to be done on a case-by-case basis.

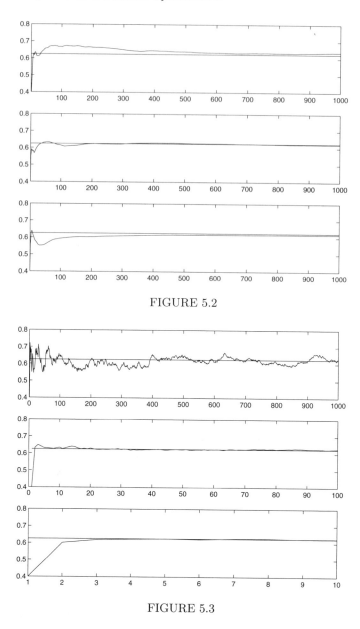

FIGURE 5.2

FIGURE 5.3

5b   A Queuing Example

Many decision problems in today's performance engineering involve the choice of routing protocols in networks. For an extremely simple example, consider a priority two-server queue. Customers of two types A, B arrive according to independent Poisson processes with rates $\lambda_A = 0.9, \lambda_B = 0.8$ and have i.i.d. standard exponential service times. The two servers work at rates μ_1, μ_2. The penalty of a type-A customers is his delay and that of a type-B customer three times his delay. Type-B customers have higher priority and are always allocated to the fast server, those of type A to the slow server w.p. $1 - \theta$ and to the fast w.p. θ. The customers at server 2 are served in the order of arrival, that is, B customers are not given any advantage. The goal is to minimize the expected steady-state penalty $z(\theta)$.

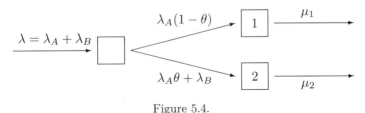

Figure 5.4.

In this model, there is of course no need for stochastic optimization: the queues at the two servers are M/M/1 queues with arrival rates $\lambda_1(\theta) \stackrel{\text{def}}{=} \lambda_A(1 - \theta)$, $\lambda_2(\theta) \stackrel{\text{def}}{=} \lambda_A \theta + \lambda_B$, and a fraction $\lambda_1(\theta)/\lambda$ of the customers go to server 1, $\lambda_2(\theta)/\lambda$ to server 2 where $\lambda \stackrel{\text{def}}{=} \lambda_A + \lambda_B = \lambda_1 + \lambda_2$. Further, a fraction $\lambda_A \theta / \lambda_2(\theta)$ of the customers at server 2 are A customers, $\lambda_B/\lambda_2(\theta)$ are B customers. Since the mean delay in the M/M/1 queue with arrival rate λ_i and service rate μ_i is $\lambda_i/\mu_i(\mu_i - \lambda_i)$, we have therefore

$$z(\theta) = \frac{\lambda_1(\theta)^2}{\lambda \mu_1 (\mu_1 - \lambda_1(\theta))} + \left(\frac{\lambda_A \theta}{\lambda_2(\theta)} + 3 \frac{\lambda_B}{\lambda_2(\theta)} \right) \frac{\lambda_2(\theta)}{\lambda \mu_2 (\mu_2 - \lambda_2(\theta))}.$$

The function $z(\theta)$ is plotted in Figure 5.5; there is a unique minimum at $\theta^* = 0.372$ and $z(\theta^*) = 1.487$.

For numerical stochastic optimization, one can use a similar regenerative representation as in Remark 2.3. The cycle τ starts by a customer entering a system with both servers idle and terminates when a customer again meets this state. Let $B_0, B_1, \ldots$ be i.i.d. Bernoulli(θ), such that customer k goes to server 1 if he is of type A and $B_k = 0$, and to server 1 if he is of type A and $B_k = 1$. A single B has density $\theta^b(1 - \theta)^{1-b}$ w.r.t. counting measure

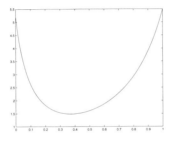

FIGURE 5.5

on $\{0, 1\}$, from which it follows easily that the score is $(B - \theta)/[\theta(1 - \theta)]$, and the score of customers $0, \ldots, n$ is therefore

$$S_n \stackrel{\mathrm{def}}{=} \sum_{k=0}^{n} \frac{B_k - \theta)}{\theta(1 - \theta)} \mathbb{1}\{\text{customer } k \text{ is of type } A\}.$$

We have $z(\theta) = z_1(\theta)/z_2(\theta)$, where $z_1(\theta) \stackrel{\mathrm{def}}{=} \mathbb{E}_\theta \sum_0^{\tau-1} W_n$, $z_2(\theta) \stackrel{\mathrm{def}}{=} \mathbb{E}_\theta \tau$, so that

$$z'(\theta) = \frac{1}{z_2(\theta)^2}\left[z_1'(\theta)z_2(\theta) - z_1(\theta)z_2'(\theta)\right],$$

where

$$z_1'(\theta) = \mathbb{E}_\theta\left[\tau S_{\tau-1}\right] = \mathbb{E}_\theta \sum_{n=0}^{\tau-1} S_n, \quad z_1'(\theta) = \mathbb{E}_\theta \sum_{n=0}^{\tau-1} W_n S_n.$$

Thus we have to search for a root of $\varphi(\cdot)$, where

$$\varphi(\theta) \stackrel{\mathrm{def}}{=} \mathbb{E}_\theta\left(\tau_1 V_2 + \tau_2 V_1 - \xi_1 W_2 - \xi_2 W_1\right),$$

where $(\tau_1, V_1, \xi_1, W_1)$, $(\tau_2, V_2, \xi_2, W_2)$ are independent copies of (τ, V, ξ, W), where

$$V \stackrel{\mathrm{def}}{=} \sum_{k=0}^{\tau-1} W_k S_k, \quad \xi \stackrel{\mathrm{def}}{=} \sum_{i=0}^{\tau-1} S_k, \quad W = \sum_{k=0}^{\tau-1} W_k.$$

5c   The Stochastic Approximation EM Algorithm

The EM algorithm is one of the main tools for performing maximum likelihood (ML) estimation in the absence of full data information (incomplete observations, lost data, etc.). An introduction is given in A4; the algorithm has two steps, the E-step, which involves computing a conditional expectation, and the M-step which is a maximization. The M-step is the same as in the presence of complete observations and therefore often easy. Thus, the main difficulty is usually the E-step, in which the relevant conditional expectations may be computed by simulation, leading to the *Monte Carlo EM algorithm* or the *stochastic EM algorithm*. We outline here a variant, the *stochastic approximation EM algorithm*, which actually can be fitted into the Robbins–Monro framework.

We will work in the framework of primary importance for the EM algorithm, exponential families with density

$$f_{\boldsymbol{\theta}}(\boldsymbol{v}) \stackrel{\mathrm{def}}{=} e^{\boldsymbol{\theta}^{\mathsf{T}} \boldsymbol{t}(\boldsymbol{v}) - \kappa(\boldsymbol{\theta})} \tag{5.1}$$

w.r.t. some reference measure $\mu(\mathrm{d}\boldsymbol{v})$, where $\boldsymbol{v}$ is the vector of (completely) observed values. The ML estimator is then often some nice explicit function

$\widehat{\boldsymbol{\theta}} \overset{\text{def}}{=} \widehat{\boldsymbol{\theta}}(\boldsymbol{t}(\boldsymbol{v}))$ of $\boldsymbol{t}(\boldsymbol{v})$, where $\boldsymbol{v}$ is the observed outcome of the r.v. $\boldsymbol{V}$ with the prescribed density $f_{\boldsymbol{\theta}}$.

To fit the stochastic approximation EM algorithm into the Robbins–Monro framework one needs to view it as an algorithm looking for a zero of a function $\boldsymbol{h}$ of the sufficient statistic $\boldsymbol{t}$ rather than the parameter $\boldsymbol{\theta}$. To this end, we will assume that the MLE $\widehat{\boldsymbol{\theta}}$ is a 1-to-1 function of the observed $\boldsymbol{t}(\boldsymbol{v})$ (as is typically the case). The EM updating (A4.4) then means that $\boldsymbol{t}_{n+1} = \mathbb{E}_{\widehat{\boldsymbol{\theta}}(\boldsymbol{t}_n)}\big[\boldsymbol{t}(\boldsymbol{V}) \,\big|\, \boldsymbol{Y}\big]$, where $\boldsymbol{Y}$ is the vector of actually observed quantities. Thus the desired fixpoint (hopefully corresponding to the MLE) is the zero of

$$\boldsymbol{h}(\boldsymbol{t}) \overset{\text{def}}{=} \mathbb{E}_{\widehat{\boldsymbol{\theta}}(\boldsymbol{t})}\big[\boldsymbol{t}(\boldsymbol{V}) \,\big|\, \boldsymbol{Y}\big] - \boldsymbol{t}.$$

To apply the Robbins–Monro updating to get $\boldsymbol{t}_{n+1}$ from $\boldsymbol{t}_n$, we need an unbiased estimate of $\boldsymbol{h}(\boldsymbol{t}_n)$, which is obtained as

$$\frac{1}{m}\sum_{1}^{m} \boldsymbol{t}(\boldsymbol{V}_{n,j}) - \boldsymbol{t}_n$$

(cf. (A4.5)) by computing $\widehat{\boldsymbol{\theta}}(\boldsymbol{t}_n)$ and simulating m replicates $\boldsymbol{V}_{n,1}, \dots, \boldsymbol{V}_{n,m}$ of $\boldsymbol{V}$ from $\mathbb{P}_{\widehat{\boldsymbol{\theta}}(\boldsymbol{t}_n)}$. The Robbins-Monro updating is thus

$$\boldsymbol{t}_{n+1} \overset{\text{def}}{=} \boldsymbol{t}_n - \varepsilon_n\Big[\frac{1}{m}\sum_{j=1}^{m} \boldsymbol{t}(\boldsymbol{V}_{n,j}) - \boldsymbol{t}_n\Big].$$

See further Cappé, Moulines, & Rydén [65], where also some numerical illustrations of the performance of the stochastic approximation EM algorithm are given.

Exercises

5.1 (A) The assignment is to experiment with the Monte Carlo EM algorithm and the stochastic approximation EM algorithm in an example of rounded data from the bivariate $\mathcal{N}(\boldsymbol{0}, \boldsymbol{\Sigma})$ distribution. There are $n = 10{,}000$ (simulated) observations $\big((Y_{11}, Y_{12}), \dots, (Y_{n1}, Y_{n2})\big)$ given by the following table:

	−4	−3	−2	−1	0	1	2	3	4
−4	0	0	1	0	2	0	1	0	0
−3	0	0	4	21	13	17	4	0	0
−2	0	2	46	142	207	166	49	2	0
−1	0	10	151	628	867	554	130	11	2
0	2	23	220	911	1465	902	249	19	0
1	0	13	134	538	950	624	182	9	1
2	1	6	32	161	259	157	36	5	0
3	0	0	5	17	21	18	6	1	0
4	0	0	0	2	0	1	0	0	0

Thus, for example, out of the n bivariate normals $\big((V_{11}, V_{12}), \dots, (V_{n1}, V_{n2})\big)$, 142 had round$(V_{i1}) = -1$ and round$(V_{i2}) = -2$.

Hints: You will need to generate $\mathcal{N}(\mathbf{0}, \boldsymbol{\Sigma})$ r.v.'s (V_1, V) conditioned on round$(V_1) = k$ and round$(V_2) = \ell$. One way is to use inversion of one-dimensional conditioned normal r.v.'s (the relevant backgorund is in II.2a), gnerating first the V_1-component according to its marginal distribution and next the V_2-component according to the conditional distribution given V_1, cf. A1. As check of your results: the $\boldsymbol{\Sigma}$ used for generating the data was

$$
\begin{pmatrix}
2 & -2 \\
-2 & 4
\end{pmatrix}.
$$

Part B:
Algorithms for Special Models

Chapter IX
Numerical Integration

All Monte Carlo computations involve expectations $z = \mathbb{E}Z$ of a r.v. Z, which can in principle be expressed as functions of our driving sequence $\{U_n\}_{n=1,2,\ldots}$ of i.i.d. uniform r.v.'s, so that

$$Z = g(U_1, U_2, \ldots)$$

for some function g. When g depends only on a (fixed) finite number d of such U_n's, the expectation

$$z = \mathbb{E}Z = \int_0^1 \cdots \int_0^1 g(x_1, \ldots, x_d)\, \mathrm{d}x_1 \cdots \mathrm{d}x_d \stackrel{\text{def}}{=} \int_{H_d} g(\boldsymbol{x})\, \mathrm{d}\boldsymbol{x}$$

can be viewed as an integral on the d-dimensional unit hypercube $H_d \stackrel{\text{def}}{=} [0,1]^d$. This raises the possibility of computing $z = \mathbb{E}Z$ by appealing to the extensive literature on (deterministic) numerical integration.

1 Numerical Integration in One Dimension

Suppose that we wish to numerically compute the integral

$$z = \int_0^1 g(x)\, \mathrm{d}x \tag{1.1}$$

via a finite number m of function evaluations. Assuming that we choose to numerically evaluate g at the points $r_1, \ldots, r_m$, a natural class of integration schemes are those that are linear in $g(r_1), \ldots, g(r_m)$, so that the numerical approximation to z takes the form

$$\hat{z}_m = \sum_{i=1}^{m} w_i g(r_i) \tag{1.2}$$

for some set of weights $w_1, \ldots, w_m$. Such an integration scheme is characterized by the choice of weights $w_1, \ldots, w_m$ and evaluation points $r_1, \ldots, r_m$.

Suppose that we have a good selection of weights $w_1, \ldots, w_m$ and points $\{r_1, \ldots, r_m\} \subset [0, 1]$. Assuming that we are prepared to do $n = km$ function evaluations, the effect of the integration rule (1.2) based on n points can be "compounded" by applying it to integrate g over each of the k subintervals $[ik, (i+1)k]$ for $0 \leq i < k$, so that $\int_{i/k}^{(i+1)/k} g(x)\, dx$ is approximated by

$$\frac{1}{k} \sum_{j=1}^{m} w_j g\big(i/k + r_j/k\big). \tag{1.3}$$

This leads to a so-called compound integration rule (using nk function evaluations) based on the weights $w_1, \ldots, w_m$ and points $r_1, \ldots, r_m$, where (1.1) is approximated by

$$\frac{1}{k} \sum_{j=1}^{m} \sum_{i=0}^{k-1} w_j g\big(i/k + r_j/k\big). \tag{1.4}$$

When k is large, each of the above subintervals has a small width $h = 1/k$. This suggests that g, over each subinterval $[ik, (i+1)k]$, should be well approximated by a low-order polynomial (assuming smoothness of g). In particular, the difference between g and its pth-order (polynomial) Taylor expansion is then of order h^{p+1}. Thus, if we can select weights $w_1, \ldots, w_m$ and points $r_1, \ldots, r_m$ such that the rule (1.2) exactly integrates all polynomials of degree p, the integration error of our compound integration rule will be of order n^{-p-1} in the number of function evaluations n. Hence, in the presence of smoothness of the integrand g, the development of integration rules with a fast rate of convergence reduces to the question of how to develop rules of type (1.2) that exactly integrate all low-order polynomials.

Assuming that the m (distinct) points $r_1, \ldots, r_m$ are fixed, we have m decision variables $w_1, \ldots, w_m$ to be chosen at our discretion, so that a good choice of weights should lead to a rule of type (1.2) capable of exactly integrating all polynomials of order m or smaller, leading to a compound convergence rate of n^{-m}. Clearly, such a choice of weights $w_1, \ldots, w_m$ can be found as the solution to the linear system of equations

$$\sum_{i=1}^{m} w_i r_i^k = \int_0^1 x^k\, dx = \frac{1}{k+1} \tag{1.5}$$

for $k = 0, 1, \ldots, m - 1$. This linear system has a unique solution, namely

$$w_i = \int_0^1 \frac{w(x)}{w'(r_i)(x - r_i)},$$

where $w(x) \overset{\text{def}}{=} \prod_{i=1}^m (x - r_i)$. When $r_i = (i - 1)/(m - 1)$ for $i = 1, \ldots, m$, this integration rule is known as a *Newton–Cotes integration rule*.

Example 1.1 Let $m = 2$. Then in the Newton–Cotes setting, (1.5) means that $w_1 + w_2 = 1$, $w_1 \cdot 0 + w_2 = 1/2$, so $w_1 = w_2 = 1/2$, and (1.4) becomes

$$\frac{1}{2k} \big(g(0) + 2g(1/k) + 2g(2/k) + \cdots + 2g((k-1)/k) + g(1) \big). \qquad (1.6)$$

This approximation is known as the *trapezoidal rule*. See further Exercise 1.1 for the case $m = 3$ (*Simpson's rule*). □

Of course, the points $r_1, \ldots, r_m$ are also decision variables that may be chosen at our discretion. When combined with the weights $w_1, \ldots, w_m$, this offers $2m$ decision variables, suggesting the possibility that a good choice of both points and weights should be capable of exactly integrating all polynomials of degree $2m - 1$ or smaller, leading to a convergence rate for the corresponding compound rule of order of n^{-2m}. Such a rule requires solving the nonlinear system of equations

$$\sum_{i=1}^m w_i r_i^k = \frac{1}{k+1}, \quad 0 \le k \le 2m - 1,$$

for the w_i and r_i. Remarkably, Gauss showed that such a selection of points is possible, leading to the integration rules known as *Gaussian quadrature rules*.

Let as usual

$$\langle g_1, g_2 \rangle \overset{\text{def}}{=} \int_0^1 g_1(x) g_2(x) \, dx$$

be the inner product in $L_2[0, 1]$, the space of square (Lebesgue) integrable functions on $[0, 1]$. By applying Gram–Schmidt orthogonalization to the sequence $1, x, x^2, \ldots$, we arrive at a sequence of orthonormal functions $p_0 = 1, p_1, p_2, \ldots$ for which $\langle p_i, p_j \rangle = \delta_{ij}$ and p_i is a polynomial of degree i. Furthermore, the roots $r_1, \ldots, r_m$ of p_m are real and distinct, and lie in the interval $[0, 1]$. Put

$$w_i \overset{\text{def}}{=} \int_0^1 \frac{1}{p'_m(r_i)(x - r_i)} p_m(x) \, dx, \quad 1 \le i \le m.$$

It turns out that the w_i are necessarily positive and sum to one (and hence can be interpreted as a probability distribution on $\{r_1, \ldots, r_m\}$). Furthermore, the rule (1.2), with this specific choice of the w_i and r_i, exactly integrates all polynomials of degree $2m - 1$ or smaller, providing a compound Gaussian quadrature rule with corresponding rate n^{-2m}. See Evans

& Swartz [114, pp. 110–113] for a more detailed discussion; numerical integration is a topic treated in virtually every textbook in numerical analysis, e.g., Press et al. [293].

Exercises

1.1 (TP) Show that the analogue of (1.6) for $m = e$ (Simpson's rule) is

$$\frac{1}{6k}\left(g(0) + 4g(1/2k) + 2g(/1k) + 4g(3/2k) + \cdots + 4g((2k-1)/2k) + g(1)\right).$$

2 Numerical Integration in Higher Dimensions

As in the one-dimensional setting, it is natural to compute an approximation to the integral

$$z = \int_{H_d} g(\boldsymbol{x}) \, d\boldsymbol{x}$$

using a linear combination of function evaluations $g(\boldsymbol{r}_1), \ldots, g(\boldsymbol{r}_m)$ at the points $\boldsymbol{r}_1, \ldots, \boldsymbol{r}_m$ lying in the hypercube $H_d = [0,1]^d$, namely via

$$\widehat{z}_m = \sum_{i=1}^{m} w_i g(\boldsymbol{r}_i) \tag{2.1}$$

for suitable weights $w_1, \ldots, w_m$. Given an integration rule of the form (2.1), its effect can be compounded by partitioning the d-dimensional hypercube into k^d subhypercubes of equal volume k^{-d} and applying the rule (2.1) to each subhypercube, analogous to (1.3) in the one-dimensional setting.

A natural means of generating integration rules in the d-dimensional context is to leverage the theory just presented in the one-dimensional setting. In particular, given a one-dimensional integration rule of the form (1.2), with weights $w_1, \ldots, w_m$ and points $\{r_1, \ldots, r_m\} \subset [0,1]$, consider the product integration rule

$$\sum_{i_1}^{m} \sum_{i_2}^{m} \cdots \sum_{i_d}^{m} w_{i_1} w_{i_2} \cdots w_{i_d} g\left(r_{i_1}, r_{i_2}, \ldots, r_{i_d}\right). \tag{2.2}$$

If the w_i and r_i are chosen to exactly integrate all single variable polynomials of degree $m - 1$ or less, then all monomials of the form $x_1^{q_1} x_2^{q_2} \cdots x_d^{q_d}$ with $0 \le j_k \le m - 1$ for $1 \le k \le d$ can be exactly integrated. Since all polynomials of degree $m - 1$ or less on H_d are linear combinations of such monomials, it follows that such a product integration rule will exactly integrate all such polynomials. Assuming one uses a compound version of such a rule (on each of k^d different subhypercubes of equal volume k^{-d}), this suggests an integration error of order k^{-m}. Of course, the number of points at which g must be evaluated is $n = (mk)^d$, so that the integration error is of order $n^{-m/d}$ when expressed in terms of n. The rapid degradation of the convergence rate as a function of the dimension d is a manifestation of the so-called *curse of dimensionality*.

Note that if the w_i and r_i are chosen according to a one-dimensional Gaussian quadrature rule, then the w_i form a probability mass function on the r_i and

$$\left(w_{i_1} w_{i_2} \cdots w_{i_d} \right)_{1 \leq i_k \leq m, \, 1 \leq k \leq d}$$

can be viewed as a probability mass function (corresponding to the product measure) on $\{r_1, r_2, \ldots, r_m\}^d$. This suggests the possibility of computing the sum (2.2) via Monte Carlo sampling, potentially significantly reducing the associated number of function evaluations; see Evans & Swartz [113] for a discussion.

An alternative approach to developing higher-dimensional integration rules is to directly attack the question of choosing weights $w_1, \ldots, w_m$ and points $r_1, \ldots, r_m$ such that (2.1) will exactly integrate all polynomials on $[0, 1]^d$ of degree less than or equal to p. Note that the points $r_1, \ldots, r_m$ need not lie on a lattice, and hence the rule need not be of product type. Exact integration of all polynomials of degree p or less is equivalent to exact integration of all monomials $x_1^{q_1} \cdots x_d^{q_d}$ with $q_1 + \cdots + q_d \leq p$. The number of such monomials is

$$\binom{d+p}{d}, \tag{2.3}$$

so in principle, such an integration rule should exist so long as the number of decision variables $2m$ is at least (2.3). To find appropriate such weights and points, one must solve a nonlinear system of equations corresponding to that associated with Gaussian quadrature in one dimension. Of course, for fixed evaluation points $r_1, \ldots, r_m$, the problem of choosing appropriate weights involves solving a linear system of equations in m unknowns and the number m^* of equations given by (2.3). Typically, one expects this to be uniquely solvable when $m = m^*$. Of course, a product integration rule involving function evaluations at $(p+1)^d$ points or fewer is capable of exactly integrating all polynomials of order p or less. However, the fact that $(p+1)^d \leq m^*$ suggests that such product integration rules are less efficient than the rules that can be potentially developed by extending Gaussian quadrature to d dimensions. Nevertheless, the exponential explosion of the number of monomials of degree less than or equal to p as a function of d causes some computational degradation in the convergence rate (even in a high-dimensional Gaussian quadrature rule) as d increases. This is another manifestation of the curse of dimensionality and is an explanation for why the theory of this section is rarely implemented in practice in high-dimensional settings $(d \geq 4)$.

3   Quasi-Monte Carlo Integration

The *quasi-Monte Carlo method* (QMCM) is a numerical integration method. It is reminiscent of Monte Carlo integration[1] in that it uses estimates of the form

$$\int_{H_d} f(\boldsymbol{x}) \, \mathrm{d}\boldsymbol{x} \approx \frac{1}{n} \sum_{i=1}^{n} f(\boldsymbol{x}_i) \qquad (3.1)$$

for suitable points $\boldsymbol{x}_1, \ldots, \boldsymbol{x}_n$ in the unit hypercube $H_d = [0,1]^d$ to estimate the integral of a function $f : H_d \to \mathbb{R}$. The difference is the way the $\boldsymbol{x}_i$ are chosen, which aims to fill H_d more regularly than can be achieved by randomness. The gain is a rate of convergence, potentially close to $\mathrm{O}(1/n)$, that is better than the $\mathrm{O}(1/\sqrt{n})$ attained by the Monte Carlo method. This is illustrated in Figures 3.1, 3.2, where we have considered estimation of

$$z = \int_{H_4} (x_1 + x_2)(x_2 + x_3)^2 (x_3 + x_4)^3 \, \mathrm{d}x_1 \, \mathrm{d}x_2 \, \mathrm{d}x_3 \, \mathrm{d}x_4 \approx 2.63$$

using the estimate (3.1) and plotting it as function of $n = 1, \ldots, 50{,}000$. In Fig. 3.1, the broken curve is the Monte Carlo estimate, where the $\boldsymbol{x}_i$ are quadruples of standard pseudorandom numbers. For the solid curve, we used instead a particular QMCM sequence, namely the Halton sequence in dimension $d = 4$ (see below). It is seen that the QMCM estimate stabilizes much faster that the Monte Carlo estimate. The slow rate at which the Monte Carlo estimate converges to the correct value is illustrated in Figure 3.2, where three independent Monte Carlo runs are plotted for $n \le 1{,}000{,}000$.

The motivation for the use of the QMCM comes to a large extent from numerical examples like this, and the method has recently become popular within the area of mathematical finance. There are of course also costs and restrictions on the applicability; we return to this in Section 3.3d.

A main reference for QMCM is Niederreiter [278]; Glasserman [133] gives a survey based on this reference, which is somewhat more extensive than our treatment. See also Caflish [64], L'Ecuyer & Lemieux [233], and Morokoff [262]. A link to the main conference series in the area is given in [w³.20].

3a   An Example: Nets

The property of regular spacing referred to above is usually termed *low discrepancy* (we discuss rigorous definitions of this in Section 3.3c). A typical example of a low-discrepancy set is a *(t, m, d) net with base b*, defined

---

[1] But not more refined quadrature rules like the ones discussed in Sections 1, 2, which will usually apply more sophisticated weights than $1/n$ in (3.1).

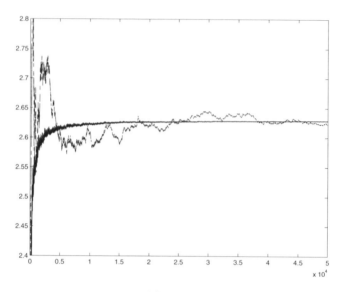

FIGURE 3.1

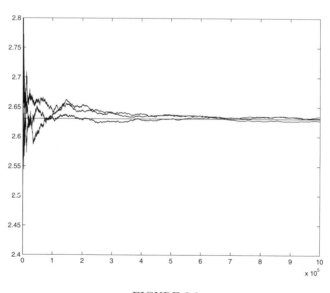

FIGURE 3.2

as a set of $n = b^m$ points in H_d such that exactly b^t points fall into a prespecified family of b-ary boxes of volume b^{t-m} (here obviously $t \leq m$). We will consider only the case $b = 2$. A binary box is then a subset of H_d of the form

$$\prod_{i=1}^{d} \left[a_i 2^{-j_i}, (a_i + 1) 2^{-j_i} \right)$$

with $j_i \in \mathbb{N}$ and $a_i \in \{0, 1, \ldots, j_i - 1\}$. For example,

$$[3/8, 1/2 = 4/8) \times [3/4 = 6/8, 7/8) \quad \text{and} \quad [1/2, 3/4) \times [3/4 = 12/16, 13/16)$$

are binary boxes of the appropriate type of volume 2^{-6} for $d = 2$ but $[5/8, 7/8) \times [3/4, 13/16)$ is not (even if the volume is 2^{-6}).

The concept of a net has the drawback of fixing the number $n = b^m$ of points in advance. This is inconvenient in particular in situations in which it is not known how large a value of n is required. To resolve these problems, one defines an infinite sequence $\{\boldsymbol{x}_k\}_{k \in \mathbb{N}}$ to be a (t, d)-sequence if for all $m > t$ the jth segment of length b^m is a (t, m, d) net in base b.

3b Specific Low Discrepancy Sequences

Nets and (t, d)-sequences have obviously desirable properties in terms of filling H_d. However, the known algorithms for generating them are quite complex and if implemented by an average user in a standard toolbox or programming language, speed will be inferior compared to built-in pseudorandom numbers. Therefore, one will usually apply standard packages (e.g., [w^3.12], [w^3.13]) for such structures; efficient generation is for the specialists. In the following, we present some different low-discrepancy sequences that have weaker regularity but are easier to generate.

For each $k \in \mathbb{N}$ and each b (typically a prime), we let $a_0(k), a_1(k), \ldots$ denote the coefficients in the (unique) expansion

$$k = \sum_j a_j(k) b^j, \quad \text{where} \quad a_j(k) \in \{0, 1, \ldots, b - 1\}, \tag{3.2}$$

of k in powers of b (the sum is of course always finite: $a_j(k) = 0$ at least for $j > \log_b k$). Here $0 \le a_j(k) < b$. We will denote by $\psi_b(k)$ the *radical inverse* of k,

$$\psi_b(k) = \sum_j a_j(k) b^{-j}.$$

This mapping "flips $a_j = a_j(k)$ around the radix to the base-b fraction $.a_0 a_1 \ldots$" For example, $\psi_{10}(1043) = 0.3401$ and $\psi_2(368) = 29/512$ since the binary representations of 368 and 29/512 are respectively 101110000 and 0.000011101, i.e.,

$$368 = 0 \cdot 2^0 + 0 \cdot 2^1 + 0 \cdot 2^2 + 0 \cdot 2^3 + 1 \cdot 2^4 + 1 \cdot 2^5 + 2^6 + 0 \cdot 2^7 + 1 \cdot 2^8,$$

$$\frac{29}{512} = \frac{16 + 8 + 4 + 1}{512} = \frac{0}{2} + \frac{0}{4} + \frac{0}{8} + \frac{0}{16} + \frac{1}{32} + \frac{1}{64} + \frac{1}{128} + \frac{0}{256} + \frac{1}{512}.$$

The *base-b Van der Corput sequence* is $\{\psi_b(k)\}_{k \in \mathbb{N}}$. Its element 0 is 0 (as for many specific low-discrepancy sequences), the next $1/2$, the two next $1/4, 3/4$, the four next $1/8, 5/8, 3/8, 7/8$, and so on. Van der Corput sequences are key ingredients in many practically relevant low-discrepancy sequences. In particular, a *Halton sequence* in dimension d is a sequence

$\{x_k\}_{k\in\mathbb{N}}$ obtained by choosing d relatively prime integers $b_1, \ldots, b_d$ and letting

$$x_k = \big(\psi_{b_1}(k), \ldots, \psi_{b_d}(k)\big).$$

Usually, $b_1, \ldots, b_d$ are the first d of the primes $2, 3, 5, 7, 11, \ldots$; in particular, this choice was made in Figure 3.3 below.

Remark 3.1 The generation of a Halton sequence is most conveniently done recursively by updating $a_0(k), a_1(k), \ldots$ (the coefficients in the expansion of k in powers of $b \in \{b_1, \ldots, b_d\}$) to $a_0(k+1), a_1(k+1), \ldots$. To this end, write $a_j = a_j(k)$ and note that if $a_0 < b - 1$, then the updating simply amounts to incrementing a_0 by 1. Otherwise, if $a_0 = b - 1$, a_0 is updated to 0 and one tries to update a_1 in the same way, going on to a_2 if $a_1 = b - 1$. Representing the a_j as a finite vector $a = (a_0, \ldots, a_m)$ with $a_m \neq 0$, the updating algorithm becomes:

(i) $i \longleftarrow 0$;

(ii) if $a_i < b - 1$: $a_i \longleftarrow a_i + 1$; go to (iv);
      else $a_i \longleftarrow 0$; $i \longleftarrow i + 1$; if $i < m + 1$, go to (ii);

(iii) $a \longleftarrow (a, 1)$;

(iv) return a.

Note that step (iii) is executed only if $k = b^m - 1$, in which case the updating results in

$$a = \big(\underbrace{b - 1, b - 1, \ldots, b - 1}_{m+1}\big) \longrightarrow \big(\underbrace{0, 0, \ldots, 0}_{m+1}, 1\big) \qquad \square$$

The low-discrepancy properties of the Halton sequence is illustrated in Figure 3.3 which shows the extent to which the bivariate marginals fill the unit square well. The number of points is 1,000 and we have taken $d = 16$; the $[2, 3]$ in (d) (say) means that it is components 2 and 3 that are plotted against each other, that is, the Van der Corput sequences with bases 3 and 5. As comparison, each of the three panels of Figure 3.4 contains 1,000 pseudorandom points in the unit square.

The figures show indeed the greater uniform coverage of the Halton sequence compared to random numbers, at least in the first components, and this is what is behind the superiority in numerical integration problems such as the one in Figure 3.1. On the other hand, some systematic deviations start to show up for the higher-dimensional components (the points get more concentrated on bands as the index of the component increases), and this is indeed a problem. The deterioration of the performance of the higher-dimensional coordinates can be seen from Figure 3.5, which is the same as Figure 3.1 (with the Monte Carlo estimates omitted), except that the x_k are now components $13, \ldots, 16$ of the Halton sequence, not $1, \ldots, 4$.

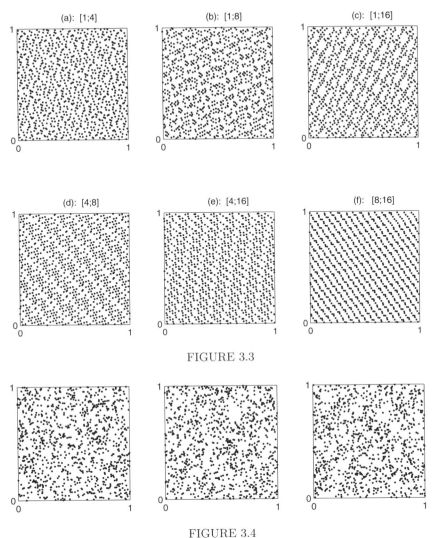

FIGURE 3.3

FIGURE 3.4

3c Discrepancy. The Koksma–Hlawka Bound

So far, we have kept the meaning of "low discrepancy" vague. The rigorous discussion in the literature involves the concept of the discrepancy

$$D\big(\boldsymbol{x}_1,\ldots,\boldsymbol{x}_n;\mathscr{A}\big) \stackrel{\text{def}}{=} \sup_{A\in\mathscr{A}} \left| \frac{\#\{i:\boldsymbol{x}_i\in A\}}{n} - |A| \right|$$

($|\cdot|$ = Lebesgue measure) of a finite set $\boldsymbol{x}_1,\ldots,\boldsymbol{x}_n \in H_d$ relative to a collection $\mathscr{A}$ of subsets of H_d.

The *ordinary discrepancy* $D(\boldsymbol{x}_1,\ldots,\boldsymbol{x}_n)$ is obtained by letting $\mathscr{A}$ consist of all rectangles $\prod_1^d[a_j,b_j)$, the *star discrepancy* $D^*(\boldsymbol{x}_1,\ldots,\boldsymbol{x}_n)$ by

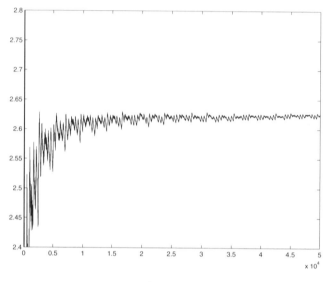

FIGURE 3.5

imposing the restriction $a_1 = \cdots = a_d = 0$ (for the connection, see Exercise 3.2). It is known that $D^*(\boldsymbol{x}_1, \ldots, \boldsymbol{x}_n)$ grows at rate at least $(\log n)^{d-1}/n$ for any sequence $\boldsymbol{x}_1, \boldsymbol{x}_2, \ldots$, and often the term "low discrepancy" is restricted to sequences attaining this lower bound.

For a smooth function f on H_d, we define the variation (associated with the names of Hardy and Krause) by

$$V(f) \stackrel{\text{def}}{=} \sum_{k=1}^{d} \sum_{1 \leq i_1 < \cdots < i_k \leq d} \int_{H_d} \left| \frac{\partial^k f}{\partial u_1 \cdots \partial u_k} (\boldsymbol{u}(i_1, \ldots, i_k)) \right| du_{i_1} \cdots du_{i_k},$$

where $\boldsymbol{u}(i_1, \ldots, i_k)$ is the point in H_d with $u(i_1, \ldots, i_k)_j = 1$ when $j \notin \{i_1, \ldots, i_k\}$ and $-u_{i_j}$ otherwise. We can now state the *Koksma–Hlawka inequality*:

$$\left| \int_{H_d} f(\boldsymbol{x}) \, d\boldsymbol{x} - \frac{1}{n} \sum_{i=1}^{n} f(\boldsymbol{x}_i) \right| \leq V(f) D^*(\boldsymbol{x}_1, \ldots, \boldsymbol{x}_n) \qquad (3.3)$$

(there is also a version for nonsmooth f that we omit).

The Koksma–Hlawka inequality justifies the statement that QMCM estimators have close-to-O$(1/n)$ convergence when implemented via sequences that are low-discrepancy in the above strict $(\log n)^{d-1}/n$ sense or close. One drawback is that it is only an upper bound, so that in practice the convergence is usually much faster (other concepts such as the L^2 discrepancy discussed in Drmota & Tichy [97] take account of this by discussing averages over a class of integrands rather than the worst case). Another is that the r.h.s. is difficult to compute; the star discrepancy has been evaluated for some specific sequences, but most often the effort is considerable, and further, the integrals in $V(f)$ are unknown and difficult to compute (note that in realistic situations we don't even know $\int_{H_d} f!$). A further restriction on the usefulness is that $V(f) = \infty$ commonly occurs.

3d   Discussion

(i) The typical application of QMCM is to evaluate integrals of the form $\int_{\Omega} \varphi(\boldsymbol{y}) \, dy$ for a suitable region $\Omega \subseteq \mathbb{R}^d$. So far, we have restricted Ω to H_d, but this is easily removed by a suitable transformation $\Omega \to H_d$; see I.5.1.

(ii) It is widely believed that QMCM is efficient only if the dimension d is fixed, so that one should avoid problems in which a stopping time is involved (as for example in r.v. generation by rejection).

To illustrate this point, we considered a first-passage problem, computing the probability z that a random walk with increments $X \stackrel{\text{def}}{=} A - B$ distributed as the independent difference between exponential r.v.'s with means 1 for A and 2 for B exits the interval $(-4, 2)$ to the right. Here

$$ z = (1 - e^2/2)(2e - e^{-2}/2) = 0.174 $$

is explicitly available via martingale stopping.

Traditional Monte Carlo confirmed this number, while in replacing pseudorandom numbers by the 2-dimensional Halton sequence with $R = 1{,}000$ runs of the random walk up to exit time τ, there appeared to be convergence to $0.118 \neq z$. A combination in which (A_1, B_1, A_2, B_2) were generated by the 4-dimensional Halton sequence but the remaining $A_3, B_3, \ldots, A_\tau, B_\tau$ via pseudorandom numbers gave convergence to the correct z, but did not improve the precision significantly compared to the run with just pseudorandom numbers.

Se further Exercise 3.3.

(iii) Concerning the dependence of the efficiency on d, the experience is that standard quadrature rules outperform both MC and QMC for small d, that QMC is the best of the three methods for moderate d, and that MC is best for large d (largely because the order of the error term does not depend on d as for the other two).

(iv) A universal advantage of MC as compared to QMC is that the error bound (the confidence interval) is easy to compute, but the one for QMC (the K-H inequality) is not. To remedy this, *randomized QMC* has been suggested. This means (in the numerical integration setting) that one selects a block size m and applies standard MC with R

replicates of the estimator

$$Z \stackrel{\text{def}}{=} \frac{1}{m} \sum_{i=1}^{m} f\left(\boldsymbol{x}_i + \boldsymbol{U}\right),$$

where $\boldsymbol{x}_1, \ldots, \boldsymbol{x}_m$ is a fixed low-discrepancy sequence (the same for all replications), $\boldsymbol{U}$ is uniform on H_d, and $\boldsymbol{x} + \boldsymbol{U}$ is computed modulo 1 componentwise (the R replications use i.i.d. $\boldsymbol{U}_1, \ldots, \boldsymbol{U}_R$). Then a confidence interval can be produced in a standard way. Randomized QMCM has the further advantage of producing estimates that are unbiased in the traditional sense.

Further randomization ideas are surveyed in L'Ecuyer & Lemieux [233].

(v) The problems associated with the higher-order components, as demonstrated by Figure 3.5 and the surrounding discussion, can sometimes be severe. One attempt to get out of this uses a repa-rameterization $(x_1, \ldots, x_d) \rightarrow (z_1, \ldots, z_d)$ such that the z_i influence on $f(z_1, \ldots, z_d)$ is monotonically decreasing in i; the low-discrepancy sequence is then used for the z_i rather than the x_i, thereby assigning the more reliable first components of the sequence to the most important variables. A notable example of this is generation of a Brownian motion $\{B(t)\}_{0 \le t \le 1}$ by bisection (see X.2b), where one may use the first coordinate for $B(1)$, the next for the normal r.v. generating $B(1/2)$ using the Brownian bridge representation, the next two for $B(1/4), B(3/4)$, and so on. Associated with this approach is the concept of *effective dimension* $d' < d$, which roughly means that the influence of $x_{d'+1}, \ldots, x_d$ on the integral of $f(x_1, \ldots, x_d)$ is minor. Another approach to high dimensions is to use *hybrid generators*, that is, sequences such that the first p components are one of the standard QMCM sequences and the last $d - p$ components are pseudorandom numbers (say $p = 50$).

Once all of this is said, there is no doubt that QMCM can be highly efficient for certain specific (and important) types of problems.

Exercises

3.1 (TP) Demonstrate why the b_i used in the Halton sequence must be relatively prime by considering the case $d = 2$, $b_1 = 2$, $b_2 = 4$.

3.2 (TP) Verify the inequality

$$D^*(\boldsymbol{x}_1, \boldsymbol{x}_n) \le D(\boldsymbol{x}_1, \boldsymbol{x}_n) \le 2^d D^*(\boldsymbol{x}_1, \boldsymbol{x}_n)$$

in the case $d = 2$.

3.3 (A) Supplement (ii) of Section 3d by further runs using larger R, different sequences of quasirandom numbers etc.

3.4 (A) Consider the same random-walk first-passage problem as in (ii) of Section 3d. Redo the calculations there, using (a) QMC naively implemented via the 2-dimensional Halton sequence, (b) standard Monte Carlo, and (c) a combination of QMC and Monte Carlo explained below. Present (as in the text) the results in terms of plots showing the convergence

Details for (c): use the 6-dimensional Halton sequence to generate $U_1, U_2, U_3, V_1,$ V_2, V_3. If $\tau > 3$, generate $U_4, \ldots, U_\tau, V_4, \ldots, V_\tau$ as traditional pseudorandom numbers.

3.5 (A) Consider an Asian option with maturity $T = 3$ years, strike price $K = 50, 100$, or 150, $r = 4\%$, and $S(t)$ geometric Brownian motion with $S(0) = 100$ and yearly volatility 0.25. Compute the price using QMC implemented via the Halton sequence. Present (as in the text) the results in terms of plots showing the convergence and containing some Monte Carlo runs to compare with.

If time permits, redo using the *Halton leap sequence* instead. This is defined as the Halton sequence sampled at $b_{d+1}, 2b_{d+1}, \ldots,$ where b_{d+1} should be taken prime to $b_1, \ldots, b_d$.

3.6 (A) Consider the NIG European call option in Exercise VII.2.3. Redo the price calculation, using QMC implemented via the 2-dimensional Halton sequence. Present (as in the text) the results in terms of plots showing the convergence and containing some Monte Carlo runs to compare with. Compare also for a fixed number R of replications the precision of the estimates using QMC, crude Monte Carlo, and Monte Carlo, where the variates in the routine for generating the NIG r.v.'s are stratified.

Chapter X
Stochastic Differential Equations

1   Generalities about Stochastic Process Simulation

Let $\{X(t)\}_{0 \le t \le T}$ be a stochastic process in discrete or continuous time. The problem that we study in this and the following chapters is to generate a sample path by simulation, where T is a fixed number (say $T = 1$) or a stopping time.

The methods that we survey are highly dependent on the type of process in question, and also on the type of application (what is the sample path to be used for?). In some cases such as Lévy or stable processes, it may even be nontrivial or impossible to generate one-dimensional distributions (i.e., $X(T)$ for a fixed T); we are then faced with a particular problem in random-variate generation. In other situations such as stationary Gaussian processes, the generation of $X(T)$ for a single T may be easy, but the dependence structure may make it difficult to generate finite-dimensional distributions (i.e., random vectors of the form $\big(X(0), X(1), \ldots, X(T)\big)$ in discrete time, or a discrete skeleton in continuous time), in particular when the dimension is high. A method that is suitable for a fixed T may not be suitable if T is a stopping time; say the method is based on generating discrete skeletons by bisection, starting with generating $X(0), X(T)$, then supplementing with $X(T/2)$, next with $X(T/4), X(3T/4)$, and so on (see further Section 2b). In continuous time, it may be straightforward to generate a discrete skeleton with the correct finite-dimensional distributions (say in the case of Brownian motion), but using a discrete skeleton may introduce errors in the specific application, such as characteristics of the first-passage time $\inf\{t > 0 : X(t) \geq x\}$.

The error criteria to be used depend on the type of application. If one is interested in just generating $X(T)$ sufficiently accurately (in the distributional sense), an appropriate error criterion may be

$$\sup_{f \in \mathscr{C}} \left| \mathbb{E}f\big(\widetilde{X}(T)\big) - \mathbb{E}f\big(X(T)\big) \right| \tag{1.1}$$

for a suitable class $\mathscr{C}$ of smooth functions, where $\widetilde{X}(\cdot)$ is the simulated path.[1] The accuracy of the sample path approximation can be measured by criteria such as

$$\mathbb{E} \sup_{0 \leq t \leq T} \left| \widetilde{X}(t) - X(t) \right|, \tag{1.2}$$

assuming that $\widetilde{X}(\cdot)$ and $X(\cdot)$ can be coupled (generated on a common probability space) in a natural way, or, for some suitably chosen p, via an L_p error criterion such as

$$\mathbb{E} \int_0^T \left| \widetilde{X}(t) - X(t) \right|^p \mathrm{d}t, \quad \mathbb{E} \sum_{n=0}^T \left| \widetilde{X}_n - X_n \right|^p \tag{1.3}$$

in continuous, respectively discrete, time.

For some continuous-time processes such as workloads in queues or queue lengths and compound Poisson processes (with a possibly added linear drift), there is a natural discrete event structure, that is, an embedded sequence of points that determines the evolution of the process as a whole. For many interesting processes such as general Lévy processes and solutions to SDEs, this is not the case, and the process is then usually generated from a discrete skeleton. Sometimes it then works quite well to define $\{\widetilde{X}(t)\}$ by linear interpolation between grid points or by taking $\{\widetilde{X}(t)\}$ piecewise constant, i.e., $\widetilde{X}(t) = \widetilde{X}(nh)$ for $nh \leq t < (n+1)h$, but in other cases such procedures may be clearly unreasonable.

Remark 1.1 Note that if, as in (1.1), one wishes to compute expectations of the form $\mathbb{E}f\big(X(T)\big)$ for an SDE in a low dimension, an alternative to simulation that is likely to be more efficient numerically is to solve the corresponding backward PDE for

$$g(x, t) \stackrel{\text{def}}{=} \mathbb{E}\big[f\big(X(T)\big) \,\big|\, X(t) = x \big]. \tag{1.4}$$

Of course, if X is d-dimensional with d large, the use of PDE methods for computing such expectations suffers from the same "curse of dimensionality" as do conventional (non-sampling-based) numerical integration

---

[1] The notation $\widetilde{X}(\cdot)$ is used only at this place. Later in this chapter, we will simulate along a discrete grid $0, h, 2h, \ldots$ and then write $X^h(\cdot)$ for the simulated path.

methods in high dimensions. This makes simulation-based methods attractive for computing either expectations of complex path-dependent functionals (regardless of dimension) or computing expectations of smooth functionals of $X(T)$ when d is (say) greater than or equal to 3.   □

2   Brownian Motion

Let $\{B(t)\}_{t\geq 0}$ be standard Brownian motion (BM). Throughout this chapter, we use the notation $t_n^h \overset{\text{def}}{=} nh$, $\Delta_n^h B \overset{\text{def}}{=} B(t_n^h) - B(t_{n-1}^h)$.

The generation of BM along the discrete skeleton $0, t_1^h, t_2^h, \ldots$ is straightforward: just generate the increments $\Delta_n^h B$ as i.i.d. $\mathcal{N}(0, h)$ variables and let

$$B_n^h \overset{\text{def}}{=} B(t_n^h) = \Delta_1^h B + \cdots + \Delta_n^h B. \qquad (2.1)$$

In view of the simplicity of this procedure, there is not much literature on the simulation of Brownian motion. A notable exception is Knuth [219].

2a   The Error from Linear Interpolation

To generate a continuous-time version of BM is intrinsically impossible because of the nature of the paths. This creates the problem that most Brownian functionals cannot be generated exactly either from a discrete skeleton or from other obvious discrete-event schemes.

An obvious procedure is to use linear interpolation between grid points. Let $T = 1$, consider the grid $t_n^h \overset{\text{def}}{=} nh$ with $h \overset{\text{def}}{=} h_N \overset{\text{def}}{=} 1/N$, and let $B^h(t)$ be generated as a Gaussian random walk at the t_n^h as in (2.1) and by linear interpolation in between; note that this introduces a particular coupling between the approximation $B^h(t)$ and the BM itself, namely, the two processes are taken to coincide at the t_n^h. The error as measured by the criteria (1.2), (1.3) is then given by the following result:.

Proposition 2.1 Let $h \overset{\text{def}}{=} h_N \overset{\text{def}}{=} 1/N$. Then

$$\mathbb{E}\int_0^1 |B^h(t) - B(t)|\, dt = c_1/N^{1/2}$$

where $c_1 = \sqrt{\pi/32}$. Further,

$$\sqrt{N/\log N}\ \mathbb{E}\sup_{0\leq t\leq 1} |B^h(t) - B(t)| \to c_2$$

as $N \to \infty$ for some $0 < c_2 < \infty$.

Proof. Let $Z^T(t) \overset{\text{def}}{=} B(t) - (t/T)B(T)$, $0 \leq t \leq T$, denote the Brownian bridge in the time interval $0 \leq t \leq T$. Then $\int_0^1 |B^h(t) - B(t)|\, dt$ has the same distribution as the sum of N independent copies of $\int_0^{1/N} |Z^{1/N}(t)| \, dt$. By standard scaling properties of Brownian motion, $\{Z^T(tT)\}_{0 \le t \le 1}$ has the same distribution as $\{\sqrt{T} Z^1(t)\}_{0 \le t \le 1}$, and so

$$
\mathbb{E} \int_0^1 \big| B^h(t) - B(t) \big| \, dt
$$

$$
= \; N \mathbb{E} \int_0^{1/N} |Z^{1/N}(t)| \, dt \;=\; \mathbb{E} \int_0^1 |Z^{1/N}(t/N)| \, dt
$$

$$
= \; N^{-1/2} \mathbb{E} \int_0^1 |Z^1(t)| \, dt \;=\; N^{-1/2} \int_0^1 \sqrt{2t(1-t)/\pi} \, dt \;=\; N^{-1/2} c_1
$$

(note that $Z^1(t)$ has a $\mathcal{N}(0, t(1-t))$ distribution, that the expected value of the absolute value of a $\mathcal{N}(0,1)$ r.v. is $\sqrt{2/\pi}$, and that Beta$(3/2, 3/2) = \pi/8$).

The second assertion follows by extreme value theory. First, as above, $\sup_{0 \le t \le 1} |B^h(t) - B(t)|$ has the same distribution as the maximum of N independent copies of $\sup_{0 \le t \le 1/N} |Z^{1/N}(t)|$, which in turn has the same distribution as the maximum of N independent copies of $\sup_{0 \le t \le 1} |Z^1(t)|/N^{1/2}$. By extreme value theory, this maximum minus $\sqrt{\log N}/N^{1/2}$ has a (scaled) Gumbel limit distribution; for uniform integrability, see Pickands [290]. □

Note that the rate of convergence in Proposition 2.1 is much slower than in the deterministic case: if $f(t)$ is a function of $t \in [0, 1]$ and f^h the function obtained by linear interpolation using the same $1/N$-grid as above, then easy calculations using Taylor expansions show that

$$
\int_0^1 \big| f^h(t) - f(t) \big| \, dt \;\sim\; \frac{1}{12N^2} \int_0^1 |f''(t)| \, dt,
$$

$$
\sup_{0 \le t \le 1} \big| f^h(t) - f(t) \big| \;\sim\; \frac{1}{4N^2} \sup_{0 \le t \le 1} |f''(t)|.
$$

This is an important point in understanding the care that must be taken in the simulation of SDEs.

2b Bisection

Assume as an example that we are interested in characteristics associated with the first-passage time $\tau(x) \stackrel{\text{def}}{=} \inf\{t \ge 0 : B(t) \ge x\}$. We can then simulate a discrete skeleton $\{B_n^h\}$, where $B_n^h \stackrel{\text{def}}{=} B^h(t_n^h) = B(nh)$, and use $\tau_n(x) \stackrel{\text{def}}{=} h \cdot \inf\{n : B_n^h \ge x\}$ as approximation. However, this obviously overestimates $\tau(x)$, so one could consider making the skeleton finer and finer so as to judge whether such a discrete approximation is sufficiently accurate. We proceed to give the details for such algorithms.

The basis is the observation that

$$B(t + h) \mid B(t) = a, \ B(t + s) = b$$
$$\sim \quad \mathcal{N}\!\left(a + (b - a)h/s, \ h - h^2/s\right), \quad 0 < h < s,$$

as follows easily from the standard formula for conditioning in the multivariate normal distribution (see A1). In particular,

$$B(t + h) \mid B(t) = a, \ B(t + 2h) = b \quad \sim \quad \mathcal{N}\!\left((a + b)/2, \ h/2\right).$$

Consider $\{B(t)\}$ in the time interval $[0, 1]$. The goal of *bisection* is to generate a set of r.v.'s

$$b_0^k, \ b_1^k, \ \ldots, \ b_{2^k - 1}^k, \ b_{2^k}^k$$

that have the same joint distribution as

$$B(0), \ B(1/2^k), \ \ldots, \ B\!\left((2^k - 1)/2^k\right), \ B(1)$$

and the sample-path consistency property

$$b_{2j}^k \ = \ b_j^{k-1}, \ j = 0, 1, \ldots, 2^{k-1}. \tag{2.2}$$

First generate b_0^0, b_1^0 by taking $b_0^0 = 0$, $b_1^0 \sim \mathcal{N}(0, 1)$. If the b_j^{k-1} have been generated, define $b^k(i)$ by (2.2) for $i = 2j$. For $i = 2j + 1$, take $b_i^k \sim \mathcal{N}\!\left(y, 2^{-k-1}\right)$, where

$$y \ = \ \frac{1}{2}\left(b_j^{k-1} + b_{j+1}^{k-1}\right).$$

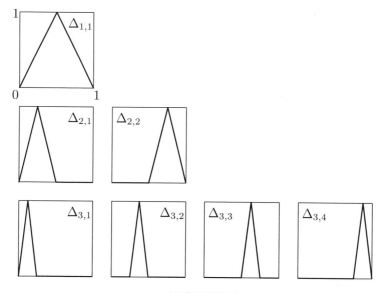

FIGURE 2.1

The construction is related to the *wavelet representation* of Brownian motion; see Steele [346]. To see this, use the representation $b_1^0 = V_0$,

$$b_i^k = \frac{1}{2}\left(b_j^{2^{k-1}} + b_{j+1}^{2^{k-1}}\right) + 2^{-(k+1)/2}V_{k,(i+1)/2}$$

for $i = 1, 3, \ldots, 2^k - 1$, where the $V_{k,i}$ are i.i.d. $\mathcal{N}(0,1)$ r.v.'s and let $\{b^k(t)\}_{0 \leq t \leq 1}$ be the process obtained by linear interpolation between the b_i^k. Then it is easy to see that

$$b^k(t) = \Delta_0(t)V_0 + \sum_{\ell=1}^{k} 2^{-(\ell+1)/2} \sum_{i=1}^{2^{\ell-1}} \Delta_{\ell,i}(t)V_{\ell,i} \,,$$

where $\Delta_0(t) = t$ and the nonzero parts of $\Delta_{1,1}(t), \Delta_{2,1}(t), \Delta_{2,2}(t), \Delta_{3,1}(t)$ are the triangle function in Figure 2.1. In fact, the wavelet representation states that

$$\Delta_0(t)Z_0 + \sum_{\ell=1}^{\infty} 2^{-(\ell+1)/2} \sum_{i=1}^{2^{\ell-1}} \Delta_{\ell,i}(t)V_{\ell,i}$$

converges a.s. and defines a Brownian motion with continuous paths.

Bisection is well suited for combining with QMCM or stratification. In both cases, one takes advantage of the fact that the Brownian path has a shape that is to first order determined by V_0, to the second by $V_{1,1}$, to the third by the $V_{2,i}$ and so on. Therefore, for QMCM one could let V_0 and the $V_{\ell,i}$, $\ell \leq k$, be generated from a low-discrepancy sequence and the $V_{\ell,i}$ with $\ell > k$ from ordinary pseudorandom numbers (cf. the discussion of hybrid generators in IX.3). For stratification, one could stratify the same V's, possibly coarser for larger ℓ.

Exercises

2.1 (A) Redo the Asian option in Exercise IX.6.2 for $N = 6$ sampling points, using bisection and stratification. The simplest way may be to generate the Brownian motion at $N' = 8$ half-yearly sampling points and ignore the last two. The stratification can be done, for example, by taking eight strata for the r.v. generating $B(8)$, four for the one for $B(4)$, and two for each of those for $B(2)$, $B(6)$.

2.2 (A) Redo Exercise III.4.2 (the Kolmogorov–Smirnov test) using bisection and stratification. The stratification can be done, for example, by taking eight strata for the r.v. generating, $B(1/2)$, and four for each of the ones for $B(1/4), B(3/4)$. What about $B(1)$?

2.3 (A) Consider an Asian option with maturity $T = 10$ years, strike price $K = 100$, and sampled monthly. That is, the price is

$$e^{-rT}\mathbb{E}^*\big[A_N(T) - K\big]^+, \quad \text{where} \quad A_N(T) \stackrel{\text{def}}{=} \frac{1}{N}\sum_{i=1}^{N} S(t_i) \,,$$

$r = 4\%$, $N = 120$, $S(t)$ is the price of the underlying asset at time t, $t_i \stackrel{\text{def}}{=} T/N$, and $\mathbb{E}^*$ refers to the risk-neutral measure. It is assumed that $S(t)$ is geometric

Brownian motion with $S(0) = 100$ and yearly volatility 0.25. Thus, $S(t)$ is distributed as

$$S(0) \exp\left\{ \left(r - \tfrac{1}{2}\sigma^2 \right)t + \sigma B(t) \right\}$$

under $\mathbb{P}^*$, where B is standard BM. Compute the price using (a) ordinary (crude) Monte Carlo (CMC), (b) CMC combined with bisection and stratification, and (c) a hybrid QMCM generator combined with bisection.

For (b), (c), it may be easier to implement the bisection by generating up to $N' = 128 = 2^7$ instead of $N = 120$. In (b), you may, for example, stratify $B(N')$ into 16 equiprobable strata and $B(N'/2)$ into 8 so that you get a total of $S = 128$ strata. The easiest way to perform the stratified sampling is probably to keep the relevant restricted normals for each stratum in an array. The arrays are then filled by sampling standard normals and moving them to the relevant array (if it is not already full) until all of the S arrays are full. In (c), "hybrid" means that the first (say) M of the needed N' uniforms are generated by something like the Halton sequence and the rest as ordinary pseudorandom numbers.

2.4 (A) The stratification in Exercise 2.3 is a naive first attempt, and a much larger variance reduction can be obtained by using a division into strata of the range $B(N')$, which is finer for large values and coarser for small values. Experiment with this, using CMC only and stratifying $B(N')$ only.

3   The Euler Scheme for SDEs

3a   Generalities

In this and the following sections, we consider the SDE $X(0) = x_0$,[2]

$$\mathrm{d}X(t) = a\big(t, X(t)\big)\,\mathrm{d}t + b\big(t, X(t)\big)\,\mathrm{d}B(t), \quad t \geq 0, \qquad (3.1)$$

subject to $X(0) = x_0$, where $\{B(t)\}_{t \geq 0}$ is standard Brownian motion. The meaning of (3.1) is that

$$X(t) = x_0 + \int_0^t a\big(s, X(s)\big)\,\mathrm{d}s + \int_0^t b\big(s, X(s)\big)\,\mathrm{d}B(s), \quad t \geq 0, \quad (3.2)$$

where the first integral is an ordinary integral and the second has to be interpreted in the Itô sense. We discuss the multidimensional case in Section 7 but assume now that everything is $\mathbb{R}$-valued. Recall that $t_n^h \overset{\text{def}}{=} nh$, $\Delta_n^h B \overset{\text{def}}{=} B(t_n^h) - B(t_{n-1}^h)$. A numerical approximation of X generated at the t_n^h and linearly interpolated in between will be denoted by $\{X^h(t)\}_{0 \leq t \leq 1}$, so that $X_n^h \overset{\text{def}}{=} X^h(t_n^h)$ is the value at the nth grid point.

---

[2]Traditionally, $\mu(t, x)$ and $\sigma(t, x)$ are used as much as $a(t, x)$ and $b(t, x)$, and we will freely change between the two notations. Also, quite a few authors have the opposite convention as here, to use a for standard deviation and b for drift.

SDEs occur in a variety of applications ranging from physics to biology to mathematical finance. Of particular interest is the *autonomous case*, in which $a(x) = a(t,x)$ and $b(x) = b(t,x)$ depend only on x but not on t. The stochastic equation (3.1) describes the time evolution of a *diffusion process* $\{X(t)\}_{0 \le t \le 1}$, with drift function $a(x)$ and diffusion coefficient $b^2(x)$, which can be thought of as a process behaving as BM with drift $a(x)$ and variance $b^2(x)$ when at x. Some particularly notable diffusions are:

Geometric Brownian motion (GBM), in which $a(x) = \alpha x$, $b(x) = \beta x$. It can be expressed explicitly in terms of the driving BM as $X(t) = X(0) \exp\{(\alpha - \beta^2/2)t + \beta B(t)\}$, and has lognormal marginals.

The **Ornstein–Uhlenbeck process,** in which $a(x) = -\alpha x$, $b(x) \equiv \beta$ with $\alpha, \beta > 0$. The Ornstein–Uhlenbeck (O–U) process is *mean-reverting*, meaning that the drift points to a certain point z (here 0) in $\mathbb{R}$. It can be expressed explicitly in terms of the driving BM as

$$X(t) = e^{-\alpha t} X(0) + \beta \int_0^t e^{-\alpha(t-s)} dB(s)$$

or, equivalently (in a form free of stochastic integrals)

$$X(t) = e^{-\alpha t} X(0) + \beta B(t) - \alpha \beta \int_0^t e^{-\alpha(t-s)} B(s) \, ds \,.$$

A d-dimensional (time-varying) version of the O–U process takes the form

$$d\boldsymbol{X}(t) = \boldsymbol{A}(t)\big(\boldsymbol{X}(t) - \boldsymbol{z}(t)\big) dt + \boldsymbol{\beta}(t) \, d\boldsymbol{B}(t) \,,$$

where $\boldsymbol{A}(t)$, $\boldsymbol{z}(t)$, $\boldsymbol{\beta}$ are (deterministic) matrix-valued functions of dimensions $d \times d$, $d \times 1$, respectively $d \times m$ and $\boldsymbol{B}$ is a m-dimensional standard BM. As for the O–U process, the solution $\boldsymbol{X}$ is Gaussian and can be explicitly computed (i.e., its mean and covariance functions are computable). The explicit solution when $d = m = 1$ is

$$X(t) = e^{H(t)} X(0) + \int_0^t e^{H(t)-H(s)} \big[\beta(s) dB(s) - A(s)z(s) \, ds\big] \,,$$

where $H(t) \stackrel{\text{def}}{=} \int_0^t A(s) \, ds$.

The **Vasicek process** is a translation $X(t) \stackrel{\text{def}}{=} z + Y(t)$ of an O–U process Y with $z > 0$. Thus $a(x) = -\alpha(x - z)$, $b(x) \equiv \beta$. The Vasicek process is often used as a model for a stochastic interest rate in finance, but a sometimes controversial feature in that connection is the property that it can attain negative values. Both the O–U and Vasicek processes are Gaussian processes (have Gaussian finite-dimensional distributions) when $X(0)$ is Gaussian, in particular when $X(0)$ is deterministic ($X(0) = x_0$).

The d-**dimensional Bessel process** has $a(x) = d$, $b(x) = 2\sqrt{x}$. It can be defined as a process with the same distribution as $\{\|\boldsymbol{B}_d(t)\|\}$, where

$\boldsymbol{B}_d(t)$ is standard d-dimensional BM (note that d is the dimension of the underlying BM, not the Bessel process itself!). The Bessel(3) process has in fact fundamental connections with the fine structure of Brownian paths; see Rogers & Williams [305].

The above examples are all special cases of a general idea. Note that because an O–U process X (a special case of which is BM) is Gaussian, it can easily be simulated on a lattice. Of course, any process that can be represented as $Y(t) = f(X(t))$ is then equally tractable. If f is 1-to-1, then such a Y satisfies an SDE of the form

$$
\begin{aligned}
\mathrm{d}Y(t) \;=\; & \left[-\alpha f'\big(g(Y(t))\big)g\big(Y(t)\big) + f''\big(g(Y(t))\big)\beta^2/2\right]\mathrm{d}t \\
& + \beta f'\big(g(Y(t))\big)\,\mathrm{d}B(t)\,,
\end{aligned} \tag{3.3}
$$

where g is the inverse to f and α,β are the parameters of the O–U process X.

The **Cox–Ingersoll–Ross process** (CIR) is another standard stochastic interest rate model. It has $a(x) = \alpha(c-x)$, $b(x) = \beta\sqrt{x}$, and $X(t) > 0$ a.s. provided that $2\alpha c \geq \beta^2$. See further Exercise 3.1.

In general, the CIR process cannot be expressed as an explicit function of the driving BM. A result appealing for simulation states that given $X(s)$ with $s < t$, $X(t)$ is distributed as a noncentral χ^2-r.v. with $f \overset{\text{def}}{=} 4\alpha c/\beta^2$ degrees of freedom (in general a noninteger) and noncentrality parameter

$$
\lambda \overset{\text{def}}{=} \frac{4\alpha e^{-\alpha(t-s)}}{\beta^2\big(1 - e^{-\alpha(t-s)}\big)}X(s)\,.
$$

Such an r.v. can be generated as a central χ^2 variate with a random (in general noninteger) degree of freedom $F = f+2N$, where N is Poisson($\lambda/2$). See also Glasserman [133, pp. 120–124].

However, if $2\alpha c = \beta$, the CIR process can be realized via the trick (3.3). In particular, $X(t) = Y(t)^2$, where Y is an O–U process with parameters $-\alpha/2, b/2$.

The **Langevin diffusion** has $b(x) \equiv 1$, $a(x) = (\mathrm{d}/\mathrm{d}x)\log\pi(x) = \pi'(x)/\pi(x)$ for a given (unnormalized) density $\pi(x)$. It has been used in continuous-time MCMC, where the target distribution Π^* is the one with density proportional to $\pi(\cdot)$; see, e.g., [301, Section 7.8.5].

Karlin & Taylor [207] is a good source for further explicit examples as well as a practically oriented discussion of diffusions. In general, additional examples of interesting SDEs and Itô integrals with a solution that can be expressed as an explicit functional of the driving BM are rare. One explicitly solvable SDE is $\mathrm{d}X(t) = 2B(t)\,\mathrm{d}B(t)$ with solution $B(t)^2 - t$. This is a special case of the class of processes X defined as the stochastic integral $X(t) = f(t, B(t))$ that can be expressed as a stochastic integral w.r.t. $\mathrm{d}B(t)$. In order that that $X(t)$ be representable explicitly as $f(t, B(t))$, the function f must satisfy the PDE $f_t + f_{xx}/2 = 0$, in which case $dX(t) = f_x(t, B(t)) \, dB(t)$.

3b   Numerical Methods for Ordinary Differential Equations

Consider the ODE $\dot{x}(t) = a(t, x(t))$ with initial condition $x(0) = x_0$ in the time interval $[0, 1]$. A numerical solution x^h is typically implemented via discrete approximations $x_n^h \overset{\text{def}}{=} x^h(t_n^h)$ at equidistant grid points $t_n^h \overset{\text{def}}{=} nh$, where $h = h_N = 1/N$. The error criterion is

$$e(h) \overset{\text{def}}{=} |x(1) - x^h(1)| \, .$$

The basic numerical method is the Euler method

$$x_0^h = x_0, \quad x_n^h = x_{n-1}^h + a(t_{n-1}^h, x_{n-1}^h)h \, . \tag{3.4}$$

Under suitable smoothness conditions (which we omit here and in the following), $e(h) = O(h) = O(1/N)$.

Here are some improvements:

(1) Take a Taylor expansion of order $p > 1$ rather than $p = 1$ as in (3.4). For $p = 2$, this gives

$$x_n^h = x_{n-1}^h + a(t_{n-1}^h, x_{n-1}^h)h + \left\{ a_t(t_{n-1}^h, x_{n-1}^h) + a_x(t_{n-1}^h, x_{n-1}^h) \right\} \frac{h^2}{2} \, ,$$

where $a_t \overset{\text{def}}{=} \partial a/\partial t$, $a_x \overset{\text{def}}{=} \partial a/\partial x$. Here $e(h) = O(h^2)$.

(2) In

$$x(h) = x(0) + \int_0^h a(t, x(t)) \, dt \, ,$$

approximate the integral by

$$\frac{h}{2} \left[a(0, x(0)) + a(h, x(h)) \right]$$

(the trapezoidal rule) rather than $a(0, x(0)) h$ as in (3.4). Here $x(h)$ is unknown but can be estimated by (3.4), i.e., predicted by $\overline{x}^h(h) \overset{\text{def}}{=} x(0) + a(0, x(0))h$. This gives $x_0^h = x_0$,

$$\overline{x}_n^h \overset{\text{def}}{=} x_{n-1}^h + a(t_{n-1}^h, x_{n-1}^h)\, h,$$

$$x_n^h = x_{n-1}^h + \left\{ a(t_{n-1}^h, x_{n-1}^h) + a(t_n^h, \overline{x}_n^h) \right\} \frac{h}{2} \, ,$$

which is an example of a *predictor–corrector* method. Again, $e(h) = O(h^2)$.

3c   The Euler Method for SDEs

The numerical methods for SDEs are modeled after those for ODEs. We will start with the Euler method, and in the next sections, we study SDE analogues of methods based on higher-order Taylor expansion. We mention for completeness that implicit methods have also been extended but shall not give the details (see Kloeden & Platen [217] and Milstein & Tretyakov [259]). In between the grid points, $X^h(t)$ may be defined either as X^h_{n-1} when $t^h_{n-1} \leq t < t^h_n$ or by linear interpolation.

The Euler scheme is $X^h(0) = x_0$,

$$X^h_n = X^h_{n-1} + a\big(t^h_{n-1}, X^h_{n-1}\big)\, h + b\big(t^h_{n-1}, X^h_{n-1}\big)\, \Delta^h_n B\,,$$

where the $\Delta^h_n B$ are i.i.d. $\mathcal{N}(0, h)$ for fixed h and $X^h_n \stackrel{\text{def}}{=} X^h(t^h_n)$. When considering the time horizon $0 \leq t \leq 1$, we take $h = 1/N$ with $N \in \mathbb{N}$.

3d   Error Criteria

For SDEs, one may be interested in two types of fit, strong and weak (which one depends on the type of application):

(s) X^h should give a good approximation of the sample path of X, that is, a good coupling. This leads to the error criterion

$$e_{\text{s}}(h) \stackrel{\text{def}}{=} \mathbb{E}\big|X(1) - X^h(1)\big| = \mathbb{E}\big|X(1) - X^h_N\big|\,.$$

(w) $X^h(1) = X^h_N$ should give a good approximation of the distribution of $X(1)$. That is, $\mathbb{E}g\big(X^h_N\big)$ should be close to $\mathbb{E}g\big(X(1)\big)$ for sufficiently many smooth functions g.

We will say that X^h converges strongly to X at time 1 with order $\beta > 0$ if $e_{\text{s}}(h) = \mathrm{O}(h^\beta)$, and weakly if

$$\big|\mathbb{E}g(X(1)) - \mathbb{E}g(X^h_N)\big| = \mathrm{O}(h^\beta)$$

for all $g \in \mathscr{C}^{2(\beta+1)}_p$, the space of functions such that $g', g'', \ldots, g^{(2(\beta+1))}$ exist [in practice, the relevant values of β are $\beta = 1, 3/2, 2, 5/2, \ldots$, so that $2(\beta + 1)$ is an integer] and that g and the first $2(\beta + 1)$ derivatives have polynomial growth (grows at most as a power of x).

The proofs of the following main results are given in Section 5:

Theorem 3.1 *The Euler scheme (3.1) converges strongly with order $\beta = 0.5$ provided a and b satisfy the technical conditions (5.1)–(5.3) stated in Section 5.*

Theorem 3.2 *The Euler scheme (3.1) converges weakly with order $\beta = 1$ under the technical conditions stated at the beginning of Section 5b.*

Remark 3.3 The use of an equidistant grid $t^h_n = nh$ is traditional and the one in common practical use. However, improvements in efficiency and accuracy can be obtained by an adaptive procedure where the step size is diminished when a and (not least) b are large; see Gaines & Lyons [124].

A related but somewhat different idea for one-dimensional diffusions is to take advantage of the fact that a diffusion can be constructed from BM by first using the natural scale $S(\cdot)$ and next a time transformation of the BM. To this end, one defines[3]

$$s(y) \overset{\mathrm{def}}{=} \exp\left\{-\int_{x^*}^{y} \left[2a(z)/b^2(z)\right] \mathrm{d}z\right\}, \quad S(x) \overset{\mathrm{def}}{=} \int_{x^*}^{x} s(y)\,\mathrm{d}y\,.$$

That is, S is the solution of the ODE $S'a + b^2 S''/2 = sa + 2s'/2 = 0$. Defining $Y(t) \overset{\mathrm{def}}{=} S(X(t))$, this ODE and Itô's formula imply that

$$\mathrm{d}Y(t) = \sigma_Y(Y(t))\,\mathrm{d}B(t) \quad \text{where } \sigma_Y(y) = s(S^{-1}(y))b(S^{-1}(y))\,.$$

Because this SDE does not contain a $\mathrm{d}t$ term, the solution can be constructed as a time change of Brownian motion B, $Y(t) = y_0 + B(\tau(t))$, where $\tau(\cdot)$ is the inverse function of

$$T(t) \overset{\mathrm{def}}{=} \int_0^t \frac{1}{\sigma_Y^2(y_0 + B(s))}\,\mathrm{d}s\,.$$

Equivalently, τ solves the ODE $\tau'(t) = \sigma_Y^2(y_0 + B(\tau(t)))$. In summary, the diffusion X can be simulated as $X(t) = x_0 + S^{-1}(B(\tau(t)))$. The difficulty is, of course, that in many cases $s(x), S(x)$, and $T^{-1}(t)$ have an intractable form. Also note that whereas $B(t)$ can be simulated exactly for any t, $T^{-1}(t)$ cannot. This representation of the solution suggests, however, potential guidelines as to how the adaptive step sizes described above should be chosen for a general SDE. See again Gaines & Lyons [124].   □

Remark 3.4 In some applications, the computation of interest involves calculating an expectation of the form $\mathbb{E}g(X(\infty))$ for an ergodic diffusion X, where $X(\infty)$ is an r.v. having the stationary distribution. Talay & Tubaro [350] show, under suitable conditions, that the Euler schemes discretization error does not degenerate at $t = \infty$. In particular, they provide conditions under which there exists $\beta \in \mathbb{R}$ such that

$$\mathbb{E}g(X(\infty)) - \mathbb{E}g(X^h(\infty)) = \beta h + \mathrm{o}(h)$$

as $h \downarrow 0$, where $X^h(\infty)$ is distributed as the steady state of X^h (shown to be ergodic for h sufficiently small).   □

The above discussion of weak and strong error focuses on the bias of the approximating process (weak error focuses on bias related to computation

---

[3]x^* is some arbitrary point in the state space or its closure. The most natural definition is to take x^* as a boundary point, but often this leads to infinite integrals, in which case x_0 needs to be taken as an interior point, thereby identifying $s(\cdot)$ only up to a constant.

of $\mathbb{E}g\big(X(t)\big)$, whereas strong error relates to the bias associated with computing the expectation of more complex path functionals). Of course, the rate of convergence of a Monte Carlo algorithm depends not only on its bias but also on its variance. To assess both effects simultaneously, consider the root-mean-square error of an SDE approximation that is known to exhibit a bias of order h^r when time increment h is used. Of course, the number R of independent replications that can be executed for a given budget c of computer time then scales as h (because R/h must be roughly proportional to c). For a scheme having bias of order h^r, an optimal trade-off of bias versus variance establishes that the best possible convergence rate for root mean square error is of order $c^{-r/(2r+1)}$, and this is attained when h is taken to be of order $c^{-1/(2r+1)}$. It follows that the order r of an SDE approximation method plays a dominant role in development of high-accuracy Monte Carlo schemes for computing solutions of SDEs; see Duffie & Glynn [99] for further details.

One additional difficulty is that many high-accuracy approximation algorithms require evaluating a large number of derivatives of the coefficient functions of the SDE at each time increment. This can be numerically costly, particularly when the SDE is high-dimensional. In part as a consequence, it is fair to say that the Euler scheme is the dominant one in practical use in higher dimensions. By working harder in the proof of Theorem 3.2, one can in fact establish that there exists $\beta \in \mathbb{R}$ such that

$$\mathbb{E}g\big(X^h(T)\big) \;=\; \mathbb{E}g\big(X(T)\big) + \beta h + \mathrm{O}(h^2)$$

as $h \downarrow 0$. As a consequence,

$$\mathbb{E}\big[2g\big(X^h(T)\big) - g\big(X^{2h}(T)\big)\big] \;=\; \mathbb{E}g\big(X(T)\big) + \mathrm{O}(h^2)\,.$$

Thus, an estimator based on replicating the r.v. $Z \overset{\text{def}}{=} 2g\big(X^h(T)\big) - g\big(X^{2h}(T)\big)$ exhibits a bias of order h^2, thereby improving the convergence rate from $c^{-1/3}$ in the simulation budget to $c^{-2/5}$. Of course, the estimator Z requires evaluating the approximation at two different step sizes h and $2h$, potentially doubling the effort per replication. Nevertheless, the improved convergence rate makes this an appealing device from a practical viewpoint. This is an example of what is called *Romberg extrapolation* in the numerical analysis literature; see Talay [349] and Kebaier [209].

Kloeden & Platen [217] is the classical reference on simulation of SDEs, but we also refer to the survey in Talay [349] and to Milstein & Tretyakov [259].

Exercises

3.1 (A) Let $p(t, T)$ be the price at time t of a zero-coupon bond expiring at time $T > t$. The return on such a bond corresponds to a continuous interest rate of

$$r(t, T) \overset{\text{def}}{=} -\frac{1}{T - t} \log p(t, T)\,.$$

Typically, $r(t, T)$ depends not only on the short rate $r(t) = r(t, t+)$ at time t but also on T, and the curve $\{r(t, t + u)\}_{u \geq 0}$ is the *term structure* at time t. Defining the instantaneous forward rate $f(t, T)$ as

$$f(t, T) \stackrel{\text{def}}{=} -\frac{\partial}{\partial T} \log p(t, T), \quad \text{we have} \quad p(t, T) = \exp\left\{-\int_t^T f(t, s)\, \mathrm{d}s\right\}.$$

The (one-factor) *Heath–Jarrow–Morton model* postulates that for any fixed T,

$$\mathrm{d}f(t, T) = \alpha(t, T)\, \mathrm{d}t + v(s, T)\, \mathrm{d}B(t), \tag{3.5}$$

where the driving BM is the same for all T.
To identify a risk-neutral measure $\mathbb{P}^*$, one combines the nonarbitrage argument with the identity

$$\exp\left\{-\int_0^T f(0, s)\, \mathrm{d}s\right\} = \mathbb{E}^* \exp\left\{-\int_0^T r(s)\, \mathrm{d}s\right\},$$

which holds because both sides must equal $p(0, T)$. After some calculations, this gives that under $\mathbb{P}^*$, the $f(t, T)$ evolve as in (3.5) with $\alpha(t, T)$ replaced by

$$\alpha^*(t, T) \stackrel{\text{def}}{=} v(t, T) \int_t^T v(t, s)\, \mathrm{d}s.$$

For these facts and further discussion, see, e.g., Björk [45].
Your assignment is to give projections) (some typical sample paths) of the risk-neutral term structure $\{r(5, 5 + u)\}_{0 \leq u \leq 10}$ after $t = 5$ years, using the Vasicek volatility structure $v(t, T) = \beta e^{-\alpha(T-t)}$ and the initial term structure $r(0, T) = \left(6 + T/30 - e^{-T}\right)/100$, which has roughly the shape of the data in Jarrow [195, p. 3]. The parameters α, β should be calibrated so that sample paths of the short rate in $[0, 5]$ look reasonable.
Generate, for example, the $r(5, 5 + u)$ at a quarter-yearly grid and use 10 yearly grid points for the $f(s, T)$. Thus, you will need to calculate the $f(i/10, 5+j/4)$ for $i = 1, \ldots, 50$, $j = 1, \ldots, 40$. Note that the initial values $f(0, T)$ are analytically available from the expression for $r(0, T)$. For calibration of a, b, use $f(t - 1/10, t)$ as approximation for $r(t)$.

4   The Milstein and Other Higher-Order Schemes

4a   The Milstein Scheme

The idea is that the approximation

$$\int_0^h b\big(t, X(t)\big)\, \mathrm{d}B(t) \sim b\big(0, X(0)\big)B(h)$$

is the main source of error for the Euler scheme. To improve it, we estimate the error by Itô's formula for $b(t, X(t))$:

$$\int_0^h b(t, X(t))\, dB(t) \; - \; b(0, X(0))B(h)$$

$$= \int_0^h \{b(t, X(t)) - b(0, X(0))\}\, dB(t)$$

$$= \int_0^h \left\{ \int_0^t \left[b_t(s, X(s)) + a(s, X(s))b_x(s, X(s)) \right. \right.$$
$$\left. + \frac{1}{2}b^2(s, X(s))b_{xx}(s, X(s)) \right] ds$$
$$\left. + \int_0^t b(s, X(s))b_x(s, X(s))\, dB(s) \right\} dB(t)$$

$$\sim \; O(h^2) + b(0, x_0)b_x(0, x_0) \int_0^h \int_0^t dB(s)\, dB(t)$$

$$\sim \; b(0, x_0)b_x(0, x_0) \int_0^h B(t)\, dB(t)$$

$$= \; b(0, x_0)b_x(0, x_0) \left\{ \frac{1}{2}B(h)^2 - \frac{1}{2}h \right\}.$$

This leads to the Milstein scheme $X_0^h = x_0$,

$$X_n^h = X_{n-1}^h + ah + b\Delta_n^h B + \frac{1}{2}bb_x\{\Delta_n^h B^2 - h\}, \qquad (4.1)$$

where $a = a(t_{n-1}^h, x_{n-1}^h)$ and similarly for b, b_x.

Theorem 4.1 *The Milstein scheme* (4.1) *converges strongly with order* $\beta = 1$.

For the proof, see Kloeden & Platen [217].

Example 4.2 Starting from the same $512 = 2^9$ i.i.d. $\mathcal{N}(0, 2^{-9})$ r.v.'s $V_1, \ldots, V_{512}$, we simulated geometric BM with $\mu = 2$, $\sigma^2 = 4$ in $[0, 1]$ using the V_i as common random numbers for the updating. We took $h = 1/n$ with $n = 4, 8, \ldots, 512$ and implemented both the Euler scheme (dashed line) and the Milstein scheme (dot-dashed line); the solid line is interpolation between the exact value of GBM at the grid point (e.g. $\exp\{(\mu - \sigma^2/2)/4 + V_1 + \cdots + V_{128}\}$ at $t = 1/4$; the normal r.v.'s used in the updating for, for example, $h = 2^{-6}$ are $V_1 + \cdots + V_8$, $V_9 + \cdots + V_{16}$, etc.). The results are given in Figure 4.1 and illustrate the better strong convergence properties of the Milstein scheme. □

It may be noticed that the weak order of the Milstein scheme is 1 as for the Euler scheme, so from this point of view there is no improvement. A numerical illustration is given in Example 5.3. Also note that if $b(t, x) \overset{\text{def}}{=}$

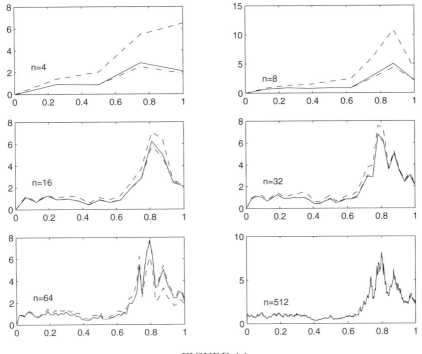

FIGURE 4.1

$b(t)$ does not depend on x (in particular, is constant), then the Euler scheme and the Milstein scheme coincide (in particular, the Euler scheme has the improved strong order $\beta = 1$).

In the view of the authors, the better strong convergence order of the Milstein scheme compared to the Euler scheme is not in itself sufficient to motivate preferring the Milstein scheme, because the concept of strong convergence order depends on a particular coupling that is seldom of intrinsic interest. Most applications (including as examples the exercises at the end of this section) call for good distributional approximations instead. However, weak error as we have defined it sofar also have its caveats because its concentrates on smooth functionals at a single point. Perez [288] takes a different direction by looking at the total variation distance between the SDE solution and its discrete approximation over the grid points.[4] In this distance measure, the Euler scheme does not even converge, whereas the Milstein scheme converges at order $h^{1/2}$. This provides one rigorous sense in which Milstein provides a better approximation than does Euler.

---

[4]To take the t.v. distance over the whole path would not make sense: any discrete approximation has quadratic variation 0, the solution itself not. Also, since t.v. distance only involves distributional properties, this definition eliminates the coupling to the coupling.

4b Itô–Taylor Expansions

We proceed by refining the estimate used for the Milstein scheme. For notational convenience, let, for example, b_x denote $b_x(0, x_0)$ when it occurs outside integrals and $b_x(s, X(s))$ when it occurs in an integral w.r.t. ds or $dB(s)$. We get

$$\int_0^h b(t, X(t))\, dB(t) \;-\; b(0, X(0)) B(h) \;=\; \int_0^h b\, dB(t) \;-\; b B(h)$$

$$= \int_0^h \left\{ \int_0^t \left[b_t + a b_x + \frac{1}{2} b^2 b_x x \right] ds \;+\; \int_0^t b b_x\, dB(s) \right\} dB(t). \quad (4.2)$$

In the last term, we expand $b b_x = b(s, X(s)) b_x(s, X(s))$ one more time by Itô's formula and note that the $dB(s)$ term dominates the ds term. Thus approximately (4.2) is

$$\left[b_t + a b_x + \frac{1}{2} b^2 b_x x \right] \int_0^h t\, dB(t) \;+\; b b_x \int_0^h B(t)\, dB(t)$$

$$+ \int_0^h dB(t) \int_0^t dB(s) \int_0^s b \frac{\partial}{\partial x}(b b_x)\, dB(u)$$

$$\sim \left[b_t + a b_x + \frac{1}{2} b^2 b_{xx} \right] (h B - V) + \frac{1}{2}(B^2 - h)$$

$$+ b(b b_{xx} + b_x^2) \int_0^h \left(\frac{1}{2} B(t)^2 - \frac{t}{2} \right) dB(t), \quad (4.3)$$

where $B \stackrel{\text{def}}{=} B(h)$,

$$V \stackrel{\text{def}}{=} V(h) \stackrel{\text{def}}{=} \int_0^h B(s)\, ds \;=\; h B(h) \;-\; \int_0^h s\, dB(s).$$

Similarly,

$$\int_0^h a(t, X(t))\, dt \;-\; a(0, X(0)) h \;=\; \int_0^h a\, dt \;-\; a h$$

$$= \int_0^h \left\{ \int_0^t \left[a_t + a a_x + \frac{1}{2} b^2 a_{xx} \right] ds \;+\; \int_0^t b a_x\, dB(s) \right\} dt$$

$$\sim \frac{1}{2} \left[a_t + a a_x + \frac{1}{2} b^2 a_{xx} \right] h^2 \;+\; b a_x \int_0^h B(t)\, dt$$

$$\sim \frac{1}{2} \left[a_t + a a_x + \frac{1}{2} b^2 a_{xx} \right] h^2 \;+\; b a_x V. \quad (4.4)$$

By evaluating the differential of $B(t)^3/3 - t B(t))$ by Itô's formula, it is seen that twice the integral in (4.3) is $B^3/3 - h B$. Hence, approximating $X(h)$

by $x_0 + ah + bW + (4.3) + (4.4)$, we arrive at the Itô–Taylor formula

$$X(h) \sim x_0 + ah + bB + \frac{1}{2}bb_x(B^2 - h) + a_x bV + \frac{1}{2}\left[a_t + aa_x + \frac{1}{2}b^2 a_{xx}\right]h^2$$

$$+ \left[b_t + ab_x + \frac{1}{2}b^2 b_{xx}\right](hB - V) + \frac{1}{2}\left[b(bb_{xx} + b_x^2)\right]\left(\frac{1}{3}B^3 - hB\right).$$

For the following, we note that the covariance matrix of $(B, V) = (B(h), V(h))$ is

$$\begin{pmatrix} \int_0^h ds & \int_0^h (h-s)\,ds \\ \int_0^h (h-s)\,ds & \int_0^h (h-s)^2\,ds \end{pmatrix} = \begin{pmatrix} h & \frac{1}{2}h^2 \\ \frac{1}{2}h^2 & \frac{1}{3}h^3 \end{pmatrix}. \tag{4.5}$$

This follows, for example, from (A6.7) by writing

$$B = \int_0^h dB(s), \quad V = \int_0^h dt \int_0^t dB(s) = \int_0^h (h-s)dB(s).$$

4c Higher Order Schemes

A scheme of strong order 1.5 is obtained directly from the Itô-Taylor expansion: $X_0^h = x_0$,

$$\begin{aligned} X_n^h &= X_{n-1}^h + ah + b\Delta_n^h B + \frac{1}{2}bb_x\left\{\Delta_n^h B^2 - h\right\} \\ &+ a_x bV_n^h + \frac{1}{2}\left[a_t + aa_x + \frac{1}{2}b^2 a_{xx}\right]h^2 \\ &+ \left[b_t + ab_x + \frac{1}{2}b^2 b_{xx}\right](h\Delta_n^h B - V_n^h) \\ &+ \frac{b}{2}\left[bb_{xx} + b_x^2\right]\left(\frac{1}{3}\Delta_n^h B^3 - V_n^h h\right), \end{aligned}$$

where $a = a(t_{n-1}^h, x_{n-1}^h)$ etc. and the $(\Delta_n^h B, V_n^h)$ are generated as i.i.d. bivariate normals with mean 0 and covariance matrix (4.5).

A scheme of weak order 2 of a slightly simpler form can be obtained by deleting the last term.

Exercises

4.1 (A) A bank wants to price its 5-year annuity loans in a market in which the short rate $r(t)$ at time t is stochastic. A loan is paid off continuously at a constant rate, say p, and thus the amount paid back is determined by

$$s(0) = 0, \quad ds(t) = (p + s(t)r(t))\,dt,$$

whereas an amount $q(0)$ kept in the bank will develop according to

$$dq(t) = q(t)r(t)\,dt.$$

Thus, for a loan of size $q(0)$ the payment rate p should be determined such that $Ec(5) = Eq(5)$ (ignoring profit and administration costs). To determine this, it suffices by an obvious proportionality argument to give estimates of the two expectations when $p = 1$, $q(0) = 1$.

Note that a short rate $r(t)$ corresponds to an interest per year of $\varepsilon = e^{r(t)} - 1$. The bank employs the Cox–Ingersoll–Ross process as model for $\{r(t)\}$. This means that we have a drift toward c, which we thus can interpret as the typical long-term interest rate and which the bank estimates corresponds to $\varepsilon = 6\%$; the interest rate at time 0 corresponds to $\varepsilon = 6.5\%$.

For your simulations of $\{r(t)\}$, use the Milstein scheme. Do first some pilot runs to determine (by sample path inspection) some values of the remaining two parameters α, β that appear to give reasonable fluctuations of $r(t)$. Compare finally your results with the deterministic values corresponding to $r(t) \equiv c$.

4.2 (A) At time $t = 0$, a submarine located at $(0, 0)$ fires a torpedo against an enemy vessel whose midpoint is currently at $(0, 4)$ (the unit is km). The vessel is 0.14 km long, its speed measured in km/h at time t is $Z_1(t)$, a Cox–Ingersoll–Ross process with parameters $\alpha_1 = 6, c_1 = 30, \beta_1 = 1$, and the direction is given by the angle 30° NW. The information available to the submarine commander is a $\mathcal{N}(c_1, \sigma^2)$ estimate $\hat{c}_1$ of c_1, where $\sigma^2 = 4$. The speed of the torpedo is another Cox–Ingersoll–Ross process $Z_2(t)$ with parameters $\alpha_2 = 60, c_2 = 60, \beta_2 = 7$, the angle (in radians!) giving the direction is $\theta(t) = \big(\theta(0) + \omega B(t)\big) \mod 2\pi$, where B is standard Brownian motion and $\omega^2 = 0.04$, and $\theta(0)$ is chosen by the submarine commander such that the torpedo would hit the midpoint of the vessel in the absence of stochastic fluctuations, that is, if the vessel moved with speed $\hat{c}_1$, and the torpedo with constant direction $\theta(0)$ and speed c_2. See Figure 4.2

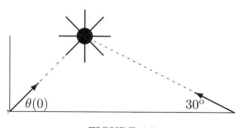

FIGURE 4.2

Compute the probability p that the torpedo hits the vessel, taking $Z_1(0) = c_1$, $Z_2(0) = c_2$. [Hint: Verify that (except in the extreme tails of $\hat{z}$), $\theta(0)$ is the arcsin of $(\hat{c}_1/2c_1)\sin 30°$.]

4.3 (A−) Complement your solution of Exercise 4.2 with an analysis of which of the parameters $\beta_1, \beta_2, \sigma^2, \omega^2$ are the major ones creating randomness.

5   Convergence Orders for SDEs: Proofs

We will give the proofs only for the autonomous case $a(t, x) = a(x)$, $b(t, x) = b(x)$; we take w.l.o.g. $T = 1$ and use the notation t_n^h, X_n^h, $\Delta_n^h B$ employed above.

5a   Strong Convergence Orders

Proof of Theorem 3.1. The conditions that are needed are the Lipschitz conditions

$$\left|a(t,x) - a(t,y)\right| \le K|x-y|, \quad \left|b(t,x) - b(t,y)\right| \le K|x-y|, \quad (5.1)$$

the linear growth condition

$$\left|a(t,x)\right| + \left|b(t,x)\right| \le K(1+|x|), \quad (5.2)$$

as well as

$$\left|a(t,x) - a(s,x)\right| + \left|b(t,x) - b(s,x)\right| \le K(1+|x|)\sqrt{t-s} \quad (5.3)$$

for $0 \le s < t \le T$, $x \in \mathbb{R}$, and some constant $K < \infty$ [here (5.1), (5.2) are in fact the standard conditions for existence of a strong solution].

Let

$$\widetilde{X}^h(t) \overset{\text{def}}{=} x_0 + \sum_{n:\, t_n^h \le t} \int_{t_{n-1}^h}^{t_n^h} a\bigl(\widetilde{X}^n(s)\bigr)\,\mathrm{d}s + \int_{t_{n-1}^h}^{t_n^h} b\bigl(\widetilde{X}^h(s)\bigr)\,\mathrm{d}B(s)$$

denote the piecewise constant extension of $\{X_n^h\}$ (i.e., the process that is X_{n-1}^h for $t \in [t_{n-1}^h, t_n^h)$) and for $t \in [t_n^h, t_{n+1}^h)$, let

$$R'(t) \overset{\text{def}}{=} \int_{t_n^h}^t a\bigl(X(s)\bigr)\,\mathrm{d}s + \int_0^{t_n^h} \bigl[a\bigl(X(s)\bigr) - a\bigl(\widetilde{X}^h(s)\bigr)\bigr]\,\mathrm{d}s\,,$$

$$R''(t) \overset{\text{def}}{=} \int_{t_n^h}^t b\bigl(X(s)\bigr)\,\mathrm{d}B(s) + \int_0^{t_n^h} \bigl[b\bigl(X(s)\bigr) - b\bigl(\widetilde{X}^h(s)\bigr)\bigr]\,\mathrm{d}B(s)\,,$$

$$r'(t) \overset{\text{def}}{=} \sup_{s \le t} \mathbb{E}R'(s)^2, \quad r''(t) \overset{\text{def}}{=} \sup_{s \le t} \mathbb{E}R''(s)^2\,.$$

Then $X(t) - \widetilde{X}^h(t) = R'(t) + R''(t)$ and hence

$$z(t) \overset{\text{def}}{=} \sup_{s \le t} \mathbb{E}\bigl(X(s) - \widetilde{X}^h(s)\bigr)^2 \le 2r'(t) + 2r''(t)\,.$$

For $s \in [t_n^h, t_{n+1}^h)$, we get

$$
\begin{aligned}
\mathbb{E}R'(s)^2 &\le 2\mathbb{E}\Bigl[\int_{t_n^h}^s a\bigl(X(r)\bigr)\,\mathrm{d}r\Bigr]^2 + 2\mathbb{E}\Bigl[\int_0^s \bigl|a\bigl(X(r)\bigr) - a\bigl(\widetilde{X}^h(r)\bigr)\bigr|\,\mathrm{d}r\Bigr]^2 \\
&\le 2K\mathbb{E}\Bigl[\int_{t_n^h}^s \bigl(1+|X(r)|\bigr)\,\mathrm{d}r\Bigr]^2 + 2K\mathbb{E}\Bigl[\int_0^s \bigl|X(r) - \widetilde{X}^h(r)\bigr|\,\mathrm{d}r\Bigr]^2 \\
&\le 2K\mathbb{E}\int_{t_n^h}^s \bigl(1+|X(r)|\bigr)^2\,\mathrm{d}r + 2K\mathbb{E}\int_0^s \bigl(X(r) - \widetilde{X}^h(r)\bigr)^2\,\mathrm{d}r \\
&\le 2K\mathbb{E}\int_{t_n^h}^s \bigl(1+|X(r)|\bigr)^2\,\mathrm{d}r + 2K\mathbb{E}\int_{t_n^h}^s \bigl(X(r) - \widetilde{X}_n^h(r)\bigr)^2\,\mathrm{d}r \\
&\le \frac{1}{n}4K(1+L) + 2K\int_0^s z(r)\,\mathrm{d}r\,,
\end{aligned}
$$

where the third step used the Cauchy–Schwarz inequality and we wrote $L \stackrel{\text{def}}{=} \sup_{r \leq 1} \mathbb{E} X(r)^2$. Since by the Itô isometry (A6.7),

$$\mathbb{E}\left[\int_0^{t_n^h} \left[b(X(s)) - b(\widetilde{X}^h(s))\right] \mathrm{d}B(s)\right]^2 = \mathbb{E}\left[\int_0^{t_n^h} \left[b(X(s)) - b(\widetilde{X}^h(s))\right] \mathrm{d}s\right]^2,$$

the same estimate holds for $\mathbb{E} R''(s)^2$, and taking the sup over $s \leq t$ we get $z(t) \leq C_1 h + C_2 \int_0^t z(r) \, \mathrm{d}r$, where $C_1 \stackrel{\text{def}}{=} 8K(1 + L)$, $C_2 \stackrel{\text{def}}{=} 4K$. By a standard analytic fact (Gronwall's inequality), this implies $z(t) \leq C_3 h$ with C_3 independent of h. Taking $t = 1$, we get

$$\mathbb{E}|X^h(1) - X(1)| = \mathbb{E}|\widetilde{X}^h(1) - X(1)| \leq \left[\mathbb{E}(\widetilde{X}^h(1) - X(1))^2\right]^{1/2}$$

$$\leq [z(1)]^{1/2} \leq C_4 h^{1/2}.$$

$\square$

5b   Weak Convergence Orders

We have to estimate

$$\left|\mathbb{E}_x f(X^h(1)) - \mathbb{E}_x f(X(1))\right| = \left|\mathbb{E}_x f(X_N^h) - \mathbb{E}_x f(X(1))\right|$$

for a given $f \in \mathscr{C}_p^{2(\beta+1)}$, where $h = 1/N$.

The differential generator of the diffusion $\{X(t)\}$ is

$$\mathscr{A} = a(x)\frac{\mathrm{d}}{\mathrm{d}x} + \frac{b^2(x)}{2}\frac{\mathrm{d}^2}{\mathrm{d}x^2}.$$

We let $u(t, x) = \mathbb{E}_x f(X(t))$, $u_t(\cdot) = u(t, \cdot)$ and will exploit the fact that u satisfies the PDE

$$\frac{\partial}{\partial t}u(t, x) = (\mathscr{A} u_t)(x) \tag{5.4}$$

for $t \geq 0$, $x \in \mathbb{R}$ subject to the initial condition $u(0, x) = f(x)$. We shall assume that $a, b \in \mathscr{C}_p^{2(\beta+1)}$. It can then be shown (see Kloeden & Platen [217, pp. 476, 473–474]) that also $u_t \in \mathscr{C}_p^{2(\beta+1)}$ for $t \geq 0$. In the following, $\beta = 1$.

Proof of Theorem 3.2. For $0 \leq s \leq 1$, set $v(s, x) = u(1 - s, x)$. A key fact is that $\{v(t, X(t))\}_{0 \leq t \leq 1}$ is a martingale [for the integrability, use the fact that $u_s \in \mathscr{C}_p^{2(\beta+1)}$ and that X has moments of all orders]. This martingale property follows from Itô's formula.

Note that

$$\mathbb{E}_x f(X^h(1)) - \mathbb{E}_x f(X(1)) = \mathbb{E}u(0, X^h(1)) - u(1, x)$$

$$= \mathbb{E}[v(1, X_N^h) - v(0, X_0^h)] = \sum_{n=0}^{N-1} [v(t_{n+1}^h, X_{n+1}^h) - v(t_n^h, X_n^h)].$$

Using the martingale property of $\{v(s, X(s)\}$ in the first step, we get

$$\mathbb{E}\left[v\left(t_{n+1}^h, X_{n+1}^h\right) - v\left(t_n^h, X_n^h\right) \mid X_n^h = y\right]$$

$$= \mathbb{E}\left[v\left(t_{n+1}^h, X_{n+1}^h\right)\right) - v\left(t_n^h, X_n^h\right) \mid X_n^h = y\right]$$
$$- \mathbb{E}\left[v\left(t_{n+1}^h, X_{n+1}^h\right)\right) - v\left(t_n^h, X_n^h\right) \mid X_n^h) = y\right]$$

$$= \mathbb{E}\left[\frac{\partial v}{\partial x}(t_n^h, y)\left(X_{n+1}^h - y\right) + \frac{1}{2}\frac{\partial^2 v}{\partial x}(t_n^h, y)\left(X_{n+1}^h - y\right)^2\right.$$
$$+ \frac{1}{6}\frac{\partial^2 v}{\partial x}(t_n^h, y)\left(X_{n+1}^h - y\right)^3 + \frac{1}{24}\frac{\partial^2 v}{\partial x}(t_n^h, y)\left(X_{n+1}^h - y\right)^4$$
$$\left. + v\left(t_{i+1}, X_{n+1}^h\right) - v\left(t_n^h, X_n^h\right) \mid X_n^h = y\right]$$

$$-\mathbb{E}\left[\frac{\partial v}{\partial x}(t_n^h, y)\left(X_{n+1}^h - y\right) + \frac{1}{2}\frac{\partial^2 v}{\partial x}(t_n^h, y)\left(X_{n+1}^h - y\right)^2\right.$$
$$+ \frac{1}{6}\frac{\partial^2 v}{\partial x}(t_n^h, y)\left(X_{n+1}^h - y\right)^3 + \frac{1}{24}\frac{\partial^2 v}{\partial x}(t_n^h, y)\left(X_{n+1}^h - y\right)^4$$
$$\left. + v\left(t_{n+1}^h, X_{n+1}^h\right) - v\left(t_n^h, X_n^h\right) \mid X_n^h = y\right].$$

The following Lemma 5.1 therefore completes the proof, provided $(\partial/\partial x)v(s, x)$ is uniformly bounded (this uniform boundedness can be relaxed).   □

Lemma 5.1 *For* $j = 1, 2, 3$,

$$\mathbb{E}_y\left(X_1^h - y\right)^j = \mathbb{E}_y\left(X(h) - y\right)^j + \mathrm{O}(h^2).$$

Further,

$$\mathbb{E}_y\left(X_1^h - y\right)^4 = \mathbb{E}_y\left(X(h) - y\right)^4 = \mathrm{O}(h^2),$$

Proof. Itô's formula implies that $\mathbb{E}_y\left(X(h) - y\right)^j$ equals

$$j\int_0^h \mathbb{E}_y a\left(X(s)\right)\left(X(s)-y\right))^{j-1}\,ds + j(j-1)\int_0^h \mathbb{E}_y \frac{b\left(X(s)\right)}{2}\left(X(s)-y\right))^{j-2}\,ds.$$

So for $j = 1$,

$$\mathbb{E}_y\left(X(h) - y\right) = \int_0^h \mathbb{E}_y a\left(X(s)\right)\,ds$$

$$= a(y)h + \mathbb{E}_y\left[\int_0^h \left[a\left(X(s)\right) - a\left(X(0)\right)\right]\,ds\right]$$

$$= a(y)h + \int_0^h \left[(\mathscr{A}a(y)s + \mathrm{o}(s)\right]\,ds$$

$$= a(y)h + \mathrm{O}(h^2),$$

where $\mathscr{A}$ is the generator. Similar but more tedious calculations yield.

$$\mathbb{E}_y\left(X(h) - y\right)^2 = b(y)h + \mathrm{O}(h^2),$$
$$\mathbb{E}_y\left(X(h) - y\right)^3 = \mathrm{O}(h^2) = \mathbb{E}_y\left(X(h) - y\right)^4.$$

Since

$$\mathbb{E}\Delta_n^h B = 0, \quad \mathbb{E}\Delta_n^h B^2 = h, \quad \mathbb{E}\Delta_n^h B^3 = O(h^2), \quad \mathbb{E}\Delta_n^h B^4 = O(h^2), \quad (5.5)$$

[in fact, $\mathbb{E}\Delta_n^h B^3 = 0$!] the moments of $X^h(h) - y$ indeed match those of $X(h) - y$ up to terms of order $O(h^2)$. $\quad\square$

Remark 5.2 An inspection of the proof shows that it is fact not necessary to make the obvious choice of $\Delta_n^h B$ as $\mathcal{N}(0, h)$; it suffices that (5.5) holds so that, for example, one could take

$$\mathbb{P}(\Delta_n^h B = \sqrt{h}) = \mathbb{P}(\Delta_n^h B = -\sqrt{h}) = \frac{1}{2}. \quad\square$$

Example 5.3 We simulated $R = 5{,}000$ replications of geometric BM with $\mu = 2$, $\sigma^2 = 2$ in the time interval $[0, 1]$. The value, say Y_r, of the log of the rth replication at $t = 1$ was transformed to $W_r \stackrel{\text{def}}{=} (Y_r - \mu + \sigma^2/2)/\sigma$, so that the weak error can be measured by how close the W_r are to standard normal r.v.'s. Figure 5.1 gives Q-Q plots with the empirical quantiles of the W_r on the horizontal axis and the normal quantiles on the vertical. The first column is the Euler scheme with normal $\Delta_n^h B$, the second the Milstein scheme with normal $\Delta_n^h B$, and the third the Euler scheme with the $\Delta_n^h B$ taken as the $\pm\sqrt{h}$-variables in Remark 5.2. The three rows correspond to $n = 16, 32, 64$ steps.

The figure illustrates that indeed the Milstein scheme does not improve the weak error of the Euler scheme and also shows a perhaps surprisingly good performance of the nonnormal $\Delta_n^h B$. $\quad\square$

Remark 5.4 A key to the above analysis of the approximation error of the Euler scheme is the so-called backward PDE (5.4). This PDE also plays an important role in the following variance reduction technique for SDEs.

Note that if $u(t, x) = \mathbb{E}_x f(X(t))$ satisfies (5.4), then

$$\mathrm{d}u(T - s, X(s)) = \frac{\partial}{\partial x} u(T - s, X(s)) b(X(s)) \, \mathrm{d}B(s) \qquad (5.6)$$

for $0 \le s \le T$. Using the initial condition for u, the integrated version of (5.6) is therefore

$$f(X(T)) - \mathbb{E}f(X(T)) = \int_0^T \frac{\partial}{\partial x} u(T - s, X(s)) b(X(s)) \, \mathrm{d}B(s).$$

Note that if v is sufficiently smooth and bounded,

$$\int_0^T \frac{\partial}{\partial x} v(T - s, X(s)) b(X(s)) \, \mathrm{d}B(s)$$

has mean zero, so it can be used as a control variate. So, consider estimating $\mathbb{E}f(X(T))$ by averaging i.i.d. replicates of an Euler approximation to the

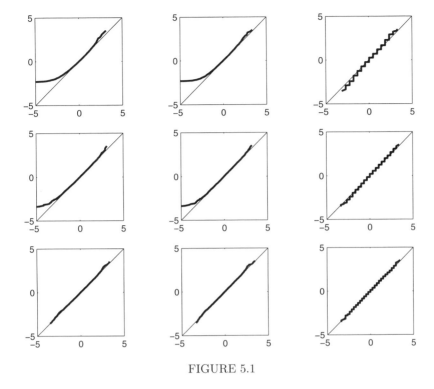

FIGURE 5.1

r.v.

$$f\big(X(T)\big) \ - \ \lambda \int_0^T \frac{\partial}{\partial x} v\big(T - s, X(s)\big) b\big(X(s)\big) \, \mathrm{d}B(s) \,,$$

where λ is an appropriately chosen value of the control coefficient. Asssuming $T = Nh$, the corresponding Euler approximation is

$$f\big(X^h(T)\big) \ - \ \lambda \sum_{n=0}^{(N-1)h} \frac{\partial}{\partial x} v\big(T - s, X_n^h\big) b\big(X_n^h\big) \, \Delta_{n+1}^h B \,.$$

If one has a good approximation to the solution u (so that $\partial v/\partial x \approx \partial u/\partial x$), one should choose $\lambda \approx 1$, in which case the degree-of variance reduction can be substantial. See Newton [277] and Henderson & Glynn [181] for further details.   □

Exercises

5.1 (TP) Verify the martingale property of $v\big(t, X(t)\big)$ and (5.4) in the proof of Theorem 3.2.

5.2 (TP) Explain why the distribution in the last column of Figure 5.1 is discrete.

6 Approximate Error Distributions for SDEs

The strong convergence order $\beta = 1/2$ for the Euler scheme (cf. Theorem 3.1) suggests a functional limit theorem with normalization constant $h^{-1/2}$ for the difference between the Euler approximation and the exact value. In this section, we heuristically derive the corresponding error distribution (see (6.2) below). We refer to Kurtz & Protter [225] and Jacod & Protter [193] for a more detailed and rigorous analysis, which also covers the multidimensional case (as well as Lévy-driven SDEs in [193]).

The general (time-homogeneous) one-dimensional SDE takes the form

$$\mathrm{d}X(t) = a\big(X(t)\big)\,\mathrm{d}t + b\big(X(t)\big)\,\mathrm{d}B(t), \qquad (6.1)$$

and its Euler approximation (for time increment h) is given by

$$X_n^h - X_{n-1}^h \stackrel{\mathrm{def}}{=} a\big(X_{n-1}^h\big)h + b\big(X_{n-1}^h\big)\Delta_n^h B$$

in our usual notation, where $X^h(t)$ is defined as X_{n-1}^h when $t_{n-1}^h \le t < t_n^n$. The normalized error process is given by

$$Z^h(t) \stackrel{\mathrm{def}}{=} h^{-1/2}\big(X^h(t) - X(t)\big).$$

Suppose that Z^h converges to a limit process Z as $h \downarrow 0$. We will now argue that the stochastic equation satified by Z should be

$$
\begin{aligned}
\mathrm{d}Z(t) = {} & a'\big(X(t)\big)Z(t)\,\mathrm{d}t + b'\big(X(t)\big)Z(t)\,\mathrm{d}B(t) \\
& + \frac{1}{\sqrt{2}}\Big(b'\big(X(t)\big)b\big(X(t)\big) + a'\big(X(t)\big)b\big(X(t)\big)\Big)\,\mathrm{d}B_1(t),
\end{aligned} \qquad (6.2)
$$

where X is the solution to the SDE (6.1) and B_1 is a standard Brownian motion independent of B.

Remark 6.1 This result provides support for the notion that the error associated with computing expectations of suitably continuous path functionals (e.g., Lipschitz) of the SDE via the Euler method is of order $h^{1/2}$, a conclusion identical to the strong error estimate in Theorem 3.1. Note, however, that the error of the Euler approximation has to do with the magnitude of $\mathbb{E}\varphi(X^h) - \mathbb{E}\varphi(X)$ (for a given functional $\varphi(\cdot)$), so that the error depends only on the marginal distributions of X^h and X. The result (6.2) is, however, specific to a particular coupling (i.e., joint distribution of X^h and X). Hence, in principle, a different coupling could lead to a different characterization of the error.   □

We now turn to the heuristic derivation of (6.2). Note that if $a(\cdot)$ and $b(\cdot)$ are smooth, then

$$
\begin{aligned}
Z_n^h - Z_{n-1}^h \;=\;& h^{-1/2}\big[b(X_{n-1}^h) - b(X_{n-1}^h)\big] \cdot \Delta_n^h B \\
&+ h^{-1/2}\big[a(X_{n-1}^h) - a(X_{n-1}^h)\big] \cdot h \\
&+ h^{-1/2}\int_{t_{n-1}^h}^{t_n^h} \big[b(X(s))) - b(X_{n-1}^h)\big]\,dB(s) \\
&+ h^{-1/2}\int_{t_{n-1}^h}^{t_n^h} \big[a(X(s)) - a(X_{n-1}^h)\big]\,ds\,.
\end{aligned}
$$

Writing

$$
\Delta_n^{h^*} \stackrel{\text{def}}{=} b'(X_{n-1}^h)Z_{n-1}^h\Delta_n^h B + a'(X_{n-1}^h)Z_{n-1}^h\, h\,,
$$

we obtain

$$
\begin{aligned}
&h^{1/2}\big[Z_n^h - Z_{n-1}^h - \Delta_n^{h^*}\big] \\
\approx\;& \int_{t_{n-1}^h}^{t_n^h} b'(X_{n-1}^h)\big[X(s) - X_{n-1}^h\big]\,dB(s) \\
&+ \int_{t_{n-1}^h}^{t_n^h} a'(X_{n-1}^h)\big[X(s) - X_{n-1}^h\big]\,ds \\
=\;& b'(X_{n-1}^h)\int_{t_{n-1}^h}^{t_n^h}\Big[\int_{t_{n-1}^h}^{s} b(X(u))\,dB(u) + \int_{t_{n-1}^h}^{s} a(X(u))\,du\Big]\,dB(s) \\
&+ a'(X_{n-1}^h)\int_{t_{n-1}^h}^{t_n^h}\Big[\int_{t_{n-1}^h}^{s} b(X(u))\,dB(u) + \int_{t_{n-1}^h}^{s} a(X(u))\,du\Big]\,ds \\
\approx\;& \big[b'(X_{n-1}^h)b(X_{n-1}^h) + a'(X_{n-1}^h)b(X_{n-1}^h)\big]I_1^h(t_n^h) \\
&+ \big[b'(X_{n-1}^h)a(X_{n-1}^h) + a'(X_{n-1}^h)a(X_{n-1}^h)\big]I_2^h(t_n^h)\,, \qquad (6.3)
\end{aligned}
$$

where

$$
I_1^h(t_n^h) \stackrel{\text{def}}{=} \int_{t_{n-1}^h}^{t_n^h}\int_{t_{n-1}^h}^{s}\,dB(u)\,dB(s)\,, \qquad I_2^h(t_n^h) \stackrel{\text{def}}{=} \int_{t_{n-1}^h}^{t_n^h}\int_{t_{n-1}^h}^{s}\,du\,dB(s)\,.
$$

Note that

$$
h^{-1/2}I_1^h(t_n^h) \;=\; h^{-1/2}\left(\frac{\big(B_n^h\big) - B_{n-1}^h\big)^2}{2} - \frac{h}{2}\right) \stackrel{\text{def}}{=} M_1^h(t_n^h) - M_1^h(t_{n-1}^h)\,.
$$

Also, let

$$
h^{-1/2}I_2^h(t_n^h) \stackrel{\text{def}}{=} M_2^h(t_n^h) - M_2^h(t_{n-1}^h)\,.
$$

Both $\{M_1^h(t_k^h)\}_{k=1,2,\ldots}$ and $\{M_2^h(t_k^h)\}_{k=1,2,\ldots}$ are square integrable martingales, with (joint) predictable variations given by

$$\langle B,B\rangle(t_n^h) = t_n^h, \qquad \langle M_1^h, M_1^h\rangle(t_n^h) = t_n^h/2,$$
$$\langle B, M_1^h\rangle(t_n^h) = 0, \qquad \langle M_2^h, M_2^h\rangle(t_n^h) = ht_n^h/3.$$

The martingale CLT implies that

$$\left(B, M_1^h, M_2^h\right) \overset{\mathscr{D}}{\to} \left(B, B_1/\sqrt{2}, 0\right)$$

in $D[0,\infty)$ as $h \downarrow 0$, where B_1 is a standard Brownian motion independent of B. Combining this with (6.3) leads to (6.2).

7   Multidimensional SDEs

The simple SDE (3.1) may be generalized to multidimensions by allowing both the solution and the driving BM to be vector-valued, say $\mathbb{R}^p$ and $\mathbb{R}^q$-valued ($p \neq q$ is allowed). Thus $\boldsymbol{X}(0) = \boldsymbol{x}_0$,

$$\mathrm{d}\boldsymbol{X}(t) = \boldsymbol{a}\big(t, \boldsymbol{X}(t)\big)\,\mathrm{d}t + \boldsymbol{b}\big(t, \boldsymbol{X}(t)\big)\,\mathrm{d}\boldsymbol{B}(t),\ 0 \le t \le 1, \qquad (7.1)$$

where $\boldsymbol{B}$ is q-dimensional Brownian motion with drift vector $\boldsymbol{0}$ and covariance matrix $\boldsymbol{I}_q$,[5] and $\boldsymbol{a} : [0,1] \times \mathbb{R}^p \to \mathbb{R}^p$, $\boldsymbol{b} : [0,1] \times \mathbb{R}^p \to \mathbb{R}^p \times \mathbb{R}^q$. Written coordinate by coordinate,

$$\mathrm{d}X_i(t) = a_i\big(t, \boldsymbol{X}(t)\big)\,\mathrm{d}t + \sum_{j=1}^{q} b_{ij}\big(t, \boldsymbol{X}(t)\big)\,\mathrm{d}B_j(t),\quad i = 1,\ldots,p. \qquad (7.2)$$

The generalization of the Euler scheme is straightforward:

$$X_{i;n}^h = a_i\big(t_{n-1}^h, \boldsymbol{X}_{n-1}^h\big)h + \sum_{j=1}^{q} b_{ij}\big(t_{n-1}^h, \boldsymbol{X}_{n-1}^h\big)\Delta_{j;n}^h B,$$

where the $\Delta_{j;n}^h B$ are i.i.d. $\mathscr{N}(0,h)$. Also the proof that the strong error order is $1/2$ and the weak error order 1 carries over with mainly notational changes.

However, the Milstein scheme gets into difficulties for $q > 1$ because the multidimensional Itô formula has a form that makes the correction term to Euler contain r.v.'s of the form

$$I_{jk} \overset{\mathrm{def}}{=} \int_0^h B_k(s)\,\mathrm{d}B_j(s),$$

---

[5]Note that the assumption of independent coordinates of $\{\boldsymbol{B}(t)\}$, i.e., covariance matrix $\boldsymbol{\Sigma} = \boldsymbol{I}$, is no restriction because it may always be made to hold by replacing $\boldsymbol{B}$ with $\boldsymbol{\Sigma}^{-1/2}\boldsymbol{B}$ if necessary.

whose density cannot be found in closed form when $j \neq k$ and for which there is no straightforward r.v. generation. Various solutions, all fairly complicated, have been suggested. For example, Rydén & Wiktorsson [323] and Wiktorsson [363] suggest using the fact that the distribution of I_{jk} is infinitely divisible with a computable Lévy measure so that general methods for Lévy processes (see Chapter XII) can be used.

In more detail, using the multidimensional Itô formula and ignoring dt terms precisely as in the derivation of the Milstein scheme with $p = 1$ in Section 4a, we get the following approximation to the correction term to the term containing d$B_j(t)$ in the ith component:

$$\int_0^h \left[b_{ij}\big(t, \boldsymbol{X}(t)\big) - b_{ij}\big(0, \boldsymbol{x}_0\big) \right] \mathrm{d}B_j(t)$$

$$\approx \int_0^h \mathrm{d}B_j(t) \int_0^t \sum_{\ell=1}^p \sum_{k=1}^q b_{\ell k} \frac{\partial b_{ij}}{\partial x_j} \, \mathrm{d}B_k(s) \approx b_{\ell k}(0, \boldsymbol{x}_0) \frac{\partial b_{ij}}{\partial x_j}(0, \boldsymbol{x}_0) I_{jk} \, .$$

It is fair to say that the Euler scheme is the dominant one in practical use in multidimensions.

Exercises

7.1 (A) The classical set of coupled ODEs for a predator–prey model is

$$\dot{\ell}(t) = \ell(t)\big(a + bh(t)\big), \quad \dot{h}(t) = h(t)\big(c - d\ell(t)\big) \, ;$$

cf. I.5.20. Khasminskii & Klebaner [216] and Chen & Kulperger [70] suggest the SDE

$$\begin{aligned} \mathrm{d}X_1(t) &= X_1(t)\big(a + bX_2(t)\big) + \sigma_1 X_1(t)\mathrm{d}B_1(t) \, , \\ \mathrm{d}X_2(t) &= X_1(t)\big(c - dX_1(t)\big) + \sigma_2 X_2(t)\mathrm{d}B_2(t) \, , \end{aligned}$$

as one among many possible stochastic versions of the model, where B_1, B_2 are independent standard Brownian motions. The linearity in $X_1(t)$ of the term $\sigma_1 X_1(t)\mathrm{d}B_1(t)$ in d$X_1(t)$ is suggested by diffusion approximations for branching processes and similarly for $\sigma_2 X_2(t)\mathrm{d}B_2(t)$.
Experiment with the sample path behavior of the solution for the set of parameters $a = -0.2$, $b = 0.001$, $c = 0.3$, $d = 0.01$ in I.5.20 and various values of σ_1, σ_2.

8 Reflected Diffusions

Brownian motion $W \stackrel{\text{def}}{=} W_{\mu,\sigma^2}$ with drift μ and variance constant σ^2 is defined as $\sigma B(t) + \mu t$, where $\{B(t)\}$ is standard Brownian motion. Since this process has the same distribution as $\{\sigma\big(B(t) + t\mu/\sigma\big)\}$, we may assume $\sigma^2 = 1$ and will just write $W \stackrel{\text{def}}{=} W_{\mu,1}$ in the following. Obviously, $\{W(t)\}$ is straightforward to simulate among a discrete skeleton by just adding a linear term to $\{B(t)\}$.

Reflected Brownian motion (RBM) $\{V(t)\}$ with drift μ and starting from $V(0) = x \geq 0$ is defined as

$$V(t) = W(t) + L_x(t), \quad \text{where } L_x(t) \stackrel{\text{def}}{=} \left(-\inf_{0 \leq s \leq t} W(s) - x\right)^+$$

(see [16, Section IX.2] for a general discussion of the reflection mapping). If $\mu = 0$, $x = 0$, then $\{V(t)\}$ has the same distribution as $\{|B(t)|\}$ and so simulating at discrete grid points $t_n^h = nh$ is easy and there is no error at the grid points. If $\mu \neq 0$, it seems natural to simulate first a discrete skeleton $\{W_n^h\}$ of $\{W(t)\}$ and let $\{V_n^h\}$ be the (discrete skeleton) of the reflected version defined by the Lindley recursion

$$V_n^h = \left(V_{n-1}^h + \Delta W_n^h\right)^+ = W_n^h + \left(-\min_{\ell=0,\dots,k} W^h - x\right)^+. \tag{8.1}$$

However, now there is a systematic error also at the grid points, because even if $\{W_n^h\}$ fits exactly, the minimum is taken over too small a set and so V_n^h is smaller than $V(t_n^h)$.

A careful analysis of this situation was given by Asmussen, Glynn, & Pitman [20], who showed that the error at a fixed point, say $T = 1$, is of order $h^{1/2}$ and gave precise asymptotics. They also discussed the following algorithm as improvement of (8.1). The key observation is that for fixed $h > 0$, the joint density of

$$\left(W(h), -\min_{0 \leq t \leq h} W(t)\right) \tag{8.2}$$

is known. For simulation purposes, a convenient representation of this distribution is to note that marginally, $W(h)$ is $\mathcal{N}(\mu h, h)$, and that by a result of Lévy,

$$F_y(x) \stackrel{\text{def}}{=} \mathbb{P}\left(-\min_{0 \leq t \leq h} W(t) - y \leq x \,\middle|\, W(h) = y\right) = 1 - e^{-2x(y+x)/h}. \tag{8.3}$$

By easy calculus,

$$F_y^{-1}(z) = \frac{-y + \sqrt{y^2 - 2h \log(1-z)}}{2}.$$

Thus, an algorithm for exact simulation of the values V_n^h of RBM at the $t_n^h = nh$ is obtained as follows:

(i) Let $t \longleftarrow 0$, $V \longleftarrow 0$.

(ii) Generate ΔW as $\mathcal{N}(\mu h, h)$ and let

$$R \longleftarrow -\frac{\Delta W}{2} + \frac{\sqrt{\Delta W^2 - 2h \log(U)}}{2},$$

where U is uniform on $(0, 1)$.

(iii) Let $L \longleftarrow (-R - V)^+$.

(iv) Let $t \longleftarrow t + h$, $V \longleftarrow V + \Delta W + L$.

(v) Return to 2.

Example 8.1 We considered RBM with $\mu = -1$, $\sigma^2 = 1$, and reflection at 0. It is known that the stationary distribution is exponential with density $\pi(x) \overset{\text{def}}{=} 2e^{-2x}$. Thus, if we generate $V(0)$ with density $\pi(x)$, $V(T)$ should again have density $\pi(x)$ and the weak error of a simulation scheme may be assessed by a Q-Q plot. This is given in Figure 8.1, with the upper row corresponding to the naive algorithm (8.1) and the lower to the improvement using (8.3); we took $T = 100$, and the columns correspond to $h = 1, 1/2, 1/4$.

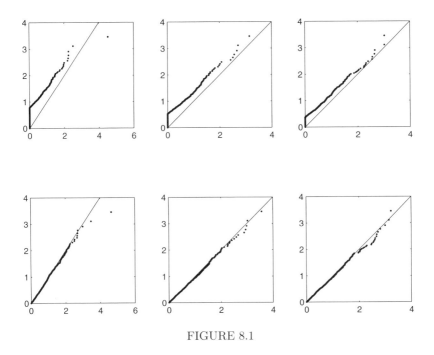

FIGURE 8.1

The improvement provided by the more sophisticated algorithm is quite clear, and one also notes the slow convergence of the naive algorithm.

□

Now consider a general diffusion given by $\mu(x)$ and $\sigma^2(x)$ and having (upward) reflection at x_0. The naive Euler scheme is then

$$X_n^h = \max\left[x_0, X_n^h + \mu(X_n^h)h + \sigma(X_n^h)\Delta_n^h B\right]. \qquad (8.4)$$

It is clear that the order of its strong error must be at least $1/2$, whereas the analysis of RBM above indicates that the weak order error is also $1/2$ (rather than 1 as in the case of no reflection). However, the exact scheme for RBM outlined above suggests that an improvement may be obtained using the approximation

$$X_n^h \;=\; \varphi\big(h, X_{n-1}^h, \mu\big(X_{n-1}^h\big), \sigma^2\big(X_{n-1}^h\big)\big), \qquad (8.5)$$

where $\varphi\big(h, x, \mu, \sigma^2\big)$ is the mapping given by the algorithm that updates RBM(μ, σ^2) from $V_{n-1}^h = x$ to $V_n^h = \varphi(\cdots)$. Lepingle [239] confirms that this approximation improves the weak error from order $1/2$ to order 1.

A similar conclusion can be drawn for one-dimensional SDEs subject to killing the process at the boundary. If a naive implementation of the Euler scheme is used, the bias of the scheme is of order $h^{1/2}$ (because of the order-$h^{1/2}$ random fluctuations exhibited in intervals of length h). On the other hand, a recursive scheme similar to (8.5) yields a bias of order h in the weak sense. The recursion depends on an exact simulation algorithm for killed BM (i.e., for an SDE with constant drift and variance functions; see Gobet [158] for details).

Example 8.2 We considered GBM with $\mu = -1/2$, $\sigma^2 = 1$, and reflection at $x_0 = 1$. Then $\log X(t)$ is RBM$(-1, 1)$. Thus, proceeding as in Example 8.1, we generate $\log X(0)$ with density $\pi(x)$, and study the weak error by an exponential Q-Q plot of the $\log X(T)$. This is given in Figure 8.2, with the upper row corresponding to the naive algorithm (8.4) and the lower to (8.5); again, $T = 100$ and the columns correspond to $h = 1, 1/2, 1/4$.

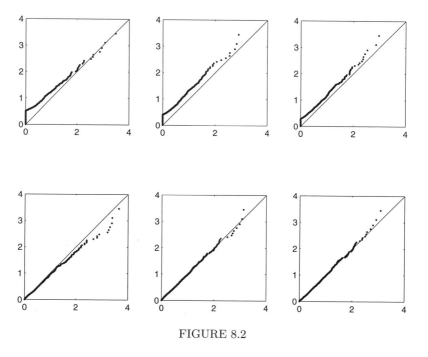

FIGURE 8.2

The conclusions are similar to Example 8.1.   □

An open problem of considerable interest is to find good algorithms for simulating reflected Brownian motion in higher-dimensional regions; see Chen & Yao [69] for a discussion of this class of models..

Exercises

8.1 (A) Produce some further illustrations for RBM like that of Example 8.1 by choosing $\mu = 0$, $V(0) = 0$, and taking advantage of the fact that then the marginal distribution of $V(t)$ is that of the absolute value of a $\mathcal{N}(0, t)$ r.v.

8.2 (TP) In [20], it is suggested that RBM be simulated on a random grid associated with a Poisson process that is run independently of the BM W by using the fact that if T is an exponential r.v. with rate λ that is independent of W, then the r.v.'s $W(T) - \min_{0 \le t \le T} W(t)$ and $-\min_{0 \le t \le T} W(t)$ are independent and exponentially distributed with rates η and ω respectively, where

$$\eta \overset{\text{def}}{=} -\mu + \sqrt{\mu^2 + 2\lambda}, \qquad \omega \overset{\text{def}}{=} \mu + \sqrt{\mu^2 + 2\lambda}.$$

More precisely, the suggested algorithm for simulation of BM W, the minimum M, and thereby RBM $V = W - M$ at the epochs t of a Poisson(λ) grid, is:

(i) Let $t \leftarrow 0$, $W \leftarrow 0$, $V \leftarrow 0$, $M \leftarrow 0$.
(ii) Generate T, S_1, S_2 as independent exponential r.v.'s with rates 1, η, ω, respectively.
(iii) Let $t \leftarrow t + T$, $M \leftarrow \min(M, W - S_2)$, $W \leftarrow B + W_1 - S_2$, $V \leftarrow W - M$.
(iv) Return to (ii).

Explain why this algorithm does not produce unbiased estimates as claimed in [20].

Chapter XI
Gaussian Processes

1 Introduction

Gaussian models are widely used not only in the setting of $\mathbb{R}$- or $\mathbb{R}^d$-valued stochastic processes with time parameter t belonging to one-dimensional sets such as $\mathbb{N}, [0, \infty), [0, T], (-\infty, \infty)$, but also in the context of random fields. We therefore allow t to vary in a general index set $\mathscr{T}$, and define $\{X(t)\}_{t \in \mathscr{T}}$ to be d-dimensional Gaussian if for any finite subset $\mathscr{T}_0$, $X(\mathscr{T}_0) \stackrel{\text{def}}{=} \{X(t)\}_{t \in \mathscr{T}_0}$ has a $d \times |\mathscr{T}_0|$-dimensional Gaussian distribution with mean vector $\boldsymbol{\mu}(\mathscr{T}_0)$ and covariance matrix $\boldsymbol{\Sigma}(\mathscr{T}_0)$ (say); when $\mathscr{T} \not\subseteq \mathbb{R}$, one often talks of a *Gaussian random field*.

For examples beyond $\mathscr{T} \subseteq \mathbb{R}$, take, e.g., $\mathscr{T} \subseteq \mathbb{R}^2$ and think of $\mathscr{T}$ as a geographical region. Then $X(t)$ could be a vector of climatic data (yearly rainfall, mean temperatures in the 12 months of the year, etc.) at location t, or geophysical information like such as strength and direction of the magnetic field, or the gravitation. Gaussian random fields evolving in time are also covered by the formalism. For example, let $S \subset \mathbb{R}^2$ be a part of an ocean and $\mathscr{T} \stackrel{\text{def}}{=} S \times [0, \infty)$. When modeling the movement of ocean waves, $X(s, t)$ could then be the deviation of the actual water level from a base level at location $s \in S$ at time t.

1a   Covariance Representation

If the mean function $\mu(s) \stackrel{\text{def}}{=} \mathbb{E}X(t)$ and the covariance function $r(s,t) \stackrel{\text{def}}{=}$ $\mathbb{C}ov\big(X(s), X(t)\big)$ are known, we also know

$$\boldsymbol{\mu}(\mathscr{T}_0) \stackrel{\text{def}}{=} \big(\mu(t)\big)_{t\in\mathscr{T}_0} \quad \text{and} \quad \boldsymbol{\Sigma}(\mathscr{T}_0) \stackrel{\text{def}}{=} \big(r(s,t)\big)_{s,t\in\mathscr{T}_0} \tag{1.1}$$

for any finite subset $\mathscr{T}_0$ and thereby the distribution of the whole process. Conversely, a pair of given μ- and r-functions will specify a consistent family of finite-dimensional distributions and thereby the distribution of the whole $\{X(t)\}_{t\in\mathscr{T}}$, provided $\boldsymbol{\Sigma}(\mathscr{T}_0)$ is symmetric and nonnegative definite for each $\mathscr{T}_0$. In summary, a Gaussian process is completely specified by its mean and covariance functions $\mu(\cdot)$, $r(\cdot,\cdot)$. When viewing $\mu(\cdot)$, $r(\cdot,\cdot)$ as the fundamental parameters of the Gaussian process, one often talks of a representation in the *time domain*. In Section 4, we will meet representations in the *frequency domain* obtained by Fourier transforming the covariance function.

If $\mathscr{T} \subseteq \mathbb{R}^d$, we call $\{X(t)\}_{t\in\mathscr{T}}$ *stationary* when $\mu(t) \stackrel{\text{def}}{=} \mu$ is independent of t and $r(s,t) \stackrel{\text{def}}{=} r(t-s)$ depends only on $t-s$. If, in addition $d > 1$, and $r(t-s)$ depends only on $\|t-s\|$ (Euclidean norm), we call the process *isotropic*. The distribution of a stationary process is invariant under shifts and the distribution of an isotropic process under rotations (orthonormal transformations).

The smoothness of the sample paths at t is determined by smoothness properties of $r(s,t)$ at $s = t$. Consider for simplicity a stationary process with $\mathscr{T} = \mathbb{R}$, $d = 1$, and covariance function $r_X(t)$. A sufficient condition for a.s. continuity is $r(h) - r(0) = \mathrm{O}\big(h/|\log h|\big)$ as $h \downarrow 0$, which is slightly weaker than differentiability from the right. For example, this covers the stationary Ornstein–Uhlenbeck process where $r_X(t) = r_X(0)\mathrm{e}^{-\alpha|t|}$ for some $\alpha > 0$. The sample functions are differentiable, with derivative say $X'(t)$, if $r_X(h)$ is twice differentiable at $h = 0$, and then the covariance function of X' is $r_{X'}(t) = -r_X''(t)$. More generally, the bivariate process (X, X') has

$$r_{X,X'}(t) = \begin{pmatrix} r_X(t) & -r_X'(t) \\ -r_X'(t) & -r_X''(t) \end{pmatrix}. \tag{1.2}$$

Similarly, the condition for existence of the kth derivative of X is essentially the existence of the $2k$th derivative of $r(t)$.

Figure 1.1 illustrates the effect of the smoothness properties of $r(\cdot)$ on the sample path. We took $r(t) = \mathrm{e}^{-4|t|^\beta}$ with $\beta = 1$ (the Ornstein–Uhlenbeck process), $\beta = 4/3$, $5/3$, and (the differentiable case) $\beta = 2$. The figure uses 2,000 time steps in $\mathscr{T} = [0,5]$, Cholesky factorization (see Section 2), and common random numbers to produce the Gaussian white noise driving the simulations.

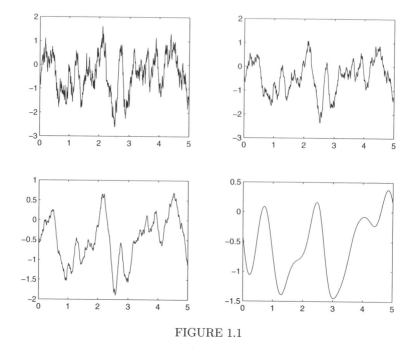

FIGURE 1.1

That the universe of Gaussian processes includes examples with smooth sample paths is definitely a main reason that Gaussian modeling is attractive in many situations.

1b   Simple Examples

A discrete-time ARMA(p, q) process has representation

$$X_{n+1} = \beta_1 X_n + \beta_2 X_{n-1} + \cdots + \beta_p X_{n-p+1} + \alpha_1 \varepsilon_n + \cdots + \alpha_q \varepsilon_{n-q+1} , \quad (1.3)$$

where the ε_n are i.i.d. standard normal variables. If $p = 0$, we talk of an MA(q) (moving average) process, and if $q = 1$, of an AR(p) (auotoregressive) process.

A set of i.i.d. $\mathcal{N}(0,1)$ r.v.'s indexed by $t \in \mathscr{T}$, where $\mathscr{T}$ is discrete, is often referred to as *white noise* on $\mathscr{T}$. An example is of course the ε_k in (1.3). The meaning of the concept is less clear for a continuous $\mathscr{T}$. It is commonly agreed, not least among engineers, to term the formal infinitesimal increments $dB(t)$ of standard Brownian motion as white noise, but the mathematical meaning of the concept in this context is unclear.

Moving averages

$$X(t) = \int_{-\infty}^{t} \beta(t - s) \, dB(s) \qquad (1.4)$$

of white noise of this type provide an obvious continuous-time and infinite-memory analogue of MA(q) processes as defined above.

If $\varepsilon_1, \varepsilon_2$ are i.i.d. $\mathcal{N}(0,1)$ and $\lambda \in (0, 2\pi)$, then

$$X(t) = \cos(t\lambda)\varepsilon_1 + \sin(t\lambda)\varepsilon_2 \qquad (1.5)$$

defines a stationary Gaussian process that we can think of as a random periodic oscillation with frequency $f = \lambda/2\pi$. The same is true for linear combinations of independent such processes. We thereby obtain a class of processes that, in a suitable sense, is dense in the class of Gaussian processses; see further Section 4.

1c   Gaussian Processes as Limits

Many stochastic models can be approximated by Gaussian processes for reasons having to do with CLT effects. A good example is given by the GI/G/∞ queue with arrival rate λ.[1] As the renewal process is accelerated (i.e., λ sent to infinity), the total number-in-system process can then be approximated by a stationary Gaussian process with a covariance that depends on the fine structure of the service-time distribution, but only on the interarrival time T_k through $\lambda = 1/\mathbb{E}T_1$ and $\mathbb{E}T_1^2$; see, e.g., Borovkov [51]. Certain characteristics of the limiting process may require simulation, for example, computation of the probability that the limiting process exceeds x in some time interval. For such computations, the ideas of this chapter are relevant. As we shall see, Gaussian processes have certain properties that lend themselves to efficiently implementable simulations. In addition, a simulation of the approximating Gaussian process is clearly far less time-consuming than a more detailed simulation of the underlying queuing dynamics themselves.

One point of caution is worth raising, however. As in the setting of the standard CLT, Gaussian approximations tend to work well when the event of interest or expectation to be computed is sampled from the "middle of the distribution." On the other hand, Gaussian approximations tend to be poor when one is computing tail events. For such computations, simulating the underlying model (with all its inherent complexities) is unavoidable.

Another significant example of a Gaussian process occurring as a limit is fractional Brownian motion, which we treat separately in Section 6.

1d   First Remarks on Simulation

An important feature in the simulation of processes such as Markov chains, regenerative processes, and solutions of SDEs is the fact that such processes are causally simulated via recursive updating. Such recursive updating has

---

[1] A queue with renewal arrivals, general service times, and infinitely many servers.

an associated complexity forf N updatings of O(N). This is not true for Gaussian processes except for special cases such as finite-order moving averages and autoregressions. In general, the complexity of simulating N correlated Gaussian r.v.'s (exactly) increases at a rate factor faster than N. On the other hand, Gaussian processes have the unusual characteristic that they do not need to be simulated causally. In particular, when simulating a Gaussian process over the time interval $\mathscr{T} = [0, 1]$, we may use bisection in a similar way as in X.2b by first simulating at times 0 and 1, then at 1/2, next at 1/4 and 3/4, and so on. If we wish, we can successively simulate the Gaussian r.v.'s that bisect the successive subintervals that are generated. This freedom to simulate noncausally can potentially be quite useful in the case of computing the distribution of the maximum of a Gaussian process for which it is known by theory (see, e.g., Mandjes [248]) that the maximizer is likely to be close to a specific time t_0. In this context, we may wish to focus most of the early sampling at time epochs close to t_0.

For a few Gaussian processes, one may in principle have the possibility to generate values at an arbitrary $t \in \mathscr{T}$ from just a finite set of r.v.'s; (1.5) is one example. In general, one has to be satisfied with generating the process on finite subsets of $\mathscr{T}$. This will then be achieved by successively generating $X(\mathscr{T}_n)$, where the sets $\mathscr{T}_n$ are finite and increasing in n. If the target is just $X(\mathscr{T}_0)$ for some finite $\mathscr{T}_0$, one may want to avoid the recursive step and generate $X(\mathscr{T}_0)$ directly. This would cover cases such as evaluating the tail of the maximum on a compact set, whereas in dealing with stopping time problems (e.g., first-passage problems) the recursive step is essential.

In dealing with random fields, the problems for going from $\mathscr{T}_{k-1}$ to $\mathscr{T}_k$ are in general just the same as for processes with a one-dimensional time scale. Therefore for most purposes the problem of simulating a Gaussian process boils down to the case $\mathscr{T} = \mathscr{T}_0 = \{0, 1, \ldots, N\}$, with the twist that in some cases N is not determined a priori.

We are therefore back to the problem of II.3 of simulating from a multivariate Gaussian distribution in $\mathbb{R}^N$ with covariance matrix $\boldsymbol{\Sigma}_N$ and (w.l.o.g) mean vector $\mathbf{0}$. One method, Cholesky factorization, has already been introduced in II.3, but we revisit it here in Section 2 from a stochastic process point of view. The method uses a factorization $\boldsymbol{\Sigma}_N = \boldsymbol{C}\boldsymbol{C}^{\mathsf{T}}$. In Section 3, we discuss another method to provide such a factorization that is potentially more efficient when N is large, as will often be the case in a stochastic-process setting. Both of these methods are exact, that is, they produce samples with exactly the required $\mathscr{N}(\mathbf{0}, \boldsymbol{\Sigma}_N)$ distribution. Section 4 gives some Fourier-inspired methods which are potentially fast but also introduce some bias. Finally, Section 6 deals with a more specialized topic, fractional Brownian motion.

Exercises

1.1 (TP) Show that the Brownian bridge $B(t) - tB(1)$ is a Gaussian process on $\mathscr{T} = [0, 1]$ and compute $r(s, t)$.

1.2 (TP) Compute r for the moving average (1.4).
1.3 (TP) Show (1.2) by taking limits in L_2.

2 Cholesky Factorization. Prediction

We consider recursive algorithms based on the covariance function, studying how to generate X_{n+1} given that $X_0, \ldots, X_n$ have been generated. Thus we need to specify the conditional distribution of X_{n+1} given $X_0, \ldots, X_n$, which is a standard problem in the multivariate normal distribution. Write $\boldsymbol{R}(n)$ for the covariance matrix of $X_0, \ldots, X_n$, and let $\boldsymbol{r}(n)$ be the $(n+1)$-column vector with $r_k(n) = r(n+1, k)$, $k = 0, \ldots, n$. Then

$$\boldsymbol{R}(n+1) \; = \; \begin{pmatrix} \boldsymbol{R}(n) & \boldsymbol{r}(n) \\ \boldsymbol{r}(n)^{\mathsf{T}} & r(n+1, n+1) \end{pmatrix}.$$

Therefore by general formulas for the multivariate normal distribution given in A1, the conditional distribution of X_{n+1} given $X_0, \ldots, X_n$ is $\mathcal{N}\!\left(\widehat{X}_{n+1}, \sigma_n^2\right)$, where

$$\widehat{X}_{n+1} \; = \; \boldsymbol{r}(n)^{\mathsf{T}} \boldsymbol{R}(n)^{-1} \begin{pmatrix} X_0 \\ X_1 \\ \vdots \\ X_n \end{pmatrix}, \quad \sigma_n^2 \; = \; r(n+1, n+1) - \boldsymbol{r}(n)^{\mathsf{T}} \boldsymbol{R}(n)^{-1} \boldsymbol{r}(n),$$

and we can just generate X_{n+1} according to $\mathcal{N}\!\left(\widehat{X}_{n+1}, \sigma_n^2\right)$.

Note that in the terminology of time series (e.g., Brockwell & Davis [58]), $\widehat{X}_{n+1}$ is the best linear predictor for X_{n+1} (in terms of minimizing the mean-square error) and σ_n^2 the corresponding prediction error.

From the point of view of simulation, the difficulty is to organize the calculations economically, say by recursive computation of $\boldsymbol{\gamma}(n)^{\mathsf{T}} \boldsymbol{\Gamma}(n)^{-1}$ to avoid matrix inversion in each step, or by some other method.

An established device is *Cholesky factorization*. This is an algorithm for writing a given symmetric $(n+1) \times (n+1)$ matrix $\boldsymbol{\Gamma} = (\gamma_{ij})_{i,j=0,\ldots,n}$ as $\boldsymbol{\Gamma} = \boldsymbol{C}\boldsymbol{C}^{\mathsf{T}}$, where $\boldsymbol{C} = (c_{ij})_{i,j=0,\ldots,n}$ is (square) lower triangular ($c_{ij} = 0$ for $j > i$), and has already been used in II.3, where it was found that $\boldsymbol{C}$ can be computed recursively: for $i = j = 0$, determine c_{00} by $\gamma(0,0) = c_{00}^2$ ($i = j = 0$); for $i = 1$ use the two equations

$$\gamma(1,0) = c_{10}c_{00}, \quad \gamma(1,1) = c_{10}^2 + c_{11}^2$$

determining c_{10}, c_{11}; in general, if $c_{i'j}$ has been computed for $i' < i$, let

$$c_{ij} \; = \; \frac{1}{c_{jj}}\left(\gamma(i,j) - \sum_{k=0}^{j-1} c_{ik}c_{jk}\right), \quad j < i, \quad c_{ii}^2 \; = \; \gamma(i,i) - \sum_{k=0}^{i-1} c_{ik}^2. \quad (2.1)$$

For simulation of $X_0, \ldots, X_n$, the implication is that we can take $Y_0, \ldots, Y_n$ to be i.i.d. standard normal, write $\boldsymbol{Y}(n) = (Y_0 \ldots Y_n)^\mathsf{T}$, $\boldsymbol{X}(n) = (X_0 \ldots X_n)^\mathsf{T}$, and define $\boldsymbol{C}(n)$ to be the Cholesky factorization of $\boldsymbol{R}(n)$. Then the X_i are generated by $\boldsymbol{X}(n) = \boldsymbol{C}(n)\boldsymbol{Y}(n)$. Component by component,

$$X_i = \sum_{k=0}^{i} c_{ik} Y_k, \quad i = 0, \ldots, n. \tag{2.2}$$

Note that we did not write $c_{ik}(n)$, because (2.1) shows that $c_{ik}(n)$ does not depend on n as long as $i, k \leq n$. This means in particular that to get $\boldsymbol{C}(n+1)$ from $\boldsymbol{C}(n)$, one needs to compute only the last row ($i = n+1$). That $\boldsymbol{X}(n)$ has the correct distribution follows from

$$\mathbb{C}\text{ov}\big(\boldsymbol{X}(n)\big) = \mathbb{C}\text{ov}\big(\boldsymbol{C}(n)\boldsymbol{Y}(n)\big) = \boldsymbol{C}(n)\boldsymbol{I}\boldsymbol{C}(n)^\mathsf{T} = \boldsymbol{R}(n).$$

Remark 2.1 The representation (2.2) shows that $Y_0, \ldots, Y_n$ form a Gram-Schmidt orthonormalization of $X_0, \ldots, X_n$. That is, (in the L_2 sense) $Y_0, \ldots, Y_n$ are orthonormal and

$$\text{span}(Y_0, \ldots, Y_k) = \text{span}(X_0, \ldots, X_k). \qquad \square$$

Remark 2.2 The number of flops needed to update element j in row i of C via(2.1) is O(j). Therefore the number of flops for row i is O(i^2), and that for all $n+1$ rows is O(n^3). $\qquad \square$

In conclusion, simulation via Cholesky factorization is exact (no approximation is involved), and one does not need to set the time horizon in advance. Note also that no matrix inversion at all is involved. The drawback of the method is that because of the complexity O(n^3), it becomes slow as n becomes large. Also, memory may present a problem (one needs to store all c_{ij}).

In general, Cholesky factorization is just a mathematical device for matrix manipulation. However, in the case of Gaussian processes the procedure can be given an interesting interpretation in terms of the standard problem of time-series analysis of *prediction* or *forecasting*: given that we have observed $X_0, \ldots, X_n$, we want a predictor of X_{n+1}. Now the best linear predictor (in terms of minimizing the mean-square error) of X_{n+1} is

$$\widehat{X}_{n+1} = \mathbb{E}\big[X_{n+1} \,\big|\, X_0, \ldots, X_n\big] = \boldsymbol{r}(n)^\mathsf{T} \boldsymbol{R}(n)^{-1} \begin{pmatrix} X_0 \\ X_1 \\ \vdots \\ X_n \end{pmatrix},$$

and thus algorithms for recursive prediction are potentially useful for simulation as well.

Two such algorithms are given in [58, Chapter 5], the *Durbin–Levinson algorithm* and the *innovations algorithm*. We consider only the latter. The key is to represent $\widehat{X}_i$ as a linear combination of the $X_k - \widehat{X}_k$ with $k < i$ rather than the X_k,

$$\widehat{X}_i \;=\; \sum_{k=0}^{i-1} \theta_{i,i-k}(X_k - \widehat{X}_k) \tag{2.3}$$

in the notation of [58]. Obviously $\widehat{X}_0 = 0$, $\sigma_n^2 = \mathbb{V}ar(X_{n+1} - \widehat{X}_{n+1})$, $\sigma_{-1}^2 = r(0,0)$. Define $Y_k \overset{\text{def}}{=} (X_k - \widehat{X}_k)/\sigma_{k-1}$.

Proposition 2.3 *The Y_k are i.i.d. standard normal.*

Proof. All that needs to be shown is independence. Let $\mathscr{H}_k$ denote the subspace of L_2 spanned by $X_0, \ldots, X_k$ and let $\langle X, Y \rangle = \mathbb{E}(XY)$ denote the usual inner product in L_2. For $i < j$ we have $X_i - \widehat{X}_i \in \mathscr{H}_{j-1}$ and $X_j - \widehat{X}_j \perp \mathscr{H}_{j-1}$ by definition of $\widehat{X}_j$. Thus the r.v.'s $Y_j \overset{\text{def}}{=} X_j - \widehat{X}_j$ are orthogonal, i.e., uncorrelated, and independence follows from properties of the multivariate normal distribution. $\qquad\square$

If we let $c_{ik} \overset{\text{def}}{=} \theta_{i,i-k}\sigma_{k-1}$, $k < i$, $c_{ii} = \sigma_{i-1}$, and write $X_i = (X_i - \widehat{X}_i) + \widehat{X}_i$, (2.3) takes the form (2.2). That is, *determining the $\theta_{i,j}$ needed for the innovation algorithm involves just the same equations as Cholesky factorization, and* (with the right choice of sign) *the Y_k in* (2.2) *can be interpreted as the* $(X_k - \widehat{X}_k)/\sigma_{k-1}$, *which in turn form a Gram–Schmidt orthonormalization of the X_k* (cf. Remark 2.1).

For a further variant of Cholesky factorization, see Hosking [185].

Exercises

2.1 (A) Let X be a stationary Gaussian process with mean zero and covariance function

$$\gamma(s) \;=\; \gamma(t, t+s) \;=\; \begin{cases} (2 + |s|)(1 - |s|)^2 & -1 < s < 1, \\ 0 & |s| \geq 1. \end{cases}$$

Simulate a discrete skeleton of X by Cholesky factorization to get a Monte Carlo estimate of

$$z \;=\; \mathbb{P}\left(\sup_{0 \leq t \leq 2} X(t) > 2 \right).$$

2.2 (A) Consider a stationary Gaussian process $\{X(t)\}_{t \geq 0}$ in continuous time with covariance function e^{-t^2}. Then the sample paths are differentiable, cf. Section 1a, so we can define the upcrossing times $u_1, u_2, \ldots$ of y by $u_1 \overset{\text{def}}{=} \inf\{t > 0 : X(t) = y, X'(t) > 0\}$, $u_{n+1} \overset{\text{def}}{=} \inf\{t > u_n : X(t) = y, X'(t) > 0\}$ and the subsequent downcrossing times by $d_n \overset{\text{def}}{=} \inf\{t > u_n : X(t) = y, X'(t) < 0\}$. The nth excursion above y is $\{X(t)\}_{u_n \leq t \leq d_n}$ and its length is $d_n - u_n$. The problem is to use Cholesky factorization to give a simulation estimate (with confidence limits) of the typical length ℓ of an excursion, as defined as the limiting time average

$$\ell = \lim_{T \to \infty} \frac{L_T}{N_T} \quad \text{where} \quad N_T = \#\{n : u_n \le T\}, \; L_T = \sum_{n : u_n \le T} (d_n - u_n).$$

Start by simulating some discrete skeletons $X(0), X(h), X(2h), \ldots$ to determine an appropriate mesh h. For the estimation of ℓ, use the Palm representation $\ell = \mathbb{E}L_T/(\lambda T)$, where $\lambda = \mathbb{E}N_T/T$; cf. IV.6e and Aberg et al. [3].

2.3 (A) Complement Exercise 2.2 by using the number $N_T(y_i)$ of upcrossings of y_i in $[0, T]$, $i = 1, \ldots, k$, as multiple control variates. The mean of N is given by *Rice's formula*: for a stationary and differentiable Gaussian process with variance σ^2, the rate of upcrossings of level $y > 0$ is $1/\mathbb{E}N_T(L)$ where

$$\mathbb{E}N_T(L) = \frac{1}{2\pi} e^{-y^2/2} \sqrt{-r''(0)}.$$

3   Circulant-Embeddings

We shall here survey an exact method for simulating a segment of a stationary process, say $(X_0, X_1, \ldots, X_N)$, in discrete time given its covariance function $r_0, r_1, \ldots, r_N$, which has the advantage over Cholesky factorization of having a far better complexity, $O(N \log N)$ compared to $O(N^3)$. References are Davis et al. [81], Ripley [300], Dietrich & Newsam [90], and Wood & Chan [367].

The method uses finite Fourier methods. Recall from A3 that the (finite) Fourier matrix $\boldsymbol{F} = \boldsymbol{F}_n$ of order n is the square matrix with rows and columns indexed by $\{0, 1, \ldots, n-1\}$ and rsth entry w^{rs}, where $w = w_n = e^{2\pi i/n}$ is the nth root of unity. The finite Fourier transform of a vector $\boldsymbol{a} = (a_0 \ldots a_{n-1})^\mathsf{T}$ is then $\widehat{\boldsymbol{a}} = \boldsymbol{F}\boldsymbol{a}/n$, and since $\boldsymbol{F}^{-1} = \overline{\boldsymbol{F}}/n$ (complex conjugate), $\boldsymbol{a}$ can be recovered from the Fourier transform $\widehat{\boldsymbol{a}}$ as $\boldsymbol{a} = \overline{\boldsymbol{F}}\widehat{\boldsymbol{a}}$. Element by element,

$$\widehat{a}_r = \sum_{s=0}^{n-1} a_s w^{rs}/n, \quad a_r = \sum_{s=0}^{n-1} \widehat{a}_s w^{-rs}.$$

A *circulant* of dimension n is a $n \times n$ matrix of the form

$$\boldsymbol{C} = \begin{pmatrix} c_0 & c_{n-1} & \cdot & c_2 & c_1 \\ c_1 & c_0 & c_{n-1} & \cdot & c_2 \\ \cdot & c_1 & c_0 & \cdot & \cdot \\ c_{n-2} & \cdot & \cdot & \cdot & c_{n-1} \\ c_{n-1} & c_{n-2} & \cdot & c_1 & c_0 \end{pmatrix};$$

note the pattern of equal entries $c_{ij} \overset{\text{def}}{=} c_k$ on $\{ij : i - j = k \bmod n\}$. Again, we label the rows and columns $0, 1, \ldots, n-1$.

Proposition 3.1 *The eigenvalues of a circulant are* $\lambda_s = \sum_{v=0}^{n-1} c_v w^{-vs}$, $s = 0, \ldots, n - 1$. *The eigenvalue corresponding to* λ_s *is the sth column*

$$(w^{st})_{t=0,\ldots,n-1} = \left(1 \;\; w^s \;\; w^{2s} \;\; \ldots \;\; w^{s(n-1)}\right)^\mathsf{T}$$

of $\boldsymbol{F}$. *That is,* $\boldsymbol{C} = \boldsymbol{F}\boldsymbol{\Lambda}\overline{\boldsymbol{F}}/n$, *where* $\boldsymbol{\Lambda}$ *is the diagonal matrix with the* λ_s *on the diagonal. In vector notation,* $\boldsymbol{\lambda} = \boldsymbol{F}\boldsymbol{c}$.

Proof. The $r$sth element of $\boldsymbol{CF}$ is

$$\sum_{t=0}^{n-1} c_{rt} f_{ts} = \sum_{t=0}^{n-1} c_{t-r} w^{ts} = \sum_{v=0}^{n-1} c_v w^{rs-vs} = \lambda_s w^{-rs},$$

which we recognize as the $r$sth element of $\boldsymbol{F}\boldsymbol{\Lambda}$. That is, $\boldsymbol{CF} = \boldsymbol{F}\boldsymbol{\Lambda}$, which is equivalent to the assertion of the proposition because of $\boldsymbol{F}^{-1} = \overline{\boldsymbol{F}}/n$. $\square$

We now turn to the algorithm. The first step is to embed the covariance matrix $\boldsymbol{\Sigma}$ of $X_0, \ldots, X_N$ as the upper left corner of a circulant of order $2M$. It is easy to see that this is possible if and only if $M \geq N$. If $M = N$, the circulant $\boldsymbol{C}$ is unique and equals

$$\begin{pmatrix} r_0 & r_1 & \cdot & r_{N-1} & r_N & r_{N-1} & r_{N-2} & \cdot & r_2 & r_1 \\ r_1 & r_0 & \cdot & r_{N-2} & r_{N-1} & r_N & r_{N-1} & \cdot & r_3 & r_2 \\ \cdot & \cdot & \cdot & \cdot & \cdot & \cdot & \cdot & \cdot & \cdot & \cdot \\ r_N & r_{N-1} & \cdot & r_1 & r_0 & r_1 & r_2 & \cdot & r_{N-2} & r_{N-1} \\ r_{N-1} & r_N & \cdot & r_2 & r_1 & r_0 & r_1 & \cdot & r_{N-3} & r_{N-2} \\ \cdot & \cdot & \cdot & \cdot & \cdot & \cdot & \cdot & \cdot & \cdot & \cdot \\ r_1 & r_2 & \cdot & r_N & r_{N-1} & r_{N-2} & r_{N-3} & \cdot & r_1 & r_0 \end{pmatrix} \cdot$$

For example, if $N = 3$, $\boldsymbol{r} = (8\ 4\ 2\ 1)^\mathsf{T}$, then

$$\boldsymbol{\Sigma} = \begin{pmatrix} 8 & 4 & 2 & 1 \\ 4 & 8 & 4 & 2 \\ 2 & 4 & 8 & 4 \\ 1 & 2 & 4 & 8 \end{pmatrix} \quad \text{and} \quad \boldsymbol{C} = \begin{pmatrix} 8 & 4 & 2 & 1 & 2 & 4 \\ 4 & 8 & 4 & 2 & 1 & 2 \\ 2 & 4 & 8 & 4 & 2 & 1 \\ 1 & 2 & 4 & 8 & 4 & 2 \\ 2 & 1 & 2 & 4 & 8 & 4 \\ 4 & 2 & 1 & 2 & 4 & 8 \end{pmatrix} \cdot$$

The idea of this circulant embedding is that the eigendecomposition of $\boldsymbol{C}$ is easy to evaluate via Proposition 3.1 and that one thereby also obtains a square root $\boldsymbol{D}$ (in the sense that $\boldsymbol{D}\overline{\boldsymbol{D}}^\mathsf{T} = \boldsymbol{C}$), namely $\boldsymbol{D} \overset{\text{def}}{=} (1/n^{1/2})\boldsymbol{F}\boldsymbol{\Lambda}^{1/2}$, *provided all entries of* $\boldsymbol{\Lambda}$ *are nonnegative, that is, provided the vector* $\boldsymbol{Fc}$ *is nonnegative.* Of course, $\boldsymbol{D}$ is typically complex, but this is no problem: we may choose $\boldsymbol{\varepsilon} = (\varepsilon_0, \ldots, \varepsilon_{2N-1})$ with components that are i.i.d. and standard complex Gaussian (i.e., the real and the imaginary parts are independent $\mathcal{N}(0, 1)$), and then the real and the imaginary parts, say $X_0, \ldots, X_{2N}$ and $X'_0, \ldots, X'_{2N}$, of $\boldsymbol{D}\boldsymbol{\varepsilon}$ will be (dependent) $2N$-dimensional

Gaussian with covariance matrix C. Therefore $X_0, \ldots, X_N$ and $X'_0, \ldots, X'_N$ have the desired distribution.

The advantage compared to Cholesky factorization is that the necessary matrix manipulations can be done via the fast Fourier transform (FFT), which has complexity $O(N \log N)$ as compared to the $O(N^3)$ for Cholesky factorization. Algorithmically, this means that $\boldsymbol{\lambda}$ is first evaulated via the FFT as $\boldsymbol{Fc}$. Next one forms the vector $\boldsymbol{\eta}$ with components $\lambda_i^{1/2} \varepsilon_i$, and $\boldsymbol{D\varepsilon}$ is computed as $\boldsymbol{F\eta}$.

The nonnegativity assumption is of course crucial for the algorithm and is discussed by Dembo et al. [86] and Dietrich & Newsam [90]. Both papers contains results and examples that show that nonnegativity is not automatic, but the overall picture is rather optimistic. A particular result of [90], which would cover many practical cases, states that nonnegativity holds if the sequence $r_0, r_1, \ldots, r_N$ is nonnegative, decreasing, and convex. A case not covered by this is the Gaussian covariance function $r_k = \mathrm{e}^{-k^2/\ell^2}$. However, it is shown in [90] that nonnegativity holds for a given ℓ if m and ℓ are sufficiently large, more precisely, $\ell \geq \pi^{-1/2}$ and $m \geq \pi \ell^2$. A further relevant reference is Craigmile [77], to which we return in Section 6.

The big advantages of the circulant-embedding method are of course its speed (the complexity $O(n \log n)$) and that it gives exact results. The drawback is that one needs to predefine the simulation horizon n, so that the method is not suitable for situations in which the time horizon is adaptively determined by the history of the realization simulated thus far.

Exercises

3.1 (TP) Verify the claim that the circulant embedding exists and is unique when $M = N$ by giving expressions for $c_0, c_1, \ldots, c_{2M-1}$ in terms of the r_k.
3.2 (TP) For $M > N$, find the number of free parameters from which to choose the circulant.

4 Spectral Simulation. FFT

We consider throughout this section a stationary Gaussian process X with discrete or continuous time, covariance function $r(t)$, and (for simplicity) mean 0.

A simple example is

$$X(t) \stackrel{\mathrm{def}}{=} \cos(\lambda t)Y_1 + \sin(\lambda t)Y_2, \tag{4.1}$$

where Y_1, Y_2 are independent $\mathcal{N}(0, \sigma^2)$. Indeed, stationarity follows from the independence of

$$\mathbb{C}ov\big(X(s+t), X(s)\big)$$
$$= \sigma^2 \cos\big(\lambda(s+t)\big)\cos(\lambda t) + \sigma^2 \sin\big(\lambda(s+t)\big)\sin(\lambda t) = \sigma^2 \cos(\lambda t)$$

of t. The process (4.1) is periodic with period $2\pi/\lambda$, or equivalently with frequency $f(\lambda) \overset{\text{def}}{=} \lambda/2\pi$, and random phase.

A superposition

$$X(t) \overset{\text{def}}{=} \sum_{i=1}^{n} \cos(\lambda_i t)Y_{i,1} + \sin(\lambda_i t)Y_{i,2}$$

of independent processes of type (4.1) is obviously again stationary with covariance function $r(t) = \sum_{1}^{n} \sigma_i^2 \cos(\lambda_i t)$. One refers to $\{\lambda_1, \ldots, \lambda_n\}$ as the *spectrum*. A common engineering interpretation is that frequency $\lambda_i/2\pi$ is represented in the signal X with average weight (energy) σ_i^2.

Example 4.1 We considered

$$X(t) = \sum_{i=1}^{9} \left[\cos(\lambda_i t)Y_{i,1} + \sin(\lambda_i t)Y_{i,2}\right] + \sum_{i=1}^{9} \left[\cos(\mu_i t)Y_{i,3} + \sin(\mu_i t)Y_{i,4}\right],$$

where the $Y_{i,k}$ are independent normals with mean 0 and $\mathbb{V}ar(Y_{k,i}) = 1/9$ for $k = 1, 2$, $\mathbb{V}ar(Y_{k,i}) = 1/9 \cdot 16$ for for $k = 3, 4$, and

$$\begin{aligned}
(\lambda_1, \lambda_2, \ldots, \lambda_9) &= (2\pi \cdot 0.92, \, 2\pi \cdot 0.94, \, \ldots, \, 2\pi \cdot 1.08), \\
(\mu_1, \mu_2, \ldots, \mu_9) &= (2\pi \cdot 4.6, \, 2\pi \cdot 4.7, \, \ldots, \, 2\pi \cdot 5.4).
\end{aligned}$$

Thus the signal X is a perturbed oscillation with frequency 1 and weight 1 superimposed by a perturbed oscillation with frequency 5 and weight $1/16$.

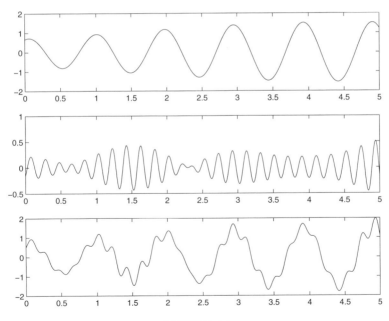

FIGURE 4.1

Figure 4.1 shows simulated paths of the two components in the first two panels, and X in the third.   □

In the general case, consider first the case of a discrete-time process $X_0, X_1, X_2, \ldots$ and write $r_k = r(k)$. Then the sequence $\{r_k\}$ is positive definite, and so by Herglotz's theorem, it can be represented as

$$r_k = \int_0^{2\pi} e^{ik\lambda}\, \nu(d\lambda) \tag{4.2}$$

for some finite real measure ν on $[0, 2\pi)$, the *spectral measure*; the condition that the process be real-valued is equivalent to

$$\int_A \nu(d\lambda) = \int_{2\pi - A} \nu(d\lambda), \quad A \subseteq [0, \pi) \tag{4.3}$$

(if the spectral density $s = d\nu/d\lambda$ exists, this simply means that s is symmetric around π, $s(\lambda) = s(2\pi - \lambda)$, $0 < \lambda < \pi$). The *spectral representation* of the process is

$$X_n = \int_0^{2\pi} e^{in\lambda}\, Z(d\lambda), \tag{4.4}$$

where $\{Z(\lambda)\}_{\lambda \in [0, 2\pi)}$ is a complex Gaussian process that is traditionally described as having increments satisfying

$$\mathbb{E}\left[(Z(\lambda_2) - Z(\lambda_1)) \overline{(Z(\lambda_4) - Z(\lambda_3))} \right] = 0,$$

$$\mathbb{E}\left| Z(\lambda_2) - Z(\lambda_1) \right|^2 = \nu(\lambda_1, \lambda_2] \tag{4.5}$$

for $\lambda_1 \le \lambda_2 \le \lambda_3 \le \lambda_4$; the integral (4.4) should be understood as the L_2 limit of approximating step functions (of course, the imaginary part in (4.4) has to vanish since X is real-valued). See, e.g., Cramér & Leadbetter [78].

For simulation, it is then appealing to simulate Z and construct X via (4.4). However, Z is not completely specified by (4.5). But:

Proposition 4.2 *Assume that X is real-valued and that $\nu(\{\pi\}) = 0$. Define Z by first taking $\{Z_1(\lambda)\}_{0 \le \pi}$, $\{Z_2(\lambda)\}_{0 \le \pi}$ to be independent real-valued Gaussian with independent increments satisfying*

$$\mathbb{V}ar\big(Z_i(\lambda_2) - Z_i(\lambda_1)\big) = \frac{1}{2}\nu(\lambda_1, \lambda_2], \quad i = 1, 2,$$

and next letting $Z(\lambda) \stackrel{\text{def}}{=} Z_1(\lambda) + iZ_2(\lambda)$ for $0 \le \lambda \le \pi$ and

$$Z(\pi + \lambda) = Z(\pi) + \overline{Z(\pi) - Z(\pi - \lambda-)} \tag{4.6}$$

$$= 2Z_1(\pi) - Z_1(\pi - \lambda-) + iZ_2(\pi - \lambda-), \quad 0 < \lambda < \pi. \tag{4.7}$$

Then (4.4) holds, i.e.,

$$X_n = 2 \int_0^{\pi} \cos(n\lambda) Z_1(d\lambda) - 2 \int_0^{\pi} \sin(n\lambda) Z_2(d\lambda). \tag{4.8}$$

Note that in the presence of a spectral density s, we may rewrite the definition of the Z_i as

$$dZ_i(\lambda) \ = \ \sqrt{\frac{1}{2}s\big(B_i(\lambda)\big)}\, dB_i(\lambda)\,,$$

where B_1, B_2 are independent standard Brownian motions.

Proof. Given Z, we define X by (4.4). By definition, $dZ(\pi+\lambda) = \overline{dZ(\pi - \lambda)}$, so that $\int_\pi^{2\pi} e^{in\lambda}\, Z(d\lambda) \ = \ = \ \int_0^\pi e^{-in\lambda}\, \overline{Z(d\lambda)}$. This implies that (4.4) can be written in the alternative form (4.8). In particular, (4.8) shows that X is real-valued (and obviously Gaussian), so it suffices to show that the covariance function is correct. But by (A6.7),

$$
\begin{aligned}
\mathbb{C}ov(X_{n+k}, X_n) \ &= \ \mathbb{E}[X_{n+k}X_n] \ = \ \mathbb{E}\left[X_{n+k}\overline{X_n}\right] \\
&= \ \mathbb{E}\left[\int_0^{2\pi} e^{i(n+k)\lambda}\, Z(d\lambda) \cdot \int_0^{2\pi} e^{-in\lambda}\, \overline{Z(d\lambda)}\right] \\
&= \ \int_0^{2\pi} e^{i(n+k)\lambda}e^{-in\lambda}\, \mathbb{E}\big|Z(d\lambda)\big|^2 \\
&= \ \int_0^{2\pi} e^{ik\lambda}\, \nu(d\lambda) \ = \ r_k.
\end{aligned}
$$

$\square$

For practical implementation, the stochastic integrals in (4.8) may be computed either by SDE schemes as in Chapter X, or by discrete approximations as follows. We say that X has *discrete spectrum* if ν has a finite support, say mass σ_k^2 at λ_k, $k = 0, \ldots, N - 1$. If X is real-valued, this means by (4.3) that N is of the form $2M$ and that we can choose $\lambda_k \le \pi$, $\lambda_{M+k} = 2\pi - \lambda_k$, $k = 0, \ldots, M - 1$. Then we can write

$$Z_i(\lambda) \ = \ \sum_{k:\, \lambda_k \le \lambda} \sigma_k Z_{k,i}, \quad 0 \le \lambda \le \pi,$$

with the $Z_{k,i}$ i.i.d. real normal with variance $1/2$, and (4.8) becomes

$$X_n \ = \ 2\sum_{k=0}^{M-1} \sigma_k\{\cos(n\lambda_k)Z_{k,1} - \sin(n\lambda_k)Z_{k,2}\}\,. \tag{4.9}$$

A process of the form (4.9) is of course straightforward to simulate. In general, spectral simulation is then performed by approximating the spectral measure by a measure with finite support, which is always possible. But note that there is no canonical way to perform this discrete approximation, and that the method is only approximative.

The great advantage of the method is, however, the speed when (4.9) is implemented via the FFT (see A3) when N is a power of 2, $N = 2^m$. One then needs to choose the λ_k of the form $2\pi k/N$, which again is always possible, let $a_k = \sigma_k Z_k$ and take X_n as $\Re\widehat{a}_n$ where $\boldsymbol{F}\widehat{a}$ with $\boldsymbol{F}$ the finite Fourier matrix and $\boldsymbol{a} = (a_0 \ldots a_{M-1})$.

In continuous time, the same method of course applies to simulating any discrete skeleton $X_0, X_h, X_{2h}, \ldots$. One needs then to compute the spectral measure ν_h of the skeleton, which is most often performed from formulas such as

$$\nu_h(\mathrm{d}\lambda) = \sum_{k=-\infty}^{\infty} \nu\big(\mathrm{d}(\lambda + 2k\pi/h)\big)$$

where ν is the spectral measure in the Bochner representation

$$r(t) = \int_{-\infty}^{\infty} \mathrm{e}^{\mathrm{i}t\lambda}\, \nu(\mathrm{d}\lambda) \tag{4.10}$$

($r(t) = r(s, s+t)$). For details and implementation issues, see for example Lindgren [240, Appendix D].

An idea somewhat related to spectral simulation is to use wavelets. See, e.g., Abry & Sellan [4] for a special case.

Exercises

4.1 (A) The *Pierson–Moskowitz spectrum* has spectral density

$$\frac{\mathrm{d}\nu}{\mathrm{d}\lambda}(\lambda) = \frac{5H_s^2}{\lambda_p(\lambda/\lambda_p)^5} \exp\left\{-\frac{5}{4}\frac{1}{(\lambda/\lambda_p)^4}\right\}, \quad -\infty < \lambda < \infty.$$

It is a standard spectrum in the ocean sciences for waves in equilibrium with the wind. Here H_s is the *significant wave height*, defined as four times the standard deviation of the surface elevation, and $\lambda_p > 0$ is the *peak frequency*, i.e., the frequency where the density has its maximum.
Use spectral simulation to generate some typical wave paths from the Pierson–Moskowitz spectrum. Experiment also with the common suggestion to truncate the spectrum to $|\lambda| \le \lambda_c$ for some $\lambda_c < \infty$, in order eliminate some of the fine ripple on the surface.

5   Further Algorithms

A stationary Gaussian discrete-time process has the ARMA form (1.3) if and only if the spectral measure is absolutely continuous with density of the form $p(\mathrm{e}^{\mathrm{i}\lambda})/q(\mathrm{e}^{\mathrm{i}\lambda})$ where p, q are polynomials; cf. Brockwell & Davis [58, Chapter 4]. For a general $\{X_n\}$, one can find polynomials p_n, q_n such that

$$\frac{p_n(\mathrm{e}^{\mathrm{i}\lambda})}{q_n(\mathrm{e}^{\mathrm{i}\lambda})}\, \mathrm{d}\lambda \xrightarrow{w} \nu(\mathrm{d}\lambda);$$

see again [58]. This suggests that one choose p, q such that the measure with density $p(\mathrm{e}^{\mathrm{i}\lambda})/q(\mathrm{e}^{\mathrm{i}\lambda})$ is close to ν, and simulate the corresponding ARMA process as an approximation to $\{X_n\}$.

The precise form of p, q is of course arbitrary and subject to choice. See Krenk & Clausen [223] and Gluver & Krenk [141] for some relevant discussion.

Some further references relevant for the general problem of simulating Gaussian processes are Wood & Chan [367] and Dietrich & Newsam [89].

6   Fractional Brownian Motion

Mandelbrot & Van Ness [247] defined fractional Brownian motion (fBM) as a mean-zero Gaussian process $\{B_H(t)\}_{t \geq 0}$ with covariance function of the form

$$r(t, s) = \frac{\sigma^2}{2} \left(|t|^{2H} + |s|^{2H} - |t - s|^{2H} \right) \tag{6.1}$$

for some $H \in (0, 1)$ (the *Hurst parameter* or self-similarity parameter). Equivalently, for any fixed $h > 0$ the increments $B_H(h), B_H(2h) - B_H(h), B_H(3h) - B_H(2h), \ldots$ form a stationary Gaussian sequence (called *fractional Brownian noise*) with variance $r(0) = \sigma^2 h^{2H}$ and covariance function

$$r(n) = \frac{\sigma^2 h^{2H}}{2} \left((n+1)^{2H} + (n-1)^{2H} - 2n^{2H} \right), n = 1, 2, \ldots. \tag{6.2}$$

For $H = 1/2$, we are back to standard Brownian motion. For $H > 1/2$, (6.2) and a convexity argument shows that the increments are positively correlated, and similarly for $H < 1/2$, they are negatively correlated. The class of fBMs can be characterized as the only self-similar[2] Gaussian processes with stationary increments. There has been a boom of interest in fractional Brownian motion in recent years, mainly because of statistical studies of phenomena such as internet traffic that show self-similarity and long-range dependence, meaning a nonintegrable covariance function violating the conditions for a CLT for the time average in $[0, t]$ with normalizing constant $t^{1/2}$; cf. IV.1.3. For this and further aspects of self-similarity and long-range dependence, see, e.g., Park & Willinger [284], Leland et al. [238], Paxson & Floyd [287], and Taqqu et al. [96]. Fractional Brownian motion is then a natural candidate as a model both because of its intrinsic properties and because it appears as (functional) limit of many more detailed descriptors of network traffic. Some selected references on performance of systems with fractional Brownian input are Norros [279], Duffield & O'Connell [100], and Narayan [270]. However, there are virtually no explicit results, and so simulation becomes an important tool.

---

[2] A stochastic process $\{X(t)\}$ is self-similar if for some H, $\{X(at)\} \overset{\mathcal{D}}{=} \{a^H X(t)\}$ for all $a \geq 0$.

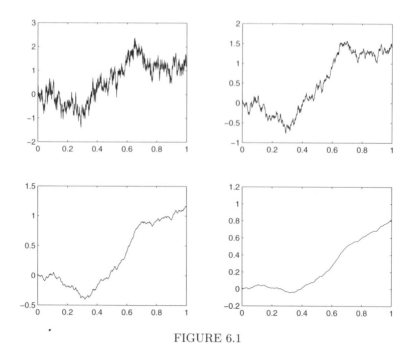

FIGURE 6.1

Figure 6.1 contains sample paths of fBM with $H = 0.25$, 0.5 (Brownian motion), 0.75, and 0.95. The figure uses $1{,}000$ time steps in $\mathcal{T} = [0, 1]$, Cholesky factorization, and common random numbers to produce the Gaussian white noise driving the simulations.

For simulation, one can use one of the methods discussed in Sections 2–5 for general stationary Gaussian processes (in this setting, the increments of $\{B_H(t)\}$). In particular, the Cholesky factorization method has been implemented by Michna [257]. In general, circulant embeddings seem to be a very attractive approach; note that, in the case of the covariance function (6.2), the required non-negativity assumption on the eigenvalues of the circulant embedding are covered by Dietrich & Newsam [90] for $1/2 < H < 1$ and by Craigmile [77] for $0 \le H < 1/2$ (more generally, [77] covers a general stationary Gaussian sequence with $r(n) \le 0$ for all $n \ge 1$). One should note that the (fast) methods of ARMA approximations and FFT are potentially dangerous because they destroy the long-range dependence.

Special algorithms for fractional Brownian motion could potentially be based on some of the many stochastic integral representations of $\{B_H(t)\}$

that are around. The classical one is

$$B_H(t) = C_H \left[\int_{-\infty}^{0} \left\{ (t-y)^{H-1/2} - |y|^{H-1/2} \right\} dW(y) \right.$$
$$\left. + \int_{0}^{t} (t-y)^{H-1/2} dW(y) \right], \qquad (6.3)$$

where

$$C_H \overset{\text{def}}{=} \sqrt{\frac{2H}{(H-1/2)\text{Beta}(H-1/2, 2-2H)}}$$

and $\{W(y)\}$ is a two-sided Brownian motion. However, simulating via (6.3) faces the same difficulty as above, because one has to truncate the integral to a finite range, thereby destroying long-range dependence. A representation without this difficulty is

$$B_H(t) = \int_{0}^{t} K(t,y) \, dW(y), \qquad (6.4)$$

where

$$K(t,s) \overset{\text{def}}{=} C_H' \, s^{1/2-H} \int_{s}^{t} u^{H-1/2}(u-s)^{H-3/2} \, du,$$

$$C_H' \overset{\text{def}}{=} \sqrt{\frac{H(2H-1)}{\text{Beta}(H-1/2, 2-2H)}}$$

(cf. Norros, Valkeila, & Virtamo [281]). However, we do not know of practical implementations of simulating via (6.4).

Another representation of relevance for simulation is

$$B_H(t) = \sum_{n=1}^{\infty} \frac{\sin(x_n t)}{x_n} X_n + \sum_{n=1}^{\infty} \frac{1-\cos(y_n t)}{y_n} Y_n, \quad 0 \le t \le 1, \qquad (6.5)$$

where $0 < x_1 < x_2 < \cdots$ are the positive real zeros of J_{-H}, the Bessel function of the first kind, and $0 < y_1 < y_2 < \cdots$ the positive real zeros of J_{1-H}, and the X_n, Y_n are independent Gaussian r.v.'s with mean zero and

$$\mathbb{V}ar X_n = \frac{2c_H^2}{x_n^2 J_{1-H}(x_n)}, \quad \mathbb{V}ar Y_n = \frac{2c_H^2}{y_n^2 J_{-H}(y_n)}, \quad c_h^2 \overset{\text{def}}{=} \frac{\Gamma(1+H)\sin(\pi H)}{\pi};$$

see Dzhaparidze & van Zanten [107]. The convergence of (6.5) (in L_2 and a.s.) follows from the known asymptotics $x_n, y_n \sim n\pi$. Kozachenko et al. [222] discuss simulation apects and give careful remainder estimates for the tail series in (6.5).

Some further references on aspects of simulating fractional Brownian motion are Mandelbrot [246] (a fast ad hoc method), Abry & Sellan [4] (wavelets), Paxson [286] (FFT), Dieker & Mandjes [91] (spectral simulation), [92] (rare-event simulation), Norros, Mannersalo, & Wang [280] (bisection with truncated memory), and Michna [257], and Huang et al. [186] (importance sampling methods).

Exercises

6.1 (TP) Derive (6.2) from (6.1).

Chapter XII
Lévy Processes

1   Introduction

An important family of stochastic processes arising in many areas of applied probability is the class of Lévy processes. A process $X = \{X(t)\}_{t \geq 0}$ is said to be a *Lévy process* if it has stationary and independent increments and satisfies $X(0) = 0$.

Standard Brownian motion B is a Lévy process, and so is a Brownian motion $\{\mu t + \sigma B(t)\}$ with general drift and variance parameters, and a further fundamental example is the counting process N_λ of a Poisson process, where λ is the rate. In fact, one of the central results in the theory of Lévy processes is that any Lévy process can be represented as an independent sum of a Brownian motion and a "compound Poisson"-like process. In particular, any Lévy process exhibiting finitely many jumps per unit time can be represented as

$$X(t) \;=\; \mu t + \sigma B(t) + \sum_{i=1}^{N_\lambda(t)} Y_i$$

for $t \geq 0$, where the Y_i are i.i.d. and independent of B, N_λ. The compound Poisson process is easily simulatable, at least when an algorithm for generating r.v.'s from the distribution of Y_1 is possible. To simulate the Brownian component, one can of course just proceed as in Chapter X.

However, there are Lévy processes for which the non-Brownian jump component $J = \{J(t)\}_{t \geq 0}$ exhibits infinitely many jumps per unit time. Dealing with such processes is the main topic of this chapter.

The jump process J is characterized by its Lévy measure $\nu(dx)$, which can be any nonnegative measure on $\mathbb{R}$ satisfying $\nu(\{0\}) = 0$ and

$$\int_{-\infty}^{\infty} (y^2 \wedge 1) \, \nu(dy) \; < \; \infty. \tag{1.1}$$

Equivalently, $\int_{|y|>\varepsilon} \nu(dy)$ and $\int_{-\varepsilon}^{\varepsilon} y^2 \, \nu(dy)$ are finite for some (and then all) $\varepsilon > 0$.

A rough description of J is that jumps of size x occur at intensity $\nu(dx)$. In particular, if ν has finite mass $\lambda \overset{\text{def}}{=} \int_{-\infty}^{\infty} \nu(dy)$, then J is a compound Poisson process with intensity λ and jump size distribution $\nu(dy)/\lambda$. In general, for any bounded interval I separated from 0, the sum of the jumps of size $\in I$ in the time-interval $[s, s+t)$ is a compound Poisson r.v. with intensity $t\lambda_I \overset{\text{def}}{=} t \int_I \nu(dy)$ and jump-size distribution $\nu(dy)\mathbb{1}_{y \in I}/\lambda_I$. Jumps in disjoint intervals are independent, and so we can describe the totality of jumps by the points in a planar Poisson process $N(dy, dt)$ with intensity measure $\nu(dy) \otimes dt$. A point of N at (Y_i, T_i) then corresponds to a jump of size Y_i at time T_i for J. If in addition to (1.1) one has

$$\int_{-\infty}^{\infty} (|y| \wedge 1) \, \nu(dy) \; < \; \infty \tag{1.2}$$

(this is equivalent to the paths of J being of finite variation), one can simply write

$$J(t) \; = \; \int_{\mathbb{R} \times [0,t]} y \, N(dy, dt). \tag{1.3}$$

If (1.2) fails, this Poisson integral does not converge absolutely, and J has to be defined by a *compensation* (centering) procedure. For example, letting

$$Y_0(t) \overset{\text{def}}{=} \int_{\{y:\, |y|>1\} \times [0,t]} y \, N(dy, dt), \quad Y_n(t) \overset{\text{def}}{=} \int_{|y| \in (y_{n+1}, y_n]} y \, N(dy, dt),$$

one can let

$$J(t) \overset{\text{def}}{=} Y_0(t) + \sum_{n=1}^{\infty} \{Y_n(t) - \mathbb{E}Y_n(t)\}, \tag{1.4}$$

where $1 = y_1 > y_2 > \cdots \downarrow 0$ and

$$\mathbb{E}Y_n(t) \; = \; t \int_{|y| \in (y_{n+1}, y_n]} y \, \nu(dy).$$

The series converges a.s. since

$$\sum_{n=1}^{\infty} \mathbb{V}ar(Y_n(t)) \; = \; \sum_{n=1}^{\infty} t \int_{|y| \in (y_{n+1}, y_n]} y^2 \nu(dy) \; = \; t \int_{-1}^{1} y^2 \nu(dy) \; < \; \infty,$$

and the sum is easily seen to be independent of the particular partitioning $\{y_n\}$. But note that since the role of the interval $[-1, 1]$ is arbitrary, a compensated Lévy jump process is given canonically only up to a drift term.

If $J(t) \geq 0$ for all $t \geq 0$, then J is called a *subordinator*. The Lévy measure for a subordinator necessarily satisfies (1.2), and any Lévy jump process satisfying (1.2) can be written as the independent difference between two subordinators, defined in terms of the restriction of ν to $(0, \infty)$, respectively the restriction of ν to $(-\infty, 0)$ reflected to $(0, \infty)$ (possibly a positive drift term has to be added).

The property of stationary independent increments implies that $\log \mathbb{E}e^{sX(t)}$ has the form $t\kappa(s)$. Here $\kappa(s)$ is called the *Lévy exponent*; its domain includes the imaginary axis $\Re s = 0$ and frequently larger sets depending on properties of ν, say $\{s : \Re s \leq 0\}$ in the case of a subordinator. Thus, $\kappa(s)$ is the cumulant g.f. of an infinitely divisible distribution, having the *Lévy–Khinchine representation*

$$\kappa(s) = cs + \frac{\sigma^2 s^2}{2} + \int_{-\infty}^{\infty} \left(e^{sy} - 1 - sy\mathbb{1}_{|y|\leq 1}\right)\nu(\mathrm{d}y), \qquad (1.5)$$

where one refers to (c, σ^2, ν) as the *characteristic triplet*.

In the finite variation case (1.2), the Lévy-Khinchine representation (1.5) is often written

$$\kappa(s) = c_1 s + \frac{\sigma^2 s^2}{2} + \int_{-\infty}^{\infty} \left(e^{sy} - 1\right)\nu(\mathrm{d}y), \qquad (1.6)$$

where $c_1 = c - \int_{-1}^{1} y\,\nu(\mathrm{d}y)$.

In the older literature, the Lévy–Khinchine representation often appears in the form

$$\kappa(s) = c_2 s + \frac{\sigma^2 s^2}{2} + \int_{-\infty}^{\infty} \left(e^{sy} - 1 - \frac{sy}{1+y^2}\right)\nu(\mathrm{d}y). \qquad (1.7)$$

A different equivalent form that is sometimes used is

$$\kappa(s) = c_2 s + \frac{\sigma^2 s^2}{2} + \int_{-\infty}^{\infty} \left(e^{sy} - 1 - \frac{sy}{1+y^2}\right)\frac{1+y^2}{y^2}\theta(\mathrm{d}y), \qquad (1.8)$$

where $\theta(\mathrm{d}y) \stackrel{\text{def}}{=} y^2/(1+y^2)\,\nu(\mathrm{d}y)$ can be any nonnegative finite measure.

Bertoin [41] and Sato [330] are the classical references for Lévy processes, but there are also some good recent texts such as Applebaum [8] and Kyprianou [226]. A good impression of the many directions into which the topic has been developed and applied can be obtained from the volume edited by Barndorff-Nielsen et al. [37].

1a Special Lévy Processes

In the examples we treat, the Lévy measure will have a density w.r.t. Lebesgue measure, which we denote by $n \stackrel{\text{def}}{=} \mathrm{d}\nu/\mathrm{d}x$. The density of $X(t)$ is denoted by $f_t(x)$ throughout.

Example 1.1 For $1 < \alpha < 2$, $\alpha \neq 1$, the *α-stable $S_\alpha(\sigma, \beta, \mu)$ distribution* is defined as the distribution with c.g.f. of the form

$$\kappa(s) \;=\; -\sigma^\alpha |s|^\alpha \left(1 - \beta \operatorname{sign}(s/i)\right) \tan \frac{\pi\alpha}{2} \;+\; s\mu, \quad \Re s = 0 \,,$$

for some $\sigma > 0$, $\beta \in [-1, 1]$, and $\mu \in \mathbb{R}$. There is a similar but somewhat different expression, which we omit, when $\alpha = 1$. The reader should note that the theory is somewhat different according to whether $0 < \alpha < 1$, $\alpha = 1$, or $1 < \alpha < 2$.

If the r.v. Y has an $S_\alpha(\sigma, \beta, \mu)$ distribution, then $Y + a$ has an $S_\alpha(\sigma, \beta, \mu + a)$ distribution and aY an $S_\alpha(\sigma|a|, \operatorname{sign}(a)\beta, \mu)$ distribution. Thus, μ is a translation parameter and σ a scale parameter. The interpretation of β is as a skewness parameter, as will be clear from the discussion of stable processes to follow.

A *stable process* is defined as a Lévy jump process in which $X(1)$ has an α-stable $S_\alpha(\sigma, \beta, 0)$ distribution. This can be obtained by choosing the Lévy density as

$$n(y) \;=\; \begin{cases} C_+/y^{\alpha+1} & y > 0, \\ C_-/|y|^{\alpha+1} & y < 0, \end{cases} \tag{1.9}$$

with

$$C_\pm \stackrel{\text{def}}{=} C_\alpha \frac{1 \pm \beta}{2} \sigma^\alpha, \quad C_\alpha \stackrel{\text{def}}{=} \frac{1 - \alpha}{\Gamma(2 - \alpha) \cos(\pi\alpha/2)}.$$

One can reconstruct β from the Lévy measure as $\beta = (C_+ - C_-)/(C_+ + C_-)$. If $0 < \alpha < 1$, then (1.2) holds and the process can be defined by (1.3). If $1 \leq \alpha < 2$, compensation is needed and care must be taken to choose the drift term to get $\mu = 0$. Stable processes have a scaling property (self-similarity) similar to Brownian motion, $\{T^{-1/\alpha} X(tT)\}_{t \geq 0} \stackrel{\mathscr{D}}{=} \{X(t)\}_{t \geq 0}$ ($\mu = 0$ is crucial for this!).

Stable processes and some of their modifications are treated in depth in Samorodnitsky & Taqqu [329]. □

Example 1.2 An important property of stable processes is that the Lévy density and hence the marginals have heavy tails. A modification with light tails corresponds to the Lévy density

$$n(x) \;=\; \begin{cases} C_+ e^{-Mx}/x^{1+Y} & x > 0, \\ C_- e^{Gx}/|x|^{1+Y} & x < 0, \end{cases}$$

where $C, G, M > 0$, $0 \leq Y < 2$. Such a Lévy process is called a *tempered stable process* (see Section 7 for a generalization). For $Y > 0$ and $C_+ = C_- = C$, the corresponding Lévy process is called the *CGMY process* (CGMY = Carr–Geman–Madan–Yor; cf. the notation for the parameters!); for $Y = 0$ and $C_+ = C_- = C$, the process is called the *variance Gamma process.* The Lévy exponent is

$$\kappa(s) = C_+\Gamma(-Y)\big[(M-s)^Y - M^Y\big] + C_-\big[(G+s)^Y - G^Y\big]. \qquad \square$$

Example 1.3 Since the Gamma distribution with density proportional to $x^{\alpha-1}e^{-\lambda x}$ is infinitely divisible, there is a Lévy process with this distribution of $X(1)$. For obvious reasons, it is called the *Gamma process*. The Lévy measure can be shown to have density $n(x) = \alpha e^{-\lambda x}/x$ for $x > 0$; note that $n(x) \sim x^{-1}$, $x \downarrow 0$, so the Lévy measure is infinite but at the borderline of being so. Hence small jumps play a relatively small role for the Gamma process. By standard properties of the Gamma distribution,

$$\kappa(s) = \log\Big(\frac{\lambda}{\lambda-s}\Big)^\alpha, \quad f_t(x) = \frac{\lambda^{\alpha t}}{\Gamma(\alpha t)}x^{\alpha t-1}e^{-\lambda x}. \qquad \square$$

The variance Gamma process in Example 1.2 is the difference between two independent Gamma processes.

Example 1.4 The *normal inverse Gaussian* (NIG) *Lévy process* has four parameters $\alpha, \delta > 0, \beta \in (-\alpha, \alpha), \mu \in \mathbb{R}$, and

$$\kappa(s) = \mu s - \delta\Big(\sqrt{\alpha^2 - (\beta+s)^2} - \sqrt{\alpha^2 - \beta^2}\Big).$$

The Lévy measure has density

$$\frac{\alpha\delta}{\pi|x|}K_1\big(\alpha|x|\big)e^{\beta x}, \quad x \in \mathbb{R}, \tag{1.10}$$

(here as usual K_1 denotes the modified Bessel function of the third kind with index 1), and the density of $X(1)$ is

$$f_1(x) = \frac{\alpha\delta}{\pi}\exp\Big\{\delta\sqrt{\alpha^2 - \beta^2} - \beta\mu\Big\}\frac{K_1\big(\alpha\sqrt{\delta^2 + (x-\mu)^2}\big)}{\sqrt{\delta^2 + (x-\mu)^2}}e^{\beta x},$$

which is called the NIG$(\alpha, \beta, \mu, \delta)$ density; the density $f_t(x)$ of $X(t)$ is NIG$(\alpha, \beta, t\mu, t\delta)$. $\qquad \square$

Example 1.5 Let X be any Lévy process with nonnegative drift. Then $T(x) \overset{\text{def}}{=} \inf\{t : X(t) > x\}$ is finite a.s., and clearly $\{T(x)\}_{x\geq 0}$ has stationary independent increments, so it is a Lévy process (in fact a subordinator, since the sample paths are nondecreasing).

The most notable example is the *inverse Gaussian Lévy process*, which corresponds to X being Brownian motion with drift $\gamma > 0$ and variance 1. Here

$$\kappa(s) = \gamma - \sqrt{\gamma^2 - 2s}, \quad f_x(t) = \frac{x}{t^{3/2}\sqrt{2\pi}}\exp\Big\{\gamma x - \frac{1}{2}\Big(\frac{x^2}{t} + \gamma^2 t\Big)\Big\},$$

and the Lévy measure has density

$$n(x) = \frac{1}{\sqrt{2\pi}\,x^{3/2}}e^{-x\gamma^2/2}, \quad x > 0. \qquad \square$$

We will see further examples in Section 5 in connection with subordination.

1b   Lévy Processes in Mathematical Finance

Using a Lévy process X instead of Brownian motion to model the log returns has recently become an extremely popular approach; see Cont & Tankov [75] and Schoutens [331]. There are two motivations, both empirically motivated. The first is statistical and originates from studies in which the geometric BM Black–Scholes model does not appear to describe reality well. One reason may be that log returns appear to have a clearly nonnormal distribution, another that the implied volatility (see further Exercise 3.2) computed from traded option prices does not appear to be a constant function of the strike price K, but exhibits a skew or a smile. Working with a Lévy process instead widens the class of possible marginals to the class of infinitely divisible distributions, which is much more flexible than the class of normals.

The second motivation is that Lévy processes allow for sample paths with jumps. Even if continuous observation is in principle impossible, one may observe atypically large changes at certain times and may then interpret these as jumps. For example, one could model crashes at the stock exchange as a Poisson process N, at the epochs of which the stock prices decrease by a factor of $1 - \delta$. In the Black–Scholes model, this leads to $X(t) = \mu t + \sigma B(t) + \log(1 - \delta)N(t)$, where B is a Brownian motion independent of N.

A difficulty is that the risk-neutral measure $\mathbb{P}^*$ is no longer unique. As discussed in I.3, the requirement is absolute continuity and that $e^{-rt+X(t)}$ be a $\mathbb{P}^*$-martingale. A popular choice is exponential change of measure (often referred to as the *Esscher transform* in this setting), where X is a Lévy process under $\mathbb{P}^*$ with Lévy exponent

$$\kappa^*(s) \;=\; \kappa(s+\theta) - \kappa(\theta)\,,$$

where (in order that the martingale property hold) θ satisfies $\kappa^*(1) = r$. Occasionally one also encounters the *minimal entropy* measure, which comes out as follows. Assuming that the jump part has finite expectation, i.e., that $\int_{|x|>1} |x|\,\nu(\mathrm{d}x) < \infty$, one can write

$$\kappa(s) \;=\; \mu s + \sigma^2 s^2/2 + \int_{-\infty}^{\infty} (e^{sy} - 1 - sy)\,\nu(\mathrm{d}y)\,. \qquad (1.11)$$

The c.g.f. under the minimum entropy measure is then of the same form, but with μ and ν replaced by

$$\mu^* \;=\; \mu + \lambda\sigma + \int_{-\infty}^{\infty} y\big(\exp\{\lambda(e^y - 1)\} - 1\big)\,\nu(\mathrm{d}y),$$
$$\nu^*(\mathrm{d}y) \;=\; \exp\{\lambda(e^y - 1)\}\,\nu(\mathrm{d}y)\,,$$

where λ is determined by $\kappa^*(1) = r$.

Of the examples in Section 1a, particularly popular choices in mathematical finance are the CGMY process and the NIG process. For further examples and comprehensive treatments of Lévy processes in finance, see Schoutens [331] and Cont & Tankov [75].

Exercises

1.1 (A) Consider the Asian option in the GBM model with the same parameters as in Exercise I.6.2, but assume in addition that events occur at the stock exchange at Poisson rate $\lambda = 1/5$, causing the stock prices to fall by 20% (that is, $\log S(t)$ has to be replaced by $\log S(t) + \log(0.8)N(t)$, where N is an independent Poisson(λ) process). Compute the option price for $K = 100$ under both the Esscher measure and the minimal entropy measure.

2   First Remarks on Simulation

Because of the property of stationary independent increments, the problem of simulating a discrete skeleton $\{J_n^h\}$, where $J_n^h \overset{\text{def}}{=} J^h(t_n^h), t_n^h \overset{\text{def}}{=} nh$, of a Lévy jump process is obviously equivalent to the problem of r.v. generation from a specific infinitely divisible distribution. One can tentatively distinguish three types of processes, for which one would apply different methods:

(a) The marginal densities $f_t(x)$ are available in a simple form, so standard methods will apply to simulate the increments.

(b) In quite a few other cases (the NIG process is one example), the $f_t(x)$ may be explicitly known but involve special functions, making r.v. generation tedious.

(c) A third group of Lévy processes consists of those for which the $f_t(x)$ are not known in closed form [the Lévy density $n(x)$ and/or Lévy exponent $\kappa(s)$ will typically be known].

Examples from (a) are the Gamma, inverse Gaussian, and Cauchy Lévy processes. The NIG process is one important example from (b); others are the hyperbolic Lévy and Mexiner processes not discussed here, and so is the minimum entropy process (cf. Section 1b) associated with a given Lévy process because of the $e^{\lambda e^y}$ occuring in the Lévy density. The CGMY process provides an example from (c).[1]

---

[1] A curious example falling between these groups is a Lévy process with $f_t(x)$ a Student t density for some $t = t_0$ but not explicitly available for $t \neq t_0$, as discussed by Hubalek [188].

The simulation of a Lévy process of type (a) is of course rather easy, though the r.v. generation may be more or less straightforward (see, e.g., Example 5.1 for the inverse Gaussian case). Also for some special processes of type (b) or (b) there exist simple algorithms. In particular, stable processes are of type (C) except for $\alpha = 1$ or $1/2$, but nevertheless, r.v. generation from stable distributions is surprisingly easy; see Section 2a below. Subordination may also provide an easy route in some cases, cf. Section 5. Methods that can be used beyond such special cases are discussed in Sections 3 and 4. Here the r.v. generation has to be based directly on the Lévy measure ν. It is obviously impossible to generate an infinity of jumps, and so invariably some truncation or limiting procedure is involved. An alternative for case (B) is course to bound the special functions by more standard ones so that acceptance–rejection methods may apply. Finally, both in (b) and (c) transform-based methods as discussed in II.2e may apply.

We proceed to some special algorithms.

2a   Stable Processes

For stable distributions, there is a standard algorithm due Chambers, Mallow, & Stuck [66] (see also Samorodnitsky & Taqqu [329]). It has a particularly simple form for a symmetric stable distribution ($\beta = 0$): if Y_1, Y_2 are independent such that Y_1 is standard exponential and Y_2 uniform on $(-\pi/2, \pi/2)$, then

$$X = \frac{\sin(\alpha Y_2)}{(\cos Y_2)^{1/\alpha}} \left(\frac{\cos((1-\alpha)Y_2)}{Y_1} \right)^{(1-\alpha)/\alpha} \tag{2.1}$$

has an $S_\alpha(1,0,0)$ distribution. Note that if $\alpha = 2$, then (2.1) reduces to

$$X = \sqrt{Y_1} \frac{\sin(2Y_2)}{\cos Y_2} = 2\sqrt{Y_1} \sin Y_2 \,,$$

which is the Box–Muller method for generating a normal r.v. with variance 2. The algorithm is also fairly simple and efficient in the asymmetric case, but is then of a somewhat more complicated form.

For asymmetric stable distributions and processes, the right-skewed case $\beta = 1$ can be viewed as a building block because of the fact that if Y_1, Y_2 are independent and $S_\alpha(1,0,0)$ distributed, then

$$Y = \mu + \sigma \left(\frac{1+\beta}{2} \right)^{1/\alpha} Y_1 - \sigma \left(\frac{1-\beta}{2} \right)^{1/\alpha} Y_2$$

has an $S_\alpha(\sigma, \beta, \mu)$ distribution.

2b   The Damien–Laud–Smith Algorithm

Early algorithms for simulation from an infinitely divisible distribution were suggested by Bondesson [50] and Damien, Laud, & Smith [80]. Bondesson's method is an early instance of ideas related to the series representations discussed below, and we return to it there. The starting point of [80] is the finite measure $\theta(dy) = y^2/(1 + y^2)\,\nu(dy)$ in the form (1.8) of the Lévy–Khinchine representation. Write $c \overset{\text{def}}{=} \int_{-\infty}^{\infty} \theta(dy)$.

Proposition 2.1 ([80]) *Let* (U_i, V_i), $i = 1, \ldots, n$, *be i.i.d. pairs such that* U *has distribution* $\theta(dy)/c$ *and the conditional distribution of* V *given* $U = y$ *is* $\text{Poisson}(c(1 + y^2)/ny^2)$, *and let*

$$
Z_n \overset{\text{def}}{=}
\begin{cases}
\displaystyle\sum_{i=1}^{n} U_i V_i & \text{in the finite variation case,} \\[2em]
\displaystyle\sum_{i=1}^{n} \left(U_i V_i - \frac{c}{nU_i} \right) & \text{in the compensated case.}
\end{cases}
$$

Then $Z_n \overset{\mathcal{D}}{\to} X(1)$ *as* $n \to \infty$.

Proof. Letting $\lambda_n(y) \overset{\text{def}}{=} c(1 + y^2)/ny^2$, we have $\mathbb{E}\big[e^{sUV} \,|\, U = y\big] = \exp\{\lambda_n(y)(e^{sy} - 1)\}$. Thus in the finite-variation case, we get for $\Re s = 0$ that

$$
\begin{aligned}
\log \mathbb{E} e^{sZ_n} &= n \log \left(\int_{-\infty}^{\infty} \exp\{\lambda_n(y)(e^{sy} - 1)\} \, \frac{y^2}{c(1 + y^2)} \nu(dy) \right) \\
&= n \log \left(\int_{-\infty}^{\infty} \{1 + \lambda_n(y)(e^{sy} - 1) + O(1/n^2)\} \frac{y^2}{c(1 + y^2)} \nu(dy) \right) \\
&= n \log \left(1 + \frac{1}{n} \int_{-\infty}^{\infty} (e^{sy} - 1)\nu(dy) + O(1/n^2) \right) \\
&= n \log \left(1 + \frac{1}{n}\kappa(s) + O(1/n^2) \right) \;\to\; \kappa(s) \,,
\end{aligned}
$$

as should be (using that the $O(1/n^2)$ term is uniform in y). For the compensated case, see [80]. □

A particularly appealing case is the $S_\alpha(1, 1, 0)$ case in which U can be generated as $\sqrt{1/W - 1}$ with W having a $\text{Beta}(\alpha/2, 1 - \alpha/2)$ distribution.

For some purposes, a disadvantage of using discrete skeletons is that one cannot precisely identify the location of the large jumps; doing so is important, for example, for reducing the uniform error X.(1.2) (note that even in the simple case of a Poisson process, X.(1.2) evaluated for a discrete skeleton does not go to zero). The methods discussed in the two next sections are better suited for this.

3   Dealing with the Small Jumps

Since we have discussed Brownian motion separately, we consider here only the case $\sigma^2 = 0$. Recall also that

$$\int_{-\infty}^{\infty} (y^2 \wedge 1)\, \nu(dy) \; < \; \infty, \tag{3.1}$$

and that the paths of J are of finite variation (no compensation needed) if and only if

$$\int_{-\infty}^{\infty} (|y| \wedge 1)\, \nu(dy) \; < \; \infty \tag{3.2}$$

(say a stable process with $0 < \alpha < 1$ or a subordinator).

For simulation, the compound Poisson case is obviously straightforward from any point of view (provided at least that it is straightforward to simulate from the probability measure proportional to $\nu(dy)$), and so we concentrate in the following on the case $\int_{-\infty}^{\infty} \nu(dy) = \infty$.

Any Lévy jump process J can be written as the independent sum

$$J(t) \;=\; J^{(1)}(t) + J^{(2)}(t), \tag{3.3}$$

where the Lévy measures of $J^{(1)}$, $J^{(2)}$ are the restrictions $\nu^{(1)}$, $\nu^{(2)}$ of ν to $(-\varepsilon, \varepsilon)$, respectively $\{y : |y| \geq \varepsilon\}$. Here $\nu^{(2)}$ is finite, so $J^{(2)}$ is a compound Poisson process, and simulation is straightforward. As a first attempt, one would then choose $\varepsilon > 0$ so small that $J^{(1)}$ can be neglected and just simulate $J^{(2)}$.

As a more refined procedure, it is often suggested (e.g., Bondesson [50], Rydberg [322]) to replace $J^{(1)}$ by a Brownian motion with the appropriate variance $\sigma_\varepsilon^2 \stackrel{\text{def}}{=} \int_{-\varepsilon}^{\varepsilon} y^2\, \nu(dy)$ and mean $\mu_\varepsilon \stackrel{\text{def}}{=} \int_{-\varepsilon}^{\varepsilon} y\, \nu(dy)$ in the finite-variation case (3.2), $\mu_\varepsilon = 0$ in the compensated case. The justification for this is the folklore that because small jumps become more and more dominant as ε becomes small, one should have

$$\{(J^{(1)}(t) - \mu_\varepsilon t)/\sigma_\varepsilon\}_{t \geq 0} \; \stackrel{\mathscr{D}}{\to} \; \{B(t)\}_{t \geq 0} \tag{3.4}$$

as $\varepsilon \downarrow 0$ where B is standard BM. However, the gap in the argument is that even if jumps of $J^{(1)}$ are small, such is not necessarily the case for the jumps of the l.h.s. of (3.4) because of the division by σ_ε, which is small. Thus standard central limit theory suggests that

$$\varepsilon/\sigma_\varepsilon \; \to \; 0 \tag{3.5}$$

is a necessary condition. A rigorous discussion was provided by Asmussen & Rosiński [26], who showed that (3.4) holds if and only if

$$\sigma_{c\sigma_\varepsilon \wedge \varepsilon} \; \sim \; \sigma_\varepsilon \tag{3.6}$$

for all $c > 0$. This is certainly implied by (3.5), and is in fact equivalent, provided the Lévy measure μ has no atoms close to 0.

The proof uses a special CLT for Lévy jump processes. Rather than going into this, we will give a direct proof of the following result, which covers most cases of practical interest:

Proposition 3.1 *Assume that ν has a density of the form $L(x)/|x|^{\alpha+1}$ for all small x, where $L(x)$ is slowly varying as $x \to 0$ and $0 < \alpha < 2$. Then (3.4) holds.*

Proof. We show only that $J^{(1)}(1)$, properly normalized, has a limiting standard normal distribution. By Karamata's theorem ([116] or [44]),

$$\sigma_\varepsilon^2 = \int_{-\varepsilon}^\varepsilon x^2\, \nu(\mathrm{d}x) = \int_{-\varepsilon}^\varepsilon |x|^{1-\alpha} L(x)\, \mathrm{d}x \sim \frac{L(\varepsilon) + L(-\varepsilon)}{2 - \alpha} \varepsilon^{2-\alpha}.$$

Since $L(\varepsilon)\varepsilon^\gamma \to 0$, $\varepsilon^\gamma/L(\varepsilon) \to 0$ for any $\gamma > 0$ and similarly for $L(-\varepsilon)$, we therefore have $\varepsilon/\sigma_\varepsilon \to 0$, so that

$$\log \mathbb{E} \exp \left\{ s\bigl(J^{(1)}(1) - \mu_\varepsilon\bigr)/\sigma_\varepsilon \right\}$$

$$= \int_{-\varepsilon}^\varepsilon (e^{sx/\sigma_\varepsilon} - 1 - sx/\sigma_\varepsilon)\, \nu(\mathrm{d}x) = \int_{-\varepsilon}^\varepsilon \left(\frac{s^2 x^2}{2\sigma_\varepsilon^2} + \mathrm{O}\!\left(\frac{|s^3 x^3|}{\sigma_\varepsilon^3} \right) \right)\, \nu(\mathrm{d}x)$$

$$= \frac{s^2}{2} + \mathrm{o}(1),$$

where the last equality follows from

$$\int_{-\varepsilon}^\varepsilon |x^3|\, \nu(\mathrm{d}x) \sim \frac{L(\varepsilon) + L(-\varepsilon)}{3 - \alpha} \varepsilon^{3-\alpha}. \qquad \square$$

Remark 3.2 The process $J^{(2)}$ of large jumps will of course be simulated at a Poisson grid $0 < t_1 < t_2 < \cdots$. The whole process along this Poisson grid is then obtained by adding a $\mathcal{N}\bigl((t_k - t_{k-1})\mu_\varepsilon, (t_k - t_{k-1})\sigma_\varepsilon^2\bigr)$ r.v. to the jump at t_k for the update from t_{k-1} to t_k. However, in some cases it is desirable to have the process represented along a deterministic equidistant grid $t_n^h \stackrel{\text{def}}{=} nh$. This is easily implementable by performing the update from t_{n-1}^h to t_n^h by adding a $\mathcal{N}\bigl(h\mu_\varepsilon, h\sigma_\varepsilon^2\bigr)$ r.v. to the jumps of $\{J^{(2)}(t)\}$ in $(t_{n-1}^h, t_n^h]$. $\qquad \square$

In summary, we have the following three approximate algorithms for generating J:

$$J(t) \approx J^{(2)}(t), \tag{3.7}$$

$$J(t) \approx \mu_\varepsilon t + J^{(2)}(t), \tag{3.8}$$

$$J(t) \approx \mu_\varepsilon t + \sigma_\varepsilon B(t) + J^{(2)}(t), \tag{3.9}$$

where $\mu_\varepsilon = 0$ for all small ε in the infinite-variation case (so that algorithms (3.7) and (3.8) are the same). Clearly, the algorithms in the list are increasingly complex but also, one would hope, increasingly accurate. Thus in particular, to achieve the same accuracy, one would believe that algorithm (3.9) could be run be with a larger ε than algorithm (3.7), which would reduce the computational effort. Indeed, generating $J^{(2)}$ is much more demanding than generating the drift or Brownian term, so that roughly the computational effort should be proportional to the Poisson parameter $\lambda_\varepsilon \overset{\text{def}}{=} \int_{|x|>\varepsilon} \nu(\mathrm{d}x)$, that goes to ∞ as $\varepsilon \downarrow 0$.

Example 3.3 We consider a stable process with index $\alpha = 1/2$ and Levy density $\frac{c}{2}x^{-3/2}$, $x > 0$, in the completely asymmetric case, and $\frac{c}{2}|x|^{-3/2}$, $x \in \mathbb{R}$, in the symmetric case, where $c = \sqrt{2/\pi}$. Then $X(1)$ can be generated as $1/U^2$ in the asymmetric case and as $1/U^2 - 1/V^2$ in the symmetric case, where U, V are independent standard normals.

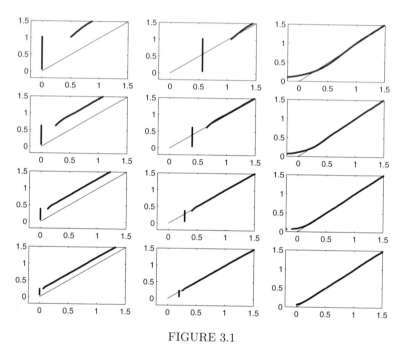

FIGURE 3.1

Figure 3.1 is a Q-Q plot restricted to $X(1) \in (0, 1.2)$ (which corresponds to 0.87% of the mass) of the quantiles of $R = 25{,}000$ simulated approximate values of $X(1)$ toward the quantiles of simulated exact values, using the above exact algorithm. The four rows correspond to $\varepsilon = 1/2, 1/3, 1/8, 1/16$. The columns are algorithms (3.7), (3.8), and (3.9). Note that in this example,

$$\mu_\varepsilon = \int_0^\varepsilon \frac{c}{2}x^{-1/2}\,\mathrm{d}x = c\varepsilon^{1/2}, \quad \sigma_\varepsilon^2 = \int_0^\varepsilon \frac{c}{2}x^{1/2}\,\mathrm{d}x = \frac{c}{3}\varepsilon^{3/2}.$$

Figure 3.2 is a similar Q-Q plot for the symmetric case in the interval $X(1) \in (-1.2, 1.2)$, which corresponds to 25% of the mass. The first column

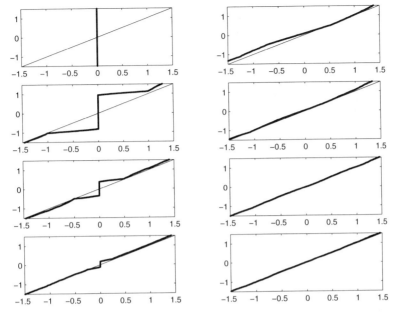

FIGURE 3.2

is the identical algorithms (3.7), (3.8), the second (3.8). The four rows correspond to $\varepsilon = 2, 1, 1/2, 1/4$, and we have

$$\mu_\varepsilon = 0, \quad \sigma_\varepsilon^2 = \int_{-\varepsilon}^{\varepsilon} \frac{c}{2} |x|^{1/2} \, dx = \frac{2c}{3} \varepsilon^{3/2}$$

(note that the mean of the small jumps is 0 by symmetry, so that pure mean compensation does not give anything new).

It is seen in the asymmetric case that mean compensation (algorithm (3.8) is indeed an improvement but has the problem of the atom at μ_ε, which is largely removed by the normal refinement (3.9). Some relevant numbers, including the computational effort

$$\lambda_\varepsilon = \begin{cases} \int_\varepsilon^\infty \frac{c}{2} x^{-3/2} \, dx = c\varepsilon^{-1/2} & \text{in the asymmetric case,} \\ \int_{|x|>\varepsilon} \frac{c}{2} x^{-3/2} \, dx = 2c\varepsilon^{-1/2} & \text{in the symmetric case,} \end{cases}$$

and the probability of simulating $X(1) = 0$, which is $e^{-\lambda_\varepsilon}$, respectively $e^{-2\lambda_\varepsilon}$, are given in the following tables, where the first corresponds to the asymmetric case and the second to the symmetric case:

ε	1/2	1/4	1/8	1/16
λ_ε	1.13	1.60	2.26	3.19
μ_ε	0.56	0.40	0.28	0.20
σ_ε	0.31	0.18	0.11	0.06
$e^{-\lambda_\varepsilon}$	0.32	0.21	0.10	0.04

ε	2	1	1/2	1/4
λ_ε	1.13	1.60	2.26	3.19
σ_ε	1.23	0.79	0.43	0.26
$e^{-2\lambda_\varepsilon}$	0.32	0.21	0.10	0.04

□

Error rates for the various algorithms discussed above have been derived independently in Signahl [342] and Cont & Tankov [75, Sections 6.3, 6.4].

Remark 3.4 In many of the parametric examples, r.v. generation from the distribution F_ε obtained by normalizing the Lévy measure ν to (ε, ∞) is nonstandard. For example, for the CGMY process of Example 1.2, we need to generate r.v.'s from the density proportional to e^{-Mx}/x^{1+Y}, $\varepsilon < x < \infty$. An obvious procedure is to use acceptance–rejection, with proposal either the exponential distribution with rate M shifted to (ε, ∞), or the Pareto density c_ε/x^{1+Y} on (ε, ∞). If M is large, one expects the first alternative to be the more efficient; for M small, the second. An alternative approach to simulation from the CGMY process is given in Example 4.5.

A general approach is to find a majorizing Lévy density $\widetilde{n}(x)$ from which the jumps are easily generated. Assuming $n(x) \leq c\widetilde{n}(x)$ with $c = 1$ (the case $c \geq 1$ is easily handled by a time-change argument), one then simulates the jumps corresponding to $\widetilde{n}(\cdot)$ and accepts a jump of size x w.p. $n(x)/\widetilde{n}(x)$. By standard properties of Poisson thinning, this makes jumps in $[x, x + \mathrm{d}x)$ occur at intensity $n(x)\,\mathrm{d}x$, as should be. See Rosiński [307] for further discussion of this approach.   □

Exercises

3.1 (TP) Carry out the time-change argument in Remark 3.4.

3.2 (A) Consider a European call option with the underlying risk-neutral asset price of the form $S(t) = S(0)e^{X(t)}$, where $X(t)$ is the CGMY process of Example 1.2. The maturity is $T = 1$ years, the short rate is 4% per year, and w.l.o.g., we take $S(0) = 1$. By a least squares fit to traded option prices, Asmussen et al. [24] calibrated the parameters $C = 1.17$, $G = 3.58$, $M = 11.5$, $Y = 0.50$. Using these parameters and any of the numerical schemes of this section, compute the option price $p(K)$ for strike prices $K \in [0.85, 1.15]$ and thereby the implied volatility $\sigma(K)$ as a function of K. That is, $\sigma(K)$ is the solution of $p(K) = \mathrm{BS}(r, T, K, \sigma(K))$, where $\mathrm{BS}(\cdot)$ is the Black–Scholes price.

4   Series Representations

We will start by deriving the most basic series representation of a subordinator $\{X(t)\}_{0 \leq t \leq 1}$ with Lévy density $n(x) \overset{\text{def}}{=} \mathrm{d}\nu/\mathrm{d}x$, $x > 0$, and no drift term.

The idea comes from the algorithm for simulation of an inhomogeneous Poissson process from i.i.d. standard exponentials given in the second half of II.4.2. More precisely, if $\beta(t)$ is the intensity function, if $B(t) \overset{\text{def}}{=} \int_0^t \beta(s)\,\mathrm{d}s < \infty$ for all $t < \infty$, and $\Gamma_1, \Gamma_2, \ldots$ are the epochs of a Poisson process with unit rate, then the epochs of the inhomogeneous Poissson process can be constructed as $B^{-1}(\Gamma_1), B^{-1}(\Gamma_2), \ldots$. In particular, if an epoch at time t gives the reward $r(t)$, we obtain the total reward as

$$\sum_{i=1}^{\infty} r\big(B^{-1}(\Gamma_i)\big).$$

The relevance for our subordinator comes from the fact that the totality of jumps of $\{X(t)\}_{0\leq t\leq 1}$ (that is, $X(1)$) can be seen as the total reward of an inhomogeneous Poisson process on $(0,\infty)$ with intensity function $n(x)$ and reward function $r(x) = x$. However, we encounter the difficulty that $\int_0^t n(x)\,dx$ is infinite, so that the above (forward) algorithm does not apply directly. To resolve this, we simply reverse time by letting it run down from ∞ to 0 rather than up the other way. This provides a backward algorithm in which, letting $\overline{\nu}(x) \stackrel{\text{def}}{=} \int_x^{\infty} n(y)\,dy$, we construct the largest jump as $\overline{\nu}^{-1}(\Gamma_1)$, the second-largest one as $\overline{\nu}^{-1}(\Gamma_2)$, and so on. In summary, we have proved the first part of the following theorem:

Theorem 4.1 *Consider a subordinator $\{X(t)\}_{0\leq t\leq 1}$ with Lévy density $n(x) = d\nu/dx$, $x > 0$, and no drift term, and let $\Gamma_1, \Gamma_2, \ldots$ be the epochs of a Poisson process with unit rate. Then*

$$X(1) \stackrel{\mathscr{D}}{=} \sum_{i=1}^{\infty} \overline{\nu}^{-1}(\Gamma_i).$$

Further, if $U_1, U_2, \ldots$ are i.i.d. uniform$(0,1)$ and independent of $\Gamma_1, \Gamma_2, \ldots$, then

$$\{X(t)\}_{0\leq t\leq 1} \stackrel{\mathscr{D}}{=} \Big\{\sum_{i=1}^{\infty} \overline{\nu}^{-1}(\Gamma_i)\mathbb{1}_{U_i\leq t}\Big\}_{0\leq t\leq 1}.$$

To obtain the second part, i.e., the representation of the whole sample path, we just have to note that the algorithm places the ith-largest jump at $t = U_i$, and that a given jump of size x in $[0,1]$ is equally likely to occur at any t.

To get an implementable algorithm, the series expansion has of course to be truncated. The two most obvious possibilities are to consider either $\sum_1^N$ with N fixed and large, or $\sum_1^{\tau}$ with ε small, where $\tau \stackrel{\text{def}}{=} \inf\{i : \overline{\nu}^{-1}(\Gamma_i) < \varepsilon\}$. In the latter case, the algorithm yields the jumps of size greater than ε, with the largest one smaller than ε added. Since this term will be negligible for ε small, one does not expect the approximations provided by the algorithm to be essentially different from the compound Poisson ones discussed in the previous section in which the Lévy measure is truncated to $[\varepsilon,\infty)$. Similarly, in considering $\sum_1^N$, the jumps smaller than $\overline{\nu}^{-1}(\Gamma_N)$ are neglected, but because of the LLN $\Gamma_N/N \to 1$, these are to the first order of approximation the same as those smaller than $\varepsilon = \overline{\nu}^{-1}(N)$. In summary, the algorithm provided by Theorem 4.1 may just be seen as the compound Poisson algorithm with the generation of the jump sizes implemented via inversion. However, more sophisticated series expansions sometimes contains genuine improvements. Also series expansions are potentially more advantageous for some aspects of variance reduction (cf. Section 6b).

In the following, we survey several variants of the representation in Theorem 4.1. Throughout, let M be a Poisson process on $[0, \infty)$ with intensity λ, and denote the nth epoch by Γ_n (thus, $\{\Gamma_n - \Gamma_{n-1}\}_{n=1,2,\dots}$ is a sequence of i.i.d. exponential r.v.'s with mean $1/\lambda$). Let further the sequences $\{U_n\}$, $\{\xi_n\}$ be independent of M and i.i.d., such that U_n is uniform$(0, 1)$ and ξ_n has some distribution varying from case to case in the following. The representations we consider typically have the form

$$\{X(t)\}_{0 \leq t \leq 1} \stackrel{\mathscr{D}}{=} \left\{ \sum_{n=1}^{\infty} G(\xi_n, \Gamma_n) \mathbb{1}_{U_n \leq t} \right\}_{0 \leq t \leq 1}$$

in the finite-variation case.

We first consider the algorithm of Bondesson [50].

Proposition 4.2 ([50]) *Assume that $\{X(t)\}$ is a subordinator. Then:*

(i) *There exist a family $\{H(\mathrm{d}y, u)\}_{u \geq 0}$ of distributions on $[0, \infty)$ and a $\lambda > 0$ such that*

$$\lambda \int_0^{\infty} H(\mathrm{d}y, u)\, \mathrm{d}u \;=\; \nu(\mathrm{d}y). \tag{4.1}$$

(ii) *For such a family $\{H(\mathrm{d}y, u)\}$, let $W_1, W_2, \dots$ be r.v.'s that are conditionally independent given M, such that W_i has distribution $H(\cdot, \Gamma_i)$ given M. Then $X = W_1 + W_2 + \cdots$ has the same distribution as X_1.*

Proof. For (i), we let $\lambda = 1$ and $\overline{H}(x, u) \stackrel{\mathrm{def}}{=} \int_{x+}^{\infty} H(\mathrm{d}y, u)$. Then (4.1) can be rewritten as

$$\int_0^{\infty} \overline{H}(x, u)\, \mathrm{d}u \;=\; \overline{\nu}(x), \quad x \geq 0. \tag{4.2}$$

Letting $H(\cdot, u)$ be the degenerate distribution at $\overline{\nu}^{-1}(u)$, we have $\overline{H}(x, u) = 1$ when $x \leq \overline{\nu}^{-1}(u)$, i.e., $u \leq \overline{\nu}(x)$, and $\overline{H}(x, u) = 0$ when $x > \overline{\nu}^{-1}(u)$, i.e., $u > \overline{\nu}(x)$, cf. II.2.2(a). This proves (4.1). For an alternative construction, see the proof of Corollary 4.4 below.

For (ii), we can use (4.1) to write the Lévy exponent $\kappa(s) = \log \mathbb{E}e^{sX_1}$ as

$$\kappa(s) \;=\; \int_0^{\infty} (e^{sy} - 1)\nu(\mathrm{d}y) \;=\; \int_{u=0}^{\infty} \int_{y=0}^{\infty} \lambda(e^{sy} - 1)H(\mathrm{d}y, u)\, \mathrm{d}u,$$

which we recognize as the c.g.f. of the total reward $X^*(\infty)$ in a time-inhomogeneous compound Poisson process $\{X^*(t)\}_{0 \leq t < \infty}$ with constant arrival rate λ and jump-size distribution $H(\cdot, u)$ at time u. From this the result follows. $\qquad\square$

As is seen from the proof, Theorem 4.1 is the special case in which $H(\cdot, u)$ is the degenerate distribution at $\overline{\nu}^{-1}(u)$.

Bondesson considers only one-dimensional distributions, not processes, but in fact, we have the following:

Corollary 4.3 *Under* (i) *of Proposition 4.2,*

$$\{X(t)\}_{0 \le t \le 1} \stackrel{\mathscr{D}}{=} \left\{\sum_{n=1}^{\infty} W_n \mathbb{1}_{U_n \le t}\right\}_{0 \le t \le 1}. \tag{4.3}$$

Proof. Let $\widetilde{X}(t) \stackrel{\text{def}}{=} \sum_{n=1}^{\infty} W_n \mathbb{1}_{U_n \le t}$. We can then think of $\widetilde{X}(t)$ as the total reward in the process obtained from $\{X^*(t)\}$ by thinning with retention probability t. Hence as in the proof of Proposition 4.2, with λ replaced by λt, we get $\log \mathbb{E} e^{s\widetilde{X}(t)} = t\kappa(s)$, and it remains only to show independence of increments. But if we split $\{X^*(t)\}$ into three processes $\{X^*(t; 1)\}$, $\{X^*(t; 2)\}$, $\{X^*(t; 3)\}$ by letting a jump go to the three processes with respective probabilities t, $t + s$, and $1 - t - s$ according to the U_n, these processes are independent and hence so are the total rewards

$$X^*(\infty; 1) = \sum_{n=1}^{\infty} W_n \mathbb{1}_{U_n \le t} = \widetilde{X}(t),$$

$$X^*(\infty; 2) = \sum_{n=1}^{\infty} W_n \mathbb{1}\{t < U_n \le t + s\} = \widetilde{X}(t + s) - \widetilde{X}(t).$$

$\square$

In the following, let $\lambda = 1$. There are several series representations of Lévy processes of similar type as (4.3) around in the literature. For example, an $S_\alpha(1, \beta, 0)$ process with $\alpha < 1$ can be represented as

$$C_\alpha^{1/\alpha} \sum_{n=1}^{\infty} \xi_n \Gamma_n^{-1/\alpha} \mathbb{1}_{U_n \le t}, \tag{4.4}$$

where

$$C_\alpha \stackrel{\text{def}}{=} \left(\int_0^\infty x^{-\alpha} \sin x \, \mathrm{d}x\right)^{-1}, \quad \mathbb{P}(\xi_n = 1) = \mathbb{P}(\xi_n = -1) = \frac{1 + \beta}{2}.$$

Letting $H(\cdot, u)$ be the distribution of an r.v. that attains the values $\pm C_\alpha^{1/\alpha} u^{-1/\alpha}$ with probabilities $(1 \pm \beta)/2$, this representation is of the same form as in Corollary 4.3. For $1 \le \alpha < 2$, there are similar expansions like (4.4) but with certain centering terms $tb_n^{(\alpha)}$ added for each term (corresponding to compensation). See [329] for details. Perhaps more surprisingly, if the process is not completely skewed (i.e., if $|\beta| \ne 1$) then such centering can be avoided. For example, for $\alpha \ne 1$ a possible representation is (4.4) with the distribution of ξ_n changed to $\mathbb{P}(\xi_n = a_\pm) = 1 - a_\pm/(a_+ + a_-)$, where

$$a_\pm \stackrel{\text{def}}{=} \pm \left[\frac{1 \pm \beta}{2}\left(\frac{1 + \beta}{1 - \beta}\right)^{\pm 1/(\alpha - 1)} + 1\right]^{1/\alpha};$$

cf. Janicki & Weron [194].

Here is one more example, random thinning of i.i.d. sequences (Rosiński [306], [307]); the thinning corresponds to allowing $H(\cdot, u)$ to have an atom at 0 in Proposition 4.2.

Corollary 4.4 *Consider the finite-variation case. Let the ξ_n have distribution F, where F is a probability distribution on $\mathbb{R}/\{0\}$ that is equivalent to ν in the Radon–Nikodym sense, and let $g = \mathrm{d}\nu/\mathrm{d}F$. Then the process can be represented as*

$$\sum_{n=1}^{\infty} \xi_n \mathbb{1}\{g(\xi_n) \geq \Gamma_n, U_n \leq t\}.$$

Proof. For $y > 0$, we get

$$H(\mathrm{d}y, u) = \mathbb{P}(\xi_n \in \mathrm{d}y; g(\xi_n) \geq u) = F(\mathrm{d}y)\mathbb{1}\{g(y) \geq u\}.$$

Thus $g = \mathrm{d}\nu/\mathrm{d}F$ yields

$$\int_0^\infty H(\mathrm{d}y, u)\,\mathrm{d}u = F(\mathrm{d}y)g(y) = \nu(\mathrm{d}y). \qquad \square$$

Example 4.5 For the specific case of tempered stable processes, in particular the CGMY process, some series expansions have recently been developed by Rosiński [308] that compared to the acceptance–rejection procedure of Remark 3.4 have the attractive feature of not needing tuning to the particular set of parameters. Consider the one-sided case with Lévy density $n(x) = Ce^{-Mx}/x^{1+\alpha}$, $x > 0$. In the finite variation case $0 < \alpha < 1$, this expansion is

$$X(t) = \sum_{n=1}^{\infty}\left[\left(\frac{C}{\alpha\Gamma_n}\right)^{1/\alpha} \wedge \left(T_n V_n^{1/\alpha}/M\right)\right]\mathbb{1}\{U_n \leq t\}, \quad 0 \leq t \leq 1,$$

where the V_n are uniform$(0,1)$ and the T_n standard exponential (with obvious independence properties). In the compensated case $1 \leq \alpha < 2$, the expansion takes instead the form

$$X(t) = bt + \sum_{n=1}^{\infty}\left\{\left[\left(\frac{C}{\alpha\Gamma_n}\right)^{1/\alpha} \wedge \left(\frac{T_n V_n^{1/\alpha}}{M}\right)\right]\mathbb{1}\{U_n \leq t\} - t\left(\frac{C}{\alpha n}\right)^{1/\alpha}\right\}$$

for $0 \leq t \leq 1$, where

$$b = \alpha^{-1/\alpha}\zeta(1/\alpha)C^{1/\alpha} - \Gamma(1-\alpha)CM^{\alpha-1}$$

($\zeta(\cdot)$ is Riemann's zeta function). We omit the formula for b for the special case $\alpha = 1$. $\qquad \square$

Exercises

4.1 (A) In earthquake modeling, let $S(t)$ be the stress potential at time t (taken to be left-continuous) and $s = 20$ a threshold value. At time $\tau = \inf\{t : S(t) > s\}$

an earthquake occurs, and the potential is then reset to 0, $S(\tau+) = 0$. Between quakes, the stress potential builds up like a subordinator with Lévy density

$$n(x) = \frac{x^{1/2} + 3}{x^{7/4}}.$$

Produce a histogram for the severity of an earthquake, as defined by the r.v. $S(\tau) - s$. You will not need to generate more complicated r.v.'s than mixtures of Paretos (Pareto r.v.'s are easily generated by inversion).

5   Subordination

Let $\{T(t)\}_{t\geq0}$ be a subordinator (a nondecreasing Lévy process) and $\{Y_t\}_{t\geq0}$ an independent general Lévy process, possibly starting from $Y_0 \neq 0$. It is then clear that the process $\{X(t)\}_{t\geq0}$ given by $X(t) \overset{\text{def}}{=} Y_{T(t)}$ has stationary independent increments, hence is a Lévy process, which is said to be *subordinated* to $\{Y_t\}$. When the process $\{X(t)\}$ can be represented this way and the increments of $\{Y_t\}$ are easily simulatable, an obvious algorithm for simulating $\{X(t)\}$ is to simulate $\{T(t)\}$ at a grid $0 < t_1 < t_2 < \cdots$ that is equidistant or Poisson, and to generate $X(t_k)$ as $Y_{T(t_k)}$.

The following two examples, in which $\{Y_t\}$ is Brownian motion, have recently become popular in finance as models for log returns of stock prices etc., due in part to a stochastic volatility interpretation: the increment $Y_{T(t+h)} - Y_{T(h)}$ can be viewed as Gaussian with the stochastic variance $T(t + h) - T(t)$, or in different terms, as a normal variance mixture.

Example 5.1 The NIG (normal inverse Gaussian) Lévy process in Example 1.4 is Brownian motion subordinated to the inverse Gaussian Lévy process. More precisely, in terms of the four parameters $\alpha, \beta, \mu, \delta$, $\{Y_t\}$ is Brownian motion with initial value $Y_0 = \mu$, drift β, and variance 1, whereas $T(t) = T_\gamma(\delta t)$, where $\{T_\gamma(t)\}$ is inverse Gaussian as in Example 1.5 with $\gamma = \sqrt{\alpha^2 - \beta^2}$.

To exploit this structure for simulation, one starts by generating the inverse Gaussian subordinator $\{T_\gamma(\delta t)\}_{t\geq0}$ along a discrete grid $h, 2h, \ldots$. This can be done, for example, by the following algorithm due to Michael, Schuchany, & Haas [256] for generating an r.v. X having the distribution of the first passage time of a BM$(\gamma, 1)$ to the level $\delta > 0$:

(i) Let $Y \sim \mathcal{N}(0,1)$, $Z \longleftarrow Y^2/\gamma$.

(ii) Let $Z \longleftarrow \left(\delta + Z/2 - \sqrt{\delta Z + Z^2/4}\right)/\gamma$, $p \longleftarrow \delta/(\delta + \gamma Z)$.

(iii) Let $U \sim \text{uniform}(0, 1)$. If $U > p$, return $X = (\delta/\gamma)^2/Z$. Else return $X = Z$.

□

Example 5.2 The variance Gamma Lévy process is Brownian motion subordinated to the Gamma process. Since Gamma r.v.'s will typically be easily available, this implies that the variance Gamma process can be simulated exactly along a discrete grid $t_n^h \overset{\text{def}}{=} nh$.   □

Simulation of Lévy processes via subordination is further discussed in Madan & Yor [243], Rubenthaler [311], and Rubenthaler & Wiktorsson [312].

Exercises

5.1 (TP) The definition of the NIG distribution amounts to sampling a BM B_1 at the first passage time of an *independent* BM B_2. Show that one does not get a larger class of distributions by allowing (B_1, B_2) to be a bivariate BM with a possibly nonzero correlation coefficient.

5.2 (TP) Show that if X is inverse Gaussian(δ, γ), then $(X - \delta/\gamma)^2/X$ has a (scaled) χ_1^2 distribution. Use this to verify the above algorithm for generating X.

6   Variance Reduction

If a Lévy process is simulated via discrete skeletons, variance reduction can be carried out much as for random walks, using importance sampling, control variates, stratification, QMC, or other methods.

6a   Importance Sampling

The most obvious choice of importance distribution for a Lévy process is of course another Lévy process. Compared to random walks, one specific feature of Lévy processes is that absolute continuity (on a time interval $[0, T]$) of two continuous-time Lévy processes is not automatic even if the marginals are absolutely continuous. More precisely:

Theorem 6.1 *Let* $\mathbb{P}_1, \mathbb{P}_2$ *be the governing probabilities for two Lévy processes with characteristic triplets* (c_1, σ_1^2, ν_1), (c_2, σ_2^2, ν_2), *and let* $T < \infty$. *If* $\sigma^2 > 0$, *then* $\mathbb{P}_1$ *and* $\mathbb{P}_2$ *restricted to* $\mathscr{F}_T \overset{\text{def}}{=} \sigma\big(X(t) : 0 \leq t \leq T\big)$ *are equivalent if and only if* (i) $\sigma_1^2 = \sigma_2^2$, (ii) $\int_{-\infty}^{\infty}(e^{\ell(x)/2} - 1)^2\, \nu_1(\mathrm{d}x) < \infty$, *where* $\ell \overset{\text{def}}{=} \log(\mathrm{d}\nu_2/\mathrm{d}\nu_1)$. *If* $\sigma^2 = 0$, *it is in addition required that* (iii) $c_2 - c_1 = \int_{-1}^{1} x\big(\nu_2(\mathrm{d}x) - \nu_1(\mathrm{d}x)\big)$.

The proof is in Sato [330, Section 33]; see also Cont & Tankov [75, pp. 307 ff].

The likelihood ratio between two compound Poisson processes is given in Example V.1.13 and is simple. The expression in case of general Lévy processes is somewhat more complicated; see again the above references.

Example 6.2 Let $\mathbb{P}_1 = \mathbb{P}$ corresponding to the characteristic triplet (c, σ^2, ν), and let $\mathbb{P}_2 = \mathbb{P}_\theta$ be defined by exponential tilting, i.e., $\kappa_\theta(s) = \kappa(s + \theta) - \kappa(\theta)$, where $\kappa(\theta) < \infty$, i.e., $e^{\theta x} \wedge x^2$ is ν-integrable. We get

$$\kappa_\theta(s) = (c + \sigma^2 \theta)s + \sigma^2 s^2/2 + \int_{-\infty}^{\infty} \left(e^{(\theta + s)x} - e^{\theta x} - sx\mathbb{1}_{x \leq 1}\right)\nu(dx)$$

$$= \left(c + \sigma^2\theta + \int_{-1}^{1}(e^{\theta x} - 1)\nu(dx)\right)s + \sigma^2 s^2/2$$

$$+ \int_{-\infty}^{\infty} e^{\theta x}\left(e^{sx} - 1 - sx\mathbb{1}_{x \leq 1}\right)\nu(dx),$$

which gives the exponentially tilted triplet as $(c_\theta, \sigma_\theta^2, \nu_\theta)$, where $\sigma_\theta^2 = \sigma^2$, $\nu_\theta(dx) = e^{\theta x}\nu(dx)$, and

$$c_\theta = c + \sigma^2\theta + \int_{-1}^{1}(e^{\theta x} - 1)\nu(dx).$$

To verify absolute continuity, note that (i) is trivial. For requirement (ii), note that $\ell(x) = \theta x$. The asymptotics of $\left(e^{\theta x/2} - 1\right)^2$ at $x = \pm\infty$ are $e^{\theta x} \vee 1$, which is ν_1-integrable away from $x = 0$. The asymptotics at $x = 0$ are $\theta^2 x^2/4$, which is ν_1-integrable away from $x = \pm\infty$. This verifies (ii) and absolute continuity when $\sigma^2 > 0$. If $\sigma^2 = 0$, we have

$$c_\theta - c = \int_{-1}^{1} x(e^{\theta x} - 1)\nu(dx) = \int_{-1}^{1} x\big(\nu_\theta(dx) - \nu(dx)\big).$$

So (iii) holds and absolute continuity.

In this example, the likelihood ratio takes a simple form,

$$\left.\frac{d\mathbb{P}}{d\mathbb{P}_\theta}\right|_{\mathscr{F}_T} = e^{-\theta X(T) + T\kappa(\theta)},$$

as may be seen, for example, by discrete-random-walk approximations.   □

6b   Variance Reduction via the Large Jumps

The wavelet representation of BM via Brownian bridges is, as discussed in X.2b and the exercises there, particularly well suited for implementing variance reduction via stratification and QMC by concentrating on the r.v.'s along a binary grid (say $B(1/4), B(1/2), B(3/4), B(1)$) that are most important for the shape of the path. However, the representation does not carry over to general Lévy processes. Instead we propose that in some situations one could concentrate on the big jumps. More precisely, it follows from Theorem 4.1 that steps (i), (ii) of the following algorithm generate the K biggest jumps of a Lévy process over the time interval $[0, 1]$ ($\overline{\nu}(x) = \int_x^\infty \nu(dy)$ denotes the tail of the Lévy measure and $\overline{\nu}^{-1}$ its inverse). The whole process can then be obtained by supplementing with the small jumps as in steps (iii), (iv):

(i) Generate $T_1, \ldots, T_K$ as standard exponentials and $U_1, \ldots, U_K$ as uniform$(0,1)$.

(ii) Let $x_1 = \overline{\nu}^{-1}(T_1)$, $x_2 = \overline{\nu}^{-1}(T_1 + T_2)$, $\ldots$, $x_K = \overline{\nu}^{-1}(T_1 + \cdots + T_K)$.

(iii) Using any of the algorithms of Sections 2–4, generate a Lévy process $\{Y(t)\}_{0 \le t \le 1}$ with Lévy measure $\nu(\mathrm{d}x)\mathbb{1}\{x \le x_K\}$;

(iv) Put $X(t) = Y(t) + \sum_{k=1}^{K} x_k \mathbb{1}_{U_k \le t}$, $\quad 0 \le t \le 1$.

In itself, this algorithm does of course not provide variance reduction. However, the point is that it identifies some crucial r.v.'s, namely $T_1, \ldots, T_K$ and $U_1, \ldots, U_K$, to which one may apply say stratification, importance sampling, or QMC.

We do not know of implementations of this idea. One would expect that the efficiency depends on a proper tuning of K to the tail of the Lévy measure (a larger K for a lighter tail).

Obvious variants of this algorithm are obtained by replacing Theorem 4.1 by some of the other series representations in Section 4.

Exercises

6.1 (TP) Show that two tempered stable processes with parameters C_+^1, C_-^1, G^1, M^1, Y^1, $C_+^2, C_-^2, G^2, M^2, Y^2$ (cf. Example 1.2) satisfy the conditions of Theorem 6.1 if and only if $C_+^1 = C_2^+$, $C_-^1 = C_2^-$.

6.2 (TP) Discuss the minimum entropy entropy measure parallel to Example 6.2 dealing with the Esscher tranform.

6.3 (TP) Show that for the NIG process, exponential tilting means changing β to $\beta + \theta$.

7 The Multidimensional Case

We shall be relatively brief, and therefore consider only the case corresponding to finite variation for a pure jump process. Let now ν be a nonnegative measure on $\mathbb{R}^d$ with $d > 1$ and satisfying $\int_{\mathbb{R}^d} \|\boldsymbol{x}\| \, \nu(\mathrm{d}\boldsymbol{x}) < \infty$, where $\|\cdot\|$ is any norm. The same construction as in (1.3) then immediately applies and gives a process $\boldsymbol{X}(t) \in \mathbb{R}^d$, the dynamics of which can informally be described by a jump $\boldsymbol{x} = \boldsymbol{X}(t) - \boldsymbol{X}(t-)$ occurring at intensity $\nu(\mathrm{d}\boldsymbol{x})$. The (trivial) case of independent components of $\boldsymbol{X}(t)$ corresponds to $\nu(\mathrm{d}\boldsymbol{x})$ being concentrated on the union of the d one-dimensional boundaries $\{\boldsymbol{x} : x_j = 0 \text{ when } i \ne j\}$.

Two main examples are as follows:

(i) Multivariate α-stable processes, which can simply be constructed by equipping the jumps of a one-dimensional α-stable process by i.i.d. random directions. That is, if $\nu_{\mathrm{P}}(\mathrm{d}r, \mathrm{d}\boldsymbol{u})$ is the Lévy measure written in polar coordinates, so that $r > 0$, $\boldsymbol{u} \in S^{d-1} \overset{\text{def}}{=} \{\boldsymbol{x} \in \mathbb{R}^d : \|\boldsymbol{x}\| = 1\}$, then

$$\nu_{\mathrm{P}}(\mathrm{d}r, \mathrm{d}\boldsymbol{u}) \;=\; \frac{C}{r^{\alpha+1}} \, \mathrm{d}r \, \sigma(\mathrm{d}\boldsymbol{u}) \,,$$

where σ is a probability measure on S^{d-1}.

(ii) Tempered multivariate α-stable processes, which are defined by a tempering function $q(r, \boldsymbol{u})$ that is completely monotonic for fixed $\boldsymbol{u}$ with $q(0+, \boldsymbol{u}) = 1$, $q(\infty, \boldsymbol{u}) = 0$ (hence $q(\cdot, \boldsymbol{u})$ is convex and decreasing on $(0, \infty)$), so that

$$\nu_{\mathrm{P}}(\mathrm{d}r, \mathrm{d}\boldsymbol{u}) \;=\; \frac{Cq(r, \boldsymbol{u})}{r^{\alpha+1}} \, \mathrm{d}r \, \sigma(\mathrm{d}\boldsymbol{u}) \,.$$

The simulation of the process can be carried out in a similar way as in Section 3 by truncating ν to C_d^c, where $C_d = [-\varepsilon, \varepsilon]^d$. The corresponding jump part is then d-dimensional compound Poisson$(\lambda_\varepsilon, F_\varepsilon)$, where $\lambda_\varepsilon = \nu(C_d^c)$ and F_ε is ν/λ_ε truncated to C_d^c. The remaining small jumps can be neglected, replaced by their expected values, or by a d-dimensional BM. See further Rosiński [308] and Cohen & Rosiński [73].

Whereas of course the case of independent components is straightforward, the dependence structure of multivariate Lévy processes presents interesting problems from the modeling, and thereby the simulation, point of view. The construction of a copula from a multivariate distribution by transforming by the inverse marginal c.d.f.'s (cf. II.3) does not carry over to multivariate Lévy measure because of the singularity at $\boldsymbol{x} = \boldsymbol{0}$. Instead, one proceeds via the marginal tails. To be more precise, we assume for simplicity that $\nu(\mathrm{d}\boldsymbol{x})$ is concentrated on $(0, \infty)^d$ with density $\nu(\boldsymbol{x})$. The Lévy tail is then

$$\overline{\nu}(x_1, \ldots, x_d) \overset{\text{def}}{=} \nu\big((x_1, \infty) \times \cdots \times (x_d, \infty)\big) \,,$$

and the kth marginal Lévy tail is

$$\overline{\nu}_k(x_k) \overset{\text{def}}{=} \nu\big((0, \infty)^{k-1} \times (x_k, \infty) \times (0, \infty)^{d-k}\big) \;=\; \overline{\nu}(0, \ldots, 0, x_k, 0, \ldots, 0) \,.$$

We define the *Lévy copula* corresponding to ν by

$$C(y_1, \ldots, y_d) \overset{\text{def}}{=} \overline{\nu}\big(\nu_1^{-1}(y_1), \ldots, \nu_1^{-1}(y_d)\big) \,, \quad 0 < y_1, \ldots, y_d < \infty \,.$$

The kth marginal is

$$\begin{aligned} C_k(y_k) &\overset{\text{def}}{=}\; C(\infty, \ldots, \infty, y_k, \infty, \ldots, \infty) \\ &=\; \overline{\nu}(0, \ldots, 0, \nu_k^{-1}(y_k), 0, \ldots, 0) \;=\; y_k \,. \end{aligned}$$

This is similar to the uniform marginals for an ordinary copula, except that y_k varies in $(0, \infty)$ rather than $(0, 1)$, and that therefore C is an infinite measure rather than a probability measure.

Now assume that we are given d Lévy measures $\nu_1^*, \ldots, \nu_d^*$ on $(0, \infty)$, and that we want to construct a Lévy measure ν^* on $(0, \infty)^d$ having the ν_k^* as marginals and the same Lévy copula as ν. This is obtained by letting

$$\overline{\nu}^*(x_1, \ldots, x_d) = C\big(\overline{\nu}_1^*(x_1), \ldots, \overline{\nu}_d^*(x_d)\big).$$

For $d = 2$, the corresponding Lévy process can be simulated as

$$\boldsymbol{X}^*(t) = \left(\sum_{i=1}^{\infty}(\overline{\nu}_1^{*-1}(\Gamma_{1,i})\mathbb{1}_{U_i \leq t}, \sum_{i=1}^{\infty}(\overline{\nu}_2^{*-1}(\Gamma_{2,i})\mathbb{1}_{U_i \leq t}\right) \qquad (7.1)$$

for $t \in [0, 1]$, provided C is absolutely continuous, where the U_i are uniform$(0, 1)$, the $\Gamma_{1,i}$ are the epochs of a standard Poisson process, and $\Gamma_{2,i}$ has c.d.f. $C_2(\Gamma_{1,i}, \cdot)$ given $\Gamma_{1,i}$, where $C_2 = (\partial/\partial y_2)C$.

A simple example is an Archimedean Lévy copula, constructed as

$$C(y_1, \ldots, y_d) = \phi^{-1}\big(\phi(y_1), \ldots, \phi(y_1)\big),$$

where $\phi : (0, \infty) \to (0, \infty)$ is strictly decreasing and convex with $\phi(0+) = \infty$, $\phi(\infty) = 0$.

For further examples, proofs, and discussion, see Cont & Tankov [75, Sections 5.4, 5.5, 6.6], Tankov [351], and Barndorff-Nielsen & Lindner [36].

Exercises

7.1 (TP) Explain that (7.1) cannot be a valid representation for independent components. Which condition is violated?

8   Lévy-Driven SDEs

An SDE driven by a Lévy process $\{X(t)\}$ has just the same form as in the case of a driving BM,

$$\mathrm{d}Y(t) = a\big(t, Y(t)\big)\,\mathrm{d}t + b\big(t, Y(t)\big)\,\mathrm{d}X(t); \qquad (8.1)$$

for the stochastic calculus needed to deal with such SDEs in the presence of jumps of the driving process, see Protter [297].

Example 8.1 One of the most notable examples is a Lévy driven Ornstein–Uhlenbeck process, $\mathrm{d}Y(t) = -\alpha Y(t)\,\mathrm{d}t + \mathrm{d}X(t)$ with $\alpha > 0$. If $\{X(t)\}$ is a subordinator, the model has the notable property that negative values cannot occur, which is one of several reasons that the Lévy-driven Ornstein–Uhlenbeck process is a popular model for stochastic interest rates in mathematical finance.

Similar ideas have been employed in stochastic volatility models. We mention in particular the model

$$\begin{aligned}
\mathrm{d}Y(t) &= \big(\mu + \beta\sigma^2(t)\big)\,\mathrm{d}t + \sigma(t)\,\mathrm{d}B(t) + \rho\,\mathrm{d}X(t), \\
\mathrm{d}\sigma^2(t) &= -\alpha\sigma^2(t)\,\mathrm{d}t + \mathrm{d}X(t),
\end{aligned}$$

for an asset price $S(t) = S(0)e^{Y(t)}$ suggested by Barndorff-Nielsen & Shephard [38]; here X is a subordinator, B an independent BM, and the interpretation of the term $\beta\sigma^2(t)$ is as a "volatility risk premium."   □

The dominant numerical method for Lévy-driven SDEs is the Euler scheme, in which an approximation $Y_n^h \overset{\text{def}}{=} Y^h(t_n^h)$ along a discrete grid $t_n^h \overset{\text{def}}{=} nh$ is generated as

$$Y_n^h = Y_{n-1}^h + a\big(t_{n-1}^h, Y_{n-1}^h\big)h + b\big(t_{n-1}^h, Y_{n-1}^h\big)\big[X\big(t_n^h\big) - X(t_{n-1}^h)\big].$$

For convergence properties, see, e.g., Jacod [191], Jacod et al. [192], and references there.

Remark 8.2 It is tempting to attempt to modify the derivation of the Milstein scheme for SDEs driven by BM to Lévy-driven SDEs. If the driving Lévy process is simulated along a discrete skeleton only, it turns out, however, that in order to implement such a scheme one has in general to be able to generate the increment of X jointly with the quadratic variation over $(t_n^h, t_{n-1}^h]$, and no simple algorithms are known for this.

The situation is somewhat different when the process is simulated as in Section 3 as sum of the process $J^{(2)}$ of large jumps and a part W that is a BM$(\mu_\varepsilon, \sigma_\varepsilon^2)$. If one is satisfied with the solution of the SDE along the Poisson grid $0 < t_1 < t_2 < \cdots$ for $J^{(2)}$, one can then just use the one-step Milstein update corresponding to replacing $\mathrm{d}X(t)$ in (8.1) by $\mathrm{d}W(t)$ in the open time interval (t_{n-1}, t_n). If one wants to have Y represented along a deterministic equidistant grid t_n^h, the update from t_{n-1}^h to t_n^h can the be done as follows. If there are no jumps of $J^{(2)}$ in $\big(t_{n-1}^h, t_n^h\big]$, just use one-step Milstein updating. If there is precisely one jump, say at time $t_{n-1}^h < s < t_n^h$ and of size $\Delta X(s)$, use one-step Milstein updating in $\big(t_{n-1}^h, s\big)$. Then add $b\big(s, Y(s)\big)\Delta X(s)$ to $Y(s-)$ and use one-step Milstein updating in (s, t_n^h). For $m > 1$ jumps, split in a similar way in $m + 1$ subintervals.

We know of neither implementations nor analysis of these ideas.   □

Chapter XIII
Markov Chain Monte Carlo Methods

1 Introduction

Markov chain Monte Carlo (MCMC) is a method for obtaining information on a distribution Π^* whose point probabilities (or density w.r.t. some reference measure μ) are typically known only up to a constant. Given a nonnegative function $\pi(x)$ on a set E, set

$$
C \stackrel{\text{def}}{=}
\begin{cases}
\displaystyle\sum_{x \in E} \pi(x) & \text{in the discrete case,} \\
\displaystyle\int_E \pi(x)\,\mu(\mathrm{d}x) & \text{in the continuous case.}
\end{cases}
$$

Then Π^* is the distribution with point probabilities $\pi^*(x) \stackrel{\text{def}}{=} \pi(x)/C$ in the discrete case and density $\pi^*(x) \stackrel{\text{def}}{=} \pi(x)/C$ w.r.t. μ in the continuous case, where the normalization constant C is assumed unknown.[1] In order not to have to differentiate between the two cases in the following, we will let μ denote counting measure on E in the discrete case. Because many arguments are basically the same in the two cases but technically more transparent in the discrete case, we will often deal just with the discrete case.

The setup of a known $\pi(\cdot)$ but an unknown C may appear quite special at first sight, but we will see some very important examples in Section 2.

---

[1] Note the notational difference from Chapter IV, where π is the target distribution itself, not its unnormalized density.

MCMC proceeds by finding an ergodic Markov chain $\xi_0, \xi_1, \ldots$ having stationary distribution Π^*. One then simulates $\xi_0, \xi_1, \ldots, \xi_n$ for some large n. As $n \to \infty$, the empirical distribution $\widehat{\Pi}_n^*$ given by

$$\widehat{\Pi}_n^*(A) \stackrel{\text{def}}{=} \frac{1}{n+1} \sum_{k=0}^{n} \mathbb{1}\{\xi_k \in A\} \tag{1.1}$$

(A is a measurable subset of E) converges to Π^* under weak conditions, and $\widehat{\Pi}_n^*$ is therefore close to Π^* if n is large. Thus, if we are interested in the Π^*-expectation $\Pi^*[f] \stackrel{\text{def}}{=} \int_E f(x)\pi^*(x)\,\mu(\mathrm{d}x)$ of a function f on E, we may use the estimator

$$\widehat{f}_n \stackrel{\text{def}}{=} \frac{1}{n+1} \sum_{k=0}^{n} f(\xi_k). \tag{1.2}$$

Similarly, an idea of the shape of Π^* can be obtained by plotting a histogram of $\xi_0, \xi_1, \ldots, \xi_n$ in the one-dimensional case, of one-dimensional marginals in the multidimensional case, or by other methods such as scatter plots.

These ideas are exactly the same as those of Chapter IV on steady-state simulation, except for the first step: whereas Chapter IV tacitly assumes that it is straightforward to generate paths of ξ, as will be the case for a queuing or inventory system, a Markov chain specified in terms of its transition mechanism rather than its stationary distribution, etc., it is not a priori clear in the MCMC setting how one can construct the transition mechanism for $\{\xi_k\}$ so that the stationary distribution is Π^*. A candidate is of course just to let $\xi_0, \xi_1, \ldots$ be i.i.d. r.v.'s with distribution Π^*; however, since C is unknown, the standard methods for r.v. generation seldom apply to Π^*, so this idea is seldom practical (see II.2.16 for a discussion of the applicability of acceptance–rejection). We return to these points later, but start the more detailed treatment of MCMC in the next section with some examples indicating how the setup of an unknown C but known $\pi(x)$ may arise.

The literature on MCMC is enormous and rapidly expanding. Established textbook treatments are Gilks, Richardson, & Spiegelhalter [129] and Robert & Casella [301]. There are, however, many surveys around, e.g., Brooks [60], Green [166], Roberts & Rosenthal [302]. Also, a number of texts that are focused on some special application areas have extensive treatments of MCMC, e.g., Häggström [189] and Cappé et al. [65].

While it is important for many applications to allow a general (non-discrete) state space, the ideas of MCMC can best be understood by thinking of the Markov chain $\{\xi_n\}$ as an ordinary discrete one, i.e., E is finite or countable. Then one traditionally specifies $\{\xi_n\}$ via the transition matrix $\boldsymbol{P} \stackrel{\text{def}}{=} \big(p(x,y)\big)_{x,y \in E}$. The point probabilities $\widehat{\pi}_n^*(x)$ of the empirical distribution and $\Pi^*[f]$ are then given by

$$\widehat{\pi}_n^*(x) \overset{\text{def}}{=} \frac{1}{n+1} \sum_{k=0}^{n} \mathbb{1}\{\xi_k = x\}\,,\ x \in E,\quad \Pi^*[f] = \sum_{x \in E} f(x)\pi(x)/C\,,$$

and so on. In the general case, the role of the transition matrix $\boldsymbol{P}$ is taken by the *transition kernel* P defined by $P(x, A) \overset{\text{def}}{=} \mathbb{P}_x(\xi_1 \in A)$. This can be viewed as an operator $f \to Pf$, $\nu \to \nu P$ acting on functions to the right by $Pf(x) \overset{\text{def}}{=} \int_E f(y)\,P(x, \mathrm{d}y)$ and on measures to the left by $\nu P(A) \overset{\text{def}}{=} \int_E \nu(\mathrm{d}x)P(x, A)$. In particular, the condition that π^* be a stationary density can be written as $\pi^* P = \pi^*$ in the general case, and $\boldsymbol{\pi}^* \boldsymbol{P} = \boldsymbol{\pi}^*$ in the discrete case, where $\boldsymbol{\pi}^*$ is the row-vector representation of π^*. Note that since C cancels, alternative formulations of the stationarity requirement are $\pi P = \pi$ and $\boldsymbol{\pi} \boldsymbol{P} = \boldsymbol{\pi}$, where $\boldsymbol{\pi} \boldsymbol{P} = \boldsymbol{\pi}$ can be written out as

$$\pi(y) = \sum_{x \in E} \pi(x)p(x, y) \tag{1.3}$$

in the discrete case, and similarly in the continuous case $\pi P = \pi$ means

$$\pi(A) = \int_E p(x, A)\pi(x)\,\mu(\mathrm{d}x)\,. \tag{1.4}$$

2 Application Areas

2a Bayesian Statistics

The point of view of Bayesian statistics is to view the parameter θ (typically a vector) as the outcome of an r.v. Θ with a specific distribution, the *prior distribution* or just the *prior*. The density (continuous case) or the probability mass function (discrete case) of Θ is denoted by $\pi^{(0)}(\theta)$ in the following, and the density of the observation vector $\boldsymbol{y} = (y_1, \ldots, y_p)$ is denoted by $L(\boldsymbol{y} \,|\, \theta)$ (the likelihood). Since now both the observations and the parameter are random, it makes sense to consider the conditional density (or p.m.f.) $\pi^*(\theta \,|\, \boldsymbol{y})$ of Θ given $\boldsymbol{Y} = \boldsymbol{y}$. The corresponding distribution is the *posterior distribution* or just the *posterior*, and is the form in which the Bayesian statistician usually gives his estimate; traditional maximum likelihood (ML) would typically report the point estimate

$$\widehat{\theta} \overset{\text{def}}{=} \operatorname{argmax}_{\theta \in \Theta} L(\boldsymbol{y} \,|\, \theta)\,,$$

possibly supplemented by a confidence interval based on a CLT for $\widehat{\theta}$.

The Bayesian procedure is certainly different from traditional statistics, one main reason being that prior information on θ is assumed available.

We will discuss this below, but we start with *Bayes's formula*

$$\pi^*(\theta \,|\, \boldsymbol{y}) \;=\; \frac{L(\boldsymbol{y} \,|\, \theta)\pi^{(0)}(\theta)}{\displaystyle\int L(\boldsymbol{y} \,|\, \zeta)\pi^{(0)}(\zeta)\,\mathrm{d}\zeta}, \tag{2.1}$$

which is the basic formula in the area.

Bayes's formula can be found in virtually any undergraduate textbook in probability or statistics. The derivation is elementary: just substitute Θ for X in the general formula

$$
\begin{aligned}
f_{X \,|\, Y=y}(x) &\;=\; \frac{f_{X,Y}(x,y)}{f_Y(y)} \;=\; \frac{f_{Y \,|\, X=x}(y)f_X(x)}{\int f_{X,Y}(z,y)\,\mathrm{d}z} \\[2mm]
&\;=\; \frac{f_{Y \,|\, X=x}(y)f_X(x)}{\int f_{Y \,|\, X=z}(y)f_X(z)\,\mathrm{d}z}.
\end{aligned}
$$

Example 2.1 Assume that $Y_1,\ldots,Y_p$ are i.i.d. $\mathscr{N}(\theta,1)$ given θ and that θ itself is $\mathscr{N}(\mu,\omega^2)$. Then the numerator in (2.1) is

$$
\prod_{i=1}^{p} \frac{1}{\sqrt{2\pi}}\mathrm{e}^{-(y_i-\theta)^2/2} \cdot \frac{1}{\sqrt{2\pi}\,\omega}\mathrm{e}^{-(\theta-\mu)^2/2\omega^2}
$$

$$
\;=\; c_1(\mu,\omega^2,\boldsymbol{y})\exp\left\{-\theta^2(p+1/\omega^2)/2 + \theta(p\overline{y}+\mu/\omega^2)\right\},
$$

where $\overline{y} \stackrel{\text{def}}{=} (y_1 + \cdots + y_p)/p$. Letting

$$
\omega^{*2} \stackrel{\text{def}}{=} \frac{1}{p+1/\omega^2}, \qquad \mu^* \stackrel{\text{def}}{=} \omega^{*2}(p\overline{y}+\mu/\omega^2),
$$

this can be written as

$$
c_1(\mu,\omega^2,\boldsymbol{y})\exp\{-\theta^2/2\omega^{*2} + \theta\mu^*/\omega^{*2}\} \;=\; c_2(\mu,\omega^2,\boldsymbol{y})\mathrm{e}^{-(\theta-\mu^*)^2/2\omega^{*2}}.
$$

This is proportional to the $\mathscr{N}(\mu^*,\omega^{*2})$ density. Since the denominator in (2.1) is just a normalizing constant ensuring that the θ-integral of (2.1) is 1, the posterior distribution is $\mathscr{N}(\mu^*,\omega^{*2})$.

Note that $\mu^*/\overline{y} \to 1$ as $p \to \infty$, so that the posterior mean is approximately $\overline{y}$ (the ML estimator $\widehat{\mu}$ in the traditional setting) for large p, whereas $\omega^{*2} \to 0$. Similarly, if $\omega^2 \to \infty$, one gets $\mu^* \to \overline{y}$, $\omega^{*2} \to 0$. □

Some easy elementary examples of the same type can be found in the exercises. Parameters in the prior distribution such as (μ,ω^2) are referred to as *hyperparameters*. Example 2.1 is a simple case of *conjugate families*: the posterior distribution belongs to the same parametric family as the prior.

The examples of explicit posteriors are, however, rather few and elementary in nature. Nevertheless, Bayes's formula (2.1) contains some essential information, since the numerator is essentially explicit as the product of the standard likelihood $L(\boldsymbol{y} \,|\, \theta)$ and the prior density $\pi^{(0)}(\theta)$. However, for purposes such as computing the posterior mean of θ (the *Bayes estimator*), it is the denominator in (2.1) that presents the major obstacle from the numerical point of view, since the integral may be high-dimensional and not available in closed form. Thus we have a case for MCMC, with Π^* the posterior and $\pi(\theta) = L(\boldsymbol{y} \mid \theta)\pi^{(0)}(\theta)$.

In contrast, the evaluation of the point at which the *maximum a posteriori probability* is attained (the point at which the posterior density or p.m.f. has its maximum) does not in itself require the numerator of (2.1). However, the optimization of the denominator may be difficult, and a possible approach is then to run an MCMC sequence $\xi_0, \dots, \xi_n$ and use ξ_{k^*} as estimator, where k^* is the maximizer of $L(\boldsymbol{y} \mid \xi_k)\pi^{(0)}(\xi_k)$. Similar remarks apply to posterior quantiles.

The Bayesian approach has up to rather recently been considered controversial in statistics, with the frequentist's point of view having been the more dominant one. The frequentist will possibly admit having some subjective beliefs on θ. Say we want to test whether a coin is fair. Then we consider it far more likely that $\theta = 0.47$ than that $\theta = 0.96$. But the frequentist's problem is how to quantify this belief to the degree that he/she is willing to postulate a very specific distributional form like the Beta with specified parameters. He/she will basically consider this as intrinsically impossible and will not be willing to accept that the estimates depend on the prior distribution.

As noted in the last paragraph of Example 2.1, the Bayesian approach does, however, often lead to answers that are basically close to traditional ML. More precisely, this is typically the case when the number of observations is large (the uncertainty on θ expressed by the variability in the prior distribution then becomes eliminated by the information in the observations as $p \to \infty$), or when the prior is *flat*, i.e., spread out over a much larger range than where the observations indicate that θ should be located (e.g., in the setting of Example 2.1, a flat prior means a large ω^2). This makes the Bayesian view more acceptable to the frequentist, and the gain is then that Bayesian computations using MCMC are much more straightforward than a corresponding likelihood maximization. This has substantially widened the class of statistical problems that are tractable, and it is now common that former frequentists take a pragmatic view and attack at least some statistical problems using Bayesian computations implemented via MCMC.

The numerical difficulties of ML are, however, not the only driver for the increasing use of Bayesian MCMC methods. One further case is empirical Bayes situations in which individual-specific data are more limited than data on the population from which the individual is sampled; see Exercise 2.2 for an example. Also, model estimation and prediction (e.g., predicting future values of a time series with unknown parameters) are "decoupled" problems in the frequentist world, but they allow a unified and simultaneous implementation via the Bayesian approach.

Example 2.2 As a typical example of a statistical problem in which Bayesian MCMC methods are used, consider one-way analysis of variance for a group of q items, where the observation vector is

$$\boldsymbol{Y} = \left(Y_{ij}\right)_{i=1,\ldots,q,\,j=1,\ldots,n_i},$$

where i denotes group index and j stands for the index of the n_i replications within the group. The two traditional statistical models are then the *fixed-effects model* with the Y_{ij} being independent such that $Y_{ij} \sim \mathcal{N}(\theta_i, \sigma^2)$, where $\theta_1, \ldots, \theta_q, \sigma^2$ are unknown parameters, and the *random-effects model*, in which one assumes $Y_{ij} = \theta_i + \sigma \varepsilon_{ij}$, where as in the fixed-effects model σ^2 is an unknown parameter but $\theta_1, \ldots, \theta_q$ are r.v.'s assumed to be i.i.d. $\mathcal{N}(\mu, \omega^2)$ r.v.'s and the ε_{ij} are i.i.d. $\mathcal{N}(0, 1)$. One can think of μ as the overall level, θ_i as the level within group i, ω^2 as the variance between groups, and σ^2 as the variance within a group.

ML estimation of the parameters in the fixed-effects model can be done explicitly in a standard way. The same is true for the random-effects model provided $n \overset{\text{def}}{=} n_i$ does not depend on i. Otherwise, the computational difficulties are at a different level, and Bayesian MCMC methods provide an appealing alternative. As an example of the models that are typically used, assume that the prior on σ^2 is inverse Gamma (IG, see A1) with parameters α_1, λ_1, that the θ_i are i.i.d. $\mathcal{N}(\mu, \omega^2)$ given μ, where μ itself has a $\mathcal{N}(\mu_0, \tau^2)$ distribution and ω^2 an IG(α_2, λ_2) distribution. Since the squared coefficient of variation of the Gamma(α, λ) distribution is $1/\alpha$, this prior is flat assuming that α_1, α_2 are small and τ^2 is large. The posterior density π of the parameters $\mu, \sigma^2, \omega^2, \theta_1, \ldots, \theta_p$ is proportional to the joint density of the parameters and the observations y_{ij}, which means that we may take

$$
\begin{aligned}
\pi\big(&\mu, \sigma^2, \omega^2, \theta_1, \ldots, \theta_p\big) \\
&= \mathrm{e}^{-(\mu-\mu_0)^2/2\tau^2} h_{\alpha_1,\lambda_1}(\sigma^2) h_{\alpha_2,\lambda_2}(\omega^2) \\
&\quad \times \prod_{i=1}^{q} \frac{1}{\omega} \mathrm{e}^{-(\theta_i-\mu)^2/2\omega^2} \times \prod_{i=1}^{q}\prod_{j=1}^{n_i} \frac{1}{\sigma} \mathrm{e}^{-(y_{ij}-\theta_i)^2/2\sigma^2},
\end{aligned}
$$

where $h_{\alpha,\lambda}(y) = y^{-\alpha-1}\mathrm{e}^{-\lambda/y}$; cf. (A1.1). □

2b Point Processes and Particle Systems

As noted in I.5.2, one often specifies point processes in a multidimensional region Ω by their unnormalized density $\pi(x)$ w.r.t. the standard Poisson process on Ω, where $x = (\boldsymbol{a}_1, \ldots, \boldsymbol{a}_m)$ with m the number of points and the $\boldsymbol{a}_i$ their locations. For example, for the Strauss process, $\pi(x) = \lambda^{m(x)} \eta^{t(x)}$ with $t(x) \overset{\text{def}}{=} \sum_{i \neq j} \mathbb{1}\{d(\boldsymbol{a}_i, \boldsymbol{a}_j) < r\}$ (Euclidean distance). MCMC is relevant because the normalizing constant is intractable, and the purpose of

MCMC may be just to see what a typical realization looks like, or to compute some characteristics $\Pi^*[f]$ such as the expected total number of points which corresponds to $f(x) = m(x)$ (see also Exercise V.1.9).

A closely related model, first used in statistical mechanics, is used for *particle systems*.[2] A particle system is a configuration of particles on a discrete set S of sites, such that a particle at site i is described by a characteristic (mark) y_i taking values in a set that for simplicity we shall assume is $\{-1, 1\}$ ($\{0, 1\}$ is more convenient in some situations). The state of the system is a configuration $x \overset{\text{def}}{=} (x_i)_{i \in S}$, and we shall think of it as random, that is, as the outcome of a r.v. $(X_i)_{i \in S}$ with values in $\{-1, 1\}^S$.

For example, in a ferromagnet a site may be an atom, and -1 and 1 may correspond to two possible directions (down and up) of the spin. In a biological population, the set of sites may be taken as a set of grid points in a region under study. Each site is assumed to be occupied by an individual, and -1 and 1 could correspond to two genetic types, to the individual carrying or not carrying a certain disease, etc. Or a site could be a square part of a region, with mark 1 if the disease is present in the area, -1 if not. In image analysis, $S = \{1, \ldots, N\} \times \{1, \ldots, M\}$ could be the set of pixels, with marks $+1$ at the black pixels and -1 at the white pixels.

There exists a huge number of suggestions for the distribution Π^* of $(X_i)_{i \in S}$. Most (but not all) exhibit attraction in the sense of biasing neighboring sites to have the same mark (certainly, this is reasonable in the biological examples above). One then needs to specify the meaning of "neighbor". This is usually done in graph-theoretic terms: think of S as the set of vertices of a graph $G = (S, \mathcal{E})$, with $\mathcal{E}$ the set of edges. The neighbors of $i \in S$ are then the i' with $ii' \in \mathcal{E}$. For example, if $S = \mathbb{Z} \times \mathbb{Z}$ (or a subset thereof), a site has the form $i = (u, v)$ with $u, v \in \mathbb{Z}$ and the common choice is

$$ii' \in \mathcal{E} \iff |u' - u| + |v' - v| = 1 .$$

Example 2.3 An important example from statistical physics is the *Ising model*, in which the density w.r.t. the uniform distribution on the set of configurations is proportional to

$$\pi(x) \overset{\text{def}}{=} \exp\left\{-\beta \sum_{ii' \in \mathcal{E}} \mathbb{1}\{x_i \neq x_{i'}\}\right\}, \tag{2.1}$$

where $\beta > 0$. Equivalently, one could take

$$\pi(x) = \exp\left\{\beta \sum_{ii' \in \mathcal{E}} x_i x_{i'}\right\} \tag{2.2}$$

(note that the r.h.s.'s of (2.1), (2.2) differ only by a constant).

---

[2]We use here the term in the static sense in which there is no time evolution.

The Ising model is a special case of the Boltzmann distribution introduced in I.5.4, and in analogy with the terminology there, β is called the *inverse temperature*. Note that $\pi(x)$ is smaller when more sites have opposite marks. Thus the configurations with either all 1's or all -1's are the most likely, and the lower the temperature gets (the larger β gets), the more unlikely other patterns become. If two patterns have the same number of 1's and -1's, the one with the larger number of neighbors of equal marks is the more likely.

In the repulsive Ising model, one has instead $\beta < 0$. This favors configurations with many neighboring sites having opposite marks.

The normalizing constant for the Ising model is

$$ C = \sum_{x \in \{-1,1\}^{|S|}} \exp\left\{-\beta \sum_{ii' \in \mathcal{E}} \mathbb{1}\{x_i \neq x_{i'}\}\right\}. $$

Thus, if S is as small as 8×8, the outer sum has already 2^{64} terms, illustrating that explicit direct computation of C is not possible even in toy models. Even if C were known, explicit computation of the characteristics of Π^* would be infeasible for similar dimensionality reasons.

Worked-out examples of MCMC for the Ising model are given in XIV.4. In the early days of image analysis, the Ising model played an important role as a prior: a black-and-white picture, say a satellite photo of a military base, would typically have large connected areas of either color (say white for the airstrip and black for the planes). A noisy image, meaning that a number of pixels are flipped to their opposite color, would blur this structure, whereas imposing an Ising prior with a relatively low temperature would restore it to some extent in the posterior. For the implementation of MCMC in this setting, see Example 5.7, and for more sophisticated and more realistic applications of MCMC to image analysis, see Green [165].   □

2c   Combinatorial Objects

In quite a few applications of MCMC, E is just a finite set and the target distribution Π^* just the uniform distribution on E. The problem is that even when finite, E is typically huge and of complicated structure, so that generating r.v.'s from the uniform distribution Π^* is not straightforward.

Three examples are given in Figure 2.1, all on an $N \times M$ lattice (with $N = 5$, $M = 6$ in the figure). Panel (a) is the *hard-core model*: E is the set of configurations with a particle present at a lattice point (bulleted in the figure) or not, such that no two neighboring points are occupied. In panel (b), the connecting edges are given directions, and E is the set of configurations such that each nonboundary vertex has two ingoing and two outgoing arrows (at the boundary, at most two arrows going in or out are permitted). This is a model for *square ice*, to which we return in XIV.4.2. Finally in panel (c), E is the set of possible allocations of q colors (here

A–G, so that $q = 7$) to the squares of the lattice, such that no two squares with a common boundary have the same color.

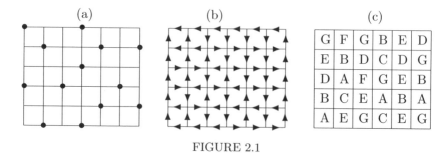

FIGURE 2.1

A simple question for which MCMC may be helpful is to give an estimate of the cardinality $|E|$ of E. We return to this in XIV.3.

2d   Hidden Markov Models

A hidden Markov model aims at describing observations $Y_0, \ldots, Y_n$ with a dependence structure governed by an underlying unobserved time-homogeneous Markov chain $X_0, \ldots, X_n$. That is,

$$\mathbb{P}(Y_0 \in A_0, \ldots, Y_n \in A_n \mid X_0 = x_0, \ldots, X_n = x_n) = G_{x_0}(A_0) \ldots G_{x_n}(A_n)$$

for suitable distributions $G_x(\cdot)$. The Markov chain $X_0, \ldots, X_n$ may be discrete with transition matrix $\boldsymbol{Q} = (q_{ij})_{i,j \in E}$ or general with transition kernel say $Q(x, F)$. We will use notation like $\boldsymbol{Y}_{0:n} \overset{\text{def}}{=} (Y_0, \ldots, Y_n)$, $\boldsymbol{X}_{0:n} \overset{\text{def}}{=} (X_0, \ldots, X_n)$ to denote segments in the two chains' histories. In the discrete case, the joint density of $(\boldsymbol{X}_{0:n}, \boldsymbol{Y}_{0:n})$ at $(\boldsymbol{x}_{0:n}, \boldsymbol{y}_{0:n})$ is

$$\nu_{i_0} q_{x_0 x_1} \cdots q_{x_{n-1} x_n} g_{x_0}(y_0) \cdots g_{x_n}(y_n), \tag{2.3}$$

where $g_x(y)$ is the density of $G_x(dy)$ w.r.t. some reference measure $\mu(dy)$ and ν is the initial distribution of X_0.

Hidden Markov models are one of the basic vehicles for modeling dependent observations, and one gets quite far by just using a finite set E of Markov states for the chain X. In fact, if the Y_i take values in, say, $\mathbb{R}^d$, then *any* distribution H on $\mathbb{R}^{(n+1)d}$ is the weak limit of a sequence of distributions H_m corresponding to hidden Markov models (typically with the size of $E = E_m$ going to ∞ with m).

MCMC has various purposes for hidden Markov models. One is the classical statistical problem of *filtering* (see I.5.5 and A5) or *smoothing* where one asks for the distribution Π^* of $\boldsymbol{X}_{0:n}$ (or suitable marginals) given the observed values $\boldsymbol{y}_{0:n}$ of $\boldsymbol{Y}_{0:n}$ (for example, one could ask for the path $\boldsymbol{x}_{0:n}$ having the maximal conditional probability given $\boldsymbol{y}_{0:n}$). Then C is obtained by integrating out the y_k in (2.3), which is typically computationally prohibitive when n is large (but note that feasible algorithms exist for a finite E or Gaussian processes, see A5). Another purpose of MCMC is parameter estimation, where the $g_x(y) = g_x(y, \theta)$ depend on some unknown parameter (vector) θ. Then, in a Bayesian setting, Π^* is the posterior of θ for which we need to integrate out the x_k in (2.3). In a traditional parametric setting, parameter estimation can be implemented via the EM algorithm (see A4). One then needs the $\mathbb{P}_\theta\big(X_k = i, X_{k+1} = j \,\big|\, \boldsymbol{y}_{0:n}\big)$, so that once again we are facing a filtering problem.

Examples of hidden Markov models in which filtering and smoothing are relevant follow.

Example 2.4 Consider a change-point problem in which E has only two states $1, 2$ such that a transition $2 \to 1$ cannot occur. This means that the Y_k have distribution $g_1(\cdot)$ up to a certain point (the time of change of state from 1 to 2) and $g_2(\cdot)$ thereafter. A basic task is to ascertain whether such a change occurs and when, which basically amounts to giving statements on the conditional distribution Π^* of $\boldsymbol{X}_{0:n}$ given $\boldsymbol{Y}_{0:n} = \boldsymbol{y}_{0:n}$. In particular, one is interested in the (conditional) probabilities of the sequences $\boldsymbol{x}_{0:n}^{(0)}$ of all 1's and $\boldsymbol{x}_{0:n}^{(k)}$, $k = 1, \dots, n-1$, with 1's at the first k places and 2's at the rest.

A more complicated and realistic change-point problem (geological layers of different types) is used as a recurrent example in [65].

□

Example 2.5 A target moves according to known transition probabilities but the observation is blurred by noise. A simple model assumes that the positions $X_0, X_1, \dots$ constitute a Gaussian random walk in $\mathbb{R}^2$ with mean zero and a known covariance matrix, and that $Y_k = X_k + \varepsilon_k$, where the ε_k are i.i.d. bivariate Gaussian with mean zero and a known covariance matrix. For an example of a more sophisticated model, assume that the acceleration $X_k^{(1)}$ of the target is white noise, whereas velocity $X_k^{(2)}$, position $X_k^{(3)}$, and observed position Y_k are blurred by noise, i.e.,

$$X_k^{(2)} = X_{k-1}^{(2)} + X_k^{(1)} + \varepsilon_k^{(1)}, \quad X_k^{(3)} = X_{k-1}^{(3)} + X_k^{(2)} + \varepsilon_k^{(2)}, \quad Y_k = X_k^{(3)} + V_k,$$

where $\{X_k^{(1)}\}$, $\{\varepsilon_k^{(1)}\}$, $\{\varepsilon_k^{(2)}\}$, $\{V_k\}$ are independent sequences of i.i.d. bivariate Gaussians with known parameters. One is interested in $\mathbb{P}\big(X_n = x \,\big|\, \boldsymbol{Y}_{0:n}\big)$ or the prediction probability $\mathbb{P}\big(X_{n+1} = x \,\big|\, \boldsymbol{Y}_{0:n}\big)$.

□

Example 2.6 In financial modeling of the log returns $Y_0, \dots, Y_n$, the simplest model (Black–Scholes) is just that the Y_k are i.i.d. $\mathcal{N}(\mu, \sigma^2)$. However, often one observes phenomena such as periods with larger variation than typical, i.e., stochastic volatility. In a hidden Markov model, one takes instead $g_x(\cdot)$ as the $\mathcal{N}(\mu_x, \sigma_x^2)$ density. A frequently used model is

$$X_k = \rho X_{k-1} + V_k, \quad Y_k = \beta e^{X_k/2} V_k', \tag{2.4}$$

with the V_k, V_k' independent $\mathcal{N}(0,1)$ r.v.'s. That is, the underlying Markov chain is an autoregressive process with Gaussian innovations. However, as noted above, an appealing alternative is to use just a finite-state Markov chain. □

Example 2.7 Let $C_0, \ldots, C_n \in \{-1,1\}$ be a sequence of bits transmitted along a noisy channel. A common model assumes that the observed sequence is $Y_0, \ldots, Y_n$, where $Y_k = W_k C_k + V_k$, the V_k are i.i.d. $\mathcal{N}(0,\sigma^2)$, and the W_k are so-called *fading coefficients*, describing time-varying properties of the channel and often modeled as an autoregressive process $W_{k+1} = \rho W_k + V_k'$ with the V_k' i.i.d. $\mathcal{N}(0,\omega^2)$. Thus the hidden Markov chain may be taken as $X_k = (C_k, W_k)$, and one is facing a filtering problem, to reconstruct the C_k from the Y_k (the W_k constitute nuisance variables, i.e. they are of no direct interest in themselves). □

A basic reference for hidden Markov models is Cappé, Moulines, & Rydén [65].

Exercises

2.1 (TP) Assume that $Y_1, \ldots, Y_n \in \{0,1\}$ are i.i.d., having probability θ for 1 and $1 - \theta$ for 0, where θ itself follows the Beta density $\theta^{\alpha-1}(1-\theta)^{\beta-1}/B(\alpha,\beta)$. Show that the posterior distribution is again Beta with parameters

$$\alpha^* = n\overline{y} + \alpha, \quad \beta^* = n - n\overline{y} + \beta.$$

Find the posterior mean and variance, and show that the mean goes to $\overline{y}$ as $n \to \infty$ with $\overline{y}$ fixed and that the variance goes to 0.

2.2 (TP) In car insurance, the numbers $Y_1, Y_2, \ldots$ of accidents by a specific driver D in consecutive years may well be assumed to be i.i.d. Poisson, with parameter say θ. Further, this θ varies from driver to driver, so a priori, driver D's θ can reasonably be viewed as the outcome of an r.v. Θ, obtained by picking an individual at random from the population of drivers.

It seems reasonable to let D's premium p per year be proportional to D's θ, $p = c\theta$. But of course θ is not observable, so we need an estimate. To this end, assume that Θ has distribution F with density $f(\theta)$. Assume that the insurance company wants to set D's premium for year $t + 1$ (thus $y_1, \ldots, y_t$ have been observed), and that at the beginning of the year it has past experience with m drivers (some of whom may still be with the company and some of whom may have died, stopped driving, switched company, etc.), such that driver j has a total of n_j accidents during a period of t_j years. Then $\widehat{\theta}_j = n_j/t_j$ is a reasonable estimator of driver j's Poisson parameter θ_j.

One now proceeds to model $\theta_1, \ldots, \theta_m$ as i.i.d. and Gamma distributed. Thus a reasonable prior for D's theta is a Gamma(α, λ), where the particular values of α, λ are those obtained as ML estimates based on $\widehat{\theta}_1, \ldots, \widehat{\theta}_m$ associated with the company's entire portfolio of insured drivers.

Show that the posterior is Gamma with parameters $\lambda^* = \lambda + \alpha$, $\alpha^* = \alpha + t\overline{y}$. Thus to set the premium the company could use

$$p^* \stackrel{\text{def}}{=} c\frac{\alpha^*}{\lambda^*} = wc\overline{y} + (1-w)c\frac{\alpha}{\lambda} \quad \text{where} \quad w \stackrel{\text{def}}{=} \frac{t}{\lambda+t}$$

(note that α^*/λ^* is the mean of the posterior distribution). This formula expresses how prior experience of typical drivers, as reflected in α^*, λ^*, is weighted with D's own driving record as summarized in $\bar{y}$; as years pass, more information on D's performance becomes available and $\bar{y}$ becomes more and more important compared to the Gamma prior.

3   The Metropolis–Hastings Algorithm

We now consider the question raised in Section 1: given a distribution Π^* with density or probability mass function proportional to $(\pi(x))_{x \in E}$, how do we simulate a Markov chain $\{\xi_n\}$ having stationary distribution Π^*? For example, in the particularly simple and not unrealistic example of a finite state space E, how do we exhibit a transition matrix $\boldsymbol{P} = (p(x,y))_{x,y \in E}$ such that $\boldsymbol{\pi P} = \boldsymbol{\pi}$, where $\boldsymbol{\pi}$ is the row-vector representation of π?

The suggestion given by Metropolis et al. [254] and later extended by Hastings [177] is remarkably simple. It involves the concept of a *proposal distribution*, just as for acceptance–rejection (A-R), but now sampled from a different p.m.f. $q(x,y)$ for each starting point x. Given that a proposal y has been generated from $q(x, \cdot)$, it is accepted w.p.

$$\alpha(x,y) \overset{\text{def}}{=} \min\bigl(1, r(x,y)\bigr) \quad \text{where} \quad r(x,y) \overset{\text{def}}{=} \frac{\pi(y)q(y,x)}{\pi(x)q(x,y)}, \qquad (3.1)$$

that is, by sampling a uniform r.v. U and accepting y if $U < \alpha(x,y)$; otherwise, one returns x. Thus, the algorithm does not (as for ordinary A-R) go on until eventually a point is accepted, but it chooses to remain at the same point x if the proposal is rejected. The crucial feature of (3.1) is that we have a ratio between two π-values, so that the unknown normalizing constant is actually not needed. We refer to $r(x,y)$ in (3.1) as the *Metropolis–Hastings* (MH) *ratio*, and to the algorithm as a whole as the *Metropolis–Hastings algorithm*.

We denote by $p(x,y)$ the probability (density) that y is returned and get

$$p(x,y) = q(x,y)\alpha(x,y), \quad y \neq x. \qquad (3.2)$$

For $y = x$ we have to add a term corresponding to rejection and get

$$\begin{aligned} p(x,x) &= q(x,x)\alpha(x,x) + \sum_{y \in E} q(x,y)\bigl(1 - \alpha(x,y)\bigr) \\ &= q(x,x)\alpha(x,x) + 1 - \sum_{y \in E} q(x,y)\alpha(x,y), \qquad (3.3) \end{aligned}$$

with the sum replaced by a μ-integral in the continuous case (but we won't need this expression).

Theorem 3.1 *The distribution Π^* is stationary for the MH chain $\xi_0, \xi_1, \ldots$ constructed above.*

Proof. The key idea is that the construction allows the stationarity equations (1.3) and (1.4) (the global balance equations) to be sharpened to the detailed (or local) balance equations, which state that

$$\pi(y)p(y,x) = \pi(x)p(x,y) \tag{3.4}$$

in the discrete case, and

$$\int_A p(x,B)\pi(x)\,\mu(\mathrm{d}x) = \int_B p(y,A)\pi(y)\,\mu(\mathrm{d}y) \tag{3.5}$$

in the continuous case. To see that this is sufficient, consider first the discrete case. Then summing (3.4) over x, the l.h.s. becomes $\pi(y)$ because $\sum_x p(y,x) = 1$, and the r.h.s. is the same as the r.h.s. of (1.3).

To verify (3.4), it suffices to take $x \neq y$, and we may then assume $\pi(y)q(y,x) \geq \pi(x)q(x,y)$ (the case $\leq$ is symmetric). Then $\alpha(x,y) = 1$, so that $p(x,y) = q(x,y)$ and we get

$$\begin{aligned}
\pi(y)p(y,x) &= \pi(y)q(y,x)\alpha(y,x) = \pi(y)q(y,x)\frac{\pi(x)q(x,y)}{\pi(y)q(y,x)} \\
&= \pi(x)q(x,y) = \pi(x)p(x,y).
\end{aligned}$$

The continuous case is similar and left as Exercise 3.1.   □

Remark 3.2 Equations (3.4), (3.5) are equivalent to $\{\xi_n\}$ being time-reversible, that is, to $(\xi_0, \dots, \xi_m)$ having the same distribution for all m as $(\xi_m, \dots, \xi_0)$ in the stationary case, where ξ_0 has distribution Π^*. For this fact and further discussion, see [16, Section II.5]. Reversibility is, however, not an intrinsically necessary property of an MCMC algorithm. Rather, the point is that the balance equations take a much simpler form for reversible chains, so that the requirements for a transition mechanism to make Π^* stationary become much simpler in the reversible case, thereby facilitating the design of suitable algorithms.   □

3a Convergence

We have immediately the following corollary:

Corollary 3.3 *If in the discrete case the MH chain is irreducible, then the empirical distribution $\widehat{\Pi}_n^*$ in (1.1) converges to Π^* and the average $\widehat{f}_n$ of the $f(\xi_k)$ in (1.2) converges to $\Pi^*[f]$.*

Remark 3.4 A transition $x \to y$ in one step is possible if and only if $q(x,y) > 0$, $\alpha(x,y) > 0$, which is the same as $q(x,y) > 0$, $q(y,x) > 0$. Therefore irreducibility is not automatic (consider for example a partitioning of E into two disjoint subsets E_1, E_2 and a proposal $q(\cdot,\cdot)$ such that $q(x,y)$ can be positive only if x, y are in the same E_i). Thus, for each choice of the proposal one has to check irreducibility. This is easily done for a discrete E by connecting points x, y with $q(x,y) > 0$ and $q(y,x) > 0$; then the resulting graph must be connected in order to guarantee irreducibility. A safe choice is to take $q(x, y) > 0$ for all x, y.

The MH chain is aperiodic under weak conditions. For example, it suffices in the discrete case that the probability $p(x, x)$ in (3.3) of staying in the same state be nonzero for at least one x, and that $p(x, x)$ not be Π^*-a.s. equal to 0 in the general case.   □

Next consider a general state space and define

$$Q(x, B) \stackrel{\text{def}}{=} \mathbb{P}_x(\xi_n \in B \text{ i.o.}), \quad L(x, B) \stackrel{\text{def}}{=} \mathbb{P}_x(\tau_B < \infty),$$

where $\tau_B \stackrel{\text{def}}{=} \inf\{n > 0 : \xi_n \in B\}$. For a measure ψ on E, one calls $\{\xi_n\}$ ψ-irreducible if $L(x, B) > 0$ for all $x \in E$ and all B with $\psi(B) > 0$, and ψ-recurrent if $Q(x, B) = 1$ for all $x \in E$ and all B with $\psi(B) > 0$. If $\{\xi_n\}$ is ψ-recurrent for some $\psi \neq 0$, the chain is called a *Harris chain* (the discussion in IV.6d focuses on somewhat different features but there is a close relation; see Nummelin [282], Meyn & Tweedie [253], and [16, Section VII.3]).

Theorem 3.5 *If, in the case of a general state space E, the MH chain is Π^*-irreducible, then it is Harris recurrent as well. In particular, the empirical distribution $\widehat{\Pi}_n^*$ in (1.1) converges to Π^* in t.v., and the average $\widehat{f}_n$ of the $f(\xi_k)$ in (1.2) converges to $\Pi^*[f]$.*

The result is due to Tierney [355]. There are quite a few variants of the result around. In particular, Athreya, Doss, & Sethuraman [31] weaken the assumption of Π^*-irreducibility.

The proof we give here is essentially a variant of that of [355]. It exploits two basic specific properties of the MH chain. The first is that existence of a stationary distribution is guaranteed in advance by Theorem 3.1. The second is that absolute continuity properties allow for short proofs that statements holding for Π^*-a.a. x in fact often hold for all $x \in E$. To make this more precise, let $\omega \stackrel{\text{def}}{=} \inf\{n : \xi_n \neq \xi_0\}$ be the first exit time of the initial state. Then $\mathbb{P}_x(\omega < \infty) = 1$ by irreducibility, and further:

Lemma 3.6 *Let $\xi_n' \stackrel{\text{def}}{=} \xi_{\omega+n}$. Then the distribution of ξ_n' is absolutely continuous w.r.t. Π^* for all $n \geq 0$ and any distribution of ξ_0.*

Proof. We may assume $\xi_0 = x$ and $n = 0$, so we have to show

$$\Pi^*(N) = 0 \quad \Rightarrow \quad \mathbb{P}_x(\xi_\omega \in N) = 0. \tag{3.6}$$

But a proposal with $\pi(y) = 0$ is always rejected, cf. (3.1). Since the proposal is absolutely continuous w.r.t. μ, so is the distribution of ξ_ω, and it follows that we can write its μ-density $g(y)$ as $\pi(y)h(y)$ where $h(y) < \infty$. Since the r.h.s. of (3.6) is $\int_N g(y)\,\mu(\mathrm{d}y)$, the result follows   □

Lemma 3.7 $Q(x, B) > 0$ *for all $x \in E$ and all B with $\Pi^*(B) > 0$.*

Proof. By the ergodic theorem.

$$\frac{1}{N+1}\sum_{n=0}^{N} \mathbb{1}\{\xi_n \in B\} \overset{\text{a.s.}}{\to} \mathbb{P}_{\Pi^*}\big(\xi_0 \in B \,\big|\, \mathscr{I}\big),$$

where $\mathscr{I}$ is the shift-invariant σ-field. The limit on the r.h.s. has expectation $\Pi^*(B) > 0$ and is therefore not 0 Π^*-a.s. This implies that $\Pi^*(B_\infty) > 0$, where $B_\infty \overset{\text{def}}{=} \{x : Q(x, B) > 0\}$. That $Q(x, B) > 0$ for any given x therefore follows since $L(x, B_\infty) > 0$ by irreducibility.   □

Lemma 3.8 *Assume $Q(x, B) < 1$ for all $x \in B$. Then $\Pi^*(B) = 0$.*

Proof. Assume $\Pi^*(B) > 0$ and let $C' \overset{\text{def}}{=} \{x \in B : L(x, B) < 1\}$. The absolute continuity in Lemma 3.6 excludes $\Pi^*(C') = 0$ since otherwise

$$\mathbb{P}_x\big(\xi_n \in B \text{ for some } n \geq n_0\big) = 1$$

by Lemma 3.6 and hence $Q(x, B) = 1$. It follows that we can choose $\delta > 0$ such that $\Pi^*(C) > 0$ where $C \overset{\text{def}}{=} \{x \in B : L(x, B) \leq 1 - \delta\}$. Since $L(x, C) \leq L(x, B) \leq 1 - \delta$, a geometric trial argument gives $Q(x, C) = 0$ for all $x \in E$, which contradicts Lemma 3.7.   □

Proof of Theorem 3.5. If E_1, E_2 are disjoint closed nonempty subsets of E with $E_1 \cup E_2 = E$, one must have positive Π^*-measure, say E_1, so that by Π^*-irreducibility, $L(x, E_1) > 0$ for $x \in E_2$, which contradicts E_2 being closed. Hence E is indecomposable, and Theorem 3.6(i) and Proposition 3.9(iv) of Nummelin [282] then give two possibilities. The first is that E is the countable union of transient sets E_k, i.e., sets with $Q(x, E_k) < 1$ for $x \in E_k$. But we must have $\Pi^*(E_k) > 0$ for some k which is impossible by Lemma 3.8. Therefore, the second possibility of [282] applies, that $Q(x, B) = 1$ for Π^*-a.a. x whenever $\Pi^*(B) > 0$. Arguments similar to those used in the proof of Lemma 3.7 then give $Q(x, B) = 1$ for all x and Harris recurrence.   □

Unfortunately, the results from [282] that were used are nonelementary, and it would be desirable to have a proof using only first principles, but we are not aware of such a one (Robert & Casella [301] outline a more elementary argument, but use an 0–1 law in [301, p. 240]) without giving a reference (that we don't know either).

3b   Convergence Rates

When using the MH algorithm in practice, it is of course important to give error estimates. For this, variance estimation methods such as batch means, stationary process methods, and the regenerative method as in Chapter IV could be used. The discussion of this issue is, however, just the same for MCMC as for general steady-state simulation and will not be repeated here.

Efficiency of an MCMC scheme entails roughly that the empirical distribution $\widehat{\Pi}^*_{n(T)}$ should approach Π^* quickly as $T \to \infty$, where $n(T)$ is the number of Markov chain steps that can be performed within T units of processing time.[3]

For the purpose of choosing the proposal kernel, $q(\cdot, \cdot)$, we therefore face two considerations:

- How fast will each updating of the Markov chain from time k to $k+1$ be, i.e., how fast is it to sample from the proposal distribution?

- How fast will the Markov chain approach its stationary distribution Π^* (we will refer to this as a question of *mixing rate* or *decorrelation rate*)?

These two concerns are often a trade-off.

One expects ξ_k to be closer in distribution to Π^* the larger k is; in fact, it is easy to see that the total variation distance defined by

$$d_k(y) \stackrel{\text{def}}{=} \sum_{x \in E} \left| \mathbb{P}_y(\xi_k = x) - \pi^*(x) \right| \tag{3.7}$$

in the discrete case and by

$$d_k(y) \stackrel{\text{def}}{=} \sup_A \left| \mathbb{P}_y(\xi_k \in A) - \Pi^*(A) \right| \tag{3.8}$$

in the continuous case (the sup extends over all measurable subsets of E) is nonincreasing in k regardless of the initial state y. Therefore the bias is particularly marked in the first part of the series $\xi_0, \ldots, \xi_n$, and a common technique is therefore to disregard the first part of the series, say $\xi_0, \ldots, \xi_m$ $(0, \ldots, m$ is the *burn-in* or *warm-up* period), and redefine the estimators $\widehat{\pi}^*_n$, $\widehat{f}_n$ according to

$$\widehat{\pi}^*_n(x) = \frac{1}{n-m} \sum_{k=m+1}^{n} \mathbb{1}\{\xi_k = x\}, \quad x \in E, \tag{3.9}$$

$$\widehat{f}_n = \frac{1}{n-m} \sum_{k=m+1}^{n} f(\xi_k). \tag{3.10}$$

A common choice is to let m be of order 10% of the total run length n.

An additional safety measure is to choose some pilot functions $f^{(1)}, \ldots, f^{(r)}$ and take m so large that the $\widehat{f}^{(j)}_k$ seem to have stabilized before $k = m$. We will present examples of this technique in XIV.4a.

Nevertheless, there exists no fail-safe general technique for assessing from the simulation output when actually the chain has (approximately) reached

---

[3]Making this statement, we deliberately neglect the human factor: one should really also take into account how long it takes to design and program the scheme (computer time is typically cheaper than human time!).

stationarity. An alternative is to use model structure to come up with an explicit k_0 (possibly dependent on y) such that $d_k(y) \leq \varepsilon$ when $k \geq k_0$. The cases in which this can be done are rather few and of special nature (but see XIV.3 for an example in which such estimates are crucial and can in fact be carried through). However, one important setting in which this can be done is that in which the chain exhibits *uniform ergodicity*.

Uniform ergodicity is defined by $d_k(y)$ going to zero uniformly in y. This is equivalent to the minorization condition

$$P^m(x, A) \geq \varepsilon \nu(A) \tag{3.11}$$

for some $m = 1, 2, \ldots$, some probability measure ν, all measurable A, and all $x \in E$, see Meyn & Tweedie [255] for further discussion. It is then straightforward to establish the bound

$$d_k(y) \leq (1 - \varepsilon)^{\lfloor k/m \rfloor} ; \tag{3.12}$$

see, e.g., [19]. Uniform ergodicity typically holds when the state space E is compact, but is less common when E is unbounded.

A closely related concept is that of exponential ergodicity (also known as *geometric ergodicity*). The Markov chain $\xi_0, \xi_1, \ldots$ is said to be *exponentially ergodic* if there exists $\rho < 1$ and a finite-valued function $h(\cdot)$ such that

$$d_k(y) \leq h(y)\rho^k \tag{3.13}$$

for $y \in E$. The standard criterion in establishing exponential ergodicity involves finding a so-called Lyapunov function $V(\cdot)$ (also known as a "test function") such that $\infty > V(\cdot) \geq 1$ and

$$\mathbb{E}_x V(\xi_1) \leq \beta V(x) + b \mathbb{1}_{x \in K} \tag{3.14}$$

for some $\beta < 1$, some $b < \infty$, and some compact set K; see [16, Section I.5] and Meyn & Tweedie [255]) (in shorthand notation, (3.14) can be written as $PV \leq \rho V + b \mathbb{1}_K$). While (3.14) is a key step in verifying (3.13), there is no simple expression for the key convergence-rate constant ρ (for example, it is not generally the case that $\rho \leq \beta$). This perhaps should come as no surprise, given that when $\xi_0, \xi_1, \ldots$ is a finite-state ergodic Markov chain, the chain is automatically exponentially ergodic and the best possible value of ρ is determined as the eigenvalue of the transition matrix $\boldsymbol{P}$ having second-largest modulus (the principal eigenvalue is 1). In general, characterizing such a second eigenvalue is difficult. However, when a finite matrix is symmetric (or, more generally, self-adjoint), there is a variational characterization of each of the eigenvalues (due to Rayleigh) that plays a key role in the mathematical and numerical analysis of the eigenvalues. Because the Lyapunov bound (3.14) applies to both reversible and irreversible Markov chains, it tends to generate bounds on ρ that are not as sharp as the bounds that can be computed for reversible chains.

For reversible Markov chains, a variety of useful bounds can be derived that are reasonably sharp in some problem settings; see, for example, Brémaud [57]. For such problems, these bounds can play a useful role from a simulation viewpoint, because some MCMC implementations lead to chains having surprisingly slow convergence rates to equilibrium, due to multi-modal behavior of the stationary distribution Π^* (and the near reducibility that this frequently implies). Nevertheless, while reversible dynamics are usually used in the MCMC setting, it is our view (which is subjective and not shared by all authors!) that in most applications, analytical bounds on the second eigenvalue are either too difficult to compute or too loose to be useful.

Despite this, we will give some related discussion when we study special samplers in the next section.

Exercises

3.1 (TP) Give the details of the proof of Theorem 3.1 in the continuous case.

4 Special Samplers

We turn to special cases of the MH algorithm, that is, specific forms of the proposal kernel $q(\cdot, \cdot)$. A class of proposals going under the name of the *Gibbs samplers* are of particular importance and treated separately in the next section.

In some situations, a proposal naturally suggests itself from properties of the model. For example:

Example 4.1 Assume that the target distribution is the Boltzmann distribution on a finite set E. That is, $\pi(x) = e^{-H(x)/T}$; cf. I.5.4. In many examples, there is given a graph structure on E, that is, a set $\mathcal{E}$ of edges (if no such structure is given a priori, it may be defined by the user). A natural proposal for a transition $x \to y$ is then to choose y as a random neighbor of x, that is, w.p. $1/n(x)$ if x has $n(x)$ neighbors. The transition probability for the corresponding MH chain becomes

$$p(x,y) = \frac{1}{n(x)} \min\left(1, \frac{\pi(y)/n(y)}{\pi(x)/n(x)}\right) = \frac{1}{n(x)} \min\left(1, \frac{n(x)}{n(y)} e^{H(x)/T - H(y)/T}\right)$$

if $xy \in \mathcal{E}$, and $p(x,y) = 0$ otherwise (unless $x = y$). $\qquad\square$

Example 4.2 For the Strauss process in I.5.2, consider the proposal that adds a uniform point w.p. $1/2$ and deletes a random point among the already present points w.p. $1/2$. $\qquad\square$

Some more general ideas are presented in the following.

4a   The Independence Sampler

Here the proposal is independent of x, $q(x, y) = q(y)$. The MH ratio becomes $\omega(y)/\omega(x)$, where $\omega(x) \overset{\text{def}}{=} \pi(x)/q(x)$, and thus

$$\alpha(x, y) = \min\left(1, \frac{\omega(y)}{\omega(x)}\right).$$

The proposal must be chosen such that $q(y) > 0$ whenever $\pi(y) > 0$.

The independence sampler has a particularly intuitively appealing form in the Bayes estimation case if one takes the proposal $q(\theta)$ equal to the prior $\pi^{(0)}(\theta)$ (replacing x by θ for notational convenience): since $\pi(\theta)$ is proportional to $L(\boldsymbol{y} \mid \theta)\pi^{(0)}(\theta)$, the acceptance probability becomes $\alpha(\theta_1, \theta_2) = \min\left(1, L(\boldsymbol{y} \mid \theta_2)/L(\boldsymbol{y} \mid \theta_1)\right)$. That is, a θ_2 with a larger likelihood is always accepted and otherwise the acceptance probability is just the likelihood ratio.

The convergence properties of the independence sampler are given by the following result (recall that $C = \int \pi(x)\,\mu(\mathrm{d}x)$):

Proposition 4.3 *The Metropolis–Hastings chain run by independent sampling is exponentially ergodic if and only if $A \overset{\text{def}}{=} \sup_x \omega(x) < \infty$. More precisely, it holds for any y that*

$$d_k(y) \leq \left(1 - C/A\right)^k. \tag{4.1}$$

Proof. We use discrete notation. Let first $A = \infty$ and write

$$E_\alpha(x) \overset{\text{def}}{=} \{y : r(x, y) \leq 1\} = \{y : \omega(y) \leq \omega(x)\}.$$

Given ε, choose $x \in E$ with $\omega(x) \geq \varepsilon^{-1}$ (this is possible because $A = \infty$). For $y \neq x$ and $y \in E_\alpha(x)$, we then get

$$p(x, y) = \alpha(x, y)q(y) = \frac{\omega(y)}{\omega(x)}q(y) = \frac{\pi(y)}{\omega(x)} \leq \varepsilon\pi(y),$$

whereas for $y \notin E_\alpha(x)$ we have $\omega(y) \geq \omega(x) \geq \varepsilon^{-1}$, so that $p(x, y) = q(y) \leq \varepsilon\pi(y)$. Summing over $y \neq x$ yields $\mathbb{P}_x(\xi_1 \neq x) \leq \varepsilon C$. It follows that for $y \neq x$ we have $\mathbb{P}(\xi_n = y) \leq (\varepsilon C)^n$. The existence of such an x for any $\varepsilon > 0$ excludes exponential ergodicity.

If $A < \infty$,

$$p(x, y) = \begin{cases} \dfrac{\omega(y)}{\omega(x)}q(y) \geq \dfrac{1}{A}\omega(y)q(y) = \dfrac{C}{A}\pi^*(y) & y \in E_\alpha(x),\ y \neq x, \\[2mm] q(y) \geq \dfrac{C}{A}\pi^*(y) & y \notin E_\alpha(x), \end{cases}$$

whereas for $y = x$, $C\pi^*(y)/A$ is a lower bound. Thus the minorization condition (3.11) holds with $m = 1$, $\varepsilon = C/A$, and (3.12) completes the proof. $\qquad\square$

It follows from the proof (or from a direct argument, summing $\pi(x) \leq Aq(x)$ over x), that $C \leq A$. The implication of this and Proposition 4.3 is that q should be chosen with A as close to C as possible. This is achieved by choosing $q(\cdot)$ close to $\Pi^*(\cdot)$, and experience shows indeed that the independence sampler works well in such cases but may otherwise behave poorly. A particularly dangerous situation arises when q is substantially lighter-tailed than Π^*.

Note that the condition $A < \infty$ in fact allows Π^* to be generated from q by acceptance–rejection, where an r.v. Y from q is accepted w.p. $\pi(Y)/\big(Aq(Y)\big)$. It is not a priori obvious whether this is more efficient than MCMC, and a difficulty is that it may be difficult to evaluate A or even get good bounds.

Example 4.4 We will consider an application of the independence sampler to population genetics.

The *Kingman coalescent* is a model for a random tree with n final branches. The description is backward in time: merging (coalescence) of two randomly chosen of the n branches occurs at an exponential$\big(n(n-1)/2\big)$ time T_n, the second coalescence occurs at an exponential$\big((n-1)(n-2)/2\big)$ time T_{n-1} after the first, and so on. Thus, all n branches have coalesced to the root at time $\tau \overset{\text{def}}{=} T_n + T_{n-1} + \cdots + T_2$ (the time unit may, for example, be the generation length). See Figure 4.1(a).

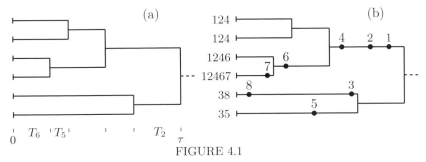

FIGURE 4.1

The Kingman coalescent arises from the Wright–Fisher model I.5.19 in population genetics by letting the population size N go to infinity and scaling (backward) time appropriately; see, e.g., Haccou et al. [172, Section 7.1]. The interpretation of τ is as the time of the most recent common ancestor, a view that has recently been pursued to provide statistical estimates of elapsed times since separation of species. We give here one example, following [172, Section 7.2] and Tavaré et al. [352], and illustrated by Figure 4.1(b). This is an infinite-sites model in which mutation at a given locus is assumed to occur at most once. The dots in Figure 4.1(b) represent mutation times, and the labels $1, 2, \ldots$ are the loci at which mutation occurred. Thus, in the current population at time $t = 0$ an individual can be characterized by a string like 1246 of the loci at which the individual carries the mutant gene. Mutations are assumed to occur at rate $\theta/2$ along any branch of the tree such that the number K of loci where a mutation has occurred is Poisson$(\theta L/2)$, where $L \overset{\text{def}}{=} nT_n + (n-1)T_{n-1} + \cdots + 2T_2$ is the total length of the tree. In Figure 4.1(b), $n = 6$, $K = 8$, and the mutations are marked by bullets and numbered in forward time.

Given a prior $\pi^{(0)}(\theta)$ on θ and the observed $K = k$, we want posterior properties of τ and θ. To this end, note that τ, L are functions of $\boldsymbol{T} \overset{\text{def}}{=} (T_n, T_{n-1}, \ldots, T_2)$, let $f(\boldsymbol{t})$ be the density of $\boldsymbol{T}$ (a product of exponential densities with rates $(n(n-1)/2), ((n-1)(n-2)/2), \ldots, 1)$, and write $q_k(\lambda) \overset{\text{def}}{=} e^{-\lambda}\lambda^k/k!$. The posterior of $(\boldsymbol{T}, \theta)$ is then proportional to

$$\pi(\boldsymbol{t}, \theta) \overset{\text{def}}{=} q_k\big(\theta L(\boldsymbol{t})/2\big) f(\boldsymbol{t})\pi^{(0)}(\theta),$$

and the suggestion of the above references is to perform MCMC for $(\boldsymbol{T}, \theta)$, using the independence sampler with proposal $f(\boldsymbol{t})\pi^{(0)}(\theta)$. $\qquad\square$

4b   Random-Walk MH and Symmetric Proposals

Here E should be a subset of $\mathbb{R}^d$ and μ Lebesgue measure, or E should be $\mathbb{Z}^d$ and μ counting measure, and the proposal is that from x we should go to y with a probability depending only on $y - x$. That is, for some distribution F with point probabilities or density $f(z)$, we let $q(x, y) = f(y - x)$; this means that from x we take a step with distribution F.

In some of the literature, the term "random-walk MH" involves a strong symmetry of F, namely that $f(z)$ depends only on z via $|z|$, the distance from the origin. In the discrete case, a main example would be a Bernoulli-type random walk, where one of the d directions is chosen w.p. $1/d$ for each and we move up or down in that direction w.p. $1/2$ for each. In the continuous case, a main example would be F being normal with i.i.d. components. Symmetry implies $q(x, y) = f(y - x) = f(x - y) = q(y, x)$ and hence $r(x, y) = \pi(y)/\pi(x)$, so that

$$\alpha(x, y) = \min\left(1, \frac{\pi(y)}{\pi(x)}\right).$$

In particular, proposals y with $\pi(y) \geq \pi(x)$ are always accepted.

Now consider the convergence properties of random walk MH, taking $d = 1$ for simplicity. Since the expected number of steps to go from x to some bounded interval $[-a, a]$ goes to ∞ as $x \to \pm\infty$, there is no hope for uniform ergodicity if E is unbounded. However, we will show that in the symmetric case geometric ergodicity holds under some regularity conditions.

Proposition 4.5 (MENGERSEN & TWEEDIE [253]) *Assume $E = \mathbb{R}$ with μ Lebesgue measure, or $E = \mathbb{Z}$ with μ counting measure, and that q is symmetric, $q(x, y) = f(y - x)$ with $f(z) = f(-z)$. Assume further that for some $\alpha > 0$ and some x_0 we have*

$$\frac{\pi(y)}{\pi(x)} \leq e^{-\alpha(|y|-|x|)}, \quad |y| \geq |x| \geq x_0, \tag{4.2}$$

and that

$$\inf_{|x| \le x_1} \pi(x) > 0 \quad \text{for all } x_1. \tag{4.3}$$

Then the random-walk MH algorithm is geometrically ergodic.

Proof. By (4.2), (4.3), and symmetry, there exists x_2 such that

$$A(x) \stackrel{\text{def}}{=} \{y : \pi(x) \le \pi(y)\} = \{y : |y| \le |x|\} \tag{4.4}$$

for $|x| \ge x_2$ (note that $A(x)$ is the set in which the proposal is accepted w.p. 1).

We apply (3.14) with test function $V(x) = e^{\beta|x|}$, where $0 < \beta < \alpha$ and $K \stackrel{\text{def}}{=} \{x : |x| \le x^*\}$, where $x^* \ge x_0 \vee x_2$ will be specified later. We verify first (3.14) for $x \ge 0$. Let first $x > x^*$. Considering proposals $y \in A(x)$ and $y \notin A(x)$ separately, applying (4.4) and using continuous-state notation, we get

$$
\begin{aligned}
s(x) & \stackrel{\text{def}}{=} \frac{1}{V(x)} PV(x) \\
& = \int_{|y| \le x} q(x, y) e^{\beta(|y|-x)} \, dy + \int_{|y| > x} q(x, y) e^{\beta(|y|-x)} \frac{\pi(y)}{\pi(x)} \, dy \\
& \quad + \int_{|y| > x} q(x, y) \left(1 - \frac{\pi(y)}{\pi(x)} \right) dy \\
& = 1 + \int_{|y| \le x} q(x, y) \left[e^{\beta(|y|-x)} - 1 \right] dy \\
& \quad + \int_{|y| > x} q(x, y) \left[e^{\beta(|y|-x)} - 1 \right] \frac{\pi(y)}{\pi(x)} \, dy \\
& \stackrel{\text{def}}{=} 1 + I_1 + I_2.
\end{aligned}
$$

In I_1, we can bound the contribution from $-x < y < 0$ by 0. In I_2, it follows by invoking (4.2) and using $\beta < \alpha$ that the contribution from $-\infty < y < -x$ can be bounded by $\int_{-\infty}^{-x} q(x, y) \, dy$. Breaking the contribution from $x < y < \infty$ into two parts coming from $x < y < 2x$, respectively $2x < y < \infty$, and using again (4.2), we get $s(x) \le 1 + J_1 + J_2 + J_3$, where

$$J_1 \stackrel{\text{def}}{=} \int_0^x q(x, y) \left[e^{\beta(y-x)} - 1 \right] dy,$$

$$J_2 \stackrel{\text{def}}{=} \int_x^{2x} q(x, y) \left[e^{\beta(y-x)} - 1 \right] e^{-\alpha(y-x)} \, dy,$$

$$J_3 \stackrel{\text{def}}{=} \int_{-\infty}^{-x} q(x, y) \, dy + \int_{2x}^{\infty} q(x, y) \, dy \le 2 \int_x^{\infty} f(z) \, dz.$$

Substituting $z = y - x$ and using the symmetry, we get

$$
\begin{aligned}
J_1 + J_2 &= \int_0^x f(z)\left[e^{-\beta z} - 1 + e^{-(\alpha-\beta)z} - e^{-\alpha z}\right] dz \\
&= -\int_0^x f(z)[1 - e^{-\beta z}]\,[1 - e^{-(\alpha-\beta)z}]\,dz\,,
\end{aligned}
$$

which has limit less than 0 as $x \uparrow \infty$. Choosing x^* sufficiently large therefore ensures that $s(x) \leq 1 + J_1 + J_2 + J_3 < 1$ uniformly in $x \geq x^*$, which together with a similar calculation for $x < 0$ implies (3.14) for $x \notin K$.

It remains only to show that $\mathbb{E}_x V(\xi_1)$ is bounded uniformly in $x \in K$, which follows from

$$
s(x) \leq 1 + 2\int_{x^*}^\infty f(z)\,dz + 2e^{\beta x^*}\int_0^{x^*} f(z)\,dz
$$

for $0 < x < x^*$, and a similar bound for $-x^* < x < 0$. □

Remark 4.6 Mengersen & Tweedie refer to (4.2) as *log-concavity* in tail. The crucial feature is that (4.2) implies that the tails of Π^* decay at least exponentially fast. Further results from [253] show that this is also close to being necessary and sufficient for exponential ergodicity, though further regularity conditions are needed for rigorous statements. □

Remark 4.7 Beyond the random-walk setting, symmetric proposals with $q(x, y) = q(y, x)$ are particularly appealing when E is finite and Π^* the uniform distribution. One reason is that the MH ratio then is just one, so that a proposal is always accepted.

Another reason is that symmetric proposals are often easy to generate. For example, in the hard-core model in Section 2c, a trivially symmetric proposal consists in (a) selecting one vertex of the graph at random, (b) deleting the object there if there is one, (b') adding an object if there is none at the vertex and the hard-core property is not violated by doing so. □

4c   Reversible-Jump MCMC

This class of samplers, for which a main reference is Green [164], is designed for the case in which $E = \bigcup_{j=1}^J E_j$ consists of J (disjoint) components of different dimensions. We can then represent $\pi(x)$ and $\mu(dx)$ by their restrictions $\pi_1(x_1), \ldots, \pi_J(x_J)$, respectively $\mu(dx_1), \ldots, \mu(dx_J)$ to the E_j. A proposal density $q(x, x')$ may be nonzero when $x \in E_j$, $x' \in E_{j'}$, corresponding to a proposed move from E_j to $E_{j'}$; typically, $q(x, x') > 0$ occurs only when $x \in E_j$, $x' \in E_{j'}$ with the dimensions of $E_j, E_{j'}$ not too different. Of course, the proposal may also allow moves within E_j.

It is worth stressing that the setup is not intrinsically different from the MH setting considered so far, since a specific form of the proposal will determine the MH ratio and thereby ensure the detailed balance condition. This is illustrated by Example 4.10 below.

The main examples occur in Bayesian statistics, where the dimension of the parameter vector x is variable. With $\boldsymbol{y}$ the observation vector, we then have a likelihood $L_j(\boldsymbol{y} \mid x_j)$ defined for $x_j \in E_j$, a prior probability ρ_j for $x \in E_j$, and a prior density $\pi_j^{(0)}(x_j)$ for $x = x_j$ given $x \in E_j$. The posterior density is therefore proportional to

$$\pi_j(x_j) \stackrel{\text{def}}{=} \rho_j \pi_j^{(0)}(x_j) L_j(\boldsymbol{y} \mid x_j), \quad x = x_j \in E_j .$$

Example 4.8 A typical example is a mixture of j exponential distributions with parameters $\lambda_1, \ldots, \lambda_j$ and weight θ_i for the ith. A typical proposal for an MH algorithm would then be to add one component in the mixture (that is, perform a transition from E_j to E_{j+1}) w.p. $q(x, x')$, where $x \in E_j$ is the current parameter vector and $x' \in E_{j+1}$ the new one, and to delete one component (perform a transition from E_j to E_{j-1}) w.p. $q(x, x')$ where $x \in E_j$, $x' \in E_{j+1}$, and possibly to perform a move within E_j w.p. $q(x, x')$, where $x, x' \in E_j$. □

Example 4.9 Further examples of variable-dimension statistical models include:

(a) Hidden Markov models in which j is the number of states for the underlying Markov chain.

(b) ARMA(p, q) models with p, q allowed to vary, say $p = 1, \ldots, P$, $q = 1, \ldots, Q$, so that $J = PQ$.

(c) Multiple regression models with covariates $t_{i1}, \ldots, t_{iM}$ for observation y_i. Here each E_j corresponds to one of the $J \stackrel{\text{def}}{=} 2^M$ subsets of $\{1, \ldots, M\}$, say S_j, and the likelihood corresponds to the statistical model

$$Y_i = \alpha + \sum_{m \in S_j} \beta_m t_{im} + \varepsilon_i ,$$

where the ε_i are i.i.d. $\mathcal{N}(0, \sigma^2)$.

(d) Change-point problems in which parameters may change at several points in time, e.g., a Poisson process on $[0, T]$ where the intensity $\lambda(t)$ is piecewise constant. □

Example 4.10 In Example 4.8 on exponential mixtures, it is convenient (but of course an overparameterization) to represent the weights in E_j as $\theta_i = a_j/(a_1 + \cdots + a_j)$, where $a_1, \ldots, a_J \in (0, \infty)$. We take a prior that makes the parameters independent, with density $f(a)$ for the a_j and $g(\lambda)$

for the λ_j. Thus with $x_j \overset{\text{def}}{=} (a_1, \ldots, a_j, \lambda_1, \ldots, \lambda_j) \in E_j$, we have

$$\pi^{(0)}(x_j) = \prod_{i=1}^{j} f(a_i)g(\lambda_i), \quad L_j(\boldsymbol{y} \,|\, x_j) = \prod_{k=1}^{n} \Big(\sum_{r=1}^{J} \frac{a_r}{a_1 + \cdots + a_j} \lambda_r e^{-\lambda_r y_k} \Big).$$

Assume that the proposal moves one up w.p. $1/2$ and then adds a, λ drawn according to the prior at position i w.p. $1/j$. This means that the proposed move is to

$$x_{j+1} \overset{\text{def}}{=} \big(a_1, \ldots, a_{i-1}, a, a_{i+1}, \ldots, a_j, \lambda_1, \ldots, \lambda_{i-1}, \lambda, \lambda_{i+1}, \ldots, \lambda_j\big) \in E_{j+1}.$$

Similarly, the downward move occurs w.p. $1/2$ and then deletes one randomly chosen component from the mixture. The MH ratio therefore becomes

$$
\begin{aligned}
r(x_j, x_{j+1}) &= \frac{1}{r(x_{j+1}, x_j)} = \frac{\rho_{j+1}\pi^{(0)}(x_{j+1})L_{j+1}(\boldsymbol{y} \,|\, x_{j+1})/2(j+1)}{\rho_j \pi^{(0)}(x_j)L_j(\boldsymbol{y} \,|\, x_j)f(a)g(\lambda)/2j} \\
&= \frac{j\rho_{j+1}L_{j+1}(\boldsymbol{y} \,|\, x_{j+1})}{(j+1)\rho_j L_j(\boldsymbol{y} \,|\, x_j)},
\end{aligned}
$$

whereby the MCMC algorithm is completely specified.

Moves up from E_J and moves down from E_1 are of course impossible. This is incorporated in the formalism for example by the convention $\rho_{J+1} = \rho_0 = 0$. $\square$

Example 4.11 A specific implementation, suggested by Green [164], is often quoted in the literature and assumes that the transitions between E_{j_1} and E_{j_2} are determined as a combination of deterministic mappings and additional randomizations. More precisely, the transition from E_{j_1} to E_{j_2} is determined by a bijection

$$\big(x_{j_2}, y_{j_2 j_1}\big) = T_{j_1 j_2}\big(x_{j_1}, y_{j_1 j_2}\big),$$

where the additional r.v.'s $Y_{j_1 j_2}, Y_{j_2 j_1}$ take values in subsets $\Omega_{j_1 j_2}, \Omega_{j_2 j_1}$ of Euclidean of space of suitable dimensions and have densities $g_{j_1 j_2}(y_{j_1 j_2})$, $g_{j_2 j_1}(y_{j_2 j_1})$. Thus,

$$T_{j_1 j_2} : E_{j_1} \times \Omega_{j_1 j_2} \rightarrow E_{j_2} \times \Omega_{j_2 j_1}, \quad T_{j_2 j_1} = T_{j_1 j_2}^{-1}.$$

For detailed balance, one can then verify that the MH ratio becomes

$$\frac{\rho_{j_2}\pi_{j_2}(x_{j_2})}{\rho_{j_1}\pi_{j_1}(x_{j_1})} \frac{p_{j_2 j_1}}{p_{j_1 j_2}} \frac{g_{j_2 j_1}(y_{j_2 j_1})}{g_{j_1 j_2}(y_{j_1 j_2})} \Big| \frac{\partial T_{j_1 j_2}(x_{j_1}, y_{j_1 j_2})}{\partial(x_{j_1}, y_{j_1 j_2})} \Big|,$$

where $|\cdot|$ is the Jacobian. In addition to Green [164], see also Robert & Casella [301, Section 11.2] and Cappé et al. [65, Section 13.2.2]. $\square$

Exercises

4.1 (TP) Compute the MH ratio for the Strauss process in Example 4.2.

4.2 (TP) Demonstrate how to fit the exponential mixture in Example 4.10 into the formalism of Example 4.11

5   The Gibbs Sampler

Assume $E \subseteq E_1 \times \cdots \times E_d$. The typical element of E is then a vector $\boldsymbol{x} = (x_1, \ldots, x_d)$, and we write

$$\boldsymbol{x}_{-i} \stackrel{\text{def}}{=} (x_1, \ldots, x_{i-1}, x_{i+1}, \ldots, x_d).$$

Identifying $\boldsymbol{x}$ with $(x_i, \boldsymbol{x}_{-i})$ and using discrete notation, we further write

$$\pi_{\boldsymbol{x}_{-i}}^{(i)}(x_i) \stackrel{\text{def}}{=} \frac{\pi(\boldsymbol{x})}{\displaystyle\sum_{y_i : (y_i, \boldsymbol{x}_{-i}) \in E} \pi(y_i, \boldsymbol{x}_{-i})}$$

for the conditional Π^*-distribution of component i given that the remaining ones equal $\boldsymbol{x}_{-i}$, using discrete notation here and in the following.

The *Gibbs sampler* generates a sequence $\boldsymbol{\xi}(k) \stackrel{\text{def}}{=} (\xi_1(k), \ldots, \xi_k(d)) \in E$, where in step k only one component is updated, say the $i = i_k$th, according to the specific rule that $\xi_i(k+1)$ is drawn according to $\pi_{\boldsymbol{x}_{-i}}^{(i)}(\cdot)$ when $\boldsymbol{\xi}(k) = \boldsymbol{x}$. There are basically two ways to choose the sequence $\{i_k\}$. The *systematic-scan* Gibbs sampler takes $\{i_k\}$ as cyclic, say $1, \ldots, d, 1, \ldots, d, \ldots$, whereas the *random-scan* sampler chooses i_k at random, typically (but not necessarily) from the uniform distribution on $\{1, \ldots, d\}$. For example, for the systematic-scan Gibbs sampler, this means that we have a movement in the state space E as illustrated in Figure 5.1 for the case $d = 2$, $E = \{x, y : 0 < y < f(x)\}$.

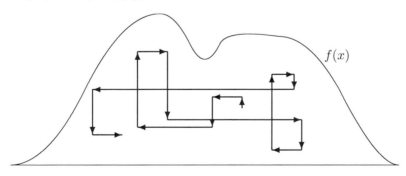

FIGURE 5.1

The idea behind the Gibbs sampler is that sampling from the $\pi_{\boldsymbol{x}_{-i}}^{(i)}(\cdot)$ in many cases is much more straightforward than sampling from Π^* itself.

Once the component i for updating has been fixed, the move performed by the Gibbs sampler can be seen as a move of the MH algorithm in E_i, where both the target and the proposal are $\pi^{(i)}_{\boldsymbol{x}_{-i}}(\cdot)$ (and therefore the proposal is accepted w.p. 1).

It is immediately clear that $\{\boldsymbol{\xi}(k)\}$ is a Markov chain in the case of the random-scan sampler. This is not the case for the systematic-scan sampler, where $\{\boldsymbol{\xi}(k)\}$ could be called a "Markov periodic chain" (not the same as a periodic Markov chain!): the transition matrix varies periodically as

$$\boldsymbol{P}^{(1)},\ldots,\boldsymbol{P}^{(d)},\boldsymbol{P}^{(1)},\ldots,\boldsymbol{P}^{(d)},\ldots,$$

where $\boldsymbol{P}^{(i)}$ is the transition matrix corresponding to the description above of how to update component i. In both cases, we have, however, the following result:

Theorem 5.1 Π^* *is stationary for* $\{\boldsymbol{\xi}(k)\}$ *in the sense that if* $\boldsymbol{\xi}(0)$ *has distribution* Π^*, *then so has* $\boldsymbol{\xi}(n)$ *for all* n.

Proof. It suffices to take $n = 1$. Write $i = i_1$ and $\boldsymbol{x} = \boldsymbol{\xi}(0)$. Then the conditional distribution of $\xi_i(0)$ given the $\xi_j(0)$ with $j \neq i$ is $\pi^{(i)}_{\boldsymbol{x}_{-i}}(\cdot)$ because $\boldsymbol{\xi}(0)$ has distribution Π^*, and by construction of the Gibbs sampler, the same is true for the conditional distribution of $\xi_i(1)$ given the $\xi_j(1)$ with $j \neq i$. Since $\xi_j(0) = \xi_j(1)$ for $j \neq i$, $\boldsymbol{\xi}(0)$ and $\boldsymbol{\xi}(1)$ must therefore have the same distribution. $\qquad\square$

The concept of irreducibility for the random-scan sampler is just the usual one for Markov chains, whereas we will call the systematic-scan sampler irreducible if the transition matrix $\boldsymbol{P} \overset{\text{def}}{=} \boldsymbol{P}^{(1)} \cdots \boldsymbol{P}^{(d)}$ (usual matrix product) is irreducible.

Corollary 5.2 *If E is finite and the Gibbs sampler is irreducible, then the empirical distribution $\widehat{\Pi}^*_n$ in (1.1) converges to Π^* and the average $\widehat{f}_n$ of the $f(\boldsymbol{\xi}(k))$ in (1.2) converges to $\Pi^*[f]$.*

Proof. This is trivial for the random-scan sampler. For the systematic scan sampler, $\boldsymbol{\pi}^*\boldsymbol{P}^{(i)} = \boldsymbol{\pi}^*$ for each i implies $\boldsymbol{\pi}^*\boldsymbol{P} = \boldsymbol{\pi}^*$, so if $\boldsymbol{P}$ is irreducible, we have $\boldsymbol{\xi}(kd) \overset{\mathscr{D}}{\to} \Pi^*$. This together with Π^* being stationary for $\boldsymbol{P}^{(1)} \cdots \boldsymbol{P}^{(\ell)}$ implies $\boldsymbol{\xi}(kd+\ell) \overset{\mathscr{D}}{\to} \Pi^*$ for $\ell = 1,\ldots,d-1$. $\qquad\square$

Example 5.3 The basic fact behind acceptance–rejection is that if $f : \mathbb{R} \to [0,\infty)$ is an unnormalized density and (X,Y) is uniform on $E \overset{\text{def}}{=} \{(x,y) : 0 < y < f(x)\}$, then X has marginal density $f^* \overset{\text{def}}{=} f/\int f$. This is the idea behind the *slice sampler* for MCMC inference on f^*: the target distribution is the uniform distribution on $E \overset{\text{def}}{=} \{(x,y) : 0 < y < f(x)\}$, and the algorithm is Gibbs sampling with uniform steps. From $(x,y) \in E$, the move is to (x,y') in (say) all even steps, where y' is uniform on the interval $\big(0, f(x)\big)$, and in odd steps, it is to (x',y), where x' is uniform on the set $I(x) \overset{\text{def}}{=} \{z : f(z) \geq f(x)\}$ (an interval if f is say log-concave but not necessarily so in general).

The slice sampler often meets the difficulty that $I(x)$ is difficult or at best cumbersome to determine. In some cases, this problem may be resolved by finding simple bounds, that is, an interval $J(x)$ with easily determined endpoints and containing $I(x)$. One can then sample x' from the uniform distribution on $J(x)$ and use acceptance–rejection, accepting the first x' such that $x' \in I(x)$. □

Example 5.4 Consider a hidden Markov model with the conditional distribution of the underlying Markov chain $\boldsymbol{X}_{0:n}$ given the observations $\boldsymbol{y}_{0:n}$ as target distribution Π^*. If S is the state space for $X_0, X_1, \ldots$, we then have $E = S^{n+1}$, and at each step k of the MCMC algorithm, the simulated value $\boldsymbol{\xi}(k)$ of the Markov chain is a sequence $\boldsymbol{\xi}(k) \stackrel{\text{def}}{=} \boldsymbol{x}_{0:n}(k)$ of $n+1$ values from S. Now consider updating of component i, that is, of x_i (for simplicity, assume $0 < i < n$). The conditional Π^*-distribution of x_i given $\boldsymbol{x}_{-i}$ equals the conditional distribution of x_i given $\boldsymbol{x}_{-i}$ and $\boldsymbol{y}_{0:n}$. According to the dependence properties of hidden Markov models, this distribution is just the conditional distribution of x_i given x_{i-1}, x_{i+1}, y_i, so that $\pi(x_i \mid \boldsymbol{x}_{-i})$ is proportional to $p(x_{i-1}, x_i)p(x_i, x_{i+1})g_{x_i}(y_i)$, where $p(x, x')$ is the transition probability $x \to x'$.

As an example, consider the stochastic volatility model (2.4). Here

$$p(x_{i-1}, x_i)p(x_i, x_{i+1})g_{x_i}(y_i)$$
$$\propto \quad \exp\left\{-\frac{(x_i - \rho x_{i-1})^2}{2(1 - \rho^2)} - \frac{(x_{i+1} - \rho x_i)^2}{2(1 - \rho^2)} - \frac{y_i^2}{\beta^2 \mathrm{e}^{x_i}}\right\}$$
$$\propto \quad \exp\left\{-\frac{x_i^2(1 + \rho^2)}{2(1 - \rho^2)} + \frac{\rho x_i(x_{i-1} + x_{i+1})}{1 - \rho^2} - \frac{y_i^2}{\beta^2 \mathrm{e}^{x_i}}\right\}.$$

R.v.'s from this density can be generated, for example, by acceptance–rejection from a $\mathcal{N}\big(\rho(x_{i-1} + x_{i+1}), 1 - \rho^2\big)$ proposal for x_i and accepting w.p.

$$\exp\left\{-\frac{x_i^2 \rho^2}{2(1 - \rho^2)} - \frac{y_i^2}{\beta^2 \mathrm{e}^{x_i}}\right\}. \qquad\qquad □$$

Example 5.5 Consider the Ising model $\boldsymbol{x} = (x_i)_{i \in S}$,

$$\pi(\boldsymbol{x}) = \exp\left\{-\beta \sum_{ii' \in \mathcal{E}} \mathbb{1}\{x_i \neq x_{i'}\}\right\}, \qquad\qquad (5.1)$$

from Section 2b. Here $d = |S|$, the number of sites, and all $E_i = \{1, -1\}$. To compute the conditional probabilities $\pi_{\boldsymbol{x}_{-i}}^{(i)}(y_i)$, it suffices to consider the case $y_i \neq x_i$, i.e., $y_i = -x_i$ (there are only two possibilities $1, -1$; a transition from $\boldsymbol{x}$ to $(y_i, \boldsymbol{x}_{-i})$ means that we flip the mark at i). Letting $n_i(\boldsymbol{x})$ be the number of neighbors i' of i with $x_{i'} \neq x_i$ and $m_i(\boldsymbol{x})$ the number with $x_{i'} = x_i$, we have $\pi(\boldsymbol{x}) = \mathrm{e}^{-\beta n_i(\boldsymbol{x})} d_i(\boldsymbol{x})$, where

$$d_i(\boldsymbol{x}) \stackrel{\text{def}}{=} \exp\left\{-\beta \sum_{j,k \neq i,\, jk \in \mathcal{E}} \mathbb{1}\{x_j \neq x_k\}\right\}.$$

Similarly, $\pi(y_i, \boldsymbol{x}_{-i}) = \mathrm{e}^{-\beta m_i(\boldsymbol{x})} d_i(\boldsymbol{x})$. Thus

$$\pi_{\boldsymbol{x}_{-i}}^{(i)}(y_i) = \frac{\pi(y_i, \boldsymbol{x}_{-i})}{\pi(y_i, \boldsymbol{x}_{-i}) + \pi(\boldsymbol{x})} = \frac{\mathrm{e}^{-\beta m_i(\boldsymbol{x})}}{\mathrm{e}^{-\beta n_i(\boldsymbol{x})} + \mathrm{e}^{-\beta m_i(\boldsymbol{x})}}. \qquad (5.2)$$

Thus, the update of site i performed by the Gibbs sampler amounts to simple binomial flipping with a probability given by (5.2).   □

The form of (5.2) is remarkably simple: all that matters is the marks at the neighbors of i. The reason behind this can be viewed as a generalized Markov property of the Ising model, that the mark at a site i and the marks at the sites that are not neighbors are conditionally independent given the marks at the neighbors. Similar features are found in a great variety of models, in particular *graphical statistical models*.

Example 5.6 Assume that $H(x) = \sum_{vv' \in \mathcal{E}} r(x_v, x_{v'})$, in a situation in which $x = (x_v)_{v \in V}$ is a configuration of values on the set V of vertices of a graph $G = (V, \mathcal{E})$ and r is a nearest-neighbour potential (as for the Ising model). We implement Gibbs sampling by updating at each vertex v, using either systematic scan or random scan. Consider updating at vertex v. The probability of changing x_v to y is then

$$\exp\left\{-\sum_{v': vv' \in \mathcal{E}} r(y, x_{v'})/T_n\right\} \bigg/ \sum_z \exp\left\{-\sum_{v': vv' \in \mathcal{E}} r(z, x_{v'})/T_n\right\}. \qquad □$$

Example 5.7 Image analysis provides an example of the product form of the state space; d is the number of pixels and each E_i is just $\{-1, 1\}$. Systematic-scan Gibbs sampling means that we update pixel by pixel in a given order, say lexicographically, and random-scan Gibbs sampling means that we update a randomly chosen pixel at each step.

As an example, consider a modification of the Ising model in which the image $\boldsymbol{x}$ itself is not observed but rather a blurred image $\boldsymbol{y}$. More precisely, we assume that the errors on the observed image $\boldsymbol{y}$ are i.i.d. with error probability $\zeta \in (0, 1)$, so that $\boldsymbol{y}$ is constructed from $\boldsymbol{x}$ by independent flipping with flipping probability ζ at each site, that is,

$$L(\boldsymbol{y} \mid \boldsymbol{x}) = \prod_{j \in S} a_j(\boldsymbol{x}) \quad \text{where} \quad a_j(\boldsymbol{x}) \stackrel{\text{def}}{=} \zeta^{1-\delta(y_j, x_j)}(1 - \zeta)^{\delta(y_j, x_j)}$$

with $\delta(y_j, x_j) = 1$ if $x_j = y_j$, $= 0$ if $x_j \neq y_j$. Thus, image analysis can be viewed as a Bayesian problem in which the purpose is to make statements on the posterior distribution of $\boldsymbol{x}$ ($=$ the parameter $=$ the actual image) given the observations $\boldsymbol{y}$. The prior $\pi^{(0)}(\boldsymbol{x})$ is given by the r.h.s. of (5.1).

We assume for simplicity that ζ is known. Letting $\pi(\boldsymbol{x})$ and $d_i(\boldsymbol{x})$, etc., be as in Example 5.5, the joint density of $(\boldsymbol{x}, \boldsymbol{y})$ is therefore proportional to $\pi^{(0)}(\boldsymbol{x}) L(\boldsymbol{y} \,|\, \boldsymbol{x})$. The posterior $\pi^*(\boldsymbol{x})$ can again be obtained by Gibbs sampling, updating site by site. Letting

$$b_j(\boldsymbol{x}) \stackrel{\text{def}}{=} \zeta^{\delta(y_j, x_j)} (1 - \zeta)^{1 - \delta(y_j, x_j)}, \quad f_i(\boldsymbol{x}) \stackrel{\text{def}}{=} \prod_{j \neq i} a_j(\boldsymbol{x}),$$

and proceeding as in Examples 5.5 and 5.6, we have with $z_i \stackrel{\text{def}}{=} -x_i$ that

$$
\begin{aligned}
\pi^{(0)}(\boldsymbol{x}) &\propto \mathrm{e}^{-\beta n_i(\boldsymbol{x})} d_i(\boldsymbol{x}), \\
\pi^*(\boldsymbol{x}) &\propto \pi^{(0)}(\boldsymbol{x}) L(\boldsymbol{y} \,|\, \boldsymbol{x}) = \mathrm{e}^{-\beta n_i(\boldsymbol{x})} d_i(\boldsymbol{x}) a_i(\boldsymbol{x}) f_i(\boldsymbol{x}), \\
\pi^*(z_i, \boldsymbol{x}_{-i}) &\propto \mathrm{e}^{-\beta m_i(\boldsymbol{x})} d_i(\boldsymbol{x}) b_i(\boldsymbol{x}) f_i(\boldsymbol{x}), \\
\pi^*(z_i \,|\, \boldsymbol{x}_{-i}) &= \frac{\mathrm{e}^{-\beta m_i(\boldsymbol{x})} b_i(\boldsymbol{x})}{\mathrm{e}^{-\beta n_i(\boldsymbol{x})} a_i(\boldsymbol{x}) + \mathrm{e}^{-\beta m_i(\boldsymbol{x})} b_i(\boldsymbol{x})}.
\end{aligned}
\tag{5.3}
$$

Thus, the update of site i performed by the Gibbs sampler amounts to simple binomial flipping with a probability given by (5.3). $\qquad\square$

5a   Componentwise Metropolis–Hastings

Componentwise MH is an extension of Gibbs sampling. As in that setting, one component at a time is updated and the target distribution in each step is the conditional Π^*-distribution $\pi_{\boldsymbol{x}_{-i}}^{(i)}(\cdot)$ of the component in question given the remaining ones. The difference is that the updating from x_i to y_i is done with a general proposal $q_{\boldsymbol{x}}^{(i)}(x_i, \cdot)$ rather than with $\pi_{\boldsymbol{x}_{-i}}^{(i)}(\cdot)$ itself as for Gibbs sampling. Componentwise MH is a special case of the MH algorithm, and the MH ratio takes the form

$$r_{\boldsymbol{x}}^{(i)}(x_i, y_i) = \frac{\pi_{\boldsymbol{x}_{-i}}^{(i)}(y_i) q_{(y_i, \boldsymbol{x}_{-i})}^{(i)}(y_i, x_i)}{\pi_{\boldsymbol{x}_{-i}}^{(i)}(x_i) q_{\boldsymbol{x}}^{(i)}(x_i, y_i)}. \tag{5.4}$$

For the Gibbs sampler,

$$q_{\boldsymbol{x}}^{(i)}(x_i, y_i) = \pi_{\boldsymbol{x}_{-i}}^{(i)}(y_i), \quad q_{(y_i, \boldsymbol{x}_{-i})}^{(i)} = \pi_{\boldsymbol{x}_{-i}}^{(i)}(x_i),$$

and thus indeed the proposal is accepted w.p. 1.

Theorem 5.8 Π^* *is stationary for the componentwise MH chain* $\{\boldsymbol{\xi}(k)\}$ *in the sense that if* $\boldsymbol{\xi}(0)$ *has distribution* Π^*, *then so has* $\boldsymbol{\xi}(n)$ *for all* n.

Proof. It suffices to take $n = 1$. Write $i = i_1$, $\boldsymbol{x} = \boldsymbol{\xi}(0)$ and let $P_{\boldsymbol{x}}^{(i)}$ be the transition kernel on E_i corresponding to MH sampling with MH ratio (5.4). Since the target distribution is $\pi_{\boldsymbol{x}_{-i}}^{(i)}(\cdot)$, we have

$$\pi_{\boldsymbol{x}_{-i}}^{(i)} P_{\boldsymbol{x}}^{(i)} = \pi_{\boldsymbol{x}_{-i}}^{(i)}. \tag{5.5}$$

The conditional distribution of $\xi_i(0)$ given the $\xi_j(0)$ with $j \neq i$ is $\pi_{\boldsymbol{x}_{-i}}^{(i)}(\cdot)$, and by (5.5), the same is true for the conditional distribution of $\xi_i(1)$ given the $\xi_j(1)$ with $j \neq i$. Since $\xi_j(0) = \xi_j(1)$ for $j \neq i$, $\boldsymbol{\xi}(0)$ and $\boldsymbol{\xi}(1)$ must therefore have the same distribution. $\qquad\square$

Corollary 5.9 *If E is finite and the componentwise MH sampler is irreducible, then the empirical distribution $\widehat{\Pi}_n^*$ in (1.1) converges to π^* and the average $\widehat{f}_n$ of the $f(\boldsymbol{\xi}(k))$ in (1.2) converges to $\Pi^*[f]$.*

Exercises

5.1 (TP) In the hidden Markov Example 5.4, derive the updating formulas if $i = 0$ or $i = n$ is a boundary value. Specialize to the stochastic volatility setting.

5.2 (A) A classsical data set (e.g. [301, p. 383, 409)] contains the numbers of cases of mastitis (an inflammatory disease) in 127 herds of dairy cattle. With n_i the number of herds having i cases, the data (adjusted for herd size) are as follows:

i	0	1	2	3	4	5	6	7	8	9	10	11	12
n_i	7	12	8	9	7	8	9	6	5	3	4	7	4
i	13	14	15	16	17	18	19	20	21	22	23	24	25
n_i	5	2	1	4	3	3	4	2	2	4	1	0	5

Assume that for a given herd j, the number of cases is Poisson(λ_j), where the λ_j are Gamma(α, β_j) and the β_j are themselves Gamma(a, b). Use Gibbs sampling with $\alpha = 0.1$, $a = b = 1$ to give histograms of the posterior distribution of λ_j for a herd with 0, 10, or 20 cases and to report the posterior means of α, a, b.

Chapter XIV
Selected Topics and Extended Examples

1 Randomized Algorithms for Deterministic Optimization

We consider the problem of finding the minimum $H(x^*)$ of a real-valued function $H(x)$ defined on a finite set S and the corresponding minimizer x^* (we ignore the problem of multiple global maxima). It is assumed that $H(x)$ is explicitly available for each x. If S is a small finite set, the problem is then easy, since we can just go through all $|S|$ possibilities to determine x^*. However, typically S is huge, so this approach is not feasible, and other methods are required.

Classical examples in the area, which also goes under the name *combinatorial optimization*, are as follows:

(a) The *traveling salesman problem*. A salesman living in city 0 has to visit every city $1, \ldots, n$ exactly once and then return to his home. The travel cost between cities i, j is c_{ij} and the problem is to find the route x that minimizes the total cost $H(x)$. The set S is then the set of all permutations $x = (x_1, \ldots, x_n)$ of $\{1, \ldots, n\}$, and

$$H(x) \;=\; c_{0x_1} + c_{x_1 x_2} + \cdots + c_{x_{n-1} x_n} + c_{x_n 0} \,. \tag{1.1}$$

(b) Consider a graph $G = (V, \mathcal{E})$. The *graph cut problem* (or *graph bisection problem*) consists in partitioning the set V of vertices into V_1, V_2, such that the number of edges connecting a vertex in V_1 with one in V_2 is minimized. Here S is the set of all partitions $x = (V_1(x), V_2(x))$

of V and

$$H(x) = \sum_{x \in S} \sum_{v_1 \in V_1(x)} \sum_{v_2 \in V_2(x)} \mathbb{1}\{v_1 v_2 \in \mathcal{E}\}.$$

The indicator function can be replaced by a cost function $c(v_1, v_2)$ that is nonzero only when $v_1 v_2 \in \mathcal{E}$.

Häggström [189] gives the following example: V is the set of all web sites containing the word "football", and two sites are connected if there is a link between them. A partitioning could serve the purpose of obtaining a rough grouping into the web sites dealing with soccer and those dealing with American football.

Though problems of this type are deterministic, randomized algorithms can perform surprisingly well compared to deterministic ones. There is a vast literature on the subject and an abundance of different approaches. We concentrate here on two methods, one classical, simulated annealing, and one more recent, the cross-entropy method. Pflug [294, Chapter 1] and Rubinstein & Kroese [318, Section 4.1] contain lists of further approaches and references.

1a   Simulated Annealing

In I.5.4, it was suggested to invoke the Boltzmann distribution π_T with point masses $\mathrm{e}^{-H(x)/T}/c(T)$, where T is the temperature, and to generate an r.v. $X(T)$ from π_T for some small T. The idea is of course that $X(T)$ will take a value with a high H-value with high probability.

Proposition 1.1 *Assume that for some $x^* \in S$ we have $a < b$, where $a \overset{\text{def}}{=} H(x^*)$, $b \overset{\text{def}}{=} \max_{x \neq x^*} H(x)$. Then $\mathbb{P}(X(T) = x^*) \to 1$ as $T \to 0$.*

Proof. By definition,

$$\mathbb{P}(X(T) = x^*) = \frac{\mathrm{e}^{-H(x^*)/T}}{\sum_{x' \in S} \mathrm{e}^{-H(x')/T}} \geq \frac{\mathrm{e}^{-a/T}}{\mathrm{e}^{-a/T} + |S - 1|\mathrm{e}^{-b/T}}$$

$$= \frac{1}{1 + |S - 1|\mathrm{e}^{(a-b)/T}},$$

which goes to 1 because $a < b$ and hence $\mathrm{e}^{(a-b)/T} \to 0$.   □

Of course, the undesirable feature of this procedure is its lack of consistency: it is not certain that $X(T) = x^*$ and hence $H(X(T)) = a$. The idea of *simulated annealing* is to use *cooling*, that is, to consider a sequence $\{T_n\}$ of temperatures such that $T_n \to 0$, $n \to \infty$. The most naive implementation of the algorithm is then to (independently) sample $X(T_n)$ for each n, whereby one obtains a consistent estimator since $X(T_n) \overset{\mathbb{P}}{\to} x^*$ by Proposition 1.1.

This simple approach is seldom feasible, since generating $X(T)$ from the Boltzmann distribution usually presents a problem (think, for example, of the Ising model). The standard implementation of simulated annealing overcomes this problem by combining with the ideas of MCMC. One then simulates a time-inhomogeneous Markov chain $\{\xi_n\}$, such that π_{T_n} is stationary for the transition matrix $(p_{xx'}(T_n))$ used at step n.

Typically, one takes T_n to be constant on intervals of increasing length. That is, $T_n = T^{(k)}$ when $n_k \leq n < n_{k+1}$, where $n_{k+1} - n_k \to \infty$. It is clear at once that if $n_{k+1} - n_k$ grows at a sufficiently large rate (slow cooling), then $\xi_n \xrightarrow{\mathbb{P}} x^*$. In fact, Geman & Geman [126] proved that if

$$T_n \geq \frac{1}{\log n} |S| \left(\max_{x \in S} H(x) - \min_{x \in S} H(x) \right)$$

then $\xi_n \xrightarrow{\mathbb{P}} x^*$. However, often this condition on slow cooling is overly careful and makes the algorithm inefficient in the sense that an unnecessarily large number of steps need to be taken before one can be sure that $\left\| \mathbb{P}(\xi_n \in \cdot) - \delta_{x^*} \right\| < \varepsilon$. On the other hand, too rapid cooling has the danger that ξ_n gets stuck in nonoptimal states, for example local minima, as is shown in Example 1.4 below.

Example 1.2 In the graph cut problem, a natural choice of an MH algorithm is to choose one vertex at random and move the vertex from its current $V_i(x)$ to the other $V_j(x)$ (unless $V_i(x)$ consists of one element only). This can be viewed as a special case of XIII.4.1, and the relevant MH ratio is given there and reproduced in (1.2) below.   □

In many examples, we have as in XIII.4.1 given a graph structure on S, and can then implement MCMC by letting the proposed move from x be to a randomly selected neighbor among the total of $n(x)$. Then as in XIII.4.1,

$$p_{xx'}(T) = \frac{1}{n(x)} \min \left(1, \frac{n(x)}{n(x')} e^{H(x)/T - H(x')/T} \right) \tag{1.2}$$

when x, x' are neighbors and $p(x, y) = 0$ when $x \neq y$ and x, x' are not neighbors. For example:

Example 1.3 Consider the traveling salesman problem. In order to design a sampler, we introduce a graph structure on the set of all permutations $x = (x_1, \ldots, x_n)$ of $\{1, \ldots, n\}$ by defining x, x' to be neighbors if x' arises from x by reversing a segment $x_i, \ldots, x_j$ with $i < j$. That is,

$$x' = (x_1, x_2, \ldots, x_{i-1}, x_j, x_{j-1}, \ldots, x_i, x_{j+1}, x_{j+2}, \ldots, x_n),$$

and each x has $n(n-1)/2$ neighbors. Then $p_{xx'}(T_n) = 0$ if x, x' not are neighbors, whereas otherwise,

$$p_{xx'}(T_n) = \frac{2}{n(n-1)} \min \left[\exp\{ (H(x) - H(x'))/T_n \}, 1 \right].$$

□

Example 1.4 Consider the setting of (1.2) and assume that x is a local minimum, i.e.,

$$c \stackrel{\text{def}}{=} \min_{x':\, xx'\in\mathcal{E}} \left[H(x') - H(x)\right] > 0\,,$$

but not a global minimum. Let $d \stackrel{\text{def}}{=} \max_{x':\, xx'\in\mathcal{E}} n(x)/n(x')$. Then for large T, the probaility of a move out of x is bounded by $de^{-c/T}$. Hence if we take $\xi_0 = x$ in the simulated annealing scheme and let $T_n \uparrow \infty$ so quickly that $d\sum_0^\infty e^{-c/T_n} < 1$, there is positive probability that $\xi_n = x$ for all x. That is, the speed of the cooling is so rapid that it makes ξ_n stick in a local minimum.   □

Some relevant references for simulated annealing are Aarts & Korst [1], Aarts & Lenstra [2], and Gidas [128].

1b   The Cross-Entropy Algorithm

Combinatorial optimization using cross-entropy was initiated by Rubinstein [317] and has since then developed into a rapidly expanding area, see [w^3.21]. A comprehensive treatment with many convincing examples is given in Rubinstein & Kroese [318]. The idea is very similar to rare-event simulation via the cross-entropy method as described in VI.8. At step n one samples R r.v.'s $X_1, \ldots, X_R$ from $f(x; \theta_n)$, a member of a parametric family $\{f(x; \theta)\}_{\theta\in\Theta}$ of probability mass functions on S. For some fixed $\delta < 1$, one next selects the $R' = \delta R$ of the X_r with the smallest H-values, say $X_1', \ldots, X_{R'}'$. Then θ_n is updated by computing θ_n^* as the θ minimizing the cross-entropy between $f(\cdot; \theta)$ and the empirical distribution of $X_1', \ldots, X_{R'}'$, and letting $\theta_{n+1} = \theta_n^*$. As in the rare-event setting, this last step is just ordinary maximum likelihood estimation.

As stopping criterion, it is suggested in [317], [318] to let $\gamma_n \stackrel{\text{def}}{=} \min\left(H(X_1'), \cdots, H(X_{R'}')\right)$ and to stop when $\gamma_n = \gamma_{n-1} = \ldots = \gamma_{n-d}$ for some d; the common value $\widehat{\gamma}$ is then the estimate of the minimum $H(x^*)$. Another implementation suggestion is to use smoothed updating, that is, to let $\theta_{n+1} = \alpha\theta_n + (1 - \alpha)\theta_n^*$ instead of $\theta_{n+1} = \theta_n^*$.

The choice of parameters needs tuning in specific applications, but some typical values are $R = 10,000$, $\delta = 0.01$, $d = 5$, $0.7 < \alpha < 1$.

We proceed to demonstrate how the parametric family of probability mass functions may be chosen in two of our main examples. Of course, the choice is by no means unique, and the efficiency of the algorithm is likely to depend crucially on the particular choice.

Example 1.5 In the graph cut problem, we can represent a cut by a vector $x = (x_v)_{v\in V}$ of 0's and 1's such that $v \in V_1(x)$ when $x_v = 0$ and $v \in V_2(x)$ when $x_v = 1$. An obvious choice is to generate the $X_r = (X_{r,v})_{v\in V}$ by Bernoulli sampling for each v. That is, $\Theta = (0, 1)^{|V|}$, $\theta = (\theta_v)_{v\in V}$, and subject to $f(\cdot, \theta)$, the $X_{r,v}$ are independent Bernoullis with $\mathbb{P}(X_{v,r} = 1) = \theta_v$. □

Example 1.6 In the traveling salesman problem, we need a parametric family of probability mass functions on the set S of all permutations $x = (x_1, \ldots, x_n)$ of $\{1, \ldots, n\}$. This can be obtained by letting Θ be the set of all transition probabilities $\theta(i, j)$ from $\{0, 1, \ldots, n\}$ to $\{1, \ldots, n\}$. Initially, x_1 is selected according to $\theta(0, \cdot)$. When $x_1, \ldots, x_k$ have been generated, the $\theta(i, j)$ are updated by removing all $j \in \{x_1, \ldots, x_k\}$ and renormalizing:

$$\theta(i, j) \longleftarrow \begin{cases} 0 & j \in \{x_1, \ldots, x_k\}, \\ \dfrac{\theta(i, j)}{\sum_{j' \notin \{x_1, \ldots, x_k\}} \theta(i, j')} & j \notin \{x_1, \ldots, x_k\}. \end{cases}$$

Then x_{k+1} is selected according to $\theta(x_k, \cdot)$.

For example, one can take all $\theta_0(i, j) = 1/(n+1)$. According to the standard formulas in maximum likelihood estimation for Markov chains, the updating formula becomes $\theta_n^*(i, j) = N_{ij}/N_i$ where

$$N_{ij} \overset{\text{def}}{=} \sum_{r=1}^{R'} \sum_{k=0}^{n-1} \mathbb{1}\{X'_{r,k} = i,\ X'_{r,k+1} = j\}, \quad N_i \overset{\text{def}}{=} \sum_{r=1}^{R'} \sum_{k=0}^{n-1} \mathbb{1}\{X'_{r,k} = i\}.$$

Note that if $\theta_n(i, j) = 1$ or 0 for some n, then such is the case for all larger n if one takes $\theta_{n+1} = \theta_n^*$, and hence the smoothing $\theta_{n+1} = \alpha\theta_n + (1 - \alpha)\theta_n^*$ serves to exclude such undesired phenomena. □

2   Resampling and Particle Filtering

Consider as in Chapter XIII on MCMC an r.v. X with distibution Π^* having density $\pi^*(x) = \pi(x)/C$ w.r.t. $\mu(\mathrm{d}x)$, which is known only up to the constant $C = \int \pi \, \mathrm{d}\mu$. If we want to compute $z = \mathbb{E}g(X)$, and $\widetilde{\pi}(x)$ is a different density (possibly also unnormalized) from which r.v.'s $\widetilde{X}_1, \ldots, \widetilde{X}_R$ are easily generated, we may then use the importance-sampling estimator

$$\widehat{z}_{\text{IS}} \overset{\text{def}}{=} \frac{\sum_{r=1}^R g(\widetilde{X}_r)\pi(\widetilde{X}_r)/\widetilde{\pi}(\widetilde{X}_r)}{\sum_{s=1}^R \pi(\widetilde{X}_s)/\widetilde{\pi}(\widetilde{X}_s)}$$

of V.1.16. In this section, we survey some implementations and extensions of this idea, for which some main references are Cappé et al. [65], Del Moral [85], Doucet et al. [95] and Liu [241]. The common terminology here is to refer to

$$w_r \overset{\text{def}}{=} \frac{(\pi/\widetilde{\pi})(\widetilde{X}_r)}{(\pi/\widetilde{\pi})(\widetilde{X}_1) + \cdots + (\pi/\widetilde{\pi})(\widetilde{X}_R)}$$

as the *importance weight of particle* r (note that $\hat{z}_{\mathrm{IS}} = \sum_1^R w_r g(\widetilde{X}_r)$).
The area is often called *sequential Monte Carlo*, except for the sampling importance resampling in Section 2b, which does not have a sequential element.

For the following, note that $\hat{z}_{\mathrm{IS}}$ has a bias of order N, as follows by the delta method, cf. III.(3.3).

2a Sequential Importance Sampling

The idea of sequential importance sampling is to break the computation of the likelihood ratio $\pi(\widetilde{X})/\widetilde{\pi}(\widetilde{X})$ (and thereby the importance weights) up into a number of steps.

To explain the idea, consider the most prominent example, a hidden Markov setup with background Markov chain $X_0, X_1, \ldots$ and observed quantities $Y_0, Y_1, \ldots$; cf. XIII.2d and A5, from which we also adapt the notation $p(\boldsymbol{x}_{0:n})$ for the density of $X_0, \ldots, X_n$, $p(\boldsymbol{x}_{0:k} \,|\, \boldsymbol{y}_{0:k})$ for the conditional density of $X_0, \ldots, X_k$ given $Y_0, \ldots, Y_k$, $p(y_k \,|\, x_k)$ for the conditional density of Y_k given X_k, etc. We consider filtering, where the goal is to provide estimates of the distribution with density $p(\boldsymbol{x}_{0:k} \,|\, \boldsymbol{y}_{0:k})$ such that the estimates are easily updated from $k-1$ to k as the observations Y_k arrive sequentially in time. The R particles observed up to time k are denoted by $\widetilde{\boldsymbol{X}}_{0:k}^{(1)}, \ldots, \widetilde{\boldsymbol{X}}_{0:k}^{(R)}$, and the problem is to find a simple recursive scheme for updating the importance weight $w_{r,k} \stackrel{\text{def}}{=} w_r\big(\widetilde{\boldsymbol{X}}_{0:k}^{(r)}\big)$ from $w_{r,k-1}$.

As proposal, we take the unconditional distribution of the Markov chain. Assuming for a fixed r that $X_0, \ldots, X_{k-1} \stackrel{\text{def}}{=} \widetilde{X}_0^{(r)}, \ldots, \widetilde{X}_{k-1}^{(r)}$ have been generated, this means that the proposal for X_k is generated according to the transition density $p(x_k \,|\, x_{k-1})$; the proposal for X_0 is $r(x_0)$, the initial distribution. Letting $\widetilde{\pi}_k(\boldsymbol{x}_{0:k})$ denote the proposal density for $X_{0:k}$, we thus have

$$\widetilde{\pi}_k(\boldsymbol{x}_{0:k}) = \widetilde{\pi}_k(\boldsymbol{x}_{0:k-1}) p(x_k \,|\, x_{k-1}) \,. \tag{2.1}$$

To obtain a similar decomposition of the target density $\pi_k^*(\boldsymbol{x}_{0:k}) \stackrel{\text{def}}{=} p(\boldsymbol{x}_{0:k} \,|\, \boldsymbol{y}_{0:k})$, we write

$$\pi_k^*(\boldsymbol{x}_{0:k}) = \frac{p(\boldsymbol{x}_{0:k}, \boldsymbol{y}_{0:k})}{p(\boldsymbol{y}_{0:k})} = \frac{p(\boldsymbol{x}_{0:k-1}, \boldsymbol{y}_{0:k-1}) p(x_k \,|\, x_{k-1}) p(y_k \,|\, x_k)}{p(\boldsymbol{y}_{0:k})}$$

$$= \frac{1}{C_k} \pi_{k-1}^*(\boldsymbol{x}_{0:k-1}) p(x_k \,|\, x_{k-1}) p(y_k \,|\, x_k) \,, \tag{2.2}$$

where $C_k \stackrel{\text{def}}{=} p(\boldsymbol{y}_{0:k})/p(\boldsymbol{y}_{0:k-1})$. Taking the ratio between (2.2) and (2.1) we get

$$w_{r,k} = w_r\big(\widetilde{\boldsymbol{X}}_{0:k}^{(r)}\big) = \frac{w_{r,k-1} p(y_k \,|\, \widetilde{X}_k^{(r)})}{\sum_{s=1}^R w_{s,k-1} p(y_k \,|\, \widetilde{X}_k^{(s)})} \,, \tag{2.3}$$

which is the desired simple updating formula.

In practice, the method does not work well for large k and needs modification, see further Section 2c.

2b Sampling Importance Resampling

An alternative to using the estimator $\widehat{z}_{\mathrm{IS}}$, called *sampling importance resampling*, is to sample $\widetilde{X}_1, \ldots, \widetilde{X}_N$ from $\widetilde{\pi}$ and compute the importance weights

$$w_n \overset{\text{def}}{=} \frac{(\pi/\widetilde{\pi})(\widetilde{X}_n)}{(\pi/\widetilde{\pi})(\widetilde{X}_1) + \cdots + (\pi/\widetilde{\pi})(\widetilde{X}_N)} \, .$$

Next the particles $X_1, \ldots, X_R$ are sampled (with replacement) from the set $\widetilde{X}_1, \ldots, \widetilde{X}_N$, such that $\widetilde{X}_n$ is sampled w.p. w_n, and one computes the estimator

$$\widehat{z}_{\mathrm{SIR}} \overset{\text{def}}{=} \frac{1}{R} \big(g(X_1) + \cdots + g(X_R) \big) \, .$$

For computing $z = \mathbb{E}g(X)$, sampling importance resampling provides asymptotically unbiased estimates, as will be shown shortly. However, $\widehat{z}_{\mathrm{SIR}}$ is not in itself an improvement of the importance-sampling estimator $\widehat{z}_{\mathrm{IS}}$ based on $\widetilde{X}_1, \ldots, \widetilde{X}_N$, because the principle of conditional Monte Carlo (cf. V.4) implies that $\widehat{z}_{\mathrm{SIR}}$ has a larger variance:

$$\mathbb{E}\big[\widehat{z}_{\mathrm{SIR}} \,\big|\, \widetilde{X}_1, \ldots, \widetilde{X}_N\big] \;=\; \mathbb{E}\big[g(X_1) \,\big|\, \widetilde{X}_1, \ldots, \widetilde{X}_N\big] \;=\; \sum_{n=1}^{N} w_n g\big(\widetilde{X}_n\big)$$

$$=\; \widehat{z}_{\mathrm{IS}} \, . \tag{2.4}$$

The potential advantage in some situations may be that one gets a sample that asymptotically behaves just like an ordinary i.i.d. sample: obviously $X_1, \ldots, X_R$ are dependent, but we will see in a moment that the dependence is weak if N is large, as well as that the marginal distribution is asymptotically correct. That is, $X_1, \ldots, X_R$ gives a sample from Π^* without the problems of MCMC associated with the unknown rate of convergence of the simulated Markov chain to the steady state. On the other hand, the bias of $\widehat{z}_{\mathrm{SIR}}$ is the same as for $\widehat{z}_{\mathrm{IS}}$, i.e., of order N, which is outperformed by MCMC when the Metropolis–Hastings chain is, say, geometrically ergodic.

The precise formulation of these statements is the following result, for which we omit a statement of regularity conditions (needed for uniform integrability properties); moreover, we just outline the proof at the heuristic level:

Proposition 2.1 *The sample* $X_1, \ldots, X_R$ *obtained by resampling from* $\widetilde{X}_1, \ldots, \widetilde{X}_N$ *satisfies the following properties as* $N \to \infty$:

(i) $\mathbb{E}g(X_1) \;\to\; \Pi^*[g] \overset{\text{def}}{=} \int g(x)\pi^*(x)\mu(\mathrm{d}x);$

(ii) a) $\mathbb{E}\big[g_1(X_1)g_2(X_2)\big] \;\to\; \Pi^*[g_1]\Pi^*[g_2]\,.$

Proof. The first statement follows immediately from (2.4) and the asymptotic unbiasedness of $\widehat{z}_{\text{IS}}$, since $\mathbb{E}g(X_1) = \widetilde{\mathbb{E}}\widehat{z}_{\text{IS}}$ by symmetry. For the second, we approximate the common numerator in the definition of w_n by

$$N\,\widetilde{\mathbb{E}}\big[(\pi/\widetilde{\pi})(\widetilde{X}_1)\big] \;=\; N \int (\pi/\widetilde{\pi})\widetilde{\pi}\,\mathrm{d}\mu \Big/ \int \widetilde{\pi}\,\mathrm{d}\mu \overset{\text{def}}{=} NC_1\,.$$

Further,

$$M_N \overset{\text{def}}{=} \max\Big((\pi/\widetilde{\pi})(\widetilde{X}_1),\,\ldots,\,(\pi/\widetilde{\pi})(\widetilde{X}_N)\Big)$$

is of order $o(N)$, as follows from general properties of i.i.d. r.v.'s with finite mean. Conditioning on $\widetilde{X}_1,\ldots,\widetilde{X}_N$, we get

$$\mathbb{E}\big[g_1(X_1)g_2(X_2)\big] \;=\; \widetilde{\mathbb{E}} \sum_{n,m=1}^{N} w_n w_m g_1(\widetilde{X}_n)g_2(\widetilde{X}_m)$$

$$=\; \sum_{n=1}^{N} \widetilde{\mathbb{E}}\big[w_n^2 g_1(\widetilde{X}_n)g_2(\widetilde{X}_n)\big] \;+\; \sum_{n\neq m} \widetilde{\mathbb{E}}\big[w_n w_m g_1(\widetilde{X}_n)g_2(\widetilde{X}_m)\big]\,.$$

In the first sum, we bound one w_n in w_n^2 by $M_N/\mathrm{O}(N) = \mathrm{o}(1)$ to conclude that the sum is of smaller order than

$$\widetilde{\mathbb{E}} \sum_{1}^{N} w_n g_1(\widetilde{X}_n)g_2(\widetilde{X}_n) \;\to\; \mathbb{E}\big[g_1(X_1)g_2(X_1)\big]\,.$$

That is, the whole sum goes to 0. The second sum can be approximated by

$$\frac{N(N-1)}{N^2 C_1^2}\widetilde{\mathbb{E}}\Big[(\pi/\widetilde{\pi}g)(\widetilde{X}_1)(\pi g/\widetilde{\pi})(\widetilde{X}_2)\Big] \;=\; \frac{N(N-1)}{N^2 C_1^2}C_1^2\Pi^*[g_1]\Pi^*[g_2]\,,$$

which goes to $\Pi^*[g_1]\Pi^*[g_2]$. $\qquad\qquad\square$

2c *Particle Filtering*

When performing sequential importance sampling for long input sequences, it turns out that the importance weights $w_{r,k}$ tend to become degenerate in the sense that for large k, just one or a few of the R particles $r = 1,\ldots,R$ have a nonvanishing importance weight. *Particle filtering*, going back to Gordon, Salmond, & Smith [162], combines sequential importance sampling with the resampling idea. The principle is simple: proceed as for sequential importance sampling, but at selected instants $0 < k_1 < k_2 < \cdots$, resample the particles according to their importance weights.

The algorithm basically allows particles with high importance weights (high fitness) to reproduce and eliminates those with low fitness. As explained in [95, p. 432] there is thereby some similarity to genetic algorithms, e.g., Goldberg [159] and Whitley [362].

Example 2.2 As an example of particle filtering, we consider a target tracking problem in which the velocity $X_n^{(1)}$ at time n of the target (in a discrete-time setup) follows an autoregressive process, and the position $X_n^{(2)}$ is observed with noise as Y_n. That is, the model is

$$
\begin{aligned}
X_n^{(1)} &= aX_{n-1}^{(1)} + \sigma\varepsilon_n' , \\
X_n^{(2)} &= X_{n-1}^{(2)} + X_{n-1}^{(1)} , \\
Y_n &= X_n^{(2)} + \omega\varepsilon_n'' .
\end{aligned}
$$

We chose a standard Gaussian setup in which the sequences $\{\varepsilon_n'\}$, $\{\varepsilon_n''\}$ are independent and Gaussian $\mathcal{N}(0,1)$. As parameters, we took $a = 0.8$, $\sigma = \sqrt{1-a^2}$ (corresponding to a stationary variance of the AR part of 1), and $\omega = 3$. We considered a sample path of length $N = 128$ with initial values $X_0^{(1)} = X_0^{(2)} = 0$.

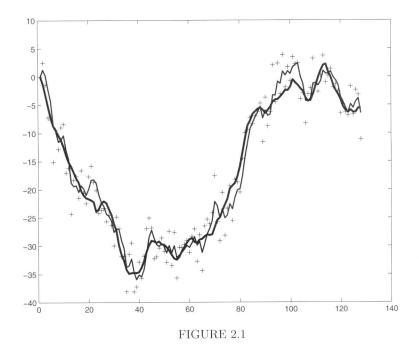

FIGURE 2.1

Figure 2.1 shows a set of data simulated from this model. The $X_n^{(2)}$ are plotted in the thick graph and the Y_n are marked with $+$'s. The thin graph is the Kalman filter that is available because of the Gaussian structure of the model. Thus, in view of the ease with which the Kalman filter is calculated, there is really no need for particle filtering in this model. However, the Kalman filter provides a benchmark against which to check the performance of particle filtering.

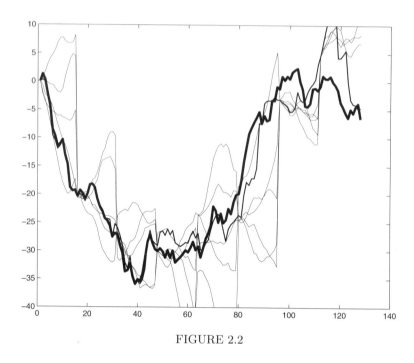

FIGURE 2.2

Figure 2.2 shows the result of a particle filtering experiment with $R = 10{,}000$ particles and resampling at times $16, 32, \ldots, 128$. The thinnest graphs show 5 particles selected at random, the thickest is the Kalman filter values, and the one in between is the Bayes estimator resulting from the particle filter, that is, the value at time k is the average of the positions of the R particle at that time. It is seen that the resampling takes the particle positions back to close to where they should be, at the Kalman filter value, In between resampling, each particle moves away from the Kalman filter, but the weighing and averaging serves to keep the Bayes estimator on track. The fit of the Bayes estimator appears to deteriorate somewhat in the last part of the plot.

$\square$

Exercises

2.1 (TP) Derive a CLT for the sampling importance resampling estimator $\widehat{z}_{\mathrm{SIR}}$. Can this be used for giving confidence intervals?

3 Counting and Measuring

One example of a counting problem, the graph-coloring problem, was presented in I.5.3. Here $G = (V, \mathcal{E})$ is a graph with $\mathcal{E}$ the set of edges and V the set of vertices, and the problem is to give each vertex one of q possible colors, such that no two neighbors (vertices connected by an edge) have the same color. With S the set of all such configurations, what is the cardinality $|S|$ of S?

Here are some further selected similar problems:

(a) Consider an 8×8 chessboard and 32 identical domino tiles, each the size of two adjacent squares of the chessboard. How many ways are there to cover the entire chessboard with the domino tiles?

(b) A famous problem, going back to Gauss in 1850, is to position eight queens on an 8×8 chessboard such that no queen can take the other. In how many ways can this be done?

(c) Given a graph $G = (V, \mathcal{E})$, how many subsets F of $\mathcal{E}$ have the property that no two sites in F are adjacent? Such a subset corresponds to a realization of a discrete hard-core model; cf. I.5.2.

(d) In statistics, $n \times m$ tables (r_{ij}), $i = 1, \ldots, n$, $j = 1, \ldots, m$, with $r_{ij} \in \mathbb{N}$ are often encountered. The marginals $r_{i\cdot} \stackrel{\text{def}}{=} \sum_{j=1}^{m} r_{ij}$, $r_{\cdot j} \stackrel{\text{def}}{=} \sum_{i=1}^{n} r_{ij}$ play an important role, and an often relevant question is, what is the number of such tables with fixed marginals $r_{i\cdot}$, $r_{\cdot j}$? See further Liu [241, p. 92].

(e) In US presidential elections, the 50 states and the District of Columbia are given certain numbers $n_1, \ldots, n_{51}$ of electors (e.g., the year 2000 numbers were 6 for Arkansas, 54 for Califonia, 25 for Florida), $N \stackrel{\text{def}}{=} \sum n_i = 538$ in total. With Θ_i defined as the indicator that candidate A beats candidate B in state i, the total scores for A, respectively B, are then

$$C_A \stackrel{\text{def}}{=} \sum_{1}^{51} n_i \Theta_i \quad C_B \stackrel{\text{def}}{=} \sum_{1}^{51} n_i (1 - \Theta_i) = N - C_A.$$

Erikson & Sigman [110] postulated that the Θ_i were i.i.d. r.v.'s with a Bernoulli($1/2$) distribution and used simulation to estimate the probability $\mathbb{P}(C_A = C_B)$ of a tie, which amounts to counting the number $|S|$ of the set S of all outcomes leading to a tie (then $\mathbb{P}(C_A = C_B) = |S|/2^{51}$; the estimate of [110] was 0.8%, corresponding to $|S| \approx 1.8 \times 10^{13}$).

Such counting problems are also closely related to tasks such as computing finite but huge sums, or measuring. For example:

(f) How do we efficiently compute the normalizing constant $\sum_{x \in S} e^{-H(x)/T}$ for the Boltzmann distribution in I.5.4 in the case of a finite but large set S of possible values for x (for example the Ising model)?

(g) How can we efficiently estimate the permanent of a $d \times d$ matrix $\boldsymbol{A} = (a_{ij})$ with 0-1 entries, defined as

$$\sum_{\sigma} \prod_{i=1}^{d} a_{i\sigma(i)} \,,$$

where the sum extends over all permutations of $\{1, \ldots, n\}$? This is a classical counting problem, with a main recent contribution being given by Jerrum, Sinclair, & Vigoda [199].

(h) (Dyer, Frieze, & Kannan [106]) How can we efficiently estimate the volume of a convex set $S \in \mathbb{R}^d$ if the available information is an oracle that for a given $x \in \mathbb{R}^d$ will return the answer whether $x \in S$?

As mentioned in I.5.3, one possible approach to counting the number of elements $|S|$ is importance sampling, where the importance distribution $\widetilde{\mathbb{P}}$ lives on a larger set T. For example, in the graph-coloring problem, T could be the set of all $|V|^q$ possible colorings and $\widetilde{\mathbb{P}}$ the uniform distribution on T. The problem we are facing is then one of rare-event simulation: we are estimating $|T|z$, where $|T| = |V|^q$ is known and z is the probability that a uniform r.v. on T takes its value in S; z is small because it can be shown under mild conditions on G that z goes to 0 exponentially fast in $|V|$ (see, e.g., [189, Problem 9.4]).

Instead of using the concepts of bounded relative error and logarithmic efficiency from Chapter VI, we follow here the tradition of computer science to distinguish between algorithms that can be run in polynomial time and those that can be run only in exponential time. More precisely, consider a computational problem parameterized with a number m such that the "size" of problem m is σ_m. A set of algorithms with running times t_m for the mth is then said to be *polynomial time* if $t_m \leq C\sigma_m^p$ for some C and some p, and *exponential time* if $t_m \geq C_1 e^{a\sigma_m}$ for some $C_1 > 0$ and some $a > 0$. Most of the above examples have the feature that naive algorithms will be exponential-time, so a major challenge is to construct polynomial-time algorithms.

A classical reference for polynomial-time counting algorithms is Sinclair [345]; and a more recent one, Jerrum [198]; an introductory treatment is given in Häggström [189]. We shall here give a sketch of the example treated in [189], the graph-coloring problem; the original treatment with a more refined analysis can be found in Jerrum [197].

The algorithm (and many others in the area) uses MCMC to produce an approximate solution, that is, an r.v. Z such that

$$\mathbb{P}\big((1 - \varepsilon)|S| < Z < (1 + \varepsilon)|S|\big) > \delta \,, \tag{3.1}$$

where $\varepsilon > 0$ and $1/2 < \delta < 1$. Given $\delta > 1/2$, one can in fact push δ arbitrarily close to 1 by replicating and taking the median:

Proposition 3.1 *Given a simulation estimator Z with the property* (3.1) *and $\eta > 0$, $1/2 < \delta < 1$, let R be so large that $2^{R+1}(1-\delta)^{R+1} < \eta/2$, and let $Z_1, \ldots, Z_{2R+1}$ be replications of Z. Then the median $Z_{(R+1)}$ satisfies*

$$\mathbb{P}\big((1-\varepsilon)|S| < Z_{(R+1)} < (1+\varepsilon)|S|\big) > 1 - \eta. \tag{3.2}$$

Proof. Assume first that Z has density f and let F be the c.d.f. Then $Z_{(R+1)}$ has density

$$\binom{2R}{R}(2R+1)f(y)F(y)^R\overline{F}(y)^R.$$

Using the bound 2^R for the binomial coefficient and 1 for $F(y)$, integration from $(1+\varepsilon)|S|$ to ∞ therefore yields

$$\mathbb{P}\big(Z_{(R+1)} > (1+\varepsilon)|S|\big) \leq 2^R(2R+1)\overline{F}\big((1+\varepsilon)|S|\big)^{R+1}/(R+1)$$
$$\leq 2^{R+1}(1-\delta)^{R+1} < \eta/2.$$

Combining with a similar lower bound gives the result in the presence of a density. The general case is handled by an easy limiting argument. □

Now turn to the precise formulation on existence of a polynomial-time algorithm for the graph-coloring problem. We take the number q of colors as fixed and consider only graphs such that each vertex has at most d neighbors, where also d is fixed. The size of the problem can then be defined as the number $m = |V|$ of vertices (note that the number $|\mathcal{E}|$ of edges is bounded by md so that it is unimportant whether we identify "size" with $|V|$ or $|\mathcal{E}|$).

Theorem 3.2 *Let q, d be fixed. Then for any $\varepsilon > 0$, $1/2 < \delta < 1$, there exists a polynomial-time algorithm with the property* (3.1).

The construction of the algorithm consists in breaking the problem into $|\mathcal{E}|$ subproblems, each of which can be handled approximately by MCMC. More precisely, we assume that the edges have been enumerated $1, \ldots, |\mathcal{E}|$ and let G_j be the graph with vertex set V and edge set $1, \ldots, j$, $j = 0, \ldots, |\mathcal{E}|$. That is, G_j is the graph obtained from G by deleting edges $j+1, \ldots, |\mathcal{E}|$. With S_j the set of q-colorings for G_j, we then have $S_{|\mathcal{E}|} = S$ and hence[1]

$$|S| = |S_0| \prod_{j=1}^{|\mathcal{E}|} \frac{|S_j|}{|S_{j-1}|}. \tag{3.3}$$

---

[1] Incidentally, note the similarity of the decomposition (3.3) with that of VI.(9.2) used in the multilevel splitting algorithm for rare-event simulation.

Now consider a q-coloring for G_{j-1}, and let v'_j, v''_j be the end vertices of edge j. Then the q-colorings of G_j are exactly the q-colorings of G_{j-1} having different colors at v'_j, v''_j. Thus, $|S_j|/|S_{j-1}|$ in (3.3) equals z_j, defined as the fraction of q-colorings of G_{j-1} having different colors at v''_j, v''_j. Equivalently,

$$z_j = \mathbb{P}\big(\xi_j(v'_j) \neq \xi_j(v''_j)\big),$$

where ξ_j has the uniform distribution Π^*_j on the q-colorings of G_{j-1} and $\xi_j(v)$ denotes the color at vertex v.

What is done is now for each j to construct an MCMC algorithm with target distribution Π^*_j and allowing an explicit bound for the rate of convergence of the corresponding Markov chain $\{\xi_{j,n}\}_{n\in\mathbb{N}}$ to Π^*_j that is tight enough to bound the number of steps needed to get within a prescribed t.v. distance to Π^*_j by a number N_j such that $N_1 + \cdots + N_{|\mathcal{E}|}$ grows at most polynomially in $|V|$ (actually, N_j will be independent of j).

The construction of $\{\xi_{j,n}\}_{n\in\mathbb{N}}$ is simple: this is just a systematic-sweep Gibbs sampler, where the vertices are sampled one by one. Given that vertex v has been sampled, the new color is then sampled uniformly in the set of colors not to be found at any neighbor (trivially, this satisfies the symmetry property noted in XIII.4.7 to be sufficient for the uniform distribution to be stationary for the MH chain). One can then prove the following lemma:

Lemma 3.3 *Let* $0 < \varepsilon < 1$ *and*

$$N(\varepsilon) \overset{\text{def}}{=} |V|\big[(\log|V| - \log\varepsilon - \log d)/a + 1\big],$$

where $a \overset{\text{def}}{=} \log(q/2d^2)$. *Then* $\big\|\mathbb{P}\big(\xi_{j,n} \in \cdot\big) - \pi^*_j(\cdot)\big\| < \varepsilon$ *for all* $n \geq N(\varepsilon)$ *and all* $j = 1, \ldots, |\mathcal{E}|$.

This is the hard part in the proof of Theorem 3.2 (e.g., [189] uses a five page coupling argument), and we will omit the proof. We shall also need another lemma:

Lemma 3.4 *For any vertices* v_1, v_2 *and all* j, *we have* $\mathbb{P}\big(\xi_j(v_1) \neq \xi_j(v_2)\big) \geq 1/2$.

Proof. We can assume that v_1, v_2 are not neighbors, since otherwise, the probability in question is 1. Let ξ^*_j denote the coloring of all vertices except v_1. Then $\xi_j(v_1)$ given ξ^*_j is uniform over all colors not attained at any of the at most d neighbors, so

$$\mathbb{P}\big(\xi_j(v_1) \neq \xi_j(v_2)\big) = 1 - \mathbb{P}\big(\xi_j(v_1) = \xi_j(v_2)\big) \geq 1 - 1/(q-d)$$

$$\geq 1 - \frac{1}{2d^2 - d} \geq 1 - \frac{1}{2} \geq \frac{1}{2}.$$

$\square$

Proof of Theorem 3.2. For simplicity, we consider only the case $\delta = 2/3$. For each $j = 1, \ldots, |\mathcal{E}|$, run $R \overset{\text{def}}{=} 48|\mathcal{E}|^3/\varepsilon^2$ replicates $\xi_{j,n,r}$ of the Gibbs sampler

$\xi_{j,n}$ up to time $N \overset{\text{def}}{=} N(\varepsilon/8|\mathcal{E}|)$ (cf. Lemma 3.3), estimate z_j by the fraction $\widehat{z}_j$ of the $\xi_{j,N}$ for which the colors at v'_j, v''_j differ, and estimate $|S|$ by $\widehat{z} \overset{\text{def}}{=} \widehat{z}_0 \widehat{z}_1 \cdots \widehat{z}_{|\mathcal{E}|}$, where $\widehat{z}_0 \overset{\text{def}}{=} q^{|V|}$. Define $z_j^* \overset{\text{def}}{=} \mathbb{P}(\xi_{j,N}(v'_j) \neq \xi_{j,N}(v''_j))$. Since $\widehat{z}_j$ is obtained by binomial sampling with success probability z_j^*, Chebyshev's inequality gives

$$\mathbb{P}(|\widehat{z}_j - z_j^*| > \varepsilon/8|\mathcal{E}|) \leq \frac{8^2 |\mathcal{E}|^2}{\varepsilon^2} \operatorname{Var} \widehat{z}_j$$

$$= \frac{64|\mathcal{E}|^2}{\varepsilon^2} \frac{z_j^*(1 - z_j^*)}{R} \leq \frac{16|\mathcal{E}|^2}{\varepsilon^2 R} \leq \frac{1}{3|\mathcal{E}|}.$$

Thus, w.p. at least $2/3$ we have $|\widehat{z}_j - z_j^*| \leq \varepsilon/8|\mathcal{E}|$ for all j, which according to the choice of N implies $|\widehat{z}_j - z_j| \leq \varepsilon/4|\mathcal{E}|$. Therefore

$$\frac{\widehat{z}_j}{z_j} \leq 1 + \frac{\varepsilon}{4|\mathcal{E}|z_j} \leq 1 + \frac{\varepsilon}{2|\mathcal{E}|},$$

where we used Lemma 3.4 in the last step. In view of $\varepsilon < 1$, this implies (assuming w.l.o.g. that $\varepsilon < 1$)

$$\frac{\widehat{z}}{z} = \prod_{j=1}^{|\mathcal{E}|} \frac{\widehat{z}_j}{z_j} \leq 1 + \varepsilon.$$

Combining with a similar $1 - \varepsilon$ lower bound, the proof is complete.   □

4   MCMC for the Ising Model and Square Ice

4a   Gibbs Sampling: a Case Study

The purpose of this section is to gain some insight into the behavior of the Ising model and the Gibbs sampling scheme in XIII.5.5 for studying this model. More precisely, we want to get an idea of what typical pictures (configurations) look like for different parameter values, what influence the intial configuration has, and most importantly, how fast the MCMC scheme appears to mix.

We used two values $\beta = 0.3$ and $\beta = 0.6$ of the inverse temperature, two initial configurations A, B described below, and a 128×128 pixel area (lattice)

$$S = \{s = (s_1, s_2) : s_1, s_2 = 1, \dots, 128\}.$$

The configuration is described by a 128×128 $\{-1, 1\}$-valued matrix with $a(s_1, s_2) = 1$ meaning that pixel s is white and $a(s_1, s_2) = -1$ that it is black. Initial configuration A has pixel colors at the different sites i.i.d. with probability $1/2$ for white and $1/2$ for black. Configuration B has the mid 16×16 (that is, $s_1, s_2 = 57, \dots, 72$) black and the rest white.

We updated pixel by pixel by systematic Gibbs sampling (using the lexicographical ordering) as in XIII.5.5 and, for simplicity of the program, froze the boundary pixels and updated only pixels with $s_1, s_2 = 2, \ldots, 127$ (thus, what we simulated is really the Ising model conditioned on a certain boundary configuration). We took 250 sweeps of the Gibbs sampler, with a sweep defined as a scan through all 128^2 pixels and ξ_k denoting the configuration after k sweeps,, and plotted (a) the initial configuration ξ_0, (b) ξ_3, (c) ξ_{10}, (d) ξ_{25}, (e) ξ_{100}, (f) ξ_{250}. In addition, plots were given of $f(\xi_k)$ as function of $k = 0, \ldots, 250$ for two functions f, namely (g) the fraction of white pixels, (h) the fraction of neighbors with marks of opposite color. The plots are in Figures 4.1-4.4.

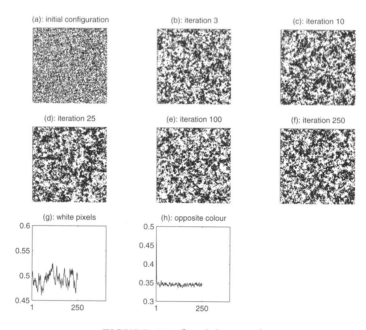

FIGURE 4.1: $\beta = 0.3$, case A

For the question of what a typical picture looks like, consider first the unrestricted Ising model in which not as here, the boundary pixels are frozen. We know the most likely pictures are either all white or all black. However, observing a picture with all pixels of the same color is (for values of β of the present order) far more unlikely than to observe one with all except one of the same color, since there are about 2×128^2 configurations of the second type and only 2 of the first. Thus, a typical picture will have a fraction of white pixels that is somewhat away from $1/2$, with say 25% and 75% being equally likely. We further expect that this fraction will move away from $1/2$ as β increases.

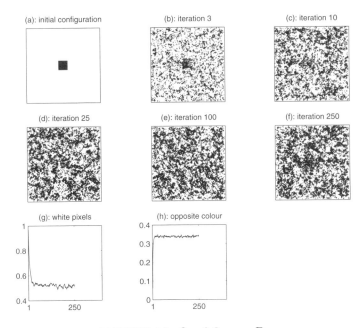

FIGURE 4.2: $\beta = 0.3$, case B

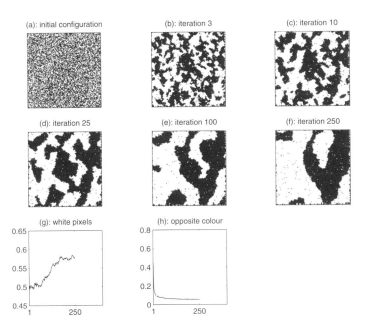

FIGURE 4.3: $\beta = 0.6$, case A

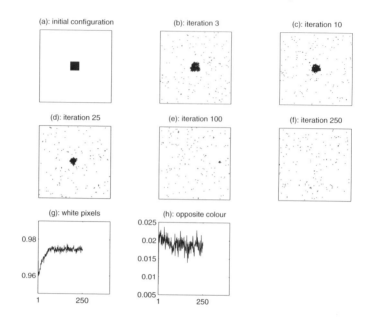

FIGURE 4.4: $\beta = 0.6$, case B

An initial configuration like B with all boundary pixels white should behave similarly, only favoring the more white configurations for the more black ones. We see this clearly in the late iterations both for $\beta = 0.3$ and $\beta = 0.6$. For $\beta = 0.3$ the typical white fraction appears from (g) to be slightly above $1/2$, for $\beta = 0.6$ it appears to be of order 97.5%.

For initial configuration A, we expect about half of the boundary pixels to be white (they are), which will preserve symmetry between black and white. Away from the boundary, one color will eventually become dominant, more so for $\beta = 0.6$ than for $\beta = 0.3$. The pictures also show this.

When not one color is predominant, the model implies clustering of black pixels and similarly for the white (this reduces the number of pairs with opposite marks), more so for $\beta = 0.6$ than for $\beta = 0.3$. The pictures for initial configuration A clearly show this. Note also that in all cases the number of pairs with opposite marks is much smaller for $\beta = 0.6$ than for $\beta = 0.3$.

Now to the question of mixing rate. Quick mixing means first of all that the initial configuration quickly is wiped out and that thereafter the configuration changes reasonably rapidly. This is certainly seen to be the case for $\beta = 0.3$ but not for $\beta = 0.6$, where the black spot in the middle in B vanishes much more slowly and the pictures for 100 and 250 steps in A have many common features. We see also that stabilization in plots (g), (h) appears to occur more slowly for $\beta = 0.6$ than for $\beta = 0.3$. Thus, there seems to be clear evidence for slower mixing for $\beta = 0.6$ than for $\beta = 0.3$.

When does the MCMC chain appear to have reached approximate stationarity (have burned in)? For $\beta = 0.3$, the figures would indicate that this is certainly so after 10 steps. For $\beta = 0.6$, the number appears rather to be 100 for initial configuration B (for initial configuration A, 250 may or may not be sufficient). However, these conclusions demonstrate a clear pitfall in burn-in tests in MCMC: we know from theory that the Ising model is symmetric in black and white pixels, and it is not clear from the figures whether the dominance of white pixels is due to the choice of having all boundary pixels white, or is an instance of quasistationary behavior. It should therefore be stressed that conclusions on decorrelation in the end are not failsafe when one is relying only on graphical and visual tests such as the ones performed.

We refer to Newman & Barkema [276] for further worked-out examples of MCMC for the Ising model, including more sophisticated pilot functions for convergence diagnostics. See also Landau [227].

4b   Wolff's Algorithm

Gibbs sampling for the Ising model, as used in Section 4a, encounters difficulties at low temperatures. The problem is that if T is small, then vertex v will with high probability be surrounded by vertices of the same spin, which will make the Gibbs sampler extremely reluctant to change the spin at vertex v. Thus, Gibbs sampling has a high rejection rate, which makes the method slow.

Swendsen & Wang developed an algorithm that improves on this situation by updating areas much larger than just one vertex, and this was later modified and refined by Wolff. We outline here Wolff's algorithm and refer to Newman & Barkema [276] for more detail both on Wolff's algorithm and the Swendsen-Wang algorithm (see also Liu [241]).

Loosely, the idea is to flip an entire cluster C of sites with equal spin and not just a single vertex. This cluster C is grown from a single vertex v picked at random, such that neighboring vertices with the same spin are added w.p. p (to be specified later) at each step. The procedure is illustrated in Figure 4.5. The initial configuration is in panel (a), and in each of the following steps, the current members of the cluster are dashboxed. The vertices marked wih a diamond $\diamond$ are those that were trials to be included in C but failed.

The algorithm is justified by a detailed balance argument, which may conveniently be phrased in terms of a state-augmentation procedure, where the state is the pair of (x, v, C), with v the vertex from which the cluster C is grown. Let k be the number of sites in C, m the number of neighbours of C of the same color as C, and n the number of neighbors of opposite color. With $x^{\#}$ the configuration obtained from x by flipping the sites in C, we then in obvious notation have $k^{\#} = k$, $m^{\#} = n$, $n^{\#} = m$, and so the

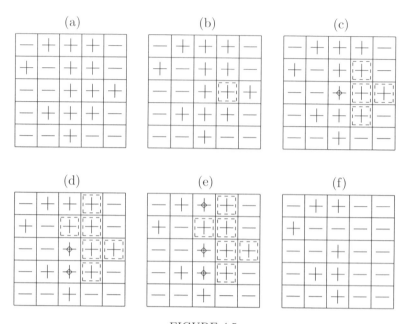

FIGURE 4.5

condition for detailed balance becomes

$$\pi(x)p^{k-1}(1-p)^m = \pi(x^{\#})p^{k-1}(1-p)^n . \tag{4.1}$$

Here $\pi(x)/\pi(x^{\#}) = e^{-\beta(n-m)}$, so taking $1-p = e^{-\beta}$ will ensure that (4.1) holds.

4c   Related Models and Algorithms

Example 4.1 An extension of the Ising model has

$$\pi(x) = \exp\left\{\alpha \sum_{i \in S} x_i - \beta \sum_{ii' \in E} \mathbb{1}\{x_i \neq x_{i'}\}\right\}. \tag{4.2}$$

If $\alpha > 0$, the first term is large if there are many 1's; if $\alpha < 0$ it is large if there are many -1's. The second term is as for the standard Ising model. Thus, compared to the standard Ising model, (4.2) has the additional feature of favoring one of the marks 1 and -1.

Another extension is the *Potts model*, having more than two types of marks (spins) than the ± 1 for the Ising model. One interpretation is different colors.   □

Example 4.2 Ice is a frozen state of water H_2O. In two dimensions, a common model for *square ice* assumes that the oxygen atoms are located at the vertices of a lattice $L \subset \mathbb{Z}^2$ and the hydrogen atoms at the edges, precisely one at each edge (the *first ice rule*). Further, a hydrogen atom is closer to one of the two oxygen atoms defining its edge, and the *second ice rule* states that there are precisely two hydrogen atoms near each oxygen atom. We represent this by directing the edges such that an arrow indicates the corresponding hydrogen atom being closer to the oxygen atom at the final point of the arrow than the one at the beginning. Thus the second ice rule means that each (nonboundary) vertex has precisely two ingoing arrows and two outgoing, and the common model asserts that all such configurations are equally likely. A pattern on (part of) the lattice compatible with both ice laws is given in panel (a) of Figure 4.6. With S the set of admissible configurations, how do we obtain information on the uniform distribution Π^* on S?

The answer is once again MCMC, and we present here the most standard algorithm, the *long loop algorithm*. See further Newman & Barkema [276].

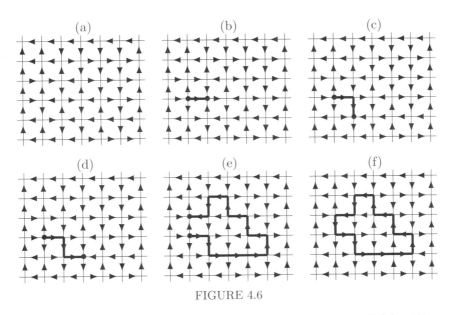

FIGURE 4.6

The initial configuration for an updating is given in panel (a) of Figure 4.6. In the first step, illustrated by panel (b), the algorithm selects an edge at random (marked by bold), chooses one of its end vertices as the first point of the loop and the other as the second (say the left is the first and the right the second), and reverses the arrow. Thereby the second ice law is violated at both points, which is marked by bullets in the points. More specifically, the first (left) point has three outgoing arrows instead of two as should be, and the second point three ingoing arrows instead of two. The algorithm next compensates for this defect at the second point (the current endpoint of the loop) by choosing one of the two of the three ingoing arrows that is not already part of the loop at random, and reversing the arrow there. The resulting configuration is in panel (c). The procedure is repeated in panel (d) and so on.

Appealing either to the finiteness of the lattice or recurrence properties of random walk in two dimensions, the loop will eventually wind its way back to one of the neighbors of the starting point as in panel (e). Here we are in the lucky situation that reversing the arrow between the starting point and the current endpoint will make the second ice rule hold at both points and hence on the whole lattice. This is done in panel (f), which then shows the final updating; the arrows around the loop have all been given opposite directions.

The transition probabilities are readily seen to be symmetric, $p(x, x') = p(x', x)$ for any $x, x' \in S$. Indeed, the probabilities of selecting the first and second points of the loop are obviously the same. In the second step as in panel (c), the two arrows to choose between for reversion are then the same, and so on. Therefore, by XIII.4.7, the resulting Markov chain on S has indeed the uniform distribution as its stationary distribution.

Note that loops may be much more complicated than in Figure 4.6. For example, the loop may return to previous vertices visited and then reverse its path. Or when it returns to a neighbor of the starting point for the first time, it is insufficient for enforcing the second ice rule that the connecting arrow be reversed etc. Loops will typically be long (random walk in two dimensions is null recurrent!), but this is not necessarily an indication of the algorithm being inefficient: as for Wolff's algorithm, the compensation is that the updating affects a potentially large part of the configuration and is not just local.   □

Exercises

4.1 (TP) In the ice model of Example 4.2 on a lattice $\{0, \ldots, N\} \times \{0, \ldots, M\}$, find a simple rule for an admissible initial configuration of an MCMC algorithm and show that the long loop algorithm is irreducible..

4.2 (TP) In the ice model on a lattice $\{0, \ldots, N\} \times \{0, \ldots, M\}$, define the set of squares as $((i-1), i) \times ((j-1), j)$, $i = 0, \ldots, N$, $j = 0, \ldots, M$. For a given coloring of the squares with colors 0,1,2 such that no adjacent squares have the same color, show that we obtain a configuration of square ice by taking pairs of adjacent squares and letting the arrow on the connecting edge point to the right if square 2 has a higher color (modulo 3) than square 1 and to the left otherwise. Show that this gives a configuration of square ice. [Hint: For the second ice law at a vertex, imagine walking around the vertex by stepping from one square to another and counting the number of times the color increases and the number of times the color decreases.] Show also that this establishes a 1-to-1 correspondance between configurations of square ice and colorings of the squares with the asserted property.

4.3 (A) In the ice model, each nonboundary vertex has $4 \cdot 3/2 = 6$ possible configurations of in- and outgoing arrows. Among these, we call a configuration *even* if both ingoing arrows are on the same horizontal or vertical line, and *odd* otherwise. Thus, out of the 6 possibilities there are 2 even and 4 odd. For a lattice $\{0, \dots, N\}^2$, give estimates of the Π^*-distribution of the frequency of odd configurations.

5 Exponential Change of Measure in Markov-Modulated Models

Consider first the discrete-time case and let $J = \{J_n\}_{n=0,1,2,\dots}$ be an irreducible Markov chain with a finite state space E. A Markov additive process (MAP) (J, S) is a generalization of a random walk, in that for the additive component $S_n = Y_1 + \dots + Y_n$ one does not equire the Y_i to be i.i.d. but rather Y_n only conditionally independent given J, such that the distribution of Y_n is $H^{(ij)}$ when $J_{n-1} = i, J_n = j$. The fundamental parameters of an MAP are thus the $H^{(ij)}$ and the transition matrix $\boldsymbol{P} = (p_{ij})_{i,j \in E}$ of J or, equivalently, the $F^{(ij)} \stackrel{\text{def}}{=} p_{ij} H^{(ij)}$; note that

$$F^{(ij)}(\infty) = p_{ij}, \quad F^{(ij)}(y) = \mathbb{P}_i(X_1 \le y, J_1 = j).$$

Of models in which discrete-time MAPs play an important role, we mention in particular Markov chains with transition matrices of GI/M/1 or M/G/1 type; see Neuts [273], [274] or [16, Section X.4]. Of course, the setup is also close to hidden Markov models.

The generalization of the m.g.f. is the $E \times E$ matrix $\widehat{\boldsymbol{F}}[\theta]$ with ijth element $\widehat{F}^{(ij)}[\theta] \stackrel{\text{def}}{=} \int e^{\theta x} F^{(ij)}(\mathrm{d}x)$, and as generalization of the cumulant g.f. one can take the logarithm $\kappa(\theta)$ of the Perron–Frobenius eigenvalue of $\widehat{\boldsymbol{F}}[\theta]$; denote the corresponding right eigenvector by $\boldsymbol{h}^{(\theta)} = \left(h_i^{(\theta)}\right)_{i \in E}$, i.e., $\widehat{\boldsymbol{F}}[\theta] \boldsymbol{h}^{(\theta)} = e^{\kappa(\theta)} \boldsymbol{h}^{(\theta)}$. The exponential change of measure (ECM) corresponding to θ is then given by

$$\widetilde{\boldsymbol{P}} = e^{-\kappa(\theta)} \boldsymbol{\Delta}_{\boldsymbol{h}^{(\theta)}}^{-1} \widehat{\boldsymbol{F}}[\theta] \boldsymbol{\Delta}_{\boldsymbol{h}^{(\theta)}}, \quad \widetilde{H}_{ij}(\mathrm{d}x) = \frac{e^{\theta x}}{\widehat{H}_{ij}[\theta]} H_{ij}(\mathrm{d}x).$$

Here $\boldsymbol{\Delta}_{\boldsymbol{h}^{(\theta)}}$ is the diagonal matrix with the $h_i^{(\theta)} e^{\theta x}$ on the diagonal. In particular, $\widetilde{p}_{ij} = e^{-\kappa(\theta)} p_{ij} h_j^{(\theta)} / h_i^{(\theta)}$ (note the similarity with the h-transform in IV.7). The likelihood ratio is

$$W_n(\boldsymbol{P}|\widetilde{\boldsymbol{P}}) = \frac{h^{(\theta)}(J_0)}{h^{(\theta)}(J_n)} e^{-\theta S_n + n\kappa(\theta)}. \tag{5.1}$$

In continuous time, an MAP with an underlying finite Markov process $J = \{J(t)\}_{t \ge 0}$ has a simple description; cf., e.g., Neveu [275]. The clue for the understanding is the structure of a Lévy process (process with stationary independent increments) as the independent sum of a deterministic drift, a Brownian component, and a pure jump process, see XII.1. Let the intensity matrix of J be $\boldsymbol{\Lambda} = (\lambda_{ij})_{i,j \in E}$. On an interval $[t, t+s]$ where

$J(t) \equiv i$, the additive part S of the MAP then evolves like a Lévy process with the drift μ_i, the variance σ_i^2 of the Brownian component, and the Lévy measure $\nu_i(\mathrm{d}x)$ depending on i. In addition, a transition of J from i to $j \neq i$ has probability q_{ij} of giving rise to a jump of S at the same time, the distribution of which has then some distribution $B^{(ij)}$.

Let $\widehat{\boldsymbol{F}}_t[\theta]$ be the matrix with ijth element $\mathbb{E}_i\big[e^{\theta S(t)}; J(t) = j\big]$. It is easy to see that $\widehat{\boldsymbol{F}}_t[\theta] = e^{t\boldsymbol{G}[\theta]}$, where

$$G^{(ij)}[\theta] = \begin{cases} q_{ij}\lambda_{ij}\widehat{B}^{(ij)}[\theta] + (1 - q_{ij})\lambda_{ij} & i \neq j, \\ \lambda_{ii} + \mu_i\theta + \frac{1}{2}\sigma_i^2\theta^2 + \int\big(e^{\theta x} - 1 - x\mathbb{1}_{|x| \leq 1}\big)\nu_i(\mathrm{d}x) & i = j. \end{cases}$$

We define $\kappa(\theta)$ as the dominant eigenvalue of $\boldsymbol{G}[\theta]$ and $\boldsymbol{h}^{(\theta)}$ as the corresponding right eigenvector. Equivalently, $e^{t\kappa(\theta)}$ is the Perron–Frobenius eigenvalue of $\widehat{\boldsymbol{F}}_t[\theta]$, and $\boldsymbol{h}^{(\theta)}$ the right eigenvector. The ECM corresponding to θ is then given by

$$\widetilde{\boldsymbol{\Lambda}} = \boldsymbol{\Delta}_{\boldsymbol{h}^{(\theta)}}^{-1}\boldsymbol{G}[\theta]\boldsymbol{\Delta}_{\boldsymbol{h}^{(\theta)}} - \kappa(\theta)\boldsymbol{I}, \quad \widetilde{\mu}_i = \mu_i + \theta\sigma_i^2, \quad \widetilde{\sigma}_i^2 = \sigma_i^2,$$

$$\widetilde{\nu}_i(\mathrm{d}x) = e^{\theta x}\nu_i(\mathrm{d}x), \quad \widetilde{q}_{ij} = \frac{\lambda_{ij}q_{ij}\widehat{B}_{ij}[\theta]}{\lambda_{ij} + \lambda_{ij}q_{ij}\big(\widehat{B}_{ij}[\theta] - 1\big)},$$

$$\widetilde{B}_{ij}(\mathrm{d}x) = \frac{e^{\theta x}}{\widehat{B}_{ij}[\theta]}B_{ij}(\mathrm{d}x).$$

In particular, the expression for $\widetilde{\boldsymbol{\Lambda}}$ means that

$$\widetilde{\lambda}_{ij} = \frac{h_j^{(\theta)}}{h_i^{(\theta)}}\lambda_{ij}\Big[1 + q_{ij}\big(\widehat{B}_{ij}[\theta] - 1\big)\Big], \quad i \neq j$$

(the diagonal elements are determined by $\widetilde{\lambda}_{ii} = -\sum_{j \neq i}\widetilde{\lambda}_{ij}$), and if $\nu_i(\mathrm{d}x)$ is compound Poisson, that is, if $\nu_i(\mathrm{d}x) = \beta_i B_i(\mathrm{d}x)$ with $\beta_i < \infty$ and B_i a probability measure, then also $\widetilde{\nu}_i(\mathrm{d}x)$ is compound Poisson with

$$\widetilde{\beta}_i = \beta_i\widehat{B}_i[\theta], \quad \widetilde{B}_i(\mathrm{d}x) = \frac{e^{\theta x}}{\widehat{B}_i[\theta]}B_i(\mathrm{d}x). \tag{5.2}$$

The likelihood ratio on $[0, T]$ is just (5.1) with n replaced by T.

Example 5.1 If all $\sigma_i^2 = 0$, $\nu_i = 0$, $q_{ij} = 0$, we have a process with piecewise linear sample paths with slope μ_i when $J(t) = i$. This process (or rather its reflected version; cf. Example 5.4) is a *Markovian fluid*, a process of considerable current interest because of its relevance for communication engineering; for some recent references, see Asmussen [13] and Rogers [303]. The ECM just means to replace λ_{ij} with $\widetilde{\lambda}_{ij} = h_j^{(\theta)}\lambda_{ij}/h_i^{(\theta)}$ for $i \neq j$ and is therefore another Markovian fluid. An important special structure that frequently arises is that the total fluid inflow is a superposition of fluid form of a large number of independent sources. The key simplifying feature of such a model is that the eigenvalue problem for the total fluid inflow can be decomposed into corresponding eigenvalue computations for each of the individual sources, each having a small state space.   □

Example 5.2 Assume that all $\sigma_i^2 = 0$, $\mu_i = -1$, $q_{ij} = 0$, and that $\nu_i(\mathrm{d}x) = \beta_i B_i(\mathrm{d}x)$ corresponds to the Poisson case with the B_i concentrated on $(0, \infty)$. This process (or rather its reflected version) corresponds to the workload process in the Markov-modulated M/G/1 queue with arrival intensity β_i and service-time distribution B_i of the customer arriving when $J(t) = i$, and the ECM just replaces these parameters by those given by (5.2).   □

Example 5.3 Assume that $\sigma_i^2 = 0$, $\mu_i = 0$ for all i, that ν_i corresponds to a Poisson(β_i) process (i.e., ν_i is a one-point measure with mass β_i at 1), and that $B^{(ij)}$ is concentrated at 1 for all i, j. Then the MAP is a counting process, in fact the same as the Markovian point process introduced by Neuts [272] and increasingly popular as a modeling tool. If $q_{ij} = 0$ for all i, j, the process is a Markov-modulated Poisson process.   □

ECM for Markov additive processes goes back to a series of papers by Keilson & Wishart and others in the 1960s, e.g., [210]. To our knowledge, the first use of the concept in simulation is Asmussen [10]. Further recent references are Bucklew [61], Bucklew et al. [63], Lehtonen & Nyrhinen [237], and Chang et al. [68]. Again, some of the most interesting applications involve combination with duality ideas, which for infinite buffer problems just means time reversal.

The approach can to some extent be generalized beyond finite E. Note, however, that if E is infinite, an MAP may be quite complicated (an example is provided by the local time of a diffusion) and that the existence of dominant eigenvalues for the relevant integral operator does not always hold. In addition, even when a dominant eigenvalue does exist, the difficulties of numerically solving the eigenvalue problem are daunting.

Example 5.4 Let S be the MAP described in Example 5.1. Then the fluid model of interest is

$$V(t) \stackrel{\mathrm{def}}{=} S(t) - \min_{0 \leq u \leq t} S(u).$$

Define the cycle as $C \stackrel{\mathrm{def}}{=} \inf \{t > 0 : S(t) = 0\}$ (this definition is interesting only if $J(0) = i$ with $\mu_i > 0$) and assume that the rare event $A(x)$ is the event $\{\sup_{0 \leq t < C} V(t) \geq x\}$ of buffer overflow within the cycle. Then (noting that $S(t) = V(t)$ for $t < C$) we can just perform the simulation by a variant of Siegmund's algorithm: determine γ by $\kappa(\gamma) = 0$, perform the corresponding ECM for the MAP (J, S), and run the MAP until it hits either x or 0. If instead $A(x) = \{V \geq x\}$ is defined in terms of the steady state, we first note the well-known representation ([14] and references there) $V \stackrel{\mathscr{D}}{=} \max_{0 \leq t < \infty} S^*(t)$, where (J^*, S^*) is the MAP we obtain by time-reversing J (replacing λ_{ij} by $\lambda_{ij}^* \stackrel{\mathrm{def}}{=} \pi_j \lambda_{ji}/\pi_i$, where π is the stationary distribution), leaving the μ_i unchanged and letting $J^*(0)$ have distribution π. Thus $z(x) = \mathbb{P}(\tau^*(x) < \infty)$, where $\tau^*(x) \overset{\text{def}}{=} \inf\{t : S^*(t) \geq x\}$ and the simulation is performed by running S until it hits x, using again ECM with $\theta = \gamma$. Similar remarks apply to Example 5.3. $\square$

A potential application of exponential change of measure or, more generally, importance sampling, for an additive process is to reduce the variance of long-run averages

$$\frac{1}{N+1} \sum_{n=0}^{N} f(X_n)$$

for a Markov chain $\{X_n\}$. It is worth pointing out that, as remarked in V.1.17, such a procedure is potentially dangerous because the variance of the likelihood ratio will typically go to infinity exponentially fast. We state and prove this for the finite case:

Proposition 5.5 *Let $\boldsymbol{P} = (p(ij))$ and $\widetilde{\boldsymbol{P}} = (\widetilde{p}(ij))$ be ergodic $d \times d$ transition matrices such that $p(x, y) = 0$ whenever $\widetilde{p}(x, y) = 0$ and that $p(x, y) \neq \widetilde{p}(x, y)$ for at least one pair x, y, and let*

$$L_N \overset{\text{def}}{=} \prod_{n=1}^{N} \frac{p(X_{n-1}, X_n)}{\widetilde{p}(X_{n-1}, X_n)}$$

be the likelihood ratio up to time N. Then there exists $\eta > 1$ such that $\lim_{N \to \infty} e^{-\eta N} \widetilde{\mathrm{Var}}_x L_N = \infty$.

Proof. We consider the Markov additive process

$$S_N \overset{\text{def}}{=} 2 \log L_N = 2 \sum_{n=1}^{N} \log p(X_{n-1}, X_n).$$

Then $\widetilde{\mathbb{E}}_x L_n^2 = \widetilde{\mathbb{E}}_x e^{S_n}$. Asymptotic results for quantities of this form are given in [16, Section XI.2] and state that $\widetilde{\mathbb{E}}_x e^{S_n} \sim Ce^{sx}$, where s is the eigenvalue of largest absolute value of the matrix $\boldsymbol{F} \overset{\text{def}}{=} \widehat{\boldsymbol{F}}[1]$ with xyth element

$$\widetilde{\mathbb{E}}_x \left[e^{S_1}; X_1 = y \right] = \widetilde{\mathbb{E}}_x \left[\frac{p(X_0, X_1)^2}{\widetilde{p}(X_0, X_1)^2}; X_1 = y \right] = \frac{p(x, y)^2}{\widetilde{p}(x, y)^2} \widetilde{p}(x, y).$$

Consider the xth row sum $(\boldsymbol{F1})_x$ of $\boldsymbol{F}$, let Y be an r.v. with $\mathbb{P}(Y = y) = \widetilde{p}(x, y)$, and let $h_x(y) \overset{\text{def}}{=} p(x, y)/\widetilde{p}(x, y)$. Then

$$(\boldsymbol{F1})_x = \widetilde{\mathbb{E}} h_x(Y)^2 \geq \left(\widetilde{\mathbb{E}} h_x(Y) \right)^2 = 1.$$

Further, $p(x, y) \neq \widetilde{p}(x, y)$ for at least one pair x, y implies that there is strict inequality for this x. Altogether, $\boldsymbol{F1} = \boldsymbol{1} + \boldsymbol{a}$, where $\boldsymbol{a} \geq \boldsymbol{0}$, $\boldsymbol{a} \neq \boldsymbol{0}$, which by Perron–Frobenius theory implies $s > 1$. Take $1 < \eta < s$. $\square$

6   Further Examples of Change of Measure

6a   *Girsanov's Formula*

Consider the SDE

$$\mathrm{d}X(t) \;=\; \mu(t)\,\mathrm{d}t \,+\, \sigma(t)\,\mathrm{d}B(t)\,, \;\; 0 \le t \le T\,,$$

where $\{B(t)\}$ is standard Brownian motion and $\{\mu(t)\}$, $\{\sigma(t)\}$ are adapted to the Brownian filtration (T may be constant or a stopping time). Assume that we want to derive characteristics of the solution $\{X(t)\}$ by instead simulating

$$\mathrm{d}\widetilde{X}(t) \;=\; \widetilde{\mu}(t)\,\mathrm{d}t \,+\, \sigma(t)\,\mathrm{d}B(t)\,, \;\; 0 \le t \le T\,.$$

What is then the likelihood ratio?

The solution is immediately suggested by the basic change-of-measure formulas for Brownian motion as derived in V.1.8 and the local Brownian character of solutions of SDEs. That is, if $\mu(t) \equiv \mu$, $\widetilde{\mu}(t) \equiv \widetilde{\mu}$, $\sigma(t) \equiv \sigma$ in an interval $(t, t+h]$, then the contribution to the likelihood from this interval is

$$\exp\{-\big(\widetilde{\mu} - \mu\big)/\sigma^2 \cdot \big(B(t+h) - B(t)\big) + h\sigma^2/2\}\,.$$

Dividing $[0, T]$ into a large number of such intervals and using the multiplicative property of the likelihood ratio therefore suggests that in the SDE setting,

$$\frac{\mathrm{d}\mathbb{P}}{\mathrm{d}\widetilde{\mathbb{P}}} \;=\; \exp\Big\{ -\int_0^T \big(\widetilde{\mu}(t) - \mu(t)\big)/\sigma^2(t)\,\mathrm{d}B(t) \,+\, \frac{1}{2}\int_0^T \sigma^2(t)\,\mathrm{d}t \Big\}\,.$$

A rigorous proof together with the technical conditions needed can be found in virtually any textbook on stochastic calculus, e.g., Steele [346]. The result is often stated in a somewhat different way and is known as *Girsanov's formula*.

For absolute continuity reasons similar to those discussed in V.1.8, it seldom makes sense to implement importance sampling for SDEs by changing $\{\sigma(t)\}$ rather than $\{\mu(t)\}$.

6b   *Many-Server Queues*

Consider a $\mathrm{GI}/\mathrm{G}/s$ queue with first-in–first-out discipline (the customers form one line and join the first server to become idle). Let U_k denote the service time of customer k and T_k between the time of his arrival and that of customer $k+1$. Assume stability, i.e., $\mathbb{E}U < s\mathbb{E}T$.

For the $\mathrm{GI}/\mathrm{G}/1$ queue with light-tailed service times, exponential change of measure obtained by exponential tilting of $X \overset{\text{def}}{=} U - T$ plays a major role in deriving asymptotics and is also a key tool for obtaining variance reduction in simulations; cf., e.g., Siegmund's algorithm in VI.2a. It is therefore natural to ask whether there is a similar analogue for many-server queues.

The answer is affirmative and is worked out in a large-deviations setting in Sadowsky [326] and Sadowsky & Spankowsky [328]. The solution is to replace $X = U - T$ by $X \stackrel{\text{def}}{=} U/s - T$. With $\kappa_U(\theta) \stackrel{\text{def}}{=} \log \mathbb{E}e^{\theta U}$, $\kappa_T(\theta) \stackrel{\text{def}}{=} \log \mathbb{E}e^{\theta T}$, this means that exponential θ-tilting is obtained by replacing the densities $f_U(u)$, $f_T(t)$ of U, T by $e^{\theta u - \kappa_U(\theta)} f_U(u)$, respectively $e^{-\theta t - \kappa_T(-\theta)} f_T(t)$. This is the same as for $s = 1$, but the difference shows in the equation that defines θ as a root. For example, in the Cramér–Lundberg exponential asymptotics $\mathbb{P}(W > x) \sim Ce^{-\gamma x}$, one should take γ as solution of $\kappa_U(\gamma) + \kappa_T(-\gamma) = 0$ for $s = 1$ but $\kappa_U(\gamma/a) + \kappa_T(-\gamma) = 0$ for $s > 1$. Another example is given below.

It is not a priori obvious that $U/s - T$ is the correct choice for the r.v. on which to perform exponential tilting. For eample, another obvious candidate is $U - T_1 - \cdots - T_s$. The justification for $U/s - T$ is the asymptotic results of [326], [328].

The Cramér–Lundberg γ is also the θ used in Siegmund's algorithm for simulating the stationary waiting-time tail when $s = 1$. It is notable, however, that the representations of GI/G/s waiting times in terms of the U_k, T_k are much more complicated when $s > 1$ than the random walk/ duality representations derived in I.1 and IV.7, so that there is no direct analogue of Siegmund's algorithm (at least no simple one; Blaszczyszyn & Sigman [49] exhibit an analogue where the state of the dual process is set-valued!). The most obvious algorithm for simulating $\mathbb{P}(W > x)$ for large x in the many-server setting is the switching regenerative method (used earlier in Exercise VI.2.6): start the system empty, run the system using the γ parameters until the waiting time of a customer exceeds x (or the system becomes empty again), and then complete the busy cycle without importance sampling.

Another example is the transient buildup of large workloads. Assume that we are interested in $z \stackrel{\text{def}}{=} \mathbb{P}(V(a) > b)$, where $\{V(t)\}$ is the workload process and both a and b are large. Exponential tilting with θ yields a drift

$$\mu_\theta \stackrel{\text{def}}{=} \frac{\mathbb{E}_\theta U}{\mathbb{E}_\theta T} - s = \frac{\kappa_U'(\theta)}{\kappa_T'(-\theta)} - s$$

of $\{V(t)\}$. The large-deviations analysis of Anantharam [7] for $s = 1$ suggests that there are two intrinsically different possibilities depending on how b/a compares to μ_γ. If $b/a > \mu_\gamma$, the importance sampling should be done by determining θ such that the drift is b/a on the whole of the time interval $[0, a]$. If on the contrary $b/a \leq \mu_\gamma$, the drift should be normal (making the system close to empty) in the beginning and then change to μ_γ, say at $t = c$, where thus c is determined by $b/(a - c) = \mu_\gamma$. See further Figure 6.1, where the corresponding large-deviations paths are illustrated.

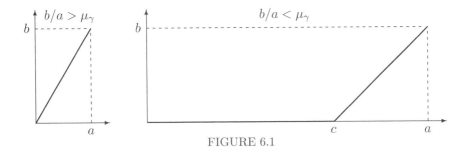

FIGURE 6.1

Remark 6.1 When $s = 1$ and the system starts empty at time $t = 0$, an alternative approach is available. Time reversal and duality permit us to express $\mathbb{P}(V(t) > x)$ as the probability that a level crossing of x for the netput process occurs prior to time t; cf. IV.7. Recall that the importance distribution used for computing the probability of a level crossing of x in finite time employs exponential twisting using parameter $\gamma > 0$, where $\kappa(\gamma) = 0$. So, if $x/\kappa'(\gamma) \le t$, this exponential twist is likely to force the level crossing to occur prior to t (and hence this importance distribution is appropriate for computing the probability of a level crossing prior to t). On the other hand, if $x/\kappa'(\gamma) > t$, then one must twist the time-reversed process using twisting parameter θ_t, where θ_t solves $\kappa(\theta_t) = x/t$. In other words, a more extreme twisting of the process is required if the time horizon t is relatively small. $\square$

6c   Local Exponential Change of Measure

We consider a stochastic process $\{X_t\}$, in discrete or continuous time, with $X_0 = 0$ and consider the problem of efficient estimation of $z \overset{\text{def}}{=} \mathbb{P}(\tau < \infty)$ in a rare-event setting where $\tau \overset{\text{def}}{=} \inf\{t > 0 : X_t > x_0\}$ for some large $x_0 > 0$.

In the light-tailed case, the standard tool for random walks and Lévy processes is Siegmund's algorithm from VI.2a. It can be motivated by a conditioned limit theorem stating that the particular change of measure used in the algorithm is close to the conditional distribution given the rare event $\{\tau < \infty\}$, cf. VI.5. We now consider the case in which $\{X_t\}$ is a more general Markov process, which, however, has the feature of behaving locally like a random walk or Lévy process. Examples are the following:

(a) A diffusion with diffusion coefficients $\mu(\cdot), \sigma^2(\cdot)$ that at level x behaves like a Brownian motion with parameters $\mu(x), \sigma^2(x)$.

(b) A compound Poisson process with drift $\lambda\mathbb{E}V - a(x)$ at level x,

$$X_t = \sum_{i=1}^{N_t} V_i - \int_0^t a(X_s)\,\mathrm{d}s\,. \tag{6.1}$$

For example, in insurance risk, consider an extended Cramér–Lundberg model (cf. I.5.14) with claim-arrival process N, claims $V_1, V_2, \ldots$, and premium rate $c(r)$ at level r. With

$$R_t = x_0 - \sum_{i=1}^{N_t} V_i + \int_0^t c(R_s)\,\mathrm{d}s \tag{6.2}$$

(the reserve at time t) and $X_t \stackrel{\text{def}}{=} x_0 - R_t$, the ruin probability with initial reserve x_0 has then the above form with $a(x) = c(x_0 - x)$.

(c) An inhomogeneous random walk on $\mathbb{Z}$ with probability p_x of going from x to $x + 1$ and $1 - p_x$ of going to $x - 1$.

In all three examples, there is an obvious local analogue $\kappa(\theta, x)$ of the cumulant function $\kappa(\theta)$ for a random walk or Lévy process, namely $\mu(x)\theta + \sigma^2(x)\theta^2/2$ in (a), $\beta\big(\mathbb{E}e^{\theta V} - 1\big) - a(x)\theta$ in (b), where β is the rate of the Poisson process N, and $\log\big(p_x e^\theta + (1 - p_x)e^{-\theta}\big)$ in (c). The idea of *local exponential change of measure* as introduced by Cottrell, Fort, & Malgoyres [76] (see also Asmussen & Nielsen [25]) is to simulate $\{X_t\}$ as a Markov process whose parameters at level x are the same as for Siegmund's algorithm based on $\kappa(\theta, x)$. That is, for each x one computes $\gamma(x)$ as the nonzero solution of $\kappa\big(\gamma(x), x\big)$ and uses the same model description as for the given model when updating from $X_n = x$ to X_{n+1}, only with $\kappa(\theta, x)$ replaced by $\widetilde{\kappa}(\theta, x) \stackrel{\text{def}}{=} \kappa\big(\theta + \gamma(x), x\big)$. In the examples:

(a) Here $\gamma(x) = -2\mu(x)/\sigma^2(x)$, which yields $\widetilde{\kappa}(\theta, x) = -\mu(x)\theta + \sigma^2(x)\theta^2/2$. That is, the importance distribution is a diffusion with the same $\sigma^2(x)$ and the sign of $\mu(x)$ reversed.

(b) Assume for simplicity that V is exponential with rate δ, so that $\mathbb{E}e^{\theta V} = \delta/(\delta - \theta)$. We then get $\gamma(x) = \delta - \beta/a(x)$, which after some algebra gives

$$\widetilde{\kappa}(\theta, x) = \widetilde{\beta}(x)\Big(\frac{\widetilde{\delta}(x)}{\widetilde{\delta}(x) - \theta} - 1\Big) - a(x)\theta\,,$$

where

$$\widetilde{\beta}(x) \stackrel{\text{def}}{=} \beta\mathbb{E}e^{\gamma(x)V} = \frac{\beta\delta}{\delta - \gamma(x)} = \delta a(x)\,, \tag{6.3}$$

$$\widetilde{\delta}(x) \stackrel{\text{def}}{=} \delta - \gamma(x) = \beta/a(x)\,. \tag{6.4}$$

That is, the level-dependent drift remains unchanged, whereas the compound Poisson part becomes level-dependent, with arrival rate and exponential jump rate at level x given by (6.3), (6.4).

(c) Here $\alpha \stackrel{\text{def}}{=} e^{\gamma(x)}$ solves $1 = p_x \alpha + (1 - p_x)/\alpha$. This is a quadratic with solutions 1 and $(1 - p_x)/p_x$, and it easily follows that $\widetilde{\mathbb{P}}$ corresponds to $\widetilde{p}_x = 1 - p_x$.

In the spatial homogeneous setting, the likelihood ratio is $\mathrm{d}\mathbb{P}/\mathrm{d}\widetilde{\mathbb{P}} = e^{-\gamma X_\tau}$. For the local ECM, one gets instead

$$
\frac{\mathrm{d}\mathbb{P}}{\mathrm{d}\widetilde{\mathbb{P}}} = \begin{cases} \displaystyle\prod_{k=1}^{\tau} \exp\{-\gamma(X_{k-1})(X_k - X_{k-1})\} & \text{in discrete time,} \\[2ex] \displaystyle\exp\left\{-\int_0^\tau \gamma(X_{t-})\,\mathrm{d}X_t\right\} & \text{in continuous time.} \end{cases}
$$

Cottrell, Fort, & Malgoyres [76] derive a logarithmic efficiency result in a *slow Markov walk* setting. This is a device from large-deviations theory and means in discrete time that $X_{k+1} = X_k + \varepsilon Y_k$, where Y_k has distribution F_x when $X_k = x$. The relevant limit for the efficiency result is then $\varepsilon \to 0$. For example in (b), this essentially means that x_0 is large compared to the typical size of a jump V.

6d   The Multidimensional Siegmund Algorithm

We now consider a multidimensional ruin problem, to estimate the probability that a random walk $\{\boldsymbol{S}_n\}$ in $\mathbb{R}^d$ hits a rare set. More precisely, we will assume that the rare set has the form $xA = \{xa : a \in A\}$, where A is convex and x is a large parameter, and that the random walk in itself would typically avoid xA. For this, the drift vector $\boldsymbol{\mu} \stackrel{\text{def}}{=} \mathbb{E}\boldsymbol{S}_1$ should as a minimum satisfy $t\boldsymbol{\mu} \notin xA$ for all t and x (technically, the existence of a separating hyperplane is sufficient). For simplicity, we take $d = 2$ in the following. Define $\tau(x) \stackrel{\text{def}}{=} \inf\{n : S_n \in xA\}$ and $z(x) \stackrel{\text{def}}{=} \mathbb{P}(\tau(x) < \infty)$.

The situation is an in Figure 6.2. Here $x\boldsymbol{\xi}(\boldsymbol{k})$ is the point at which the line with slope $\boldsymbol{k}$ hits xA. The mean drift vector $\boldsymbol{\mu}$ points away from A, so an obvious possibility is to use a change of measure changing the drift to some $\boldsymbol{k}$ pointing towards A. This change of measure would naturally be taken as exponential, given what we have already seen on optimality properties of this procedure.

Exponential change of measure for $d > 1$ has already been introduced in V.1b. Recall that for $d = 2$, we define

$$
\kappa(\boldsymbol{\theta}) = \kappa(\theta_1, \theta_2) = \log \mathbb{E}e^{\theta_1 X_1 + \theta_2 X_2}
$$

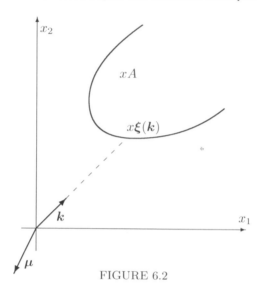

FIGURE 6.2

where $(X_1, X_2) \stackrel{\text{def}}{=} \boldsymbol{S}_1$, and that the exponentially tilted measure is a random walk with increment distribution satisfying

$$\mathbb{E}_{\theta_1,\theta_2} h(X_1, X_2) = \mathbb{E}\left[h(X_1, X_2) \exp\left\{\theta_1 X_1 + \theta_2 X_2 - \kappa(\theta_1, \theta_2)\right\}\right]. \quad (6.5)$$

It easily follows from (6.5) that the changed drift under $\mathbb{P}_{\theta_1,\theta_2}$ is given by

$$\boldsymbol{\mu}_{\theta_1,\theta_2} \stackrel{\text{def}}{=} \mathbb{E}_{\theta_1,\theta_2}(X_1, X_2) = (\kappa_1, \kappa_2) = \nabla\kappa(\theta_1, \theta_2) \quad (6.6)$$

where ∇ denotes the gradient.

The Siegmund algoritm for $d = 1$ now suggests to performing an exponential change of measure with $\boldsymbol{\gamma} = \boldsymbol{\theta}$ where $\kappa(\boldsymbol{\theta}) = 0$. However, in contrast to $d = 1$, the solution $\boldsymbol{\gamma}$ is no longer unique: the set of solutions typically form a closed curve K with $\boldsymbol{0} \in K$; cf. Figure 6.3. The arrows pointing outward from K are the gradient. The gradient is orthogonal to K and at any given point, its length is twice the radius of curvature at the given point of K.

Thus, we are faced with the problem of which $\boldsymbol{\gamma} \in K$ to work with. Now, a lower bound for $z(x)$ is given by the probability of the path following the $\mathbb{P}_{\boldsymbol{\gamma}}$-description, i.e., by

$$z(x) = \mathbb{P}\big(\tau(x) < \infty\big) = \mathbb{E}_{\boldsymbol{\gamma}} e^{-\boldsymbol{\gamma} \cdot \boldsymbol{S}_{\tau(x)}} \approx e^{-x \boldsymbol{\gamma} \cdot \boldsymbol{\xi}(\boldsymbol{\mu}_{\boldsymbol{\gamma}})}. \quad (6.7)$$

This suggests to take $\boldsymbol{\gamma} = \boldsymbol{\gamma}^*$ where

$$\boldsymbol{\gamma}^* \stackrel{\text{def}}{=} \operatorname{argmin} \boldsymbol{\gamma} \cdot \boldsymbol{\xi}(\boldsymbol{\mu}_{\boldsymbol{\gamma}}).$$

It can be shown that (under appropriate conditions) indeed the correct logarithmic asymptotics corresponds to taking $\boldsymbol{\gamma} = \boldsymbol{\gamma}^*$ in (6.7) and that the corresponding exponential change of measure with $\boldsymbol{\theta} = \boldsymbol{\gamma}^*$ leads to logarithmic efficiency; see Collamore [74].

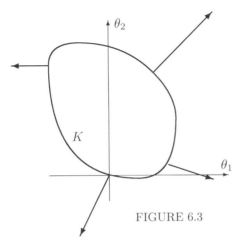

FIGURE 6.3

A more intricate but highly relevant version of the same problem occurs in FIFO single classs queuing networks where the state space is restricted to $[0, \infty)^d$. The dynamics of such as network can be represented, in great generality, as a "free" d-dimensional random walk on $\mathbb{R}^d$ that is subject to certain boundary behavior (to be precise, a d-dimensional version of the so-called Skorokhod problem) that forces the network state to remain in the non-negative orthant. One can represent the most likely path to a rare set (such as the total network population exceeding a large level) either in terms of the dynamics of the underlying "free" process (i.e., the random walk) or in terms of the queuing model itself. The major problem that arises in the network setting is that the most likely path, in fluid scale, can be a quite complex piecewise-linear trajectory, whether expressed in terms of the underlying free process or in terms of the "reflected" queueing model. A complete description of the range of possible rare event path possibilities has been developed by Avram, Dai, & Hasenbein [32] for two-dimensional reflected Brownian motion (RBM) on the non-negative orthant. Such a RBM can be viewed as a mathematically stylized version of a two-dimensional FIFO queuing network. The piecewise linear structure arises from the possibility, for example, that a large queueing population at a given station in the network may be a consequence of first filling the buffer of an upstream station with a large number of customers, followed by draining the upstream station and simultaneously slowing the server at the given (downstream) station. This type of path would have two piecewise linear components, when expressed either in terms of the free process or the reflected queueing process. Since it is unlikely that a closed-form mathematical theory for these complex piecewise linear paths exists for d-dimensional networks, such rare-event network simulation algorithms will need to algorithmically compute the appropriate paths. Some general discussion on large-deviations paths can be found in Atar & Dupuis [30] and

Dupuis & Ellis [101] (see also Foley & McDonald [119] and McDonald [250] for some special cases). Further, it seems likely that the implementation has to be via state-dependent importance sampling, as in important ongoing work of Dupuis & Wang [104], [105], [w³.5].

6e   Polymers and Self-Avoiding Random Walks

Polymers are significant objects of interest in chemistry and molecular biology. We will here consider a two-dimensional setting, even if in real life polymers are of course three-dimensional.

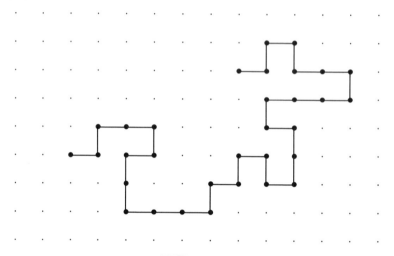

FIGURE 6.4

A polymer of length n is loosely speaking a string of $n+1$ molecules. In our planar setting, we will model the distribution of polymers of length n as the uniform distribution Π_n^* on the set E of *self-avoiding random walks* on $\mathbb{Z}^2$. That is, if $X_0, \ldots, X_n$ is a random walk on $\mathbb{Z}^2$ with probability $1/4$ for moving to each of the neighbors, then Π^* is the distribution of $X_0, \ldots, X_n$ conditioned to have no two points in common (i.e., the cardinality of the set $\{X_0, \ldots, X_n\}$ is $n+1$). Interesting questions are, for example, what is the cardinality of E, and what is the average distance between the two endpoints of a polymer of length n? Explicit formulas are not available, so simulation is one possible approach.

The most obvious simulation scheme is to simulate $X_0, \ldots, X_n$ unconditionally and reject sample path violating the self-avoidance criterion. That is, if

$$\zeta \stackrel{\text{def}}{=} \inf\{k : |\{X_0, \ldots, X_k\}| = k\}$$

is the first time the random walk returns to a previously occupied site, then only sample paths with $\zeta > n$ are accepted.

However, since the set

$$E_0 \stackrel{\text{def}}{=} \{\boldsymbol{x}_{0:n} : x_0 = (0,0), |x_k - x_{k-1}| = 1, k = 1, \ldots, n-1\}$$

of all 4^n possible sample paths for the random walk (assuming throughout w.l.o.g. that $X_0 = (0,0)$) is much larger than E, the method becomes slow for large n. Instead, we define an importance distribution $\widetilde{\mathbb{P}}$ by choosing X_k uniformly among the neighbors of X_{k-1} that have not been visited in steps $0, \ldots, k-1$; if there are no such neighbors, the sample path cannot be in E and the simulation can be stopped (the path is rejected).

For $\boldsymbol{x}_{0:n} \in E_0$, define $d_k \stackrel{\text{def}}{=} d_k(\boldsymbol{x}_{0:n})$ as the number of neighbors of x_k that have not previously been visited.

Proposition 6.2 $\dfrac{\mathrm{d}\Pi^*}{\mathrm{d}\widetilde{\mathbb{P}}}(\boldsymbol{x}_{0:n}) \propto \varphi(\boldsymbol{x}_{0:n}) \stackrel{\text{def}}{=} d_1 d_2 \cdots d_{n-1}.$

Proof. We modify the definition of d_k by letting $d_k = 4$ if x_k has no unvisited neighbors; since this occurs w.p. 0 for both Π^* and $\widetilde{\mathbb{P}}$, this does not affect the statement of Proposition 6.2. We then let $\widetilde{\mathbb{P}}'$ be the modification of $\widetilde{\mathbb{P}}$ that after ζ chooses any of the $d_k = 4$ neighbors of X_{k-1} w.p. $1/4$. With $\widetilde{p}(\boldsymbol{x}_{0:n})$, $\widetilde{p}'(\boldsymbol{x}_{0:n})$ the probabilities of a path $\boldsymbol{x}_{0:n} \in E$ w.r.t $\widetilde{\mathbb{P}}$, respectively $\widetilde{\mathbb{P}}'$, we then get

$$\frac{\mathrm{d}\widetilde{\mathbb{P}}}{\mathrm{d}\widetilde{\mathbb{P}}'}(\boldsymbol{x}_{0:n}) = \frac{\widetilde{p}(\boldsymbol{x}_{0:n})}{\widetilde{p}'(\boldsymbol{x}_{0:n})} = \frac{1}{\widetilde{\mathbb{P}}'(\boldsymbol{X}_{0:n} \in E)},$$

$$\widetilde{p}(\boldsymbol{x}_{0:n}) \propto \widetilde{p}'(\boldsymbol{x}_{0:n}) = \frac{1}{d_1}\frac{1}{d_2} \cdots \frac{1}{d_{n-1}},$$

$$\frac{\mathrm{d}\Pi^*}{\mathrm{d}\widetilde{\mathbb{P}}}(\boldsymbol{x}_{0:n}) = \frac{1/|E|}{\widetilde{p}(\boldsymbol{x}_{0:n})} \propto d_1 d_2 \cdots d_{n-1}.$$

$\square$

The normalizing constant $\widetilde{\mathbb{E}}\varphi(\boldsymbol{X}_{0:n})$ is not computable, so one has to use either MCMC or the techniques of V.1.16 and Section 2 to compute characteristics of Π^*.

Exercises

6.1 (TP) In the setting of Section 6e on self-avoiding random walks, calculate for $n = 4$ the $\widetilde{\mathbb{P}}$-densities of the paths

$$\boldsymbol{x}' = ((0,0), (1,0), (1,1), (0,1), (0,2)),$$
$$\boldsymbol{x}'' = ((0,0), (1,0), (2,0), (3,0), (4,0)).$$

[You should get $\widetilde{p}(\boldsymbol{x}') < \widetilde{p}(\boldsymbol{x}'')$, illustrating that the importance distribution favors more dispersed paths rather than more compact ones.]

7   Black-Box Algorithms

Roughly speaking, a black-box algorithm is a device for producing a desired output from an input using only minimal information on its structure. A prototype example is the Propp–Wilson algorithm in IV.8 for perfect sampling, where the desired output is an r.v. sampled from the stationary distribution of a finite Markov chain and the input a stream of simulated values of the Markov chain, where no particular property of the Markov chain such as the form of the transition probabilities is required.

We shall not attempt a general definition of black-box algorithms but rather present some examples where existence/nonexistence is an issue.

7a   Exponential Tilting

Exponential change of measure (ECM) is one of the main tools of importance sampling. In some cases such as the normal or exponential distributions, random number generation is straightforward because one stays within the same parametric class. When this is not the case, the situation may be more cumbersome, so one may ask for an automated way to perform ECM.

Let $X_1, X_2, \ldots$ be i.i.d. r.v.'s with distribution G; for simplicity we assume $X \geq 0$. Define

$$\mathscr{F}_n \stackrel{\mathrm{def}}{=} \sigma\left(X_1, \ldots, X_n, U_1, \ldots, U_n\right),$$

where $U_1, U_2, \ldots$ are independent $(0, 1)$ uniforms. A *randomized stopping time* for $X_1, X_2, \ldots$ is then a stopping time w.r.t. the filtration $\{\mathscr{F}_n\}$. Let further $\mathscr{G}$ be a class of distributions G on $[0, \infty)$ with $\kappa_G \stackrel{\mathrm{def}}{=} \int_0^\infty \mathrm{e}^{\theta x}\, G(\mathrm{d}x) < \infty$ for each $G \in \mathscr{G}$, and for $G \in \mathscr{G}$, define G_θ by $\mathrm{d}G_\theta/\mathrm{d}G(x) = \mathrm{e}^{\theta x - \kappa_G}$. We define a black-box algorithm for ECM for the given θ and the given class $\mathscr{G}$ of distributions as a pair of a randomized stopping time τ and an $\mathscr{F}_\tau$-measurable r.v. Z such that Z is distributed according to G_θ for any $G \in \mathscr{G}$. Formally, we write this as

$$\mathbb{P}_G(Z \leq x) = G_\theta(x) \quad \text{for all } x.$$

In other words, the rules of the game are that the algorithm may use simulated values from G and additional uniforms[2] to produce an r.v. from G_θ.

Proposition 7.1 *Let θ be given and let $\mathscr{G}$ be the class of all distributions on $[0, \infty)$ with $\kappa_G < \infty$. Then:*

---

[2]Note that it is no restriction to consider only a single U_k in introducing additional randomness; U_k can be transformed to a sequence of i.i.d. uniforms (or even to a whole Brownian motion).

(i) *No black-box algorithm for ECM exists when $\theta > 0$.*

(ii) *A black-box algorithm for ECM exists when $\theta < 0$.*

Proof. (i) Let θ be given and assume that a black-box algorithm exists. Choose an arbitrary distribution G on $[0, \infty)$ with infinite support such that $\mathbb{P}_G(Z \leq 1) = G_\theta(1) = 1/2$. Define $M_\sigma \overset{\text{def}}{=} \max_{k \leq \sigma} X_k$ and choose $x_0 > 1$ so large that $\mathbb{P}_G(Z \leq 1, M_\sigma \leq x_0) > 1/4$. Next define H by $H(\mathrm{d}x) = G(\mathrm{d}x)$ for $x \leq x_0$, and let H have at atom of size $1 - G(x_0)$ at some $B < \infty$. By choosing B large enough, we can make κ_H arbitrarily large and hence $H_\theta(1)$ arbitrarily small, say $H_\theta(1) < 1/8$. Then

$$\begin{aligned} 1/8 \ &> \ H_\theta(1) \ = \ \mathbb{P}_H(Z \leq 1) \\ &\geq \ \mathbb{P}_H(Z \leq 1, M_\sigma \leq x_0) \ = \ \mathbb{P}_G(Z \leq 1, M_\sigma \leq x_0) \ > \ 1/4, \end{aligned}$$

a contradiction. Here in the fourth step we used the fact that (σ, Z) can simply be viewed as an $\mathbb{N} \times \mathbb{R}$-valued function on the set of infinite sequences $(x_1, u_1, x_2, u_2, \ldots)$, such that Z can be computed from $(x_1, u_1, \ldots, x_n, u_n)$ when $\sigma = n$.

In (ii), we may just use acceptance–rejection, accepting an $X_k = x$ w.p. $e^{\theta x}$. □

For $\theta > 0$, one should not conclude that full knowledge of G is necessary for r.v. generation from G_θ. Less may do, as the following proposition demonstrates:

Proposition 7.2 *Let $\mathcal{G}$ be the class of distributions concentrated on $[0, A]$ for some $A < \infty$. Then a black-box algorithm for ECM exists for all θ.*

Proof. Acceptance–rejection with acceptance probability $e^{\theta x}$ for $x < 0$ and $e^{\theta(x-A)}$ for $x > 0$. □

7b Perfect Sampling of Countable Markov Chains

By a black-box algorithm for the stationary distributions $(\pi_P)_{P \in \mathscr{P}}$ for a class $\mathscr{P}$ of ergodic transition matrices on E (i.e., a class of ergodic Markov chains), we understand a (randomized) stopping time σ for $\{X_n\}$ and an r.v. Z, measurable w.r.t. $\mathscr{F}_\sigma$, where

$$\mathscr{F}_n \overset{\text{def}}{=} \sigma(X_0, \ldots, X_n, U_0, \ldots, U_n)$$

with $U_0, U_1, \ldots$ uniform$(0, 1)$ and independent of $\{X_n\}$, such that

$$\mathbb{P}_P(Z \in A) \ = \ \pi_P(A) \quad \text{for all } A \subseteq E \text{ and all } P \in \mathscr{P}, \qquad (7.1)$$

where $\mathbb{P}_P$ indicates that $\{X_n\}$ is simulated according to P and π_P is the stationary distribution for P. A different terminology is to call π_P *simulatable* within $\mathscr{P}$.

Remark 7.3 The "rules of the game" are thus to use nothing more than a simulated version of $\{X_t\}$ and some possible additional randomization. In particular, the algorithm is not allowed to use analytic information on the p_{ij}. The purpose of this restriction is twofold: First, if the p_{ij} are analytically available, one can argue that there exist deterministic algorithms for computing π by solving linear equations. Next, the natural description of a Markov chain is most often in terms of an updating rule rather than the p_{ij}, and one would simulate directly from the updating rule rather than use it to compute the p_{ij}. For an example, consider the Kiefer–Wolfowitz vector $\boldsymbol{W}_n \stackrel{\text{def}}{=} \left(W_n^{(1)}, \ldots, W_n^{(s)}\right)$ in a GI/G/c queue (the components give the workloads at the s servers in nondescending order at the nth arrival). Here the updating rule is

$$\boldsymbol{W}_{n+1} \;=\; \mathscr{R}\left(\left[W_n^{(1)} + U_n - T_n\right]^+, \left[W_n^{(2)} - T_n\right]^+, \ldots, \left[W_n^{(s)} - T_n\right]^+\right),$$

where U_n is the service time of customer n, T_n the nth interarrival time, and $\mathscr{R} : [0,\infty)^s \to [0,\infty)^s$ the operator rearranging the components in nondescending order. □

From IV.8 we obtain the following result:

Theorem 7.4 *If E is finite, then the stationary distribution for the class $\mathscr{P}_E$ of all ergodic transition matrices on E is simulatable.*

However (Asmussen, Glynn, & Thorisson [19]), we also have the following:

Theorem 7.5 *If E is countably infinite, then the stationary distribution for the class $\mathscr{P}_E$ of all ergodic transition matrices on E is not simulatable.*

Proof. We argue by contradiction by assuming that (7.1) holds for $\mathscr{P} = \mathscr{P}_E$. Assume w.l.o.g. that $E = \{0,1,2,\ldots\}$.

Let $\boldsymbol{P}^{(0)} = (p_{ij}^{(0)})_{i,j\in E}$ be arbitrary, write $\mathbb{P}_0 = \mathbb{P}_{\boldsymbol{P}^{(0)}}$, $\pi^{(0)} = \pi_{\boldsymbol{P}^{(0)}}$, and choose $K < \infty$ such that $\mathbb{P}_0(Z \leq K, M \leq K) > 1 - \varepsilon$, where $M \stackrel{\text{def}}{=} \max_{n \leq \sigma} X_n$ and $0 < \varepsilon < 1/2$. For $\alpha \in (0,1)$, define

$$p_{ij}^{(\alpha)} \stackrel{\text{def}}{=} \begin{cases} p_{ij}^{(0)} & i \leq K, \\ \alpha + (1-\alpha)p_{ij}^{(0)} & i = j > K, \\ (1-\alpha)p_{ij}^{(0)} & i > K,\, i \neq j. \end{cases}$$

That is, $\boldsymbol{P}^{(\alpha)}$ is obtained from $\boldsymbol{P}^{(0)}$ by adding a geometric number (with parameter α) of "self-loops" in states $i > K$; in states $i \leq K$, the transitions are just the same, and hence

$$\mathbb{P}_\alpha(Z \leq K, M \leq K) \;=\; \mathbb{P}_0(Z \leq K, M \leq K) \;>\; 1 - \varepsilon,$$

where $\mathbb{P}_\alpha \stackrel{\text{def}}{=} \mathbb{P}_{\boldsymbol{P}^{(\alpha)}}$, $\pi^{(\alpha)} \stackrel{\text{def}}{=} \pi_{\boldsymbol{P}^{(\alpha)}}$.

Let $\tau \overset{\text{def}}{=} \inf \{n > 0 : X_n = 0 \mid X_0 = 0\}$ and recall the regenerative formula IV.(6.3),

$$\pi_i = \frac{1}{\mathbb{E}\tau} \mathbb{E} \sum_{n=0}^{\tau-1} \mathbb{1}_{X_n = i}$$

for the stationary distribution of a Markov chain. For $i \leq K$, this yields

$$\pi_i^{(\alpha)} = \frac{1}{\mathbb{E}_\alpha \tau} \mathbb{E}_\alpha \sum_{n=0}^{\tau-1} \mathbb{1}_{X_n = i} = \frac{1}{\mathbb{E}_\alpha \tau} \mathbb{E}_0 \sum_{n=0}^{\tau-1} \mathbb{1}_{X_n = i} = \frac{\mathbb{E}_0 \tau}{\mathbb{E}_\alpha \tau} \pi_i^{(0)} . \quad (7.2)$$

From the self-loop property it follows that

$$\mathbb{E}_\alpha \tau \geq \frac{1}{1-\alpha} \mathbb{P}_\alpha \left(\max_{0 \leq n < \tau} X_n > K \right) = \frac{1}{1-\alpha} \mathbb{P}_0 \left(\max_{0 \leq n < \tau} X_n > K \right) . \quad (7.3)$$

As $\alpha \uparrow 1$, the r.h.s. of (7.3) goes to ∞, and hence (7.2) goes to 0. Hence with $A = \{1, \dots, K\}$, we have $\pi^{(\alpha)}(A) < \varepsilon$ for all α close enough to 1, and get

$$\begin{aligned} \mathbb{P}_\alpha(Z \in A) - \pi^{(\alpha)}(A) &\geq \mathbb{P}_\alpha(Z \leq K, M \leq K) - \pi^{(\alpha)}(A) \\ &\geq 1 - \varepsilon - \varepsilon > 0, \end{aligned}$$

contradicting (7.1). $\qquad\qquad\qquad\qquad\qquad\qquad\qquad\qquad\qquad\qquad\qquad\square$

The implication of Theorem 7.5 is not necessarily that one should consider exact simulation impossible when faced with a particular nonfinite Markov chain $\{X_n\}$. Rather, Theorem 7.5 says that exact simulation cannot be based on simulated values of $\{X_n\}$ alone, but one needs to combine with some specific properties of $\{X_n\}$, i.e., to involve knowledge of the form of $\boldsymbol{P} \in \mathscr{P}_0$, where $\mathscr{P}_0 \subset \mathscr{P}_E$. Examples are in Foss & Tweedie [121] in the framework of Harris chains, in Section 8 in a regenerative setting, and in Section 13a for the GI/G/1 waiting time.

7c From Bernoulli(p) to Bernoulli$\big(f(p)\big)$

Assume that we have access to a stream $X_1, X_2, \dots$ of Bernoulli(p) variables with the numerical value of p unknown except that we have the information that $p \in S \subset [0,1]$. For a given function $f : S \to [0,1]$, we then want to generate Bernoullis$\big(f(p)\big)$.

Some cases are obviously doable, say $S = (0,1)$ and $f(p) = \theta p$ with $0 \leq \theta \leq 1$. Then standard acceptance–rejection with acceptance probability θ will do. But what if $\theta > 1$? For example,[3] if $S = (0, 1/2)$, can we generate Bernoullis$(2p)$?

---

[3] This was the form in which the problem was raised by one of the authors, SA, at a meeting in 1991.

The general solution to this problem was obtained by Keane & O'Brien in a 1994 paper [208], where the following was shown:

Proposition 7.6 *Simulation of Bernoullis$(f(p))$ from Bernoullis(p) is possible if and only if f is constant, or if f is continuous and for all $p \in S$ and some $n \geq 1$ one has*

$$\min\big(f(p), 1 - f(p)\big) \ \geq \ \min(p, 1 - p)^n . \qquad (7.4)$$

Example 7.7 Consider the $p \to 2p$ problem with $S = (0, 1/2)$. Then $p < 1 - p$ for all $p \in S$, whereas $\min(2p, 1 - 2p)$ equals $1 - 2p$ when $p \geq 1/4$ and $2p$ otherwise. Thus, for $p \leq 1/4$ the condition of Proposition 7.6 becomes $2p \geq p^n$ for some $n \geq 1$, which trivially holds with $n = 1$. In the interval $1/4 \leq p < 1/2$, one must have $1 - 2p \geq p^n$ for some $n \geq 1$ and all p, which is impossible, as is seen by considering the limit $p \uparrow 1/2$. However, if the known information is $0 < p < 1/2 - \varepsilon$ for some $\varepsilon > 0$, the desired inequality will hold by taking n so large that $2\varepsilon \geq 2^{-n}$.

We conclude that a black-box algorithm for $p \to 2p$ exists if the available information is $0 < p < 1/2 - \varepsilon$, but not when we just know $0 < p < 1/2$. □

Nacu & Peres [269] generalized the problem one step further by asking for the number $N\big(S, p, f(p)\big)$ of Bernoullis(p) r.v.'s that one needs to generate a Bernoulli$\big(f(p)\big)$. More precisely, when does $\mathbb{P}\big(N\big(S, p, f(p)\big) > n\big)$ decay at least exponentially fast or when does $N\big(S, p, f(p)\big)$ at least have a finite kth moment? The answers given in [269] involve further smoothness properties of f.

8   Perfect Sampling of Regenerative Processes

Continuing and generalizing the studies of IV.8 and Section 7b, we now consider the problem of perfect sampling of a regenerative process X as defined in IV.6b; the exposition is basically an extract from Asmussen, Glynn, & Thorisson [19], with some improvements in Example 8.6. We know already from Section 7b that no black-box algorithm, using as input simulated zero-delayed cycles and additional uniform r.v.'s, can exist in the class of all regenerative processes, since such an algorithm would in particular apply to countable ergodic Markov chains. Thus, perfect sampling algorithms will have in one way or another to exploit some known feature of the regenerative process, and in particular we will see that some partial information on the tail $\mathbb{P}(\tau > x)$ of the distribution of the cycle length τ is potentially useful.

We use continuous-time notation (the discrete time case is entirely similar). We allow a delay, i.e. an initial cycle $C_0 \overset{\text{def}}{=} \{X(t)\}_{0 \leq t < \tau_0}$ with a distribution possibly differing from the common one of the remaining cycles

$C_k \stackrel{\text{def}}{=} \left\{ X_{t+T(0)+T(k-1)} \right\}_{0 \le t < T(k)-T(k-1)}$, $k = 1, 2, \ldots$. Formally, the cycles can be viewed as independent random elements of the Skorokhod D-space of E-valued functions with a finite lifetime such that $C_1, C_2, \ldots$ are i.i.d. We write $X \stackrel{\text{def}}{=} (C_0, C_1, \ldots)$ and C for the generic cycle, τ for its length. The starting point for the discussion is the following result on the structure of a stationary regenerative process, for which we refer to [16, Section VI.2e] or Thorisson [354]:

Proposition 8.1 (a) *Assume that the distribution of C_0 has density τ/μ w.r.t. the distribution of C, that is,*

$$\mathbb{P}(C_0 \in \cdot) = \frac{1}{\mu} \mathbb{E}\left[\tau; C \in \cdot \right].$$

Let U be uniform$(0,1)$ and define $X^*(t) \stackrel{\text{def}}{=} X(t + UT(0))$. Then X^* is a stationary version of X;
(b) *Assume that the distribution of $T(0)$ has density $\mathbb{P}(\tau > t)/\mu$ and that*

$$\mathbb{P}(C_0 \in \cdot \,|\, T(0) = t) = \mathbb{P}(\theta_t C \in \cdot \,|\, \tau > t),$$

where $\theta_t C \stackrel{\text{def}}{=} (X(s+t))_{0 \le s < \tau - t}$. Then X is stationary.

The statement of (a) is intuitively that a stationary version of X is obtained by letting the initial cycle be obtained from the generic cycle C by first length-biasing with τ and next placing the time origin uniformly within the cycle. This leads to the following corollary:

Corollary 8.2 *A black-box algorithm for simulating a stationary regenerative process exists within the class where $\tau \le a$ for some known a.*

Proof. The length-biasing can be performed by simulating a generic cycle C and performing acceptance–rejection with acceptance probability τ/a. $\square$

Example 8.3 The situations in which τ is bounded are rare. One example is an (s, S) inventory system in which the demand in $[0, t]$ has the form $bt + \sum_1^{N_t} Y_i$, where $\{N_t\}$ is Poisson and the $Y_i > 0$ i.i.d. When the inventory downcrosses level s, it is immediately replenished to S. Taking the downcrossings of s as regeneration points, one then has $\tau \le a$, where $a \stackrel{\text{def}}{=} (S - s)/b$ is the length of a cycle with no arrivals. $\square$

Part (b) of Proposition 8.1 leads to the following algorithm, in which the stationary cycle r.v. τ^* is defined as a r.v. with density $\mathbb{P}(\tau > x)/\mu$:

Proposition 8.4 *In the class of regenerative processes with τ^* simulatable, a stationary version can be generated by simulating C_0 as follows: Generate τ^* and successive cycles $C_1', C_2', \ldots$. Let σ be the first k with $\tau_k' > \tau^*$, and take $C_0 \stackrel{\text{def}}{=} \theta_{\tau^*} C_\sigma$.*

Example 8.5 To generate a stationary version of an M/G/1 queue is an easy problem under the FIFO (first-in–first-out) discipline, since there the stationary characteristics are known in closed form or at least explicitly enough to allow simulation; cf., e.g., the Pollaczeck–Khinchine formula I.(5.3) for the common stationary distribution of the workload or the delay. For other queuing disciplines the problem may be more difficult, for example for processor sharing or priority queuing. These disciplines are, however, like many others, work-conserving, meaning that the workload process has the same distribution as in the FIFO case. Therefore the simulation of a stationary workload process does not present new problems. Suppose, however, as an example that we want to simulate a stationary version $X_0^*, X_1^*, X_2^*, \ldots$ of the sequence of delays of customers $0, 1, 2, \ldots$ in the processor-sharing case (this sequence has characteristics that are genuinely different from the FIFO case). As regeneration points, we take customers arriving to an empty system, and these then are the same as for the FIFO case. Therefore, we can generate the stationary FIFO delay by the Pollaczeck–Khinchine formula and run the FIFO delay sequence according to the Lindley recursion I.(1.1) until a customer sees an empty system upon arrival. The desired stationary cycle length τ^* is then the number of customers served until then, and to complete the construction of C_0, one simulates $C_1', C_2', \ldots$ using the processor-sharing discipline. □

Example 8.6 A general criterion for the ability to simulate a stationary version of a regenerative process is the availability of an integrable function $b(t)$ of explicit form such that $\mathbb{P}(\tau > t) \le b(t)$, complemented with a (typically more easily available!) lower bound $\mathbb{P}(\tau > t) \ge q(t) > 0$. An algorithm for generating τ^* is then the following:

(i) Generate T from the density $h \stackrel{\text{def}}{=} b/\int b$. Fix $\delta > 1$ and write $t \stackrel{\text{def}}{=} T$, $g \stackrel{\text{def}}{=} g(t) \stackrel{\text{def}}{=} \delta b(t)$, $q \stackrel{\text{def}}{=} q(t)$, $p \stackrel{\text{def}}{=} \mathbb{P}(\tau > t)$.

(ii) Choose $n \ge 1$ such that $\delta(1 - g^n) \ge 1$ when $g < 1$, and $g \le f_n(q)$ when $g \ge 1$, where $f_n(p) \stackrel{\text{def}}{=} p/(1 - p)^n$ [note that $f_n(q) \uparrow \infty$ as $n \to \infty$].

(iii) From simulated values of τ, generate a Bernoulli(p/g) r.v. V using the Keane–O'Brien algorithm (see [208] and Section 7c) with n as in step (ii).

(iv) If $V = 1$, return $\tau^* \stackrel{\text{def}}{=} t$. Otherwise, return to step (i).

The algorithm is thus an acceptance–rejection algorithm, accepting an r.v. from $h(t)$ w.p. $\mathbb{P}(\tau > t)/g(t)$. Since both $h(t)$ and $g(t)$ are proportional to $b(t)$, the output τ^* therefore has density proportional to $\mathbb{P}(\tau > x)$ as desired.

A proof is needed that step (iii) is feasible, that is, that the requirement (7.4) for the Keane–O'Brien algorithm is fulfilled. In the present context, this means that we should have

$$\min(p/g, 1 - p/g) \geq \min(p, 1 - p)^n. \tag{8.1}$$

We first show that p/g is at least the r.h.s., i.e., at least either p^n or $(1-p)^n$. If $g \leq 1$, then $p/g \geq p^n$ because of $n \geq 1$. If $g > 1$, then

$$\frac{p}{g} = \frac{f_n(p)(1-p)^n}{g} \geq \frac{f_n(p)(1-p)^n}{f_n(q)} \geq (1-p)^n,$$

since $f_n(\cdot)$ is increasing and $p \geq q$. For the similar lower bound for $1 - p/g$, note first that $1 - p/g \geq 1 - p \geq (1-p)^n$ when $g \geq 1$. If $g < 1$, we will show that $1 - p/g \geq p^n$. This will hold if $1 - p/g \geq g^n$, which in turn is equivalent to $p \leq g - g^{n+1}$. The truth of this follows from

$$g - g^{n+1} = \delta b(t)(1 - g^n) \geq b(t) \geq p. \qquad \square$$

Example 8.7 For a simple yet interesting example in which an upper bound on $\mathbb{P}(\tau > t)$ can be obtained, consider an (s, S) inventory system with state space $\{0, 1, \ldots, S\}$. Items are removed at Poisson(λ) times, and when the inventory level makes a transition $s + 1 \to s$, a supplier is called that will arrive after a random time Z (the *lead time*) to replenish the inventory to S items. As regeneration times, we take the times of transitions $s + 1 \to s$. If Z is stochastically smaller than the $\Gamma(\alpha, \lambda')$ distribution for some α and some $\lambda' \geq \lambda$, an upper bound on $\mathbb{P}(\tau > t)$ is then the tail probability $b(t)$ of a $\Gamma(\alpha + S - s, \lambda)$ distribution. A (trivial) lower bound is $q(t) \stackrel{\text{def}}{=} \mathbb{P}(Z > t)$. If instead $\lambda' \leq \lambda$, use the $\Gamma(\alpha + S - s, \lambda')$ distribution as upper bound. $\qquad \square$

Example 8.8 For another example in which the stationary cycle length τ^* is simulatable, consider a Harris chain satisfying the uniform minorization condition

$$P^m(x, A) \geq \varepsilon \nu(A) \tag{8.2}$$

for some $m = 1, 2, \ldots$, some $0 < \varepsilon < 1$, some probability measure ν, all measurable A, and all $x \in E$. As explained in IV.6d in more detail, regeneration points (in the wide sense) can then be constructed by flipping a coin w.p. ε for heads at times $n = 0, m, 2m, \ldots$ and letting a wide-sense regeneration occur at time $n + m$ by taking the distribution of X_{n+m} to be ν. In this way, the distribution of the generic cycle becomes geometric on the lattice $m, 2m, \ldots$. That is, $\mathbb{P}(\tau = km) = \varepsilon(1 - \varepsilon)^{k-1}$, $k = 1, 2, \ldots$, which yields

$$\mathbb{P}(\tau^* = km + j) = \frac{\varepsilon(1 - \varepsilon)^{k-1}}{m}, \quad k = 1, 2, \ldots, \ j = 0, 1, \ldots, m - 1.$$

Of course, such a τ^* is straightforward to simulate as $mN + M$, where N is geometric on $\{1, 2, \ldots\}$ with success parameter ε and M is uniform on $\{0, \ldots, m - 1\}$. Note that when $m > 1$ and we want a whole stationary process, we cannot simulate cycles $C_1, C_2, \ldots$ as independent as in Proposition 8.4, but need to continue the simulation of the Harris chain according to the rules described in IV.6d.   □

9 Parallel Simulation

Multi-processor computing systems are playing an increasingly important role in many areas of scientific computing. The additional computing power afforded by such machines can be used to advantage in at least two different and distinctive ways to enhance the reach and efficiency of Monte Carlo simulations:

(i) For computations in which the model to be simulated has modest memory requirement, one can assign different streams of random numbers to each of the individual processors, and execute independent simulations of the same system on each of the available processors, thereby speeding up the overall calculation.

(ii) For complex and massive simulations, it may be advantageous to distribute the simulation over multiple processors with the goal of creating a computing architecture in which all the available processors efficiently and cooperatively subdivide the effect involved in simulating a given realization of the stochastic process of interest.

The situation (ii) is exemplified by the simulation of an interacting particle system on a finite grid in which individual processors are assigned to simulate the state of each site (see, for example, Newman & Barkema [276] for the Ising model). If the particle system has only "local interactions" so that each site's state is determined only by its immediate neighbors, this massively parallel architechture has a modest communication overhead (in that the processor corresponding to a given site needs to communicate the site's state only to those processors associated with the site's immediate neighbors). This leads to a simulation environment that can potentially simulate the particle system much faster than could a single processor computer. A more challenging example of such a simulation is the context of an automobile traffic simulation within a city. To implement such a simulation in a single processor context would be extremely slow or perhaps even infeasible. However, in a multi-processor setting, one can potentially subdivide the calculations across the available processors by, for example, assigning the detailed traffic simulation for the streets in a particular neighborhood to a given processor. As long as vehicles do not cross neighborhood boundaries, the neighborhood traffic simulations can proceed independently. Of course, vehicles do occasionally cross neighborhood boundaries. When an outgoing vehicle crosses a neighborhood boundary, the processor corresponding to the first neighborhood must pass a message to the processor corresponding to the neighborhood that the outgoing vehicle is entering. Since the simulations at the different processors are typically not synchronized, it may be that the second neighborhood's simulation has already been executed to a time of day at which the vehicle from the outgoing neighborhood enters the second neighborhood. This necessitates a "rollback" in the second neighborhood's simulation to a time of the day later than the instance at which the vehicle under discussion entered. Development of efficient algorithms for executing such rollbacks is an active area of research within the distributed simulation community; see Fujimoto [123].

The most straightforward application of parallel processing power arises in computations of type (i), in which independent realizations can be simulated on each of the individual processes. Suppose, for example, that our goal is to compute $z = \mathbb{E}Z$. Suppose that processor i simulates i.i.d. copies $Z_{i1}, Z_{i2}, \ldots$ of the r.v. Z, and that each independent realization takes a time units to simulate. In the presence of p parallel processors, the error associated with the grand average over all realizations simulated in c time units is

$$\frac{1}{p}\sum_{i=1}^{p}\frac{1}{\lfloor c/a \rfloor}\sum_{j=1}^{\lfloor c/a \rfloor}Z_{ij} - z \overset{\mathscr{D}}{\approx} \sqrt{\frac{a}{pc}}\mathscr{N}(0,1)\,.$$

The time c required to decrease the error to level ε is therefore of order $a/p\varepsilon^2$, so that the completion time for the computation is p^{-1} times that associated with a single processor computation. In other words, the "speed-up" associated with a p processor computer system is equal to p. There are very few other scientific computing settings in which parallel computing offers a speed-up factor of p. The more typical situation is that the speed-up is roughly of order βp with $\beta \in (0,1)$ or is even sublinear in p. Thus, the Monte Carlo method is a computational approach that becomes increasingly attractive as the degree of parallelization increases.

However, there are subleties that deserve further discussion. The first such issue is that massive parallelization magnifies the effect of any bias that is present in the estimators that are used. For exaple, suppose that z is a steady-state expectation associated with an underlying Markov process. Because of initial transient effects, the time-average $\overline{Z}(c)$ available after simulating for c time units is therefore biased. In particular, as argued in Chapter IV,

$$\mathbb{E}\overline{Z}(c) - z = \mathbb{E}\frac{1}{c}\int_0^c Z(s)\,\mathrm{d}s = \frac{b}{c} + \mathrm{o}(1/c)$$

as $c \to \infty$, for some constant b. Suppose, as is typical in the Markov setting, that the sample average $\overline{Z}(c)$ satisfies a CLT with associated time-average variance constant σ^2. It follows that if the computation is independently replicated on each of p processors, thereby producing estimators $\overline{Z}_1(c), \ldots, \overline{Z}_p(c)$, the grand average satisfies

$$\frac{1}{p}\sum_{i=1}^{p}\overline{Z}_i(c) \stackrel{\mathscr{D}}{\approx} z + \frac{b}{c} + \frac{\sigma}{\sqrt{pc}}\mathscr{N}(0,1)$$

for p and c large. Consequently, if p is of larger order than c, the bias term b/c becomes the dominant effect in the error. In particular, the introduction of additional parallel processors does not reduce the error further. The key point is that in the presence of massive parallelization, elimination or reduction of bias becomes critical to the development of effective algorithms.

A second subtlety concerns our prior assumption that each realization takes exactly a units of computer time to simulate. For most computations, this is at most approximately true. The more typical situation is that the computer time required to simulate a realization is itself a r.v. (that is correlated with the observations collected from that realization). As a consequence, the amount of computer time T_i required to simulate a fixed number of replications on processor i is random, as is the completion time $\max_{1 \le i \le p} T_i$ at which the last processor completes its final simulation. Two problems then arise. The first is that the total execution time of the computation may exceed the time limit that has been assigned to that user of the parallel computer system. The second issue is that if the processors that finish their simulations early are left idle, a significant amount of potential computing power is left unused. The alternative is to permit such processors to continue simulating new realizations until a specified time limit for completion of the overall calculation is met. Of course, in this case, the total number of realizations simulated on processor i is an r.v. and, as a result, the sample mean associated with averaging over the observations that have been completed prior to the time limit is a biased estimator. As noted earlier, bias can cause serious problems in the parallel processing context when p is large. It is therefore of significant interest to study the bias associated with such sample means, and to investigate means of mitigating such bias. These issues are addressed in Glynn & Heidelberger [146]–[150].

10   Branching Processes

10a   The Galton–Watson Process

The process in question is a Markov chain $\{X_n\}$ on $\mathbb{N}$, modeling the sizes of different generations in a one-sex population. It is defined by the recursion

$$X_{n+1} = \sum_{i=1}^{X_n} Z_{n,i},$$

where $Z_{n,i}$ is the number of children of individual i in the nth generation. The $Z_{n,i}$ are assumed i.i.d. and we write $f_j = \mathbb{P}(Z_{n,i} = j)$, $j = 0, 1, 2, \ldots$. When $X_0 = 1$, a graphical illustration is often given in terms of a family tree, see Figure 10.1, where $X_0 = 1$, $X_1 = 2$, $X_2 = 6$, $X_3 = 3$. In general, a similar illustration consists of X_0 family trees that are not connected with each other.

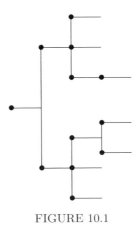

FIGURE 10.1

The initial motivation of Galton and Watson in the late nineteenth century was to compute the probability of survival of heir names. Here, X_0 is the number of present male carriers of the name, X_1 the number of their male children, X_2 the number of sons' sons, and so on. The name will die out (the family become extinct) if $X_n = 0$ for some n (then also $X_m = 0$ for all $m \geq n$), so the extinction probability is

$$\mathbb{P}(X_n \text{ for some } n) \;=\; \mathbb{P}(X_n \text{ for all large } n). \tag{10.1}$$

The extinction probability when $X_0 = 1$ is traditionally denoted by q, so that by independence, the conditional probability of extinction given $X_0 = i$ is q^i.

Another obvious application of the Galton–Watson process is biological populations in which one counts females only, so that X_n is the number of females in the nth generation. In an epidemic in a large population in its early stages, one can ignore the fact that the number of susceptibles changes slightly in the first generations of the epidemic and model the number of infectiuous as a Galton–Watson process. In cell kinetics (say cancer models), one often uses the special case $f_0 + f_2 = 1$ of *binary splitting* in which a cell in each step either dies or is divided into two. Examples of models with a clear branching structure but incorporating features not appropriately described by the Galton–Watson process appear in the next subsections and the exercises.

A standard fact that immediately follows by conditioning on $X_1 = j$ is that the extinction probability q is a solution of

$$q = \widehat{f}[q] = \sum_{n=0}^{\infty} q^j f_j. \tag{10.2}$$

More detailed properties of the Galton–Watson process involve the *offspring mean*

$$m \overset{\text{def}}{=} \mathbb{E}[X_1 \mid X_0 = 1] = \mathbb{E}Z_{1,1} = \sum_{j=0}^{\infty} j f_j.$$

From $\mathbb{E}[X_n \mid X_{n-1} = x] = xm$, it follows immediately that $W_n = X_n/m^n$ is a martingale. Being nonnegative, it has an a.s. limit W. In the *subcritical case* $m < 1$ or the *critical case* $m = 1$ (excluding $f_1 = 1$), this is possible only if $W = 0$, which shows that here $q = 1$. In the *supercritical case* $m > 1$, $q = 0$ if $f_0 = 0$, whereas $q \in (0, 1)$ is the root < 1 of (10.2) when $m > 1$, $f_0 > 0$. An easy Markov chain argument then gives $X_n \to \infty$ on the set of nonextinction, and in fact, this set coincides with $W > 0$ under weak regularity conditions, so that the process either dies out or grows exponentially at geometric rate m.

10b   Multitype Galton–Watson Processes

An extension of the GW process allows the individuals to be of several types. The simplest case is that of a finite number of types; say the types are $1, \ldots, q$ (or any other finite set). The process is then vector-valued, $\boldsymbol{X}_n = (X_n^1, \ldots, X_n^q)$, where X_n^j is the number of type-j individuals at time n, and the children of the ith type-j individual is a vector $\boldsymbol{Z}_{n,i}^j = (Z_{n,i}^{j,1}, \ldots, Z_{n,i}^{j,q})$. Thus

$$X_{n+1}^j = \sum_{k=1}^{q} \sum_{i=1}^{X_n^k} Z_{n,i}^{k,j}.$$

Examples:

(a) The type is the age of the individual, say $0, 1, 2, \ldots, 100$. Children are always of type 0 and in each type step the individual moves up one step or dies. Thus $Z_{n,i}^{j,k}$ is 0 unless $k = 0$ or $k = j$.

(b) The type of an individual is the *parity*, i.e., the number of children already born. Say only $1, \ldots, q$ are possible and type 0 corresponds to not having had children.

(c) As combination, the type may be the pair (a, p) of age and parity.

The expected offspring production is given by the $m_{jk} \overset{\text{def}}{=} \mathbb{E}Z_{n,i}^{j,k}$. Let $\boldsymbol{M}$ be the $q \times q$ matrix with elements m_{jk}. One can then show under an irreducibility assumption on M that the largest eigenvalue λ of M is strictly positive and that

$$\mathbb{E}^{(j)}\left[X_n^k\right] \sim c_j d_k \lambda^n,$$

where $\mathbb{E}^{(j)}$ refers to one initial individual of type j. One calls c_j the *reproductive value* of a type-j individual, $(d_1, \ldots, d_q)$ is the *stable type distribution*, and λ (or sometimes $\log \lambda$) is the *Malthusian rate of growth*.

In the next subsection we will see an example of an infinite-type GW process.

10c   Neutron Transport Processes

In a body of radioactive material such as Uranium 236, atoms spontaneously split and neutrons are emitted. An emitted neutron may hit other atoms, causing them also to split and emit a random number of new neutrons.

We represent this as a branching process in which the type $\in \mathbb{R}^6$ of an individual is the pair of its position $(x, y, z) \in \Omega$ in the body $\Omega \subset \mathbb{R}^3$ when born and the direction in which it starts moving, represented as a point (a, b, c) on the unit sphere $a^2 + b^2 + c^2 = 1$ (assuming the velocities to be all equal, say to α).

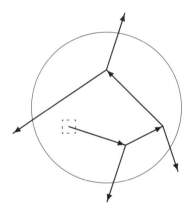

FIGURE 10.2

Figure 10.2 shows a *cascade* of neutrons initiated by a neutron created by a spontaneous (primary) splitting at the dash-boxed location. The secondary splits are binary on the figure, and once the neutron leaves the body, it cannot return. We can view this as a multitype branching process in which the children of a neutron hitting an atom are the neutron itself and the new neutrons created at a split. An individual dies once it leaves the body. Thus in the figure the ancestor gives birth to two children at the first split, one of which is himself at the new position and the other the neutron emitted. The ancestor is present in the population in generations $n = 0, 1$. The generation sizes are $1, 2, 2, 2, 0$. We have assumed for simplicity that only one new neutron is created by a hit and that the hitting neutron chooses its direction uniformly on the unit sphere. For a more realistic description, see Harris [175].

Assume for a while that the body Ω is the whole of $\mathbb{R}^3$ and that a neutron will hit an atom in the time interval $[t, t + \mathrm{d}t]$ w.p. $\alpha\,\mathrm{d}t$ (it travels at a constant speed). This means that the hazard rate of the distribution of T is α, which implies that T has an exponential distribution with density $\alpha e^{-\alpha t}$. Thus, if the speed is unity, the position (x_1, y_1, z_1) at the first hit is the initial position (x_0, y_0, z_0) shifted by T in the direction of motion. The cascade will be infinite because $\Omega = \mathbb{R}^3$, so that deaths cannot occur.

If the body is finite, we can still view (x_1, y_1, z_1) as the ficitious position of the first split in an infinite body. If $(x_1, y_1, z_1) \in \Omega$, a split actually occurs, and if $(x_1, y_1, z_1) \notin \Omega$, the particle dies. A cascade may be finite or infinite. It is clear that the cascade will be finite if the body is very small, whereas in the other extreme we get close to the case $\Omega = \mathbb{R}^3$ if the body is very large. Thus somewhere in between is a critical size at which the cascade starts being infinite with positive probability. That is, if the body represents a bomb, it goes off because eventually one of the cascades will be infinite.

To calculate the critical size, we need the extinction probability of our branching process. However, since it has an infinity of types, there is no simple way to do this, and simulation can become useful. In fact, this was done (a prominent name was von Neumann) when the first A-bomb was constructed in 1945. This was a historical milestone in applications of simulation. The problem in itself is an obvious case for simulation, but also many general methods and algorithms in the whole area arose as a by-product.

10d Simulation Aspects

There appear to be no specific problems or methodological aspects in the simulation of a branching process. We have included the topic more because it gives rise to some nice applications. So, as preparation for the exercises, we will just give a few comments on relevant problems and how to interpret the simulation output.

One purpose of simulating a branching process may be determining whether the process is subcritical or supercritical as in the A-bomb example. The obvious difficulty is that this is a limiting property. From a simulation terminating at generation N one cannot deduce with certainty from $X_N > 0$ that the process is supercritical, and similarly R replications that all become extinct before N will not tell for sure that $q = 1$, only that q is probably close to 1. Guidelines may, however, be obtained from looking at the whole sample path $X_0, \ldots, X_N$. In the subcritical case with

$X_N > 0$, this should exhibit a *quasistationary* behavior, that is, behave somewhat as a positive recurrent Markov chain (the transition probabilities are $\mathbb{P}(X_1 = j \,|\, X_0 = i)$, $i, j > 0$, in the Galton–Watson case). In the supercritical case with $X_N > 0$, the sample path should rather show exponential growth, which may be assessed by plotting $\log X_M, \ldots, \log X_N$ with say $M = N/2$. The exponential growth rate may be estimated either by doing regression of the same logs, or simply as X_N / X_{N-1}.

Exercises

10.1 (A) We consider an epidemic in Denmark initiated by one person coming from Asia and just infected by a disease that so far had not occurred in the country. As time unit, we take the time from infection to outbreak; at outbreak, the infected is taken into quarantine and cannot infect others from then on. We assume a constant population size N and let X_n denote the number of susceptibles at time n, Y_n the number of infectious, and Z_n the number of removed (thus $N = X_n + Y_n + Z_n$, $Y_0 = 1$, $Z_0 = 0$). As model, we assume that each individual has $m = 10$ contacts with other individuals in a period, that the probability that any of these is infected is Y/N, the percentage of infected, and that the person avoids infection after contact with an infected w.p. $\alpha = 0.92$.
You are to give histograms of the length T of the epidemic and its size S, that is, of

$$T \stackrel{\text{def}}{=} \inf\{n : Y_n = 0\}, \quad S \stackrel{\text{def}}{=} Y_0 + Y_1 + \cdots + Y_T = Z_T.$$

Use appropriate model simplifications to simulate the process as a Galton–Watson process with Poisson offspring distribution. Redo with $\alpha = 0.85$ [0.92 will be changed to 0.85 if 8% of the population have been vaccinated!].
10.2 (A) In the neutron-transport model for the A-bomb, assume that Ω is the ball $x^2 + y^2 + z^2 \leq r^2$, that the position of the neutron initiating a cascade is uniform on Ω, that only one new particle is created by a split and takes a direction that is uniform on $a^2 + b^2 + c^2 = 1$, and finally that $\alpha = 1$. Determine $R_- < R_+$ such that you feel fairly confident that the critical radius R_c satisfies $R_- < R_c < R_+$
10.3 (A) A rod of fuel in a nuclear reactor is represented as a line segment of length x. Spontaneous emissions of neutrons occur according to a Poisson process on $[0, x] \times [0, \infty)$ with unit rate, with the second coordinate representing time. Each split creates one new neutron, and the old and the new neutron choose independently to move left or right w.p. $1/2$ for each direction. Give simulation estimates of the temperature $T = T(X)$ of the rod as a function of x, where $T(x)$ is defined as the expected number of splits within one time unit. Take again $\alpha = 1$. Hint: Reduce again to the study of characteristics of cascades.
10.4 (A+) You are given a female population of size X_n in year n, divided into the age groups $0, \ldots, 100$. The mortality is given by the Gompertz–Makeham distribution corresponding to the G82 parameters used in Exercise II.4.1 (you will need to discretize here), and the probability of giving birth at age a and parity p is

$$b_{a,p} = \begin{cases} z_0 \beta_a & p = 0, \\ z_1 z_2^{p-1} \beta_a & p = 1, 2, \ldots, \end{cases} \tag{10.3}$$

where $\beta_a = (a-15)^3(46-a)^3$ for $a = 16, \ldots, 45$ and $\beta_a = 0$ otherwise. Further, a baby is a boy w.p. 51% and a girl w.p. 49%.

(a) Determine z_0, z_1 such that the probability of a woman giving birth to at least one child is 95% and 75% of getting at least two. Use these parameters in the following.

(b) By varying z_2, illustrate the connection between the expected number m of children born to a woman and (1) the Malthusian parameter λ, (2) the stable age distribution, using what you consider a relevant range for m.

(c) In a population with $m = 3$, the government wants to stop the population exploding by introducing drastic sanctions on women giving birth to more than two children. The government expects that this will change (10.3) to

$$b_{a,p} = \begin{cases} z_0 \beta_a & p = 0, \\ z_1 \beta_a & p = 1, \\ \frac{1}{10} z_1 z_2^{p-1} \beta_a & p = 2, 3, \ldots. \end{cases} \qquad (10.4)$$

Find the new λ and give projections for the age distribution $5, 10, 25$, and 50 years ahead.

Comments

Part (a) can be done by simulation or, alternatively, by fairly simple nonstochastic numerical analysis.

Use a multitype branching process for the woman only, but note that parity also counts the number of sons.

Age distributions may be illustrated by histograms or one-dimensional characteristics such as the proportion of old (66 or more) people to working (18–65) people, or children (17 or less) to working people.

Estimates for one-dimensional quantities (λ, the proportions above, etc.) should be accompanied by confidence intervals. In some cases, you may need the delta method.

You should of course start with a small X_0 such as 1,000, but aim for a population of several million.

11   Importance Sampling for Portfolio VaR

The present section follows two papers [135], [136] by Glasserman, Heidelberger, & Shahabuddin, who considered a portfolio exposed to d (dependent) risk factors $X_1, \ldots, X_d$ in a certain time horizon h, say one day or two weeks. The initial value of the portfolio is denoted by v and the (discounted) terminal value by $V = V(h, X_1, \ldots, X_d)$; for example, in the case of d European call options with maturity $T_i > h$, strike price K_i, normal log return X_i, and initial asset price s_i in position i,

$$v = \sum_{i=1} \mathrm{BS}(s_i, r, T_i, K_i), \quad V = \mathrm{e}^{-rh} \sum_{i=1} \mathrm{BS}(s_i \mathrm{e}^{X_i}, r, T_i - h, K_i), \quad (11.5)$$

where (in obvious notation) $\mathrm{BS}(s, r, T, K)$ is the Black–Scholes price I.(3.4).

The loss is $L = v - V$, where r is the risk-free interest rate and the VaR is a quantile of L (say the 99% one or the 99.97% one). The model for $(X_1, \ldots, X_d)$ in [135] is $\boldsymbol{X} \sim \mathcal{N}(\boldsymbol{0}, \boldsymbol{\Sigma})$, whereas [136] considers a multivariate Student t distribution, that is, the distribution of $\boldsymbol{U}/\sqrt{W/f}$, where $\boldsymbol{U} \sim \mathcal{N}(\boldsymbol{0}, \boldsymbol{\Sigma})$ and W is an independent χ^2 r.v. with f degrees of freedom.

The first step in both [135] and [136] is to note the *delta–gamma approximation*, which is based on the Taylor expansion

$$L \approx -\frac{\partial V}{\partial h} h - \sum_{i=1}^{d} \delta_i X_i - \frac{1}{2} \sum_{i,j=1}^{d} \gamma_{ij} X_i X_j, \qquad (11.6)$$

where $\delta_i \stackrel{\text{def}}{=} \partial V/\partial x_i$, $\gamma_{ij} \stackrel{\text{def}}{=} \partial^2 V/\partial x_i \partial x_j$. For example, in (11.5),

$$\delta_i = \delta_{\mathrm{BS}}(s_i, r, T_i, K_i) s_i \mathrm{e}^{X_i}, \quad \gamma_{ii} = \delta_i + \gamma_{\mathrm{BS}}(s_i, r, T_i, K_i) s_i^2 \mathrm{e}^{2X_i},$$

whereas the mixed partial derivatives vanish; here δ_{BS}, γ_{BS} are the delta, respectively the gamma, of the BS price; cf. the discussion of Greeks in I.4 and Chapter VII. For brevity, we rewrite the r.h.s. of (11.6) as $a_0 + Q$, where $Q \stackrel{\text{def}}{=} -\boldsymbol{\delta} \boldsymbol{X} - \boldsymbol{X}^{\mathsf{T}} \boldsymbol{\Gamma} \boldsymbol{X}/2$, $a_0 \stackrel{\text{def}}{=} -h \, \partial V/\partial h$. The delta–gamma approximation is then $\mathbb{P}(L > x) \approx \mathbb{P}(Q > x - a_0)$.

In the $\mathcal{N}(\boldsymbol{0}, \boldsymbol{\Sigma})$ case in [135], we can represent $\boldsymbol{X}$ as $\boldsymbol{CY}$, where $\boldsymbol{Y} \sim \mathcal{N}(\boldsymbol{0}, \boldsymbol{I})$ and $\boldsymbol{CC}^{\mathsf{T}} = \boldsymbol{\Sigma}$; standard results on diagonalization of quadratic forms show that it is possible to choose $\boldsymbol{C}$ to make $\boldsymbol{C}^{\mathsf{T}} \boldsymbol{\Gamma} \boldsymbol{C}$ diagonal, say with diagonal elements $\lambda_1, \ldots, \lambda_d$, so that

$$Q = \sum_{i=1}^{d} \left(b_i Y_i + \lambda_i Y_i^2 \right)$$

for certain b_i, λ_i. This is a sum of independent noncentral χ^2 r.v.'s; the tail is not explicitly available, but the characteristic function $\mathbb{E}\mathrm{e}^{\mathrm{i}sQ}$ is and can be inverted numerically to evaluate the delta–gamma approximation. However, evaluating $\mathbb{P}(L > x)$ and thereby the VaR in this way is quite imprecise, so that simulation comes in. Since the VaR is small, we have a rare-event problem in which variance reduction is essential. However, implementing the first idea that comes to mind, importance sampling using an exponential change of measure for L identified via $\mathbb{E}_\theta L = x - a_0$, is difficult because the distribution of L is typically too complicated to allow for identification of θ and the associated $\mathbb{P}_\theta$-distribution of L.

Instead, the idea of [135] is to use exponential change of measure for the approximating Q. The distribution $\widetilde{F}$ on $\mathbb{R}^d$ defined by $\mathrm{d}\widetilde{F}/\mathrm{d}F = \mathrm{e}^{\theta Q}/\mathbb{E}\mathrm{e}^{\theta Q}$, where $F = \mathcal{N}(\boldsymbol{0}, \boldsymbol{I})$, makes $Y_1, \ldots, Y_d$ independent, as follows from Q being a sum of terms each depending on just one Y_i. Further, $\widetilde{F}$ makes Y_i normal with mean and variance given respectively by

$$\mu_i(\theta) = \frac{\theta b_j}{1 - \theta \lambda_i} \quad \text{and} \quad \sigma_i^2(\theta) = \frac{1}{1 - \theta \lambda_i};$$

cf. Exercise V.1.2. Thus we are led to use an importance distribution $\widetilde{\mathbb{P}}$ in which the Y_i are independent normals with the means and variances just stated and θ determined by

$$x - a_0 = \mathbb{E}^* Q = \sum_{i=1}^{d} \left[b_i \mu_i(\theta) + \lambda_i \left(\mu_i(\theta)^2 + \sigma_i^2(\theta) \right) \right].$$

Indeed, the empirical finding of [135] is that this typically (i.e., in the case of a number of selected test portfolios) reduces the variance by a factor of 20–50. This is theoretically supported by a logarithmic efficiency result on a relative error defined by normalizing the variance of the importance-sampling estimator $\mathrm{d}F/\mathrm{d}\widetilde{F}\mathbb{1}\{Q > x - a_0\}$ by $\mathbb{P}(Q > x - a_0)^2$. Note, however, that this differs from the definition of logarithmic efficiency that we have used in VI.1, where the normalization would have been by $\mathbb{P}(L > x)^2$; the difference is small exactly when the delta–gamma approximation is good.

Now turn to the (heavy-tailed) multivariate Student t example considered in [136]. That is, $\boldsymbol{X} = \boldsymbol{U}/\sqrt{W/f}$, where $\boldsymbol{U} \sim \mathcal{N}(\boldsymbol{0}, \boldsymbol{\Sigma})$ and W is an independent χ^2 r.v. Replacing the representation $\boldsymbol{X} = \boldsymbol{C}\boldsymbol{Y}$ above by $\boldsymbol{U} = \boldsymbol{C}\boldsymbol{Y}$, the Q in the delta–gamma approximation $\mathbb{P}(L > x) \approx \mathbb{P}(Q > x - a_0)$ takes the form

$$Q = \frac{1}{\sqrt{W/f}} \sum_{i=1}^{d} b_i Y_i + \frac{1}{W/f} \sum_{i=1}^{d} \lambda_i Y_i^2$$

(with the same a_0, b_i, λ_i). However, since (say) each $Y_i/\sqrt{W/f}$ is t distributed and therefore heavy-tailed, $\mathbb{E}e^{\theta Q} = \infty$ for $\theta \neq 0$, so that the above method of exponential change of measure for Q is infeasible. Instead, [136] works with

$$Q_x = (W/f)(Q - x + a_0) = (W/f)(a_0 - x) + \sqrt{W/f} \sum_{i=1}^{d} b_i Y_i + \sum_{i=1}^{d} \lambda Y_i^2$$

(note that $Q > x - a_0$ if and only if $Q_x > 0$) and a corresponding importance distribution $\widetilde{F}$ defined by $\mathrm{d}\widetilde{F}/\mathrm{d}F = e^{\theta Q_x}/\mathbb{E}e^{\theta Q_x}$, where as above, $F = \mathcal{N}(\boldsymbol{0}, \boldsymbol{I})$.

We need first to identify $\widetilde{F}$. Calculations identical to those above combined with some extra steps involving a conditioning on W shows that under $\widetilde{F}$, W has a $\mathrm{Gamma}\big(f/2, \lambda(\theta)\big)$ distribution, where $\lambda(\theta) = \big(1 - 2\alpha(\theta)\big)/2$ with

$$\alpha(\theta) = -\frac{\theta(x - a_0)}{f} + \frac{\theta^2}{2f} \sum_{i=1}^{d} \frac{b_i^2}{1 - 2\theta\lambda_i}.$$

Further, $Y_1, \ldots, Y_d$ become conditional independent normals with conditional means and variances

$$\mu_i(\theta) = \frac{\theta b_j \sqrt{W/f}}{1 - \theta \lambda_i}, \quad \text{respectively} \quad \sigma_i^2(\theta) = \frac{1}{1 - \theta \lambda_i}.$$

This description makes it possible in a straightforward way to simulate from $\widetilde{F}$ as well as to find the θ similar to the one used above, that is, the θ solving $\mathbb{E}^* Q_x = 0$.

The findings of [136] concerning the efficiency of this simulation scheme are much the same as in [135]. In [136], the idea of combining with a stratification of the Y_i is also considered. The resulting variance reduction is in some cases up to a factor of 6, in other cases it is insignificant.

12   Importance Sampling for Dependability Models

In this section, we briefly discuss the use of importance sampling in computing various performance measures associated with what are commonly known as *dependability models*. Such dependability models do not contain embedded random walk structure, so much of the discussion of Chapter VI does not directly apply. Nevertheless, importance sampling has become a widely applied tool in this modeling context.

A dependability model is an effort to mathematically describe a system that is made up of individual components, each of which is potentially subject to failure. In contrast to a reliability model, dependability models typically incorporate an ability on the part of the system to repair or replace failed components. This is relevant to systems that are intended to provide a high level of availability (i.e., the percentage of time during which the system is operating). Such systems include airline reservation systems, air traffic control networks, and many other applications. Such models frequently assume that the component failure times are exponentially distributed, as are the repair durations. As a consequence, the state of the dependability model is therefore described by a continuous-time Markov pure jump process.

This continuous-time Markov process typically has an enormous state space.[4] For example, if the system consists of n components, the state descriptor typically needs to include information relating to whether each of the n components is "up" (i.e., operational) or "down" (i.e., in failed mode). This leads to a state space having 2^n states, without even considering the additional state information that might be needed to track the locations of the repair persons that are assigned to the maintenance of the system. This

---

[4]Nonexponential failure and repair times can in principle be incorporated by modeling them as phase-type distributions, see [16, Section III.4]. Of course, this easily leads to even bigger state spaces.

huge state space makes simulation an attractive solution methodology for computing associated performance measures.

The state of the art for importance sampling of such systems is best developed in the context of systems in which the individual components are highly reliable.[5] This corresponds, in the continuous-time Markov process setting, to a model in which the exponential rate parameters for the failure-time distributions are orders of magnitude smaller than the rate parameters for the repair times.

Such a highly available dependability system tends to spend a significant proportion of its time in the state in which all components are operational, and no repairs are underway; denote this state as $\vec{0}$. The performance measure that is perhaps of most interest to dependability modelers is the socalled mean time to failure (MTTF). More precisely, this is the expected time $\mathbb{E}_{\vec{0}} T_F$ for the system to enter the subset F of states in which the system is non-operational (i.e., "failed"), starting from state $\vec{0}$. Since importance sampling has been developed in the context of rare-event computations, the first step is to reformulate the computation of MTTF as a problem involving a rare-event probability. Because the continuous-time Markov process regenerates at returns to $\vec{0}$, it follows that

$$\mathbb{E}_{\vec{0}} T_F \;=\; \mathbb{E}_{\vec{0}}\big[\tau;\, \tau < T_F\big] + \mathbb{E}_{\vec{0}} T_F \mathbb{P}_{\vec{0}}(\tau < T_F) + \mathbb{E}_{\vec{0}}\big[T_F;\, \tau > T_F\big]$$

where τ is the first time that the process returns to $\vec{0}$. Consequently,

$$\mathbb{E}_{\vec{0}} T_F \;=\; \frac{\mathbb{E}_{\vec{0}} \min(\tau, T_F)}{\mathbb{P}_{\vec{0}}(T_F < \tau)}\,. \tag{12.1}$$

The r.v. appearing in the numerator of (12.1) is, with high probability, equal to τ. Given the propensity of the process to spend time in $\vec{0}$, this numerator tends to have a low coefficient of variation, and the numerator of (12.1) is easy to compute by crude Monte Carlo. On the other hand, the denominator of (12.1) is a rare-event probability and can not be efficiently evaluated by crude Monte Carlo.

To use importance sampling on the denominator, it is instructive to consider the case in which the failure rates are all of order ε, while the repair rates are of order 1. Note that the event $\{T_F < \tau\}$ depends only on the sequence of states visited, and hence can be computed by applying importance sampling to the discrete-time Markov chain embedded in the continuous-time Markov process. As $\varepsilon \downarrow 0$, it becomes more likely that the path followed by the embedded chain on the event $\{T_F < \tau\}$ includes no repair events, for otherwise an additional failure event would be required to "fail" the system (reducing the probability of the corresponding path by a factor of order ε relative to the paths in which no repair occur). Since the

---

[5] Another way to build high dependability into a system is to build a massive amount of component redundancy into the system design.

general principle of rare-event simulation is to attempt to simulate from the conditional distribution given $\{T_F < \tau\}$, this suggests generating paths for the embedded chain from a transition matrix for which the likelihood, in any state, of a failure event (i.e., a state transition corresponding to a component failure) is much higher than the likelihood of a repair event. The difficulty with such an importance-sampling algorithm is that such an importance distribution "starves" the secondary paths associated with the event $\{T_F < \tau\}$ having lower, but nonnegligible, probability relative to the primary path to system failure in which no repairs occur. This underweights the paths on which single or even multiple repairs are exhibited along the path to system failure.

An approach that works well in practice is to balance the likelihood, in any state other than $\vec{0}$, that the next event is a failure event against that of a repair event. Assume, for example, that one builds a transition matrix for the importance sampling algorithm in which the probability that the next event is a failure is $1/2$. This failure probability of $1/2$ needs to be distributed across the different types of component failures that are possible in that state. One standard way of allocating this probability is in proportion to the likelihoods under the transition matrix associated with the original model. For example, suppose that in a given state, only two possible component failures are possible, with probabilities εp_1 and εp_2, respectively, under the original transition matrix. The above allocation strategy suggest that the two failure transitions should have probabilities

$$\frac{1}{2}\frac{p_1}{p_1 + p_2} \quad \text{and} \quad \frac{1}{2}\frac{p_2}{p_1 + p_2},$$

respectively, under the importance distribution. When this importance sampling approach is followed, significant variance reduction can be achieved; see Shahabuddin [334] for a complete discussion of this and related dependability system algorithms.

Some further selected related references are Goyal et al. [163], Heidelberger [178], Heidelberger et al. [179], and Nakayama [267], [268].

13 Special Algorithms for the GI/G/1 Queue

We close this book by returning to the model that was our very first example in I.1, the single-server queue.

13a Perfect Sampling of W

We consider a random-walk maximum M in the notation of the Siegmund algorithm in VI.2a. Recall in particular that F_γ denotes the probability measure with density $e^{\gamma x}$ w.r.t. F, where $\gamma > 0$ is the solution of $\widehat{F}[\gamma] = 1$, that $\tau(x) = \inf\{n : S_n > x\}$, $\xi(x) = S_{\tau(x)} - x$, and that we have the representation $\mathbb{P}(M > x) = \mathbb{E}_\gamma e^{-\gamma S_{\tau(x)}}$, where $\mathbb{P}_\gamma$ is the probability measure such that the X_n are i.i.d. with distribution F_γ w.r.t. $\mathbb{P}_\gamma$.

An algorithm for perfect sampling of M (which has the same distribution as the GI/G/1 waiting time W subject to a suitable choice of F) was suggested by Ensor & Glynn [109]. It uses an exponential r.v. V with rate γ (independent of $\{S_n\}$) and the ladder heights $S_{\tau_+(n)}$, where

$$\tau_+(0) \stackrel{\text{def}}{=} 0, \quad \tau_+(n+1) \stackrel{\text{def}}{=} \inf\{k > \tau_+(n) : S_k > S_{\tau_+(n)}\}.$$

Cf. Figure 13.1 where the ladder heights are marked with a $\bullet$. The r.v. generated by the simulation is the last ladder height

$$Z = \sup\{S_{\tau_+(n)} : S_{\tau_+(n)} \le V\}$$

not exceeding V, see again Figure 13.1.

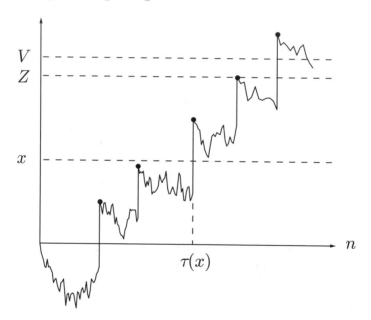

FIGURE 13.1

Proposition 13.1 *The $\mathbb{P}_\gamma$-distribution of Z is the same as the $\mathbb{P}$-distribution of M.*

Proof. First note that $S_{\tau(x)}$ is necessarily a ladder height. By sample path inspection, $Z > x$ if and only $S_{\tau(x)} \le V$, so that

$$\begin{aligned}
\mathbb{P}_\gamma(Z > x) &= \mathbb{P}_\gamma(S_{\tau(x)} \le V) = \mathbb{E}_\gamma\left[\mathbb{P}_\gamma\left(S_{\tau(x)} \le V \mid S_{\tau(x)}\right)\right] \\
&= \mathbb{E}_\gamma e^{-\gamma S_{\tau(x)}} = \mathbb{P}(M > x).
\end{aligned}$$

□

From perfect sampling of the waiting time $W \stackrel{\mathscr{D}}{=} M$, one can easily perform perfect sampling also of the steady-state queue length Q. This follows from the *distributional Little's law*, stating that Q has the same distribution as $N^*(W + U)$, where W, U, $\{N(t)\}$ are independent, U has the service-time distribution and $\{N^*(t)\}$ is a stationary version of the renewal arrival process; cf. [16, Section X.4]. Thus, if W is generated as above and $T_1, T_2, \ldots$ are independent, such that $T_2, T_3, \ldots$ follow the interarrival-time distribution and T_1 has density $\mathbb{P}(T_2 > t)/\mathbb{E}T_2$, the r.v.

$$\sup\{n = 1, 2, \ldots : T_1 + \cdots + T_n < W + U\}$$

(0 if no such n exists) has the same distribution as Q.

For the M/G/1 queue, an obvious alternative perfect simulation estimator for W comes from the Pollaczeck–Khinchine formula: W has the same distribution as $U_1^* + \cdots + U_N^*$, where $N, U_1^*, U_2^* \ldots$ are independent, N is geometric with $\mathbb{P}(N = n) = (1 - \rho)\rho^n$, $n = 0, 1, 2, \ldots$, and the U_k^* have the equilibrium service-time distribution with density $\mathbb{P}(U > x)/\mathbb{E}U$. In contrast to the Ensor–Glynn estimator, this estimator can also be used in the case of heavy tails.

13b   The Minh–Sorli Algorithm

A classical formula for the mean delay due to Marshall is

$$EW = \frac{\mathbb{E}U^2 + \mathbb{E}T^2 - 2\mathbb{E}U\mathbb{E}T - \mathbb{E}I^2}{2(\mathbb{E}T - \mathbb{E}U)}, \tag{13.1}$$

where U, T are generic service and interarrival times and I the idle period. This is obtained from the Lindley recursion $W_{n+1} = (W_n + U_n - T_n)^+$ by squaring: since

$$\mathbb{E}(W_n + U_n - T_n)^2 = \mathbb{E}\left[(W_n + U_n - T_n)^+ - (W_n + U_n - T_n)^-\right]^2$$
$$= \mathbb{E}(W_n + U_n - T_n)^{+2} + \mathbb{E}(W_n + U_n - T_n)^{-2}$$

and $(W_n + U_n - T_n)^-$ can be identified with I, we get (assuming $\{W_n\}$ to be stationary)

$$EW^2 = \mathbb{E}(W + U - T)^2 - \mathbb{E}I^2$$
$$= \mathbb{E}W^2 + \mathbb{E}U^2 + \mathbb{E}T^2 - 2\mathbb{E}U\mathbb{E}T + 2\mathbb{E}W(\mathbb{E}U - \mathbb{E}T) - \mathbb{E}I^2 .$$

From this (13.1) follows.

Minh & Sorli [260] (see also Wang & Wolff [359]) suggested that one use (13.1) for simulation by noting that everything is known except for $\mathbb{E}I^2$. Thus, one can simply simulate busy cycles and estimate $\mathbb{E}I^2$ by the empirical second moment. Methodologically, we have an instance of indirect estimation as discussed in V.8.

As ρ approaches 1, the first term in (13.1) becomes dominant, and hence one expects the possible variance reduction to be most substantial when ρ is close to 1. Indeed, in the case of the M/M/1 queue with traffic intensity $\rho = 0.9$, the variance reduction compared to regenerative simulation is reported in [11] to be about a factor of 2,000!

13c   A Control-Variate Method for $\mathbb{E}W$

The problem is to estimate $z = \mathbb{E}W$ by simulation. Of course, standard methods such as regenerative simulation apply in a straightforward way to this problem. For a more sophisticated algorithm, suggested in Asmussen [11], note first the formula

$$\mathbb{E}M = \int_0^\infty \mathbb{P}(M > x)\,\mathrm{d}x = \int_0^\infty \mathbb{P}(\tau(x) < \infty)\,\mathrm{d}x\,,$$

where $\tau(x) = \inf\{n : S_n > x\}$. Here an extremely efficient (at least for large x) estimator for $\mathbb{P}(\tau(x) < \infty)$ is provided by Siegmund's algorithm, viz. $\mathrm{e}^{-\gamma x}\mathrm{e}^{-\gamma\xi(x)}$ simulated from $\mathbb{P}_\gamma$, where $\xi(x) \overset{\text{def}}{=} S_{\tau(x)}$ is the overshoot, so that it is appealing to try the estimator

$$\int_0^\infty \mathrm{e}^{-\gamma x}\mathrm{e}^{-\gamma\xi(x)}\,\mathrm{d}x$$

for $\mathbb{E}M$. The obvious difficulty is that evaluating $\xi(x)$ for all x would require an infinitely long simulation. This can be circumvented by truncating the integral and suitably compensating. More precisely, let $V > 0$ be independent of $\{S_n\}$ and define

$$Z \overset{\text{def}}{=} \int_0^V \frac{1}{\mathbb{P}(V > x)}\mathrm{e}^{-\gamma x}\mathrm{e}^{-\gamma\xi(x)}\,\mathrm{d}x\,.$$

Then indeed

$$\begin{aligned}
\mathbb{E}_\gamma Z &= \mathbb{E}_\gamma \int_0^\infty \mathbb{1}_{x<V}\frac{1}{\mathbb{P}(V > x)}\mathrm{e}^{-\gamma x}\mathrm{e}^{-\gamma\xi(x)}\,\mathrm{d}x \\
&= \int_0^\infty \mathbb{P}(x < V)\frac{1}{\mathbb{P}(V > x)}\mathrm{e}^{-\gamma x}\mathbb{E}\mathrm{e}^{-\gamma\xi(x)}\,\mathrm{d}x \\
&= \int_0^\infty \mathrm{e}^{-\gamma x}\mathbb{E}_\gamma \mathrm{e}^{-\gamma\xi(x)}\,\mathrm{d}x = \int_0^\infty \mathbb{P}\big(\tau(x) < \infty\big)\,\mathrm{d}x = \mathbb{E}M\,.
\end{aligned}$$

Simulation experiments indicate that the variance of Z is reasonably small but not extremely small. We improve this by introducing

$$C \overset{\text{def}}{=} \int_0^V \frac{1}{\mathbb{P}(V > x)}\mathrm{e}^{-\gamma x}\,\mathrm{d}x$$

as control variate; note that by the same calculation as for $\mathbb{E}_\gamma Z$,

$$\mathbb{E}_\gamma C = \int_0^\infty \mathrm{e}^{-\gamma x}\,\mathrm{d}x = \frac{1}{\gamma}\,.$$

The control-variate estimator is

$$\overline{Z} - \overline{\alpha}(\overline{C} - \mathbb{E}_\gamma C),\tag{13.2}$$

where

$$\overline{Z} \stackrel{\text{def}}{=} \frac{1}{R}(Z_1 + \cdots + Z_R),\quad \overline{C} \stackrel{\text{def}}{=} \frac{1}{R}(C_1 + \cdots + C_R),$$

$$\overline{\alpha} \stackrel{\text{def}}{=} \frac{\sum(Z_r - \overline{Z})(C_r - \overline{C})}{\sum(C_r - \overline{C})^2}.$$

Indeed, the estimator (13.2) is a huge improvement: the observed variance reduction is often a factor of 1,000–25,000!

The relevant choice of the distribution of V turns out to be the exponential distribution with rate γ, $\mathbb{P}(V > x) = e^{-\gamma x}$, and it can be proved that this choice is asymptotically optimal in a suitable sense.

Figure 13.2 gives an example of the high linear dependence between Z and C: the correlation is 0.999!

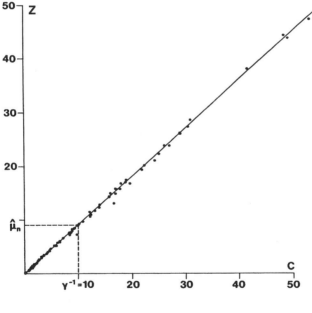

FIGURE 13.2

In the setting of the Ensor–Glynn algorithm in Section 1a, a related idea is to apply V as control for the r.v. Z generated by perfect sampling.

Appendix

A1 Standard Distributions

We give here a list of standard distributions, for the main purpose of introducing the terminology and parameterization used in the book. For a discrete distribution on $\mathbb{N}$, $p(k)$ denotes the probability mass function and $\widehat{p}[z] \overset{\text{def}}{=} \sum_0^\infty z^k p(k)$ the probability generating function (p.g.f.). In the continuous case, $f(x)$ is the density, $F(x) \overset{\text{def}}{=} \int_{-\infty}^x f(y)\,\mathrm{d}y$ the c.d.f., $\widehat{F}[s] \overset{\text{def}}{=} \int_{-\infty}^\infty e^{sy} f(y)\,\mathrm{d}y$ the m.g.f. (defined for any $s \in \mathbb{C}$ such that $\int_{-\infty}^\infty e^{\Re s\, y} f(y)\,\mathrm{d}y < \infty$), and the cumulant function (c.g.f.) is $\log \widehat{F}[s]$.

The Bernoulli(p) distribution is the distribution of $X \in \{0,1\}$, such that $\mathbb{P}(X = 1) = 1$. Here the event $\{X = 1\}$ can be thought of as heads coming up when throwing a coin w.p. 1 for heads. The **binomial (n,p) distribution** is the distribution of the sum $N \overset{\text{def}}{=} X_1 + \cdots + X_n$ of n i.i.d. Bernoulli(p) r.v.'s. The p.g.f.'s are $\mathbb{E}z^X = 1 - p + pz$, $\mathbb{E}z^N = (1 - p + pz)^n$.

The geometric distribution with success parameter $1 - \rho$ is the distribution of the number N of tails before a head comes up when flipping a coin w.p. ρ for tails, $\mathbb{P}(N = n) = (1-\rho)\rho^n$, $n \in \mathbb{N}$. Also $N' = N+1$, the total number of flips including the final head, is said to have a geometric$(1 - \rho)$ distribution, and one has $\mathbb{P}(N' = n) = (1 - \rho)\rho^{n-1}$, $n = 1, 2, \ldots$.

The Gamma(α, λ) **distribution** has

$$f(x) \;=\; \frac{\lambda^\alpha}{\Gamma(\alpha)} x^{\alpha-1} e^{-\lambda x}, \quad x > 0.$$

The m.g.f. is $\big(\lambda/(\lambda - s)\big)^\alpha$, the mean is α/λ, and the variance is α/λ^2. The Gamma$(1, \lambda)$ distribution is the exponential distribution with rate parameter λ, and the Gamma(n, λ) distribution with $n = 1, 2, 3, \ldots$ is commonly called the Erlang(n, λ) distribution or just the Erlang(n) distribution.

The **inverse Gamma**(α, λ) distribution is the distribution of $Y = 1/X$ where X has a Gamma(α, λ) distribution. The density is

$$\frac{\lambda^\alpha}{y^{\alpha+1}\Gamma(\alpha)} e^{-\lambda/y}. \tag{A1.1}$$

The inverse Gamma distribution is popular as Bayesian prior for normal variances (see, e.g., XIII.2.2).

The Inverse Gauss(c, ξ) **distribution** has density

$$\frac{c}{x^{3/2}\sqrt{2\pi}} \exp\left\{\xi c - \frac{1}{2}\left(\frac{c^2}{x} + \xi^2 x\right)\right\}, \quad x > 0. \tag{A1.2}$$

The c.g.f. is $\xi c - c\sqrt{\xi^2 - 2s}$, the mean is c/ξ, and the variance is c/ξ^3. The inverse Gaussian distribution can be interpreted as the time needed for Brownian motion with drift ξ to get from level 0 to level c. The distribution is also popular in statistics as a flexible two-parameter distribution (in fact, one of the nice examples of a two-parameter exponential family). An important extension is the NIG family of distributions; see XII.1.4 and XII.5.1.

The Weibull(β) **distribution** has tail $\overline{F}(x) = e^{-x^\beta}$. It originates from reliability. The failure rate is ax^b, where $a = \beta$, $b = \beta - 1$, which provides one of the simplest parametric alternatives to a constant failure rate as for the exponential distribution corresponding to $\beta = 1$.

The multivariate normal distribution $\mathscr{N}(\boldsymbol{\mu}, \boldsymbol{\Sigma})$ in $\mathbb{R}^d$ has density

$$\frac{1}{(2\pi)^{d/2}\det(\boldsymbol{\Sigma})^{1/2}} \exp\{-(\boldsymbol{x} - \boldsymbol{\mu})^\mathsf{T} \boldsymbol{\Sigma}^{-1}(\boldsymbol{x} - \boldsymbol{\mu})/2\}.$$

The m.g.f. for $d = 1$ is $e^{\mu s - s^2 \sigma^2/2}$.

An important formula states that conditioning in the multivariate normal distribution again leads to a multivariate normal: if

$$\begin{pmatrix} \boldsymbol{X}_1 \\ \boldsymbol{X}_2 \end{pmatrix} \sim \mathscr{N}\left(\begin{pmatrix} \boldsymbol{\mu}_1 \\ \boldsymbol{\mu}_2 \end{pmatrix}, \begin{pmatrix} \boldsymbol{\Sigma}_{11} & \boldsymbol{\Sigma}_{12} \\ \boldsymbol{\Sigma}_{21} & \boldsymbol{\Sigma}_{22} \end{pmatrix}\right),$$

then

$$\boldsymbol{X}_1 \,|\, \boldsymbol{X}_2 \sim \mathscr{N}\big(\boldsymbol{\mu}_1 + \boldsymbol{\Sigma}_{12}\boldsymbol{\Sigma}_{22}^{-1}(\boldsymbol{X}_2 - \boldsymbol{\mu}_2), \boldsymbol{\Sigma}_{11} - \boldsymbol{\Sigma}_{12}\boldsymbol{\Sigma}_{22}^{-1}\boldsymbol{\Sigma}_{21}\big).$$

A2   Some Central Limit Theory

Proposition A.1 (ANSCOMBE'S THEOREM) *Let* $X_1, X_2, \ldots$ *be i.i.d. with mean* μ *and variance* σ^2, *let* $S_n \stackrel{\text{def}}{=} X_1 + \cdots + X_n$, *and let* $T_n \in \mathbb{N}$ *be a sequence of random times satisfying* $T_n/t_n \stackrel{\text{P}}{\to} 1$ *for some deterministic sequence* $\{t_n\}$ *with* $t_n \to \infty$. *Then* $(S_{T_n} - T_n\mu)/t_n^{1/2} \stackrel{\mathscr{D}}{\to} \mathscr{N}(0, \sigma^2)$, $n \to \infty$.

For a proof, see, e.g., Chung [72].

A3   FFT

The (finite) Fourier matrix of order n is

$$
\boldsymbol{F} \stackrel{\text{def}}{=} \boldsymbol{F}_n \stackrel{\text{def}}{=} \begin{pmatrix} 1 & 1 & 1 & \cdot & 1 \\ 1 & w & w^2 & \cdot & w^{n-1} \\ 1 & w^2 & w^4 & \cdot & w^{2(n-1)} \\ \cdot & \cdot & \cdot & \cdot & \cdot \\ 1 & w^{n-1} & w^{2(n-1)} & \cdot & w^{(n-1)^2} \end{pmatrix}
$$

where $w \stackrel{\text{def}}{=} w_n \stackrel{\text{def}}{=} e^{2\pi i/n}$ is the nth root of unity. That is, labeling the rows and columns $0, 1, \ldots, n-1$, the rsth entry is w^{rs} (note that $w^0 = 1$). The finite Fourier transform of a vector $\boldsymbol{a} \stackrel{\text{def}}{=} (a_0 \ldots a_{n-1})^\mathsf{T}$ is then $\widehat{\boldsymbol{a}} = \boldsymbol{F}\boldsymbol{a}/n$, and since $\boldsymbol{F}^{-1} = \overline{\boldsymbol{F}}/n$ (complex conjugate), $\boldsymbol{a}$ can be recovered from the Fourier transform $\widehat{\boldsymbol{a}}$ as $\boldsymbol{a} = \overline{\boldsymbol{F}}\widehat{\boldsymbol{a}}$. Element by element,

$$
\widehat{a}_r = \sum_{s=0}^{n-1} a_s w^{rs}/n\,, \quad a_r = \sum_{s=0}^{n-1} \widehat{a}_s w^{-rs}\,.
$$

This procedure works in principle for any n. If the matrix multiplications in $\widehat{\boldsymbol{a}} = \boldsymbol{F}\boldsymbol{a}/n$ and/or $\boldsymbol{a} = \overline{\boldsymbol{F}}\widehat{\boldsymbol{a}}$ are implemented in the naive way, the complexity is then $O(n^2)$. The crux of the fast Fourier transform (FFT) is that when taking n of the form $n = 2^k$ for some integer k, some clever manipulations using a recursive reduction to 2^{k-1} will allow one to reduce the complexity to $O(n\log n)$ (we do not describe these manipulations here since the FFT is implemented in practically any software package for numerical calculations). See, e.g., Press et al. [293].

If the given n is not of the form 2^k, one can of course just choose $n^* = 2^{k^*}$ such that $2^{k^*-1} < n < 2^{k^*}$ and perform FFT in dimension n^*, supplementing $\boldsymbol{a}$ and/or $\widehat{\boldsymbol{a}}$ with the necessary number of zeros.

A4   The EM Algorithm

The EM algorithm is one of the main tools for performing maximum likelihood (ML) estimation in the absence of full data information (incomplete observations, lost data, etc.). The main example is exponential families with density

$$f_{\boldsymbol{\theta}}(\boldsymbol{v}) \overset{\text{def}}{=} e^{\boldsymbol{\theta}^{\mathsf{T}}\boldsymbol{t}(\boldsymbol{v})-\kappa(\boldsymbol{\theta})} \tag{A4.1}$$

w.r.t. some reference measure $\nu(d\boldsymbol{v})$, where $\boldsymbol{v}$ is the observation vector. The ML estimator is then often some nice explicit function $\widehat{\boldsymbol{\theta}} \overset{\text{def}}{=} \widehat{\boldsymbol{\theta}}(\boldsymbol{t}(\boldsymbol{v}))$ of $\boldsymbol{t}(\boldsymbol{v})$, where $\boldsymbol{v}$ is the observed outcome of the r.v. $\boldsymbol{V}$ with the prescribed density $f_{\boldsymbol{\theta}}$.

Example A.1 Let $V_1, \ldots, V_n$ be i.i.d. exponential(θ) and $\boldsymbol{V} \overset{\text{def}}{=} \left(V_1 \ \ldots \ V_n \right)^{\mathsf{T}}$. Then (A4.1) holds with $t(\boldsymbol{V}) = -V_1 - \cdots - V_n$, $\kappa(\theta) = n \log \lambda$, and we have $\widehat{\theta} = n/(V_1 + \cdots + V_n) = -n/t(\boldsymbol{V})$, as can be obtained by straightforward differentiation of the log likelihood $-\lambda(V_1 + \cdots + V_n) + n \log \lambda$. □

Example A.2 In a normal mixture problem, the observations $Y_1, \ldots, Y_n$ are i.i.d. with a common density $f_{\boldsymbol{\theta}}(y)$, which is a mixture

$$\sum_{i=1}^{d} \alpha_i \frac{1}{\sqrt{2\pi}} e^{-(y-\mu_i)^2/2}$$

of $\mathcal{N}(\mu, 1)$ densities (for simplicity, we assume that the variance is known and equal to 1), so that the unknown parameters are $(\alpha_i, \mu_i)_{i=1,\ldots,d}$, which are constrained by $\alpha_1 + \cdots + \alpha_d = 1$. One can interpret the model as a given observation Y being assigned type i w.p. α_i and then having distribution $\mathcal{N}(\mu_i, 1)$.

To obtain the representation (A4.1), one performs a data augmentation and defines

$$\boldsymbol{V} \overset{\text{def}}{=} \left(J_1, \ldots, J_n, Y_1, \ldots, Y_n \right), \quad \boldsymbol{t}(\boldsymbol{V}) \overset{\text{def}}{=} \left(N_1, \ldots, N_d, S_1, \ldots, S_d \right),$$

$$\boldsymbol{\theta} \overset{\text{def}}{=} \left(\theta_1, \ldots, \theta_d, \theta_{d+1}, \ldots, \theta_{2d} \right) \overset{\text{def}}{=} \left(\log \alpha_1, \ldots, \log \alpha_d, \mu_1, \ldots, \mu_d \right),$$

where J_k is the type of Y_k and

$$N_i \overset{\text{def}}{=} \sum_{k=1}^{n} \mathbb{1}_{J_k=i}, \quad S_i \overset{\text{def}}{=} \sum_{k: J_k=i} Y_k = \sum_{k=1}^{n} Y_k \mathbb{1}_{J_k=i}.$$

Then (A4.1) can be seen to hold for an appropriate choice of $\boldsymbol{t}$ and ν.

After conditioning on the J_k, it is readily guessed that the MLE estimator $\widehat{\boldsymbol{\theta}}$ is given by

$$\widehat{\alpha}_i = \frac{N_i}{N}, \quad \widehat{\mu}_i = \frac{S_i}{N_i}, \quad i = 1, \ldots, d, \tag{A4.2}$$

i.e., the ML estimator of α_i is the empirical mean N_i/N of the Y_k with $J_k = i$, and similarly the ML estimator of μ_i is the empirical mean N_i/N of the same set of Y_k. For a formal verification of (A4.2), introduce a Lagrangian multiplier λ and consider the minimization of

$$\sum_{j=1}^{2d} \theta_j t_j(v) \; + \; \lambda\big(e^{\theta_{d+1}} + \cdots + e^{\theta_{2d}} - 1\big). \tag{A4.3}$$

□

Now consider the general exponential family setting, and assume that only Y is observed and not the whole of V. For example in Example A.1, there could be censoring at t_0 such that only the $V_1 \wedge t_0, \ldots, V_n \wedge t_0$ are observed, and in Example A.2 only the Y_k but not their types J_k could be observed. The statistical estimation problem remains explicitly tractable in the first example but not in the second, where we are left with a $(2d-1)$-dimensional optimization problem without an explicit solution. The EM algorithm proceeds iteratively. When updating from $\boldsymbol{\theta}_m$ to $\boldsymbol{\theta}_{m+1}$, $t(V)$ is replaced by its $\mathbb{P}_{\boldsymbol{\theta}_m}$-conditional expectation t_m given Y:

$$t_m \overset{\text{def}}{=} \mathbb{E}_{\boldsymbol{\theta}_m}\big[t(V) \,\big|\, Y\big], \quad \boldsymbol{\theta}_{m+1} \overset{\text{def}}{=} \widehat{\boldsymbol{\theta}}(t_m). \tag{A4.4}$$

It can be shown that the $\boldsymbol{\theta}_m$-likelihood is nondecreasing in m and hence that $\boldsymbol{\theta}_m \to \widehat{\boldsymbol{\theta}}$ under suitable regularity conditions.

Example A.3 Consider the normal mixture example A.2. Write

$$\boldsymbol{\theta}_m = \big(\log\alpha_{m,1}, \ldots, \log\alpha_{m,d}, \mu_{m,1}, \ldots, \mu_{m,d}\big).$$

Then

$$\mathbb{P}_{\boldsymbol{\theta}_m}\big(J_k = i \,\big|\, y\big) = \mathbb{P}_{\boldsymbol{\theta}_m}\big(J_k = i \,\big|\, y_k\big)$$

$$= \frac{\alpha_{m,i}(2\pi)^{-1/2}e^{-(y_k-\mu_{m,i})^2/2}}{\sum_1^d \alpha_{m,j}(2\pi)^{-1/2}e^{-(y_k-\mu_{m,j})^2/2}} \overset{\text{def}}{=} \psi(\boldsymbol{\theta}_m, y_k),$$

$$t_m = \big(N_{m,1}, \ldots, N_{m,d}, S_{m,1}, \ldots, S_{m,d}\big),$$

where

$$N_{m,i} \overset{\text{def}}{=} \mathbb{E}_{\boldsymbol{\theta}_m}\big[N_i \,\big|\, Y\big] = \sum_{k=1}^{n} \psi(\boldsymbol{\theta}_m, Y_k),$$

$$S_{m,i} \overset{\text{def}}{=} \mathbb{E}_{\boldsymbol{\theta}_m}\big[S_i \,\big|\, Y\big] = \sum_{k=1}^{n} Y_k \psi(\boldsymbol{\theta}_m, Y_k).$$

From (A4.2), it therefore follows that $\boldsymbol{\theta}_{m+1} = \widehat{\boldsymbol{\theta}}(t_m)$ is given by $\widehat{\alpha}_{m+1,i} = N_{m,i}/N$, $\widehat{\mu}_{m+1,i} = S_{m,i}/N_{m,i}$.

It should be noted that the example of normal mixtures is one in which the likelihood has local maxima; see, e.g., Robert & Casella [301, Example

5.19]. If started in the domain of attraction of a local but not global maximum, the EM algorithm will then never escape to the global maximum. A practical solution to this problem is to run the algorithm several times with different starting points, say $\boldsymbol{\theta}_{0,1}, \ldots, \boldsymbol{\theta}_{0,L}$, and as estimator use $\lim_m \boldsymbol{\theta}_{m,\ell}$, where ℓ is the index with the highest limiting value of the likelihood.   □

The difficulty in applying the EM algorithm is usually the computation of the conditional expectation $\boldsymbol{t}_m$ (the E-step) rather than the computation of $\widehat{\boldsymbol{\theta}}(\boldsymbol{t}_m)$ (the M-step), which involves just the same calculations as in computating the MLE in the presence of full observations. It is therefore tempting to perform the E-step by Monte Carlo, which means that $\boldsymbol{t}_m$ is redefined as

$$\boldsymbol{t}_m \stackrel{\text{def}}{=} \frac{1}{R}\big(\boldsymbol{t}(\boldsymbol{V}_{m,1}) + \cdots + \boldsymbol{t}(\boldsymbol{V}_{m,R})\big), \tag{A4.5}$$

where $\boldsymbol{V}_{m,1}, \ldots, \boldsymbol{V}_{m,R}$ are simulated replications from the conditional $\mathbb{P}_{\boldsymbol{\theta}_m}$-distribution of $\boldsymbol{V}$ given $\boldsymbol{Y}$. This algorithm is known under names such as the *Monte Carlo EM algorithm* and the *stochastic EM algorithm* (outside the exponential family setting to which we have restricted ourselves, these names may actually cover slightly different algorithms). The Monte Carlo EM algorithm obviously has the property that $\{\boldsymbol{\theta}_m\}_{m \in \mathbb{N}}$ becomes a time-homogeneous Markov chain with no state being absorbing (in particular not the MLE $\widehat{\boldsymbol{\theta}}$!), so that the best one can hope for is oscillations around $\widehat{\boldsymbol{\theta}}$ that are small given that R has been chosen large enough, not convergence in probability or a.s. To obtain this, one needs to let $R \stackrel{\text{def}}{=} R_m$ go to ∞ with m, which is closely related to the *stochastic approximation EM algorithm* discussed in VIII.5.

Exercises

4.1 (TP) Verify that the minimization of (A4.3) produces the claimed result.
4.2 (TP) In Example A.1, find the explicit form of the MLE estimator for θ in the presence of censoring, and also write up the EM recursion expressing θ_{m+1} in terms of θ_m.

A5   Filtering

We summarize here the classical formulas for filtering as introduced in I.5.5.

We consider a hidden Markov model in which the unobserved Markov chain $X_0, \ldots, X_n$ has state space E (possibly uncountable) and transition density $q(x, x')$ w.r.t. $\nu(\mathrm{d}x')$. Further, the (observed) Y_k, $k = 0, \ldots, n$, depends only on $X_0, \ldots, X_n$ and the remaining Y_ℓ via X_k, such that the density is $g(y \,|\, x)$ when $X_k = x$. We use notation such as

$$\boldsymbol{x}_{0:k} \stackrel{\text{def}}{=} (x_0, \ldots, x_k), \quad p(\boldsymbol{x}_{0:k}) \stackrel{\text{def}}{=} r_0(x_0)q(x_0, x_1)q(x_1, x_2)\cdots q(x_{k-1}, x_k),$$

where r_0 is the density of X_0 w.r.t. ν. Thus $p(\boldsymbol{x}_{0:k})$ is the density of $X_{0:k}$ w.r.t. $\nu^{\otimes n}$.

The problem is to update the filtering density $p(x_k \mid \boldsymbol{y}_{0:k})$ from $k-1$ to k, motivated largely by settings such as target tracking and signal processing, where the (noisy) observations arrive sequentially in time.

The recursion involves the auxiliary quantities $p(x_{k+1} \mid \boldsymbol{y}_{0:k})$ (the predicition density) and $p(y_k \mid \boldsymbol{y}_{0:k-1})$, and is given by

$$p(y_k \mid \boldsymbol{y}_{0:k-1}) = \int p(x_k \mid \boldsymbol{y}_{0:k-1})g(y_k \mid x_k)\,\nu(\mathrm{d}x_k)\,, \qquad (A5.1)$$

$$p(x_k \mid \boldsymbol{y}_{0:k}) = p(x_k \mid \boldsymbol{y}_{0:k-1})g(y_k \mid x_k)/p(y_k \mid \boldsymbol{y}_{0:k-1}) \qquad (A5.2)$$

$$p(x_{k+1} \mid \boldsymbol{y}_{0:k}) = \int p(x_k \mid \boldsymbol{y}_{0:k})q(x_k, x_{k+1})\,\nu(\mathrm{d}x_k)\,, \qquad (A5.3)$$

as follows by obvious conditioning arguments (for (A5.2), note that $g(y_k \mid x_k) = g(y_k \mid x_k, \boldsymbol{y}_{0:k-1})$). The initialization corresponds to replacing $p(y_k \mid \boldsymbol{y}_{0:k-1})$ in (A5.1) by

$$p(y_0) = \int r_0(x_0)g(y_0 \mid x_0)\,\nu(\mathrm{d}x_0) \qquad (A5.4)$$

when $k = 0$.

The difficulty in the evaluation is the integrals in (A5.1), (A5.3), (A5.4). One solvable case is a Gaussian linear state-space model

$$\boldsymbol{X}_k = \boldsymbol{A}\boldsymbol{X}_{k-1} + \boldsymbol{R}U_k\,, \quad \boldsymbol{Y}_k = \boldsymbol{B}\boldsymbol{X}_k + \boldsymbol{S}V_k\,,$$

where $\boldsymbol{X}_k \in \mathbb{R}^p$, $\boldsymbol{Y}_k \in \mathbb{R}^q$, $\boldsymbol{U}_k \in \mathbb{R}^s$, $\boldsymbol{V}_k \in \mathbb{R}^t$, the matrices $\boldsymbol{A}, \boldsymbol{B}, \boldsymbol{R}, \boldsymbol{S}$ have the appropriate dimensions (e.g., $\boldsymbol{B} : q \times p$), and the $\boldsymbol{U}_k$, $\boldsymbol{V}_k$ are independent Gaussians with mean $\boldsymbol{0}$ and covariance matrices $\boldsymbol{I}_s$, $\boldsymbol{I}_t$. Here all relevant joint and conditional distributions are multivariate Gaussian, so (A5.1)–(A5.4) can be written out in terms of means and covariances. The resulting formulas are the classical *Kalman filter*. Another solvable case is a finite state space E, where the integrals become sums and one only has to store a finite number of values. Here (A5.1)–(A5.4) is known as the *forward algorithm*. For computing the smoothing density $p(\boldsymbol{x}_{0:n} \mid \boldsymbol{y}_{0:n})$, a further step, the *backward algorithm*, is needed. Note, however, that even the case of a finite E may create problems if E is huge, as is the case in some applications.

We refer further to Cappé et al. [65].

A6   Itô's Formula

If $\{x(t)\}_{t \geq 0}$ is a solution of the ODE

$$\dot{x}(t) = a(t, x(t))\,, \quad \text{i.e.,} \quad \mathrm{d}x(t) = a(t, x(t))\,\mathrm{d}t\,,$$

and f is a smooth function of two variables (a is assumed smooth as well), then the chain rule gives

$$
\begin{aligned}
\mathrm{d}f(t, x(t)) &= f_t(t, x(t))\,\mathrm{d}t \;+\; f_x(t, x(t))\,\mathrm{d}x(t) \\
&= f_t(t, x(t))\,\mathrm{d}t \;+\; f_x(t, x(t))a(t, x(t))\,\mathrm{d}t\,,
\end{aligned}
$$

where $f_t(t, x) = \frac{\partial}{\partial t}f(t, x)$, $f_x(t, x) = \frac{\partial}{\partial x}f(t, x)$ denotes the partial derivatives.

Itô's formula is a similar expression for a function $f(t, X(t))$ of the solution X of the SDE

$$
\mathrm{d}X(t) \;=\; a(t, X(t))\,\mathrm{d}t + b(t, X(t))\,\mathrm{d}B(t)\,, \tag{A6.5}
$$

where B is standard Brownian motion, and states that

$$
\mathrm{d}f(X(t)) \;=\; \left\{ af_x + f_t + \frac{1}{2}b^2 f_{xx} \right\}\,\mathrm{d}t \;+\; bf_x\,\mathrm{d}B(t) \tag{A6.6}
$$

where a, b, f_t, f_x, and f_{xx} (the second partial derivative w.r.t. x) are evaluated at $(t, X(t))$. The precise meaning of this statement is that (A6.5), (A6.6) should be interpreted respectively as

$$
X(t) - X(0) \;=\; \int_0^t a(s, X(s))\,\mathrm{d}s \;+\; \int_0^t b(s, X(s))\,\mathrm{d}B(s)
$$

and

$$
\begin{aligned}
&f(X(t)) - f(X(0)) \\
&= \int_0^t \left\{ a(s, X(s))f_x(s, X(s)) + f_t(s, X(s)) + b^2(s, X(s))f_{xx}(s, X(s)) \right\}\,\mathrm{d}s \\
&\quad + \int_0^t b(s, X(s))f_x(s, X(s))\,\mathrm{d}B(s),
\end{aligned}
$$

where $\int_0^t b(s, X(s))\,\mathrm{d}B(s)$, etc., denotes the Itô integral.

The proof of (A6.6) can be found in any standard textbook in stochastic calculus, e.g., Steele [346]. The heuristics is the expression $(\mathrm{d}B(t))^2 = \mathrm{d}t$ (compare to $(\mathrm{d}t)^2 = 0$!), which is motivated from quadratic variation properties of B. Thus, compared to ODEs, one needs to take into account also the term containing f_{xx} in the second-order Taylor expansion to correctly include all terms of order $\mathrm{d}t$.

A formula that is often used is the *Itô isometry*

$$
\mathrm{Cov}\left(\int_0^t f(s)\,\mathrm{d}B(s),\; \int_0^t g(s)\,\mathrm{d}B(s) \right) \;=\; \int_0^t f(s)g(s)\,\mathrm{d}s\,. \tag{A6.7}
$$

A7   Inequalities

Proposition A.1 (THE INFORMATION INEQUALITY) *Let f, g be densities. Then*

$$\int \log g(x) f(x)\, \mathrm{d}x \;\leq\; \int \log f(x)\, f(x)\, \mathrm{d}x\,,$$

where $\log 0 \cdot y = 0$, $0 \leq y < \infty$. If equality holds, then f and g define the same probability measure.

Proof. Let X be an r.v. with density $f(x)$ and write $\mathbb{E}_f$ for the corresponding expectation. Then by Jensen's inequality,

$$\int \log g(x) f(x)\, \mathrm{d}x \;-\; \int \log f(x)\, f(x)\, \mathrm{d}x$$

$$= \quad \mathbb{E}_f \log g(X) - \mathbb{E} \log f(X) \;=\; \mathbb{E}_f \log \frac{g(X)}{f(X)}$$

$$\leq \quad \log\left(\mathbb{E}_f \frac{g(X)}{f(X)} \right) \;=\; \log\left(\int_{\{f>0\}} \frac{g(x)}{f(x)} f(x)\, \mathrm{d}x \right)$$

$$= \quad \log\left(\int_{\{f>0\}} g(x)\, \mathrm{d}x \right) \;\leq\; \log 1 \;=\; 0\,,$$

where $\{f > 0\} = \{x : f(x) > 0\}$. If equality holds, then $\int_{\{f>0\}} g(x)\, \mathrm{d}x = 1$ and $g(X) = f(X)$ a.s. Hence for any A,

$$\int_A g(x)\, \mathrm{d}x \quad = \quad \int_{A \cap \{f>0\}} g(x)\, \mathrm{d}x$$

$$= \quad \mathbb{E}_f\left[\frac{g(X)}{f(X)}; X \in A \cap \{f > 0\} \right]$$

$$= \quad \mathbb{P}(X \in A \cap \{f > 0\}) \;=\; \int_A f(x)\, \mathrm{d}x\,.$$

$\square$

Proposition A.2 (CHEBYSHEV'S COVARIANCE INEQUALITY) *If X is a real-valued r.v. and f, g nondecreasing functions, then*

$$\mathbb{E}[f(X)g(X)] \;\geq\; \mathbb{E}f(X)\, \mathbb{E}g(X)\,.$$

For an elegant coupling proof, see Thorisson [354].

A8   Integral Formulas

The derivative of $\int_a^\theta g(x)\, \mathrm{d}x$ w.r.t. θ is $g(\theta)$. If g depends not only on x but also one θ, the relevant extension is the following, which is easily proved from first principles:

$$\frac{\mathrm{d}}{\mathrm{d}\theta} \int_a^\theta g(x,\theta)\,\mathrm{d}x \;=\; g(\theta,\theta) \;+\; \int_a^\theta \frac{\mathrm{d}}{\mathrm{d}\theta} g(x,\theta)\,\mathrm{d}x\,. \qquad (\mathrm{A}8.1)$$

A convenient formulation of the standard integration-by-parts formula for Lebesgue–Stieltjes integrals when dealing with r.v.'s is the following: for $X \geq 0$ an r.v. and $g(x) \geq 0$ differentiable, in the sense that $g(x) = g(0) + \int_0^x g'(y)\,\mathrm{d}y$ for some function $g'(y)$,

$$\mathbb{E}g(X) \;=\; g(0) \;+\; \int_0^\infty g'(x)\mathbb{P}(X > x)\,\mathrm{d}x\,. \qquad (\mathrm{A}8.2)$$

Similar formulas exist in multidimensions; see V.3.2 for an example.

Bibliography

[1] E.H.L. Aarts & J.H.M. Korst (1989) *Simulated Annealing and Boltzmann Machines.* Wiley.

[2] E.H.L. Aarts & J.K. Lenstra (1989) *Local Search in Combinatorial Optimization.* Wiley.

[3] S. Aberg, M.R. Leadbetter, & I. Rychlik (2006) Palm distributions on wave characteristics in encountering seas. *Submitted.*

[4] P. Abry & D. Sellan (1996) The wawelet-based synthesis for the fractional Brownian motion proposed by F. Sellan and Y. Meyer: remarks and fast implementation. *Appl. Comp. Harmonic Anal.* **3**, 337–383.

[5] R.J. Adler, R. Feldman & M.S. Taqqu, eds. (1998) *A User's Guide to Heavy Tails.* Birkhäuser.

[6] J. Albert, J. Bennett, & J.J. Cochran (2005) *Anthology of Statistics in Sports.* SIAM.

[7] V. Anantharam (1988) How large delays build up in a GI/GI/1 queue. *Queueing Systems* **5**, 345–368.

[8] D. Applebaum (2004) *Lévy Processes and Stochastic Calculus.* Cambridge University Press.

[9] S. Asmussen (1982) Conditioned limit theorems relating a random walk to its associate, with applications to risk reserve processes and the GI/G/1 queue. *Adv. Appl. Prob.* **14**, 143–170.

[10] S. Asmussen (1989) Risk theory in a Markovian environment. *Scand. Actuarial J.,* 69–100.

[11] S. Asmussen (1990) Exponential families and regression in the Monte Carlo study of queues and random walks. *Ann. Statist.* **18**, 1851–1867.

[12] S. Asmussen (1992) Queueing simulation in heavy traffic. *Math. Oper. Res.* **17**, 84–111.

[13] S. Asmussen (1995) Stationary distributions for fluid flow models with or without Brownian noise. *Stochastic Models* **11**, 21–49 (1995).

[14] S. Asmussen (1995) Stationary distributions via first passage times. *Advances in Queueing: Models, Methods & Problems* (J. Dshalalow, ed.), 79–102. CRC Press.

[15] S. Asmussen (2000) *Ruin Probabilities*. World Scientific.

[16] S. Asmussen (2003) *Applied Probability and Queues* (2nd ed.). Springer-Verlag.

[17] S. Asmussen & K. Binswanger (1997) Simulation of ruin probabilities for subexponential claims. *Astin Bull.* **27**, 297–318.

[18] S. Asmussen, K. Binswanger, & B. Højgaard (2000) Rare events simulation for heavy–tailed distributions. *Bernoulli* **6**, 303–322.

[19] S. Asmussen, P.W. Glynn, & H. Thorisson (1992) Stationarity detection in the initial transient problem. *ACM TOMACS* **2**, 130–157.

[20] S. Asmussen, P.W. Glynn, & J. Pitman (1996) Discretization error in the simulation of one–dimensional reflecting Brownian motion. *Ann. Appl. Probab.* **5**, 875–896.

[21] S. Asmussen, P. Fuckerrieder, M. Jobmann, & H.–P. Schwefel (2000) Large deviations and fast simulation of buffer overflow probabilities for queues. *Stoch. Proc. Appl.* **102**, 1–23.

[22] S. Asmussen, D.P. Kroese, & R.Y. Rubinstein (2004) Heavy tails, importance sampling and cross-entropy. *Stoch. Models* **21**, 57–76.

[23] S. Asmussen & D.P. Kroese (2006) Improved algorithms for rare event simulation with heavy tails. *Adv. Appl. Probab.* **38**, 545–558.

[24] S. Asmussen, D. Madan, & M. Pistorius (2005) Pricing equity default swaps under the CMGY model. *Submitted.*

[25] S. Asmussen & H.M. Nielsen (1995) Ruin probabilities via local adjustment coefficients. *J. Appl. Probab.* **32**, 736–755.

[26] S. Asmussen & J. Rosiński (2001) Approximations of small jumps of a Lévy process with a view towards simulation. *J. Appl. Probab.* **38**, 482–493.

[27] S. Asmussen & R.Y. Rubinstein (1995) Steady–state rare events simulation in queueing models and its complexity properties. *Advances in Queueing: Models, Methods & Problems* (J. Dshalalow ed.), 429–466. CRC Press.

[28] S. Asmussen & R.Y. Rubinstein (1999) Sensitivity analysis of insurance risk models via simulation. *Management Science* **45**, 1125–1141.

[29] S. Asmussen & K. Sigman (1996) Monotone stochastic recursions and their duals. *Probab. Eng. Inf. Sci.* **10**, 1–20.

[30] R. Atar & P. Dupuis (1999) Large deviations and queueing networks: methods for rate function identification. *Stoch. Proc. Appl.* **84**, 255–296.

[31] K.B. Athreya, H. Doss, & J. Sethuraman (1996) On the convergence of the Markov chain simulation method. *Ann. Statist.* **24**, 69–100.

[32] F. Avram, J.G. Dai, & J.J. Hasenbein (2001) Explicit solutions for variational problems in the quadrant. *QUESTA* **37**, 261–291.

[33] S. Axsäter (2000) *Inventory Control*. Kluwer.

[34] F. Baccelli & P. Brémaud (2002) *Elements of Queueing Theory. Palm–Martingale Calculus and Stochastic Recursions* (2nd ed.). Springer-Verlag.

[35] J. Banks, J.S. Carson, II, & B.L. Nelson (1996) *Discrete Event–Systems Simulation* (2nd ed.). Prentice-Hall.

[36] O. Barndorff-Nielsen & A. Lindner (2007) Lévy copulas and the Υ-transform. *Bernoulli* (to appear).

[37] O. Barndorff-Nielsen, T. Mikosch, & S. Resnick, eds. (2001) *Lévy Processes. Theory and Applications*. Birkhäuser.

[38] O. Barndorff-Nielsen & N. Shephard (2001) Non-Gaussian Ornstein-Uhlenbeck based models and some of their uses in financial econometrics. *J.R. Statist. Soc.* **B63**, 167–241.

[39] A.J. Bayes (1970) Statistical techniques for simulation models. *Austr. Comp. J.* **2**, 180–184.

[40] P. Bertail & S. Clémencon (2006) Regeneration-based statistics for Harris recurrent Markov chains. *Dependence in Probability and Statistics* (P. Bertail, P. Doukan, & P. Soulier, eds.), 3–54. Lecture Notes in Statistics, Springer-Verlag.

[41] J. Bertoin (1996) *Lévy Processes.* Cambridge University Press.

[42] A. Beskos & G.O. Roberts (2005) Exact simulation of diffusions. *Ann. Appl. Probab.* **15**, 2422–2444.

[43] R.N. Bhattacharya & J.K. Ghosh (1978) On the validity of the formal Edgeworth expansion. *Ann. Statist.* **6**, 434–451.

[44] N.H. Bingham, C.M. Goldie, & J.L. Teugels (1987) *Regular Variation.* Encyclopedia of Mathematics and Its Applications **27**. Cambridge University Press.

[45] T. Björk (2005). *Arbitrage Theory in Continuous Time* (2nd ed.). Oxford University Press.

[46] J. Blanchet & P.W. Glynn (2007) Strongly efficient estimators for light-tailed sums. *Proceedings of the 2006 Valuetools Conference, Pisa* (to appear).

[47] J. Blanchet & P.W. Glynn (2006) Efficient rare-event simulation for the maximum of heavy-tailed random walk. *Submitted.*

[48] J. Blanchet, J.C. Liu, & P.W. Glynn (2006) State-dependent importance sampling and large deviations. *Proceedings of the 6th International Workshop on Rare Event Simulation* (W. Sandmann, ed.), 154-161. Bamberg, Germany.

[49] B. Blaszczyszyn & K. Sigman (1999) Risk and duality in multidimensions. *Stoch. Proc. Appl.* **83**, 331–356.

[50] L. Bondesson (1982) On simulation from infinitely divisible distributions. *Adv. Appl. Probab.* **14**, 855–869.

[51] A.A. Borovkov (1984) *Asymptotic Methods in Queueing Theory.* Wiley.

[52] A.A. Borovkov (1998) *Mathematical Statistics.* Gordon & Breach.

[53] M. Bossy, E. Gobet, & D. Talay (2004) A symmetrized Euler scheme for an efficient approximation of reflected diffusions. *J. Appl. Probab.* **41**, 877–889.

[54] B. Bouchard & N. Touzi (2004) Discrete time approximation and simulation of backward stochastic differential equations. *Stoch. Proc. Appl.* **11**, 175–206.

[55] P. Boyle, M. Broadie, & P. Glasserman (1997) Monte Carlo methods for security pricing. *Journal of Economic Dynamics and Control* **21**, 1267–1321.

[56] P. Bratley, B.L. Fox, & L. Schrage (1987) *A Guide to Simulation.* Springer-Verlag.

[57] P. Brémaud (1999) *Markov Chains. Gibbs Fields, Monte Carlo Simulation and Queues.* Springer-Verlag.

[58] P.J. Brockwell & R.A. Davis (1991) *Time Series: Theory and Methods* (2nd ed.). Springer-Verlag.

[59] P.J. Brockwell & R.A. Davis (1996) *Introduction to Time Series and Forecasting.* Springer-Verlag.

[60] S.P. Brooks (1998) Markov chain Monte Carlo and its application. *The Statistician* **47**, 69–100.

[61] J.A. Bucklew (1990) *Large Deviation Techniques in Decision, Simulation, and Estimation.* John Wiley & Sons, New York.

[62] J.A. Bucklew (2004) *Introduction to Rare Event Simulation.* Springer-Verlag.

[63] J.A. Bucklew, P. Ney, & J.S. Sadowsky (1990) Monte Carlo simulation and large deviations theory for uniformly recurrent Markov chains. *J. Appl. Prob.* **27**, 44–59.

[64] R.E. Caflish (1998) Monte Carlo and quasi–Monte Carlo methods. *Acta Numer.* **7**, 1–49.

[65] O. Cappé, E. Moulines, & T. Rydén (2005) *Inference in Hidden Markov Models.* Springer-Verlag.

[66] J.M. Chambers, C.L. Mallows, & B.W. Stuck (1976) A method for simulating stable random variables. *J. Amer. Statist. Ass.* **71**, 340–344.

[67] T.F. Chan, G. Golub, & R.J. LeVeque (1983) Algorithms for computing the sample variance: analysis and recommendations. *Amer. Statist.* **37**, 242–247.

[68] C.S. Chang, P. Heidelberger, P. Juneja, & P. Shahabuddin (1994) Effective bandwidth and fast simulation of ATM intree networks. *Performance Evaluation* **20**, 45–65

[69] H. Chen & D.D. Yao (2000) *Fundamentals of Queueing Networks.* Springer-Verlag.

[70] Z. Chen & R. Kulperger (2005) A stochastic competing-species model and ergodicity. *J. Appl. Probab.* **42**, 738–753.

[71] Y.S. Chow & H. Robbins (1965) On the asymptotic theory of fixed-width sequential confidence intervals for the mean. *Ann. Math. Statist.* **36**, 457–462.

[72] K.L. Chung (1974) *A Course in Probability Theory* (2nd ed.). Academic Press.

[73] S. Cohen & J. Rosiński (2006) Gaussian approximations of multivariate Lévy processes with applications to simulation of tempered and operator stable processes. *Bernoulli* (to appear).

[74] J.F. Collamore (2002) Importance sampling techniques for the multidimensional ruin problem for general Markov additive sequences of random vectors. *Ann. Appl. Probab.* **12**, 382–421.

[75] R. Cont & P. Tankov (2004) *Financial Modelling with Jump Processes.* Chapman & Hall/CRC.

[76] M. Cottrel, J.C. Fort, & G. Malgouyres (1983) Large deviations and rare events in the study of stochastic algorithms. *IEEE Trans. Automatic Control* **AC–28**, 907–918.

[77] P.F. Craigmile (2003) Simulating a class of stationary Gaussian processes using the Davies–Harte algorithm, with applications to long memory processes. *J. Time Series Anal.* **24**, 505–511.

[78] II. Cramér & M.R. Leadbettcr (1967) *Stationary and Related Stochastic Processes.* Wiley.

[79] D.J. Daley, H. Stoyan, & D. Stoyan (1999) The volume fraction of a Poisson germ model with maximally non-overlapping spherical grains. *Adv. Appl. Probab.* **31**, 610–624.

[80] P. Damien, P.W. Laud, & A.F.M. Smith (1995) Approximate random variate generation from infinitely divisible distributions with applications to Bayesian inference. *J.R. Statist. Soc.* **B57**, 547–563.

[81] B.M. Davis, R. Hagan, & L.E. Borgman (1981) A program for the finite Fourier transform simulation of realizations from a one-dimensional random function with known covariance. *Comp. Geosciences* **7**, 199–206.

[82] A.C. Davison & D.V. Hinkley (1997) *Bootstrap Methods and Their Applications.* Cambridge University Press.

[83] P.T. de Boer, D.P. Kroese, S. Mannor, & R.Y. Rubinstein (2005) A tutorial on the cross-entropy method. *Ann. Oper. Res.* **134**, 19–67.

[84] L. de Haan & A. Ferreira (2006) *Extreme Value Theory. An Introduction.* Birkhäuser.

[85] P. Del Moral (2005) *Feynman-Kac Formulae. Genealogical and Interacting Particle Systems with Applications.* Springer-Verlag.

[86] A. Dembo, C.L. Mallows, & L.A. Shepp (1989) Embedding nonnegative Toeplitz matrices in nonnegative definite circulant matrices, with applications to covariance estimation. *IEEE Trans. Inf. Th.* **IT-35**, 1206–1212.

[87] A. Dembo & O. Zeitouni (1993) *Large Deviation Techniques.* Jones & Bartlett, Boston.

[88] L. Devroye (1986) *Non-Uniform Random Variate Generation.* Springer-Verlag.

[89] C.R. Dietrich & G.N. Newsam (1993) A fast and exact method for multi-dimensional Gaussian simulation. *Water Resources Res.* **29**, 2861–2869.

[90] C.R. Dietrich & G.N. Newsam (1997) Fast and exact simulation of stationary Gaussian processes through circulant embedding of the covariance matrix. *SIAM J. Sci. Comput.* **18**, 1088–1107.

[91] A.B. Dieker & M. Mandjes (2003) On spectral simulation of fractional Brownian motion. *Prob. Eng. Inf. Sci.* **17**, 417–434.

[92] A.B. Dieker & M. Mandjes (2006) Fast simulation of overflow probabilities in a queue with Gaussian input. *ACM TOMACS* **16**, 119–151.

[93] Y. Dodge (1996) A natural random number generator. *Int. Statist. Review* **64**, 329–344.

[94] J.L. Doob (1957) Conditional Brownian motion and the boundary of harmonic functions. *Bull. Soc. Math. France* **85**, 431–458.

[95] A. Doucet, N. de Freitas, & N. Gordon, eds. (2001) *Sequential Monte Carlo Methods in Practice.* Springer-Verlag.

[96] P. Doukhan, G. Oppenheimer, & M.S. Taqqu, eds. (2003) *Long-Range Dependence: Theory and Applications.* Birkhäuser.

[97] M. Drmota & R. Tichy (1997) *Sequences, Discrepancies and Applications.* Lecture Notes in Mathematics **1651**. Springer-Verlag.

[98] R.M. Dudley (1999) *Uniform Central Limit Theorems.* Cambridge University Press.

[99] D. Duffie & P.W. Glynn (1995) Efficient Monte Carlo simulation of security prices. *Ann. Appl. Probab.* **5**, 897–905.

[100] N.G. Duffield & N. O'Connell (1995) Large deviations and overflow probabilities for the general single-server queue. *Math. Proc. Camb. Philos. Soc.* **118**, 363–374.

[101] P. Dupuis & R.S. Ellis (1995) The large deviation principle for a general class of queueing systems, I. *Trans. Amer. Math. Soc.* **347**, 2689–2751.

[102] P. Dupuis & H. Wang (2004) Importance sampling, large deviations, and differential games. *Stochastics and Stochastics Reports* **76**, 481–508.

[103] P. Dupuis & H. Wang (2005) Dynamic importance sampling for uniformly recurrent Markov chains. *Ann. Appl. Prob.* **15**, 1-38.

[104] P. Dupuis & H. Wang (2006) Subsolutions of an Isaacs equation and efficient schemes of importance sampling: Convergence analysis. *Submitted.*

[105] P. Dupuis & H. Wang (2006) Subsolutions of an Isaacs equation and efficient schemes of importance sampling: Examples and numerics. *Submitted.*

[106] M. Dyer, A. Frieze, & R. Kannan (1991) A random polynomial-time algorithm for approximating the volume of convex bodies. *J. ACM* **38**, 1–17.

[107] K.O. Dzhaparidze & J.H. van Zanten (2004) A series expansion of fractional Brownian motion. *Prob. Th. Rel. Fields* **130**, 39–55.

[108] P. Embrechts, C. Klüppelberg & T. Mikosch (1997) *Modelling Extremal Events for Finance and Insurance.* Springer, Heidelberg.

[109] K.B. Ensor & P.W. Glynn (2000) Simulating the maximum of a random walk. *J. Stat. Plann. Inference* **85**, 127–135.

[110] R.S. Erikson & K. Sigman (2000) A simple stochastic model for close U.S. presidential elections. Unpublished manuscript, Columbia University.

[111] S.M. Ermakov & V.B. Melas (1995) *Design and Analysis of Simulation Experiments.* Kluwer.

[112] S. Ethier & T. Kurtz (1986) *Markov Processes: Characterization and Convergence.* Wiley.

[113] M. Evans & T. Swartz (1988) Sampling from Gauss rules. *SIAM J. Sci. Stat. Comput.* **9**, 950–961.

[114] M. Evans & T. Swartz (2000) *Approximating Integrals via Monte Carlo and Deterministics Methods.* Oxford University Press.

[115] W.J. Ewens (2004) *Mathematical Population Genetics* (2nd ed.). Springer-Verlag.

[116] W. Feller (1971) *An Introduction to Probability Theory and Its Applications* (2nd ed.) **II**. Wiley.

[117] J.A. Fill (1998) An interruptible algorithm for perfect sampling via Markov chains. *Ann. Appl. Probab.* **8**, 131–162.

[118] G. Fishman (1996) *Monte Carlo. Concepts, Algorithms and Applications.* Springer-Verlag.

[119] R. Foley & D. McDonald (2001) Join the shortest queue: stability and exact asymptotics. *Ann. Appl. Probab.* **11**, 569–607.

[120] E. Fournié, J.-M. Lasry, J. Lebouchoux, P.-L. Lions, & N. Touzi (1999) Applications of Malliavin calculus to Monte Carlo methods in finance. *Finance and Stochastics* **3**, 391–412.

[121] S.G. Foss & R.L. Tweedie (1998) Perfect simulation and backwards coupling. *Stochastic Models* **14**, 187–203.

[122] B.L. Fox & P.W. Glynn (1989) Estimating discounted costs. *Management Science* **35**, 1297-1325.

[123] R.M. Fujimoto (2000) *Parallel and Distributed Simulation Systems.* Wiley.

[124] J. Gaines & T.J. Lyons (1997) Variable step size control in the numerical solution of stochastic differential equations. *SIAM J. Appl. Math.* **57**, 1455–1484.

[125] M.J.J. Garvels & D.P. Kroese (1998) A comparison of RESTART implementations. *Proceedings of the Winter Simulation Conference* **1998**, 601–609. IEEE Press.

[126] S. Geman & D. Geman (1984) Stochastic relaxation, Gibbs distributions and the Bayesian restoration of images. *IEEE Trans. Pattern Anal. Mach. Intell.* **6**, 721–741.

[127] J.E. Gentle (2003) *Random Number Generation and Monte Carlo Methods* (2nd ed.). Springer-Verlag.

[128] B. Gidas (1994) Metropolis-type Monte Carlo simulation algorithms and simulated annealing, *Topics in Contemporary Probability* (J.L. Snell, ed.), 159–232. CRC Press.

[129] W.R. Gilks, S. Richardson & D.J. Spiegelhalter (1996) *Markov Chain Monte Carlo in Practice.* Chapman & Hall.

[130] R.D. Gill (1989) Non- and semi-parametric maximum likelihood estimators and the von Mises method (Part I). *Scand. J. Statist.* **16**, 97–128.

[131] P. Glasserman (1991) *Gradient Estimation via Perturbation Analysis.* Kluwer.

[132] P. Glasserman (1992) Smoothing complements and randomized score functions. *Ann. Oper. Res.* **39**, 41–67.

[133] P. Glasserman (2004) *Monte Carlo Methods in Financial Engineering.* Springer-Verlag.

[134] P. Glasserman, P. Heidelberger, & P. Shahabuddin (1999) Asymptotically optimal importance sampling and stratification for pricing path-dependent options. *Mathematical Finance* **9**, 117–152.

[135] P. Glasserman, P. Heidelberger, & P. Shahabuddin (2000) Variance reduction techniques for estimating Value-at-Risk. *Management Science* **46**, 1349–1364.

[136] P. Glasserman, P. Heidelberger, & P. Shahabuddin (2002) Portfolio Value-at-Risk with heavy-tailed risk factors. *Mathematical Finance* **12**, 239–269.

[137] P. Glasserman, P. Heidelberger, P. Shahabuddin, & T. Zajic (1996) Multilevel splitting for estimating rare events probabilities. *Oper. Res.* **47**, 585–600.

[138] P. Glasserman, P. Heidelberger, P. Shahabuddin, & T. Zajic (1996) Splitting for rare event simulation: Analysis of simple cases. *Proceedings of the Winter Simulation Conference* **1996**, 302–308. IEEE Press.

[139] P. Glasserman, P. Heidelberger, P. Shahabuddin, & T. Zajic (1997) A large deviations perspective on the efficiency of multilevel splitting. *IEEE Trans. Aut. Contr.* **43**, 1666–1679.

[140] P. Glasserman & Y. Wang (1997) Counterexamples in importance sampling for large deviations probabilities. *Ann. Appl. Probab.* **7**, 731–746.

[141] H. Gluver & S. Krenk (1990) Robust calibration and rescaling of ARMA processes for simulation. Danish Center for Applied Mathematics and Mechanics. Report **409**.

[142] P.W. Glynn (1987) Limit theorems for the method of replications. *Stochastic Models* **3**, 343–354.

[143] P.W. Glynn (1989) A GSMP formalism for discrete event systems. *Proc. IEEE* **77**, 14–23.

[144] P.W. Glynn (1990) Likelihood ratio gradient estimation for stochastic systems. *Comm. ACM* **33**, 75–84.

[145] P.W. Glynn (1994) Some topics in regenerative steady-state simulation. *Acta Applicandae Mathematicae* **34**, 225–236.

[146] P.W. Glynn & P. Heidelberger (1990) Bias properties of budget constrained Monte Carlo simulations. *Oper. Res.* **38**, 801–814.

[147] P.W. Glynn & P. Heidelberger (1991) Analysis of parallel, replicated simulation under a completion time constraint. *ACM TOMACS* **1**, 3–23.

[148] P.W. Glynn & P. Heidelberger (1992) Analysis of initial transient deletion for parallel steady-state simulation. *SIAM J. Scientific Stat. Computing* **13**, 904–922.

[149] P.W. Glynn & P. Heidelberger (1992) Experiments with initial transient deletion for parallel, replicated steady-state simulations. *Management Science* **13**, 400–418.

[150] P.W. Glynn & P. Heidelberger (1992) Jackknifing under a budget constraint. *ORSA J. Comput.* **4**, 226–234

[151] P.W. Glynn & D.L. Iglehart (1987) A joint central limit theorem for the sample mean and regenerative variance estimator. *Ann. Oper. Res.* **8**, 41–55.

[152] P.W. Glynn and D.L. Iglehart (1988) Simulation methods for queues. An overview. *Queueing Systems* **3**, 221-255.

[153] P.W. Glynn & D.L. Iglehart (1989) Importance sampling for stochastic simulations. *Management Science* **35**, 1367–1392.

[154] P.W. Glynn & D.L. Iglehart (1990) Simulation output analysis using standardized time series. *Math. Oper. Res.* **15**, 1–16. .

[155] P.W. Glynn & W. Whitt (1989) Indirect estimation via $L = \lambda W$. *Oper. Res.* **37**, 82–103.

[156] P.W. Glynn & W. Whitt (1992) The asymptotic efficiency of simulation estimators. *Oper. Res.* **40**, 505–520.

[157] P.W. Glynn & W. Whitt (1992) The asymptotic validity of sequential stopping rules for stochastic simulations. *Ann. Appl. Probab.* **2**, 180–198.

[158] E. Gobet (2000) Weak convergence of killed diffusion using Euler schemes. *Stoch. Proc. Appl.* **87**, 167-197.

[159] D.E. Goldberg (1989) *Genetic Algorithms in Search, Optimization and Machine Learning.* Addison-Wesley.

[160] W.B. Gong (1987) *Smoothed Perturbation Analysis of Discrete Event Dynamic Systems.* PhD Dissertation, Harvard University.

[161] W.B. Gong & Y.C. Ho (1987) Smoothed (conditional) perturbation analysis of discrete event dynamic systems. *IEEE Trans. Autom. Control* **32**, 858–866.

[162] N.J. Gordon, D.J. Salmond, & A.F.M. Smith (1993) Novel approach to nonlinear/non-Gaussian Bayesian state space estimation. *IEE Proc.-F, Radar Signal Process.* **140**, 107–113.

[163] A. Goyal, P. Shahabuddin, P. Heidelberger, V.F. Nicola, & P.W. Glynn (1992) A unified framework for simulating Markovian models of highly dependable systems. *IEEE Trans. Computers* **41**, 36–51.

[164] P.J. Green (1995) Reversible jump Markov chain Monte Carlo computation and Bayesian model determination. *Biometrika* **82**, 711–732.

[165] P.J. Green (1996) MCMC in image analysis. *Markov Chain Monte Carlo in Practice* (W.R. Gilks, S. Richardson, & D.J. Spiegelhalter, eds.), 381–399. Chapman & Hall.

[166] P.J. Green (2001) A primer on Markov chain Monte Carlo. *Complex Stochastic Systems* (C. Klüppelberg, O.E. Barndorff-Nielsen, & D.R. Cox, eds.), 1–62. Chapman & Hall.

[167] U. Grenander & M. Rosenblatt (1956) *Statistical Analysis of Stationary Time Series*. Almqvist & Wiksell, Stockholm.

[168] R.C. Griffith & S. Tavaré (1999) The age of mutations in gene trees. *Ann. Appl. Probab.* **9**, 567–590.

[169] G. Grimmett (1999) *Percolation* (2nd ed.). Springer-Verlag..

[170] A. Gut (1995) *An Intermediate Course in Probability*. Springer–Verlag.

[171] P.J. Haas (2002) *Stochastic Petri Nets: Modelling, Stability, Simulation*. Springer-Verlag

[172] P. Haccou, P. Jagers, & V.A. Vatutin (2005) *Branching Processes. Variation, Growth and Extinction of Populations*. Cambridge University Press.

[173] J.M. Hammersley & D.C. Handscomb (1964) *Monte Carlo Methods*. Methuen.

[174] Z. Haraszti & J.K. Townsend (1998) The theory of direct probability redistribution and its application to rare event simulation. *ACM TOMACS* **9**, 105–140.

[175] T.E. Harris (1963) *The Theory of Branching Processes*. Springer-Verlag.

[176] J. Hartinger & D. Kortschak (2007) On the efficiency of the Asmussen–Kroese-estimator and its application to stop-loss transforms. *J. Appl. Probab.* (to appear).

[177] W.K. Hastings (1970) Monte Carlo sampling methods using Markov chains. *Biometrika* **89**, 731–743.

[178] P. Heidelberger (1995) Fast simulation of rare events in queueing and reliability models. *ACM TOMACS* **6**, 43–85.

[179] P. Heidelberger, P. Shahabuddin & V. Nicola (1994) Bounded relative error in estimating transient measures of highly dependable non–Markovian systems. *ACM TOMACS* **4**, 137–164.

[180] P. Heidelberger & D. Towsley (1989) Sensitivity analysis from sample paths using likelihood ratios. *Management Science* **35**, 1475–1488.

[181] S.G. Henderson & P.W. Glynn (2001) Approximating martingales for variance reduction in Markov process simulation. *Math. Oper. Res.* **27**, 253-271.

[182] S.G. Henderson & P.W. Glynn (2001) Regenerative steady-state simulation of discrete-event systems. *ACM TOMACS* **11**, 313–345.

[183] Y.C. Ho & X.-R. Cao (1983) Optimization and perturbation analysis of queueing networks. *J. Optim. Th. Appl.* **40**, 559–582.

[184] F. den Hollander (2000) *Large Deviations*. Fields Institute Monographs **14**. American Mathematical Society.

[185] J.R.M. Hosking (1984) Modeling persistence in hydrological time series using fractional differencing. *Water Resources Res.* **20**, 1898–1908.

[186] C. Huang, M. Devetsikiotis, I. Lambadaris, & A.R. Kaye (1991) Fast simulation of queues with long-range dependent traffic. *Stochastic Models* **15**, 429–460.

[187] Z. Huang & P. Shahabuddin (2003) Rare-event, heavy-tailed simulations using hazard function transformations, with applications to value-at-risk. *Proceedings of the Winter Simulation Conference* **2004**, 276–284.

[188] F. Hubalek (2005) On the simulation from the marginal distribution of a Student t and generalized hyperbolic Lévy process. Working paper.

[189] O. Häggström (2002) *Finite Markov Chains and Algorithmic Applications.* Cambridge University Press.

[190] W. Hörmann, J. Leydold, & G. Derflinger (2004) *Automatic Nonuniform Random Variate Generation.* Springer-Verlag.

[191] J. Jacod (2004) The Euler scheme for Lévy driven stochastic diffferential equations: Limit theorems. *Ann. Probab.* **32**, 1830–1872.

[192] J. Jacod, T.G. Kurtz, S. Méléard, & P. Protter (2005) The approximate Euler method for Lévy driven stochastic differential equations. *Ann. Inst. Henri Poincaré* **41**, 523–558.

[193] J. Jacod & P. Protter (1996) Asymptotic error distribution for the Euler method for stochastic differential equations. *Ann. Probab.* **26**, 267–307

[194] A. Janicki & A. Weron (1994) *Simulation and Chaotic Behaviour of α–Stable Stochastic Processes.* Marcel Dekker.

[195] J.A. Jarrow (2002) *Modeling Fixed-Income Securities and Interest Rate Options.* Stanford University Press.

[196] J.L. Jensen (1995) *Saddlepoint Approximations.* Clarendon Press, Oxford.

[197] M. Jerrum (1995) A very simple algorithm for estimating the number of k-colorings of a low-degree graph. *Random Structures and Algorithms* **7**, 157–165.

[198] M. Jerrum (2003) *Counting, Sampling and Integrating: Algorithms and Complexity.* Birkhäuser.

[199] M. Jerrum, A. Sinclair, & E. Vigoda (2004) A polynomial-time algorithm for the permanent of a matrix with nonnegative entries. *J. ACM* **51**, 671–697.

[200] H. Joe (1997) *Multivariate Models and Dependence.* Chapman & Hall.

[201] M.E. Johnson (1982) *Multivariate Statistical Simulations.* Wiley.

[202] N.L. Johnson & S. Kotz (1972) *Distributions in Statistics: Continuous Multivariate Distributions.* Wiley.

[203] S. Juneja & P. Shahabuddin (2002) Simulating heavy tailed processes using delayed hazard rate twisting. *ACM TOMACS* **12**, 94–118.

[204] S. Juneja & P. Shahabuddin (2006) Rare event simulation techniques. *Simulation* (S.G. Henderson & B.L. Nelson, eds.), 291–350. Handbooks in Operations Research and Management Science. Elsevier.

[205] R. Kaas, M. Goovaerts, J. Dhaene, & M. Denuit (2001) *Modern Actuarial Risk Theory.* Kluwer.

[206] H. Kahn & T.E. Harris (1951) Estimation of particle transmission by random sampling. *National Bureau of Standard Applied Mathematics Series* **12**, 27–30.

[207] S. Karlin & H.G. Taylor (1981) *A Second Course in Stochastic Processes.* Academic Press.

[208] M.S. Keane & G.L. O'Brien (1994) A Bernoulli factory. *ACM TOMACS* **4**, 213–219.

[209] A. Kebaier (2005) Statistical Romberg extrapolation: a new variance reduction technique and applications to option pricing. *Ann. Appl. Probab.* **15**, 2681–1705.

[210] J. Keilson & D.M.G. Wishart (1964) A central limit theorem for processes defined on a finite Markov chain. *Proc. Cambridge Philos. Soc.* **60**, 547–567.

[211] W.D. Kelton, R.P. Sadowski, & D.T. Sturrock (2004) *Simulation with Arena* (3rd ed.). McGraw Hill.

[212] F. Kelly (1977) *Reversibility and Stochastic Networks*. Wiley.

[213] W.S. Kendall, F. Liang, & J.-S. Wang, eds. (2005) *Markov Chain Monte Carlo. Innovations and Applications*. World Scientific.

[214] W.S. Kendall (2005) Notes on perfect simulation. [213], 93–146.

[215] H. Kesten (1973) Random difference equations and renewal theory for products of random matrices. *Acta Mathematica* **131**, 207–248.

[216] R.Z. Khasminskii & F.C. Klebaner (2001) Long term behaviour of solutions of the Lotka-Volterra system under small random perturbations. *Ann. Appl. Probab.* **11**, 952–963.

[217] P.E. Kloeden & E. Platen (1992) *Numerical Solution of Stochastic Differential Equations*. Springer–Verlag.

[218] P.E. Kloeden, E. Platen, & H. Schurz (1994) *Numerical Solution of Stochastic Differential Equations through Computer Experiments*. Springer–Verlag.

[219] D.E. Knuth (1984) An algorithm for Brownian zeroes. *Computing* **33**, 89–94.

[220] D.E. Knuth (1998) *The Art of Computer Programming* (3rd ed.). Addison-Wesley.

[221] A. Kohatsu-Higa & M. Montero (2004) Malliavin calculus in finance. *Handbook of Numerical and Computational Methods in Finance* (S.T. Rachev, ed.), 111–174. Birkhäuser.

[222] Y. Kozachenko, T. Sottinen, & O. Vasylyk (2005) Simulation of weakly self-similar $\mathrm{Sub}_\varphi(\Omega)$-processes: a series expansion approach. *Methodology and Computing in Applied Probability* **7**, 379–400.

[223] S. Krenk & J. Clausen (1987) On the calibration of ARMA processes for simulation. In *Reliability and Optimization of Structural Systems* (P. Thoft-Christensen ed.). Springer-Verlag.

[224] H.J. Kushner & G.G. Yin (2003) *Stochastic Approximation and Recursive Algorithms and Applications*. Springer-Verlag.

[225] T.G. Kurtz & P. Protter (1991) Wong-Zakai corrections, random evolutions and numerical schemes for SDEs. *Stochastic Analysis*, 331–346. Academic Press.

[226] A. Kyprianou (2006) *Introductory Lectures on Fluctuations of Lévy Processes with Applications*. Springer-Verlag.

[227] D.P. Landau (2005) An introdiction to Monte Carlo methods in statistical physics. [213], 53–91.

[228] A.M. Law & W.D. Kelton (1991) *Simulation Modeling and Analysis* (2nd ed.). McGraw Hill.

[229] P. L'Ecuyer (1994) Uniform random number generation. *Ann. Oper. Res.* **53**, 77–120.

[230] P. L'Ecuyer (1999) Good parameters and implementations for combined multiple recursive random number generation. *Oper. Res.* **47**, 159–164.

[231] P. L'Ecuyer (1999) Tables of maximally equidistributed LFSR generators *Math. Comp.* **68**, 261–269.

[232] P. L'Ecuyer (2004) Uniform random number generation. *Stochastic Simulation*. Handbooks of Operations Research and Management Science. Elsevier.

[233] P. L'Ecuyer & C. Lemieux (2002) Recent advances in randomized quasi-Monte Carlo metods. *Modeling Uncertainty: An Examination of Stochastic Theory, Methods and Applications* (M. Dror et al., eds.), 419–474. Kluwer.

[234] P. L'Ecuyer & F. Panneton (2006) Fast recurrences based upon linear recurrences modulo 2: an overview. *Proceedings of the Winter Simulation Conference* **2006** (to appear). IEEE Press.

[235] P. L'Ecuyer & G. Perron (1994) On the convergence rates of IPA and FDC derivative estimators. *Oper. Res.* **42**, 643–656.

[236] T. Lehtonen & H. Nyrhinen (1992) Simulating level–crossing probabilities by importance sampling. *Adv. Appl. Probab.* **24**, 858–874.

[237] T. Lehtonen & H. Nyrhinen (1992) On asymptotically efficient simulation of ruin probabilities in a Markovian environment. *Scand. Actuarial J.*, 60–75.

[238] W. Leland, M. Taqqu, W. Willinger, & D. Wilson (1994) On the self-similar nature of ethernet traffic. *IEEE/ACM Trans. Netw.* **2**, 1–15.

[239] D. Lepingle (1995) Euler scheme for reflected stochastic differential equations. *Mathematics and Computers in Simulation* **38**, 119–126.

[240] G. Lindgren (2006) *Lecture Notes on Stationary and Related Stochastic Processes.* Lund University. Available from [$\mathbf{w}^3$.8].

[241] J.S. Liu (2001) *Monte Carlo Strategies in Scientific Computing.* Springer-Verlag.

[242] W. Loh (1996) On latin hypercube sampling. *Ann. Statist.* **24**, 2058–2080.

[243] D. Madan & M. Yor (2005) Representing the CGMY and Meixner Lévy processes as time changed Brownian motions. Manuscript.

[244] N. Madras (2002) *Lectures on the Monte Carlo Method.* American Mathematical Society.

[245] N. Madras & G. Slade (1993) *The Self-Avoiding Walk.* Birkhäuser.

[246] B.B. Mandelbrot (1971) A fast fractional Gaussian noise generator. *Water Resources Res.* **7**, 543–553.

[247] B.B. Mandelbrot & J.W. Van Ness (1968) Fractional Brownian motions, fractional noises and applications. *SIAM Review* **10**, 422–437.

[248] M. Mandjes (2007) *Large Deviations of Gaussian Queues.* Wiley (to appear).

[249] M. Matsumoto & T. Nishimura (1998) Mensenne twister: a 623-dimensionally equidistributed uniform pseudo-random number generator. *ACM TOMACS* **8**, 3-30.

[250] D. McDonald (1999) Asymptotics of first passage times for random walks in a quadrant. *Ann. Appl. Probab.* **9**, 110–145.

[251] M. McKay, R. Beckman & W. Conover (1979) A comparison of three methods for selecting values of output variables in the analysis of output from a computer code. *Technometrics* **21**, 239–245.

[252] A. McNeil, R. Frey, & P. Embrechts (2005) *Quantitative Risk Management. Concepts, Techniques and Tools.* Princeton University Press.

[253] K.L. Mengersen & R.L. Tweedie (1996) Rates of convergence of the Hasting and Metropolis algorithms. *Ann. Statist.* **24**, 101–121

[254] N. Metropolis, A. Rosenbluth, M. Rosenbluth, A. Teller, & E. Teller (1953) Equations of state calculations by fast computing machines. *J. Chem. Phys.* **21**, 1087–1091.

[255] S. Meyn & R.L. Tweedie (1993) *Markov Chains and Stochastic Stability.* Springer-Verlag.

[256] J.R. Michael, W.R. Schuchany, & R.W. Haas (1976) Generating random variates using transformations with multiple roots. *American Statistician* **30**, 88–90.

[257] Z. Michna (1999) On tail probabilities and first passage times for fractional Brownian motion. *Math. Methods of Oper. Res. 49*, 335–354.

[258] T. Mikosch (2006) Copulas: tales and facts (with discussion). *Extremes* **9**, 3–20, 55–62.

[259] G.N. Milstein & M.V. Tretyakov (2004) *Stochastic Numerics for Mathematical Physics.* Springer-Verlag.

[260] D.L. Minh & R.M. Sorli (1983) Simulating the GI/G/1 queue in heavy traffic. *Oper. Res.* **31**, 966–971.

[261] B.J.T. Morgan (1984) *Elements of Simulation.* Chapman & Hall.

[262] W.J. Morokoff (1998) Generating quasi-random paths for stochastic processes. *SIAM Review* **40**, 765–788.

[263] R. Motwani & P. Raghavan (1995) *Randomized Algorithms.* Cambridge University Press.

[264] J. Møller (1999) Perfect simulation of conditionally specified models. *J. R. Statist. Soc.* **B61**, 251–264.

[265] J. Møller & R.P. Waagepetersen (2000) *Statistical Inference and Simulation for Spatial Point Processes.* Chapman & Hall/CRC.

[266] M. Nakayama (1994) Two-stage stopping procedures based on standardized time series. *Management Science* **40**, 1189–1206.

[267] M.K. Nakayama (1996) A characterization of the simple failure biasing method for simulations of highly reliable Markovian systems. *ACM TOMACS* **4**, 52–88.

[268] M.K. Nakayama (1996) General conditions for bounded relative error in simulations of highly reliable systems. *Adv. Appl. Probab.* **28**, 687–727.

[269] S. Nacu & Y. Peres (2005) Fast simulation of new coins from old. *Ann. Appl. Probab.* **15**, 93–115.

[270] O. Narayan (1998) Exact asymptotic queue length distribution for fractional Brownian traffic. *Advances in Performance Evaluation* **1**, 39–65.

[271] R. Nelsen (1999) *An Introduction to Copulas.* Springer-Verlag.

[272] M.F. Neuts (1977) A versatile Markovian point process. *J. Appl. Probab.* **16**, 764–779.

[273] M.F. Neuts (1981) *Matrix-Geometric Solutions in Stochastic Models.* Johns Hopkins University Press.

[274] M.F. Neuts (1989) *Structured Stochastic Matrices of the M/G/1 Type and Their Applications.* Marcel Dekker.

[275] J. Neveu (1961) Une generalisation des processus a accroissementes positifs independantes. *Abh. Math. Sem. Hamburg* **23**, 36–61.

[276] M.E.J. Newman & G.T. Barkema (2000) *Monte Carlo Methods in Statistical Physics.* Oxford University Press.

[277] N.J. Newton (1994) Variance reduction for simulated diffusions. *SIAM J. Appl. Math.* **54**, 1780–1805.

[278] H. Niederreiter (1992) *Random Number Generation and Quasi–Monte Carlo Methods.* SIAM.

[279] I. Norros (1994) A storage model with self–similar input. *Queueing Systems* **16**, 387–396.

[280] I. Norros, P. Mannersalo & J.L. Wang (1999) Simulation of fractional Brownian motion with conditionalized random midpoint displacement. *Advances in Performance Analysis* **5**, 77–101.

[281] I. Norros, E. Valkeila & J. Virtamo (1999) An elementary approach to a Girsanov type formula and other analytical results for fractional Brownian motion. *Bernoulli* **5**, 571–587.

[282] E. Nummelin (1984) *General Irreducible Markov Chains and Non-Negative Operators.* Cambridge University Press.

[283] F. Panneton, P. L'Ecuyer, & M. Matsumoto (2006) Improved long-period generators based on linear recurrences modulo 2. *ACM Trans. Math. Software* **32** (to appear).

[284] K. Park & W. Willinger, eds. (2000) *Self-Similar Network Traffic and Performance Evaluation.* Wiley.

[285] B.N. Parlett (1980) *The Symmetric Eigenvalue Problem.* Prentice-Hall.

[286] V. Paxson (1997) Fast, approximate synthesis of fractional Gaussian noise for generating self–similar network traffic. *Computer Communication Review* **27**, 5–18

[287] V. Paxson & S. Floyd (1995) Wide area traffic: the failure of Poisson modelling. *IEEE/ACM Trans. Netw.* **3**, 226–244.

[288] J.A. Perez (2004) *Convergence of Numerical Schemes in the Total Variation Sense.* PhD dissertation, New York University.

[289] V.V. Petrov (1965) On the probabilities of large deviations for sums of independent random variables. *Th. Probab. Appl.* **10**, 287–298.

[290] J. Pickands, III (1968) Moment convergence of sample extremes. *Ann. Math. Statist.* **39**, 881–889.

[291] B.T. Polyak (1990) New stochastic approximation type procedures. *Automat. Remote Control* **7**, 98–107.

[292] B.T. Polyak & A.B. Juditsky (1992) Acceleration of stochastic approximation by averaging. *SIAM J. Control Optim.* **30**, 838–855.

[293] W.H. Press, B.F. Flannery, S.A. Teukolsky, & W.T. Vetterling (1986) *Numerical Recipes. The Art of Scientific Computing.* Cambridge University Press.

[294] G. Pflug (1997) *Stochastic Optimization.* Kluwer.

[295] J.G. Propp & D.B. Wilson (1996) Exact sampling with coupled Markov chains and applications to statistical mechanics. *Random Structures and Algorithms* **9**, 223–252.

[296] J.G. Propp & D.B. Wilson (1998) How to get a perfectly random sample from a generic Markov chains and generate a random spanning tree of a directed random graph. *J. Algs.* **27**, 170–217.

[297] P. Protter (2004). *Stochastic Integration and Differential Equations* (2nd ed.) Springer-Verlag.

[298] S.I. Resnick (1997) Heavy tail modeling and teletraffic data. *Ann. Statist.* **25**, 1805–1869.

[299] A. Ridder (1996) Fast simulation of Markov fluid models. *Adv. Appl. Probab.* **28**, 786–803.

[300] B. Ripley (1987) *Stochastic Simulation.* Wiley, New York.

[301] C.P. Robert & G. Casella (2004) *Monte Carlo Statistical Methods* (2nd ed.). Springer-Verlag.

[302] G.O. Roberts & J.S. Rosenthal (2004) General state space Markov chains and MCMC algorithms. *Probability Surveys* **1**, 20–71.

[303] L.C.G. Rogers (1994) Fluid models in queueing theory and Wiener–Hopf factorisation of Markov chains. *Ann. Appl. Probab.* **4**, 390–413.

[304] L.C.G. Rogers & D. Talay, eds. (1997) *Numerical Methods in Finance.* Cambridge University Press.

[305] L.C.G. Rogers & D. Williams (2000) *Diffusions, Markov Processes and Martingales.* Cambridge University Press.

[306] J. Rosiński (1990) On series representations of infinitely divisible random vectors. *Ann. Probab.* **18**, 405–430.

[307] J. Rosiński (2001) Series representations of Lévy processes from the perspective of point processes. [37], 401–415.

[308] J. Rosiński (2007) Tempered stable distributions. *Stoch. Proc. Appl.* (to appear).

[309] S.M. Ross (1985) *Introduction to Probability Models* (3rd ed.). Academic Press.

[310] S.M. Ross (1991) *A Course in Simulation.* Macmillan.

[311] S. Rubenthaler (2003) Numerical simulation of the solution of a stochastic differential equation driven by a Lévy process. *Stoch. Proc. Appl.* **103**, 311–349.

[312] S. Rubenthaler & M. Wiktorsson (2003) Improved convergence rate for the simulation of subordinated Lévy processes. *Stoch. Proc. Appl.* **103**, 311–349.

[313] R.Y. Rubinstein (1981) *Simulation and the Monte Carlo Method.* Wiley.

[314] R.Y. Rubinstein (1986) *Monte Carlo Optimization Simulation and Sensitivity of Queueing Networks.* Wiley.

[315] R.Y. Rubinstein (1992) Sensitivity analysis of discrete event systems by the "push-out" method. *Ann. Oper. Res.* **39**, 229–250.

[316] R.Y. Rubinstein (1997) Optimization of computer simulation models with rare events. *Europ. J. Oper. Res.* **99**, 89–112.

[317] R.Y. Rubinstein (1999) The cross-entropy method for combinatorial and continuous optimization. *Method. Comp. Appl. Prob.* **2**, 127–190.

[318] R.Y. Rubinstein & D.P. Kroese (2005) *The Cross-Entropy Method. A Unified Approach to Combinatorial Optimization, Simulation and Machine Learning.* Springer-Verlag.

[319] R.Y. Rubinstein & B. Melamed (1998) *Classical and Modern Simulation.* Wiley.

[320] R.Y. Rubinstein & A. Shapiro (1993) *Discrete Event Systems: Sensitivity Analysis and Stochastic Optimization via the Score Function Method.* John Wiley & Sons.

[321] D. Ruppert (1991) Stochastic approximation. *Handbook in Sequential Analysis* (B.K. Ghosh & P.K. Sen, eds.), 503–529. Marcel Dekker.

[322] T.H. Rydberg (1997) The normal inverse Gaussian Lévy process: simulation and approximation. *Stochastic Models* **13**, 887–910.

[323] T. Rydén & M. Wiktorsson (2001) On the simulation of iterated Itô integrals. *Stoch. Proc. Appl.* **91**, 151–168.

[324] J.S. Sadowsky (1991) Large deviations theory and efficient simulation of excessive backlogs in a $GI/GI/m$ queue. *IEEE Trans. Automat. Contr.* **AC–36**, 1383–1394.

[325] J.S. Sadowsky (1993) On the optimality and stability of exponential twisting in Monte Carlo simulation. *IEEE Transaction on Information Theory* **IT-39**, 119–128.

[326] J.S. Sadowsky (1995) The probability of large queue lengths and waiting times in a heterogeneous multiserver queue. Part II: Positive recurrence and logarithmic limits. *Advances in Applied Probability* **27**, 567–583.

[327] J.S. Sadowsky & J.S. Bucklew (1990) On large deviations theory and asymptotically efficient Monte Carlo estimation. *IEEE Trans. Inform. Theory* **IT-36**, 579–588.

[328] J.S. Sadowsky & W. Szpankowsky (1995) The probability of large queue lengths and waiting times in a heterogeneous multiserver queue. Part I: Tight limits. *Adv. Appl. Probab.* **27**, 532-566.

[329] G. Samorodnitsky & M.L. Taqqu (1994) *Non–Gaussian Stable Processes.* Chapman & Hall.

[330] K. Sato (1999) *Lévy Processes and Infinite Divisibility.* Cambridge University Press.

[331] W. Schoutens (2003) *Lévy Processes in Finance. Pricing Financial Derivatives.* Wiley.

[332] L.W. Schruben (1986) Sequential simulation run control using standardized times series. *ASA-ACM Interface Meeting*, Colorado. *PCmpScSt18*, 257–260.

[333] R.J. Serfling (1980) *Approximation Theorems of Mathematical Statistics.* Wiley.

[334] P. Shahabuddin (1994) Importance sampling for the simulation of highly reliable Markovian systems. *Management Science* **40**, 333–352.

[335] G.S. Shedler (1993) *Regenerative Stochastic Simulation.* Academic Press.

[336] R. Shrinivasan (2002) *Importance Sampling. Applications in Communication and Detection.* Springer-Verlag.

[337] A. Shwartz & A. Weiss (1995) *Large Deviations for Performance Analysis: Queues, Communication and Computers.* Chapman & Hall.

[338] D. Siegmund (1976). Importance sampling in the Monte Carlo study of sequential tests. *Ann. Statist.* **4**, 673–684.

[339] D. Siegmund (1976) The equivalence of absorbing and reflecting barrier problems for stochastically monotone Markov processes. *Ann. Probab.* **4**, 914–924.

[340] D. Siegmund (1985) *Sequential Analysis.* Springer-Verlag.

[341] K. Sigman (1994) *Stationary Marked Point Processes: An Intuitive Approach.* Chapman & Hall.

[342] M. Signahl (2003) On error rates in normal approximations and simulation schemes for Lévy processes. *Stochastic Models* **19**, 287–298.

[343] M. Signahl (2004) Sensitivity analysis via simulation in the presence of discontinuities. *Math. Meth. Oper. Res.* **60**, 29–51.

[344] E.A. Silver, D.F. Pyke, & R. Peterson (1998) *Inventory Management and Production Planning and Scheduling* (3rd ed.). Wiley.

[345] A. Sinclair (1993) *Algorithms for Random Generation and Counting.* Birkhäuser.

[346] M. Steele (2001) *Stochastic Calculus and Financial Applications.* Springer-Verlag.

[347] R. Suri (1989) Perturbation analysis: the state of the art and research issues explained via the GI/G/1 queue. *Proceeedings of the IEEE* **77**, 114–137.

[348] R. Suri & M. Zazanis (1988) Perturbation analysis gives strongly consistent estimates for the M/G/1 queue. *Management Science* **34**, 39–64.

[349] D. Talay (1995) Simulation of stochastic differential systems. *Probabilistic Methods in Applied Physics* (P. Krée & W. Wedig, eds.), 54–96. Lecture Notes in Physics **451**. Springer-Verlag.

[350] D. Talay & L. Tubaro (1990) Expansion of the global error for numerical schemes solving stochastic differential equations. *Stochastic Analysis and Applications* **8**, 94–120.

[351] P. Tankov (2005) Simulation and option pricing in Lévy copula models. In *Mathematical Modelling of Financial Derivatives* (M. Avellaneda & R. Cont, eds.). Springer-Verlag.

[352] S. Tavaré, D.J. Balding, R.C. Griffiths, & P. Donnelly (1997) Inferring coalescence times for molecular sequential data. *Genetics* **145**, 505–518.

[353] J.R. Thompson (2000) *Simulation. A Modeler's Approach.* Wiley.

[354] H. Thorisson (2000) *Coupling, Stationarity and Regeneration.* Springer-Verlag.

[355] L. Tierney (1994) Markov chains for exploring posterior distribution (with discussion). *Ann. Statist.* **22**, 1701–1786.

[356] A.W. van der Vart (1998) *Asymptotic Statistics.* Cambridge University Press.

[357] M. & J. Villén-Altamirano (1991) RESTART: A method for accelerating rare events simulations. *Queueing Performance and Control in ATM* (J.W. Cohen & C.D. Pack, eds.). *Proceedings of ITC* **13**, 71–76.

[358] M. & J. Villén-Altamirano (1994) RESTART: A straightforward method for fast simulation of rare events. *Proceedings of the Winter Simulation Conference* **1994**, 282–289. IEEE Press.

[359] C.L. Wang & R.W. Wolff (2003) Efficient simulation of queues in heavy traffic. *ACM TOMACS* **13**, 62–81.

[360] W. Whitt (1989) Planning queueing simulations. *Management Science* **35**, 1341–1366.

[361] W. Whitt (2002) *Stochastic-Process Limits.* Springer-Verlag.

[362] D. Whitley (1994) A genetic algorithm tutorial. *Statistics and Computing* **4**, 63–85.

[363] M. Wiktorsson (2001) Joint characteristic function and simultaneous simulation of iterated Itô integrals for multiple independent Brownian motions. *Ann. Appl. Probab.* **11**, 470–487.

[364] M. Wiktorsson (2002) Simulation of stochastic integrals with respect to Lévy processes of type G. *Stoch. Proc. Appl.* **101**, 113–125.

[365] D. Wilson (2000) How to couple from the past using a read-once source of randomness. *Random Structures and Algorithms* **16**, 85–113.

[366] W.L. Winston (1991) *Introduction to Mathematical Programming. Applications & Algorithms.* PWS-Kent.

[367] A. Wood & A. Chan (1994) Simulation of stationary Gaussian processes in $[0,1]^d$. *J. Comp. Graph. Statist.* **3**, 409–432.

[368] P.H. Zipkin (2000) *Foundations of Inventory Management.* McGraw-Hill.

[369] S. Yakowitz, J. Krimmel & F. Szidarovszky (1978) Weighted Monte Carlo integration. *SIAM J. Numer. Anal.* **15**, 1289–1300.

Web Links

[w³.1] http://home.imf.au.dk/asmus/books/agtypos.tex
List of typos, comments etc. for the book.

[w³.2] http://home.imf.au.dk/asmus/
Søren Asmussen.

[w³.3] http://www.stanford.edu/dept/MSandE/people/faculty/glynn/
Peter W. Glynn.

[w³.4] http://cgm.cs.mcgill.ca/~luc/
Luc Devroye; random number generation.

[w³.5] http://www.cfm.brown.edu/people/dupuis/home.html/
Paul Dupuis.

[w³.6] http://www2.gsb/columbia.edu/faculty/pglasserman/Other/
Paul Glasserman.

[w³.7] http://www.iro.umontreal.ca/~lecuyer
Pierre L'Ecuyer; pseudorandom and quasirandom numbers.

[w³.8] http://www.maths.lth.se/matstat/staff/georg
Georg Lindgren; aspects of Gaussian process simulation.

[w³.9] http://random.mat.sbg.ac.at
Random number generation.

[w³.10] http://www.t-rs.co.jp/eng/products/index.htm
A physical random number generator.

[w³.11] http://www.protego.se/sg100_en.htm
A physical random number generator.

[w³.12] http://random.mat.sbg.ac.at/links/quasi.html
Quasirandom numbers.

[w³.13] http://sourceforge.net/projects/ldsequences
Quasirandom numbers.

[w³.14] http://www.wintersim.org
Winter Simulation Conferences

[w³.15] http://www.informs-cs.org/wscpapers.html
Papers from [w³.14].

[w³.16] http://www.uni-bamberg.de/wiai/ktr/html/tagungen/
resim06/resim06.html
Last conference in the Rare Event Simulation conference series.

[w³.17] http://vega@math.spbu.ru
St. Petersburg Workshops on Simulation.

[w³.18] http://www.cs.rutgers.edu/~imacs
IMACS: The International Association for Mathematics and Computers
in Simulation.

[w³.19] http://www.informs-sim.org/
INFORMS Simulation Society.

[w³.20] http://www-sop.inria.fr/omega/MC2QMC2004/
Last conference in QMC series.

[w³.21] http://iew3.technion.ac.il/mailman/listinfo/cemethod
The cross-entropy method.

[w³.22] http://www.stats.bris.ac.uk/MCMC/
MCMC Preprint Service.

[w³.23] http://www.mrc-bsu.cam.ac.uk/bugs/
Bayesian Inference Using Gibbs Sampling; software.

[w³.24] http://research.microsoft.com/~dbwilson/exact
Perfect sampling.

Index

Stochastic Modelling and Applied Probability
formerly: Applications of Mathematics

(continued from page II)